College Algebra

DeMYSTiFieD®

DeMYSTiFieD® Series

Accounting Demystified
Advanced Calculus Demystified
Advanced Physics Demystified
Advanced Statistics Demystified
Algebra Demystified
Alternative Energy Demystified
Anatomy Demystified
asp.net 2.0 Demystified
Astronomy Demystified
Audio Demystified
Biology Demystified
Biotechnology Demystified
Business Calculus Demystified
Business Math Demystified
Business Statistics Demystified
C++ Demystified
Calculus Demystified
Chemistry Demystified
Circuit Analysis Demystified
College Algebra Demystified
Corporate Finance Demystified
Databases Demystified
Data Structures Demystified
Differential Equations Demystified
Digital Electronics Demystified
Earth Science Demystified
Electricity Demystified
Electronics Demystified
Engineering Statistics Demystified
Environmental Science Demystified
Everyday Math Demystified
Fertility Demystified
Financial Planning Demystified
Forensics Demystified
French Demystified
Genetics Demystified
Geometry Demystified
German Demystified
Home Networking Demystified
Investing Demystified
Italian Demystified
Java Demystified
JavaScript Demystified
Lean Six Sigma Demystified
Linear Algebra Demystified
Macroeconomics Demystified
Management Accounting Demystified
Math Proofs Demystified
Math Word Problems Demystified
MATLAB® Demystified
Medical Billing and Coding Demystified
Medical Terminology Demystified
Meteorology Demystified
Microbiology Demystified
Microeconomics Demystified
Nanotechnology Demystified
Nurse Management Demystified
OOP Demystified
Options Demystified
Organic Chemistry Demystified
Personal Computing Demystified
Pharmacology Demystified
Physics Demystified
Physiology Demystified
Pre-Algebra Demystified
Precalculus Demystified
Probability Demystified
Project Management Demystified
Psychology Demystified
Quality Management Demystified
Quantum Mechanics Demystified
Real Estate Math Demystified
Relativity Demystified
Robotics Demystified
Sales Management Demystified
Signals and Systems Demystified
Six Sigma Demystified
Spanish Demystified
sql Demystified
Statics and Dynamics Demystified
Statistics Demystified
Technical Analysis Demystified
Technical Math Demystified
Trigonometry Demystified
uml Demystified
Visual Basic 2005 Demystified
Visual C# 2005 Demystified
xml Demystified

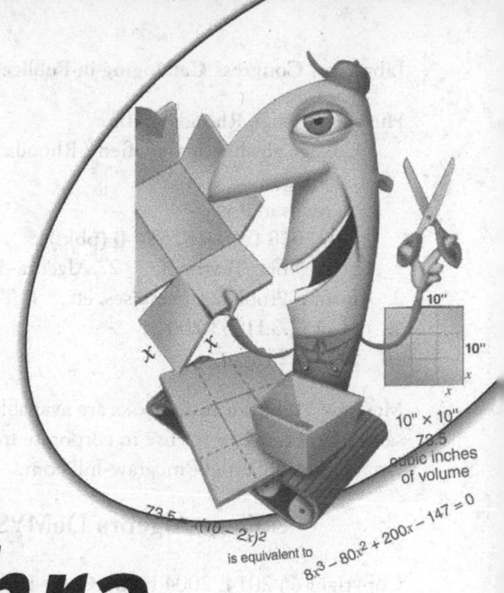

College Algebra

DeMYSTiFieD®

Rhonda Huettenmueller

Second Edition

New York Chicago San Francisco Athens London Madrid Mexico City
Milan New Delhi Singapore Sydney Toronto

Library of Congress Cataloging-in-Publication Data

Huettenmueller, Rhonda, author.
College algebra demystified / Rhonda Huettenmueller. – Second edition.
pages cm
Includes index.
ISBN 978-0-07-181584-0 (pbk.)
1. Algebra–Textbooks. 2. Algebra–Programmed instruction.
3. Algebra–Problems, exercises, etc. I. Title.
QA152.3.H853 2004
512–dc23 2013041551

McGraw-Hill Education books are available at special quantity discounts to use as premiums and sales promotions, or for use in corporate training programs. To contact a representative please e-mail us at bulksales@mcgraw-hill.com.

College Algebra DeMYSTiFieD®, Second Edition

11 12 13 14 15 16 QVS/QVS 23 22 21 20 19

ISBN 978-0-07-181584-0
MHID 0-07-181584-8

Sponsoring Editor
Judy Bass

Acquisitions Coordinator
Amy Stonebraker

Editing Supervisor
David E. Fogarty

Project Manager
Sheena Uprety,
Cenveo® Publisher Services

Copy Editor
Cenveo Publisher Services

Proofreader
Cenveo Publisher Services

Indexer
Cenveo Publisher Services

Production Supervisor
Pamela A. Pelton

Composition
Cenveo Publisher Services

Art Director, Cover
Jeff Weeks

Cover Illustration
Lance Lekander

To my friends
from the Pohl Rec Center:
Jessica Carmona, Whitney Cook, Melissa Jenkins, Christin Ledford, Angela McGuire, Angela Plata, Kristen Resendez, Samantha Rota, Mary Ann Teel, and Kia Williams.

About the Author

Rhonda Huettenmueller has been teaching at the college level since 1990 and earned a PhD in mathematics in 2001. She is the author of several books in the *Demystified* series: *Algebra Demystified, College Algebra Demystified, Precalculus Demystified*, and *Business Calculus Demystified*.

Contents

	Introduction	*xiii*
CHAPTER 1	**Fundamentals**	**1**
	The Distributive Property	2
	Rational Expressions	7
	Simplifying Rational Expressions	8
	Multiplying Rational Expressions	8
	Adding Rational Expressions	9
	Exponents and Roots	13
	Summary	17
	Quiz	18
CHAPTER 2	**Linear Equations and Inequalities**	**21**
	Basic Linear Equations	22
	Equations Leading to Linear Equations	25
	Absolute Value Equations	30
	Linear Inequalities	35
	Compound Inequalities	36
	Absolute Value Inequalities	39
	Summary	47
	Quiz	49
CHAPTER 3	**Quadratic Equations**	**51**
	Solving Quadratic Equations by Factoring	52
	The Quadratic Formula	53

	Completing the Square	56
	Summary	63
	Quiz	64
CHAPTER 4	**The *xy*-Coordinate Plane**	**65**
	Plotting Points	66
	The Distance Between Two Points	67
	The Midpoint Formula	76
	Circles	77
	Summary	88
	Quiz	89
CHAPTER 5	**Lines and Parabolas**	**91**
	Introduction to Lines	92
	Intercepts	95
	The Slope of a Line	102
	Horizontal and Vertical Lines	105
	Finding an Equation for a Line	107
	The Slope-Intercept Form of a Line	112
	Graphing the Line Using the Slope and y-Intercept	113
	Parallel and Perpendicular Lines	116
	Linear Applications	123
	Parabolas	134
	Sketching the Graph of a Parabola	135
	Locating the Vertex by Completing the Square	141
	Summary	147
	Quiz	149
CHAPTER 6	**Nonlinear Inequalities**	**153**
	Solving Nonlinear Inequalities Graphically	154
	Solving Nonlinear Inequalities	157
	Sign Graphs	159
	Rational Inequalities	165
	Summary	173
	Quiz	174

CHAPTER 7 **Functions** **175**
Introduction to Functions 176
Evaluating Functions 178
Evaluating Piecewise Functions 180
Domain and Range 182
Functions and Their Graphs 188
Finding the Domain and Range Graphically 193
Increasing Intervals and Decreasing Intervals 197
The Graph of a Piecewise Function 200
More on Evaluating Functions 209
The Difference Quotient 212
Summary 217
Quiz 220

CHAPTER 8 **Quadratic Functions** **223**
A Review of a Parabola's Vertex 224
The Range of a Quadratic Function 226
The Maximum/Minimum of a Quadratic Function 228
Applied Maximum/Minimum Problems 230
Revenue-Maximizing Price 239
Maximizing/Minimizing Other Functions 242
Summary 243
Quiz 244

CHAPTER 9 **Transformations and Combinations** **247**
Transformations 248
Reflections, and Vertical Stretching and Compressing 253
Sketching the Graph of a Transformation 259
Special Functions 264
Even/Odd Functions 281
Combining Functions 286
Function Composition 287
Function Composition for a Single Value 290
The Domain for the Composition of Functions 295
Summary 298
Quiz 300

CHAPTER 10 **Polynomial Functions** 305
Introduction to Polynomial Functions 306
Sketching Graphs of Polynomials 314
Polynomial Division 317
Synthetic Division 324
Synthetic Division and Factoring 331
Rule of Signs and Upper and Lower Bounds Theorem 340
Complex Numbers 346
Complex Solutions to Quadratic Equations 354
The Fundamental Theorem of Algebra 357
Summary 370
Quiz 372

CHAPTER 11 **Systems of Equations and Inequalities** 375
Systems of Linear Equations 376
Elimination by Addition 380
Applications for Systems of Equations 390
Systems Containing Nonlinear Equations 396
Inequalities and Systems of Inequalities 401
Systems of Inequalities 408
Summary 419
Quiz 421

CHAPTER 12 **Exponents and Logarithms** 425
Compound Growth 426
The Number *e* 430
Increasing Population 431
Logarithms 434
Properties of Logarithms 436
Three More Important Logarithm Properties 439
Using Multiple Logarithm Properties 441
Equations Involving Exponents and Logarithms 445
Exponent and Logarithm Functions 453
The Domain of a Logarithm Function 459
Summary 460
Quiz 463

Final Exam 467
Answers to Quizzes and Final Exam 483
Index 487

Introduction

This book is meant to help you *understand* college algebra. While we will cover most of what a typical college algebra student must learn, we will cover it more carefully than an instructor can do so in class. I have found that most college algebra students struggle with the course because the material progresses too quickly.

So that you do not have to absorb too much at once, each subsection contains exactly one new idea. You will not be distracted by missing algebra steps because I have included many of the algebra steps that most authors and instructors skip. The explanations are brief but clear and the examples are worked out in detail. I have used my more than 20 years of teaching experience to anticipate the questions you might have.

You'll get the most from this book if you work at it a little at a time. Because the topics build on each other, make sure that you understand the material from the previous sections before beginning a new section. If you have trouble working the Practice problems, solutions are worked out in detail so that you can self-correct. At the end of each chapter is a summary and a quiz. You should take each quiz as if you were in a classroom, that is, without notes and with a time limit. This will help you decide how well you understand the chapter. Try to prepare for the final exam at the end of the book as if it really were a comprehensive exam. Study the reviews at the end of each chapter before attempting the final. In fact, instead of answering all 90 questions at once, you might treat the final exam as three separate 30-question exams, trying to improve your score each time.

With steady work and patience, I think you will surprise yourself with success. Good luck.

Rhonda Huettenmueller

chapter 1

Fundamentals

Success in any math class depends on a solid foundation in fundamentals. For college algebra, this means the ability to do the basics: arithmetic, factoring, solving equations, and working with rational expressions, exponents, and roots. The first two chapters are meant to dust off your algebra skills. If you find anything in this chapter (or the next) that is covered too fast, you might consider using my book *Algebra Demystified*, which covers these topics more carefully. If you are already comfortable with the basics, then you can safely skip this chapter.

CHAPTER OBJECTIVES

In this chapter, you will

- Use the Distributive Property to expand and factor expressions
- Use the FOIL method to expand expressions
- Simplify rational expressions (fractions containing a variable)
- Perform arithmetic on rational expressions
- Work with exponent and radical properties

The Distributive Property

A *term* is a quantity separated by a plus or minus sign. For example, the terms in the expression $3x^2y + 10xy + 4xy^2 + 9$ are $3x^2y$, $10xy$, $4xy^2$, and 9. The number in a term is called the *coefficient*. A term without a variable is called a *constant*. The constant in this example is 9, and the coefficients are 3, 10, 4, and 9. Two terms are *alike* if they have the same variables to the same powers. We combine *like* terms by adding/subtracting coefficients on terms that are alike.

EXAMPLE 1-1

Combine like terms.

$$14x^2y + 8y + 3x + 2x^2y - 5y + 7x$$

We begin by rewriting the expression so that like terms are next to each other. After that, we simply add their coefficients.

$$14x^2y + 8y + 3x + 2x^2y - 5y + 7x = (14x^2y + 2x^2y) + (8y - 5y) + (3x + 7x) = 16x^2y + 3y + 10x$$

We use the Distributive Property a lot in algebra. This property allows us to write expressions both in *expanded form* and in *factored form*.

$$\underset{\text{Factored form}}{a(b \pm c)} = \underset{\text{Expanded form}}{ab \pm ac}$$

EXAMPLE 1-2

Use the Distributive Property to expand the expression.

- $3(2xy - 5xy^2)$

$$3(2xy - 5xy^2) = \overset{\text{3 is distributed here.}}{3(2xy) - 3(5xy^2)} = 6xy - 15xy^2$$

- $10x(4y + 6xy - 7x)$

$$10x(4y + 6xy - 7x) = 10x(4y) + 10x(6xy) - 10x(7x)$$
$$= 40xy + 60x^2y - 70x^2$$

Distributing a minus sign or a negative number changes the sign of every term inside the parentheses.

EXAMPLE 1-3

Use the Distributive Property to expand the expression.

- $-4x(3x - 2y + 5)$

$$-4x(3x - 2y + 5) = -4x(3x) - 4x(-2y) - 4x(5)$$
$$= -12x^2 + 8xy - 20x$$

- $3(x^2 - 5x + 2y) - 6x(2x - 4)$

$$3(x^2 - 5x + 2y) - 6x(2x - 4) \overset{\text{Distribute 3 and } -6x}{=} 3x^2 - 15x + 6y - 12x^2 + 24x$$
$$\overset{\text{Combine like terms}}{=} -9x^2 + 9x + 6y$$

We use the *FOIL method* to expand expressions such as $(2x + 3)(x - 4)$. The letters in FOIL help us to keep track of four individual products: First × first + Outer × outer + Inner × inner + Last × last. For example,

$$(\overset{F}{2x} + 3)(\overset{F}{x} - 4) \quad (\overset{O}{2x} + 3)(x - \overset{O}{4}) \quad (2x + \overset{I}{3})(\overset{I}{x} - 4) \quad (2x + \overset{L}{3})(x - \overset{L}{4}).$$

EXAMPLE 1-4

Use the FOIL method to expand the expression.

- $(2x + 3)(x - 4)$

$$(2x + 3)(x - 4) = \overset{F}{2x(x)} - \overset{O}{2x(4)} + \overset{I}{3(x)} + \overset{L}{3(-4)}$$
$$= 2x^2 - 8x + 3x - 12 = 2x^2 - 5x - 12$$

Factoring is the process of using the Distributive Property in reverse. We begin the factorization process by identifying any factor that is in common with each term. For example, each coefficient in $6x^2 - 9y + 15xy + 12$ is divisible by 3, so we can factor 3 from each term.

$$6x^2 - 9y + 15xy + 12 = \overset{\text{Divide 3 from each term}}{3(2x^2) - 3(3y) + 3(5xy) + 3(4)}$$
$$= \overset{\text{3 is now factored}}{3(2x^2 - 3y + 5xy + 4)}$$

Many expressions having three terms can be factored so that the FOIL method gives us the original expression. When factoring $x^2 - 4x + 3$, for example, we begin with *F* in FOIL.

$$x^2 - 4x + 3 = (x \quad\quad)(x \quad\quad)$$

We now concentrate on two factors (*L* in FOIL) that give us the last term. Here, we either want $(-1)(-3)$ or $(1)(3)$. Because the first sign is minus, we choose $(-1)(-3)$ so that O + I in FOIL gives us the middle term.

$$x^2 - 4x + 3 = (x - 1)(x - 3)$$

Let us use the FOIL method to make sure the factorization is correct.

$$(x - 1)(x - 3) = x(x) + x(-3) + (-1)(x) + (-1)(-3)$$
$$= x^2 - 3x - x + 3 = x^2 - 4x + 3$$

If the last term has a lot of factorizations, we can use a shortcut, provided the first term is x^2. Here is the shortcut: if the second sign is "+," we want the *sum* of the factors to be the middle term's coefficient (as in the above example, the sum of 3's factors -1 and -3 added to -4, the middle coefficent). If the second sign is "$-$," we want the *difference* of the factors to give the middle coefficient.

EXAMPLE 1-5

Factor the expression.

- $\mathbf{x^2 - 7x + 10}$

Because the first term is x^2, we can use the shortcut. The second sign is a plus sign, so we want the sum of 10's factors to be -7: $10 = (-2)(-5)$.

$$\mathbf{x^2 - 7x + 10 = (x - 2)(x - 5)}$$

- $x^2 + 6x + 9$

Because the second sign is a plus sign, we want the sum of 9's factors to be 6: 9 = (3)(3).

$$x^2 + 6x + 9 = (x + 3)(x + 3) = (x + 3)^2$$

- $x^2 + 2x - 8$

The second sign is a minus sign, so we want the difference of 8's factors to be +2: $8 = (4)(-2)$.

$$x^2 + 2x - 8 = (x + 4)(x - 2)$$

When the first term is not simply x^2 but something such as $6x^2$, the factoring process takes a little longer. Again, we begin with F and then L (in FOIL), checking the possibilities until we find the factors that give us O + I = middle term.

EXAMPLE 1-6

Factor the expression.

- $4x^2 - 4x - 15$

Because the first term is $4x^2$ and the last term is -15, we have several factorizations to check. The fact that the second sign is a minus sign makes the situation a little worse. Below are the candidates that give us $4x^2$ as the first term and -15 as the last term.

$(4x + 3)(x - 5)$	$(2x + 5)(2x - 3)$	$(4x + 5)(x - 3)$
$(4x - 3)(x + 5)$	$(2x - 5)(2x + 3)$	$(4x - 5)(x + 3)$
$(4x + 1)(x - 15)$	$(2x + 15)(2x - 1)$	$(4x + 15)(x - 1)$
$(4x - 1)((x + 15)$	$(2x - 15)(2x + 1)$	$(4x - 15)(x + 1)$

The factorization that gives us $-4x$ as the middle term is $(2x - 5)(2x + 3)$.

Not every expression in the form $ax^2 + bx + c$ can be factored with this method. For example, $x^2 + 5x + 1$ does not even have a factorization with real numbers. We will see in Chap. 10 how to handle these.

The *difference of two squares* can be factored with the formula $a^2 - b^2 = (a-b)(a+b)$. Once we have decided what quantities are being squared, we can simply use the formula.

EXAMPLE 1-7

Factor the expression.

- $x^2 - 4 = x^2 - 2^2 = (x-2)(x+2)$
- $9x^2 - 1 = (3x)^2 - 1^2 = (3x-1)(3x+1)$
- $\frac{1}{4}x^2 - y^2 = \left(\frac{1}{2}x\right)^2 - y^2 = \left(\frac{1}{2}x - y\right)\left(\frac{1}{2}x + y\right)$

PRACTICE

Expand the expression in Problems 1-5.

1. $4x(5y - 2x + 3)$
2. $-3y(8x - 9y - 1)$
3. $(x-3)(x-4)$
4. $(x-6)(x+6)$
5. $(2x+9)(x-1)$

Factor the expression in Problems 6-13.

6. $24x^2y + 12xy^2 - 30xy$
7. $x^2 + 5x - 14$
8. $x^2 + 5x + 4$
9. $x^2 + 4x + 4$
10. $x^2 - 10x + 25$
11. $x^2 - 64$
12. $4x^2 - 9$
13. $6x^2 + 7x - 20$

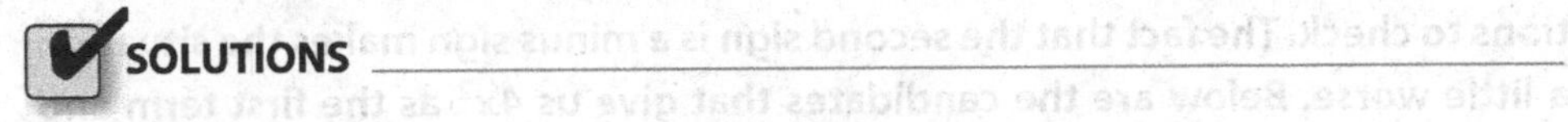

SOLUTIONS

1. $4x(5y - 2x + 3) = 20xy - 8x^2 + 12x$
2. $-3y(8x - 9y - 1) = -24xy + 27y^2 + 3y$
3. $(x-3)(x-4) = x^2 - 4x - 3x + 12 = x^2 - 7x + 12$
4. $(x-6)(x+6) = x^2 - 6x + 6x - 36 = x^2 - 36$
5. $(2x+9)(x-1) = 2x^2 - 2x + 9x - 9 = 2x^2 + 7x - 9$
6. $24x^2y + 12xy^2 - 30xy = 6xy(4x) + 6xy(2y) + 6xy(-5)$
   $= 6xy(4x + 2y - 5)$
7. $x^2 + 5x - 14 = (x+7)(x-2)$
8. $x^2 + 5x + 4 = (x+4)(x+1)$
9. $x^2 + 4x + 4 = (x+2)(x+2) = (x+2)^2$
10. $x^2 - 10x + 25 = (x-5)(x-5) = (x-5)^2$

11. $x^2 - 64 = x^2 - 8^2 = (x - 8)(x + 8)$

12. $4x^2 - 9 = (2x)^2 - 3^2 = (2x - 3)(2x + 3)$

13. $6x^2 + 7x - 20 = (3x - 4)(2x + 5)$

Rational Expressions

Most of the *rational expressions* we will see in this book look like a fraction whose numerator and/or denominator contains a variable.

$$\frac{x}{x+3} \qquad \frac{2}{x^2-5} \qquad \frac{4x-9}{x^2+3x+1}$$

Because we treat a rational expression the same as a fraction, we should be comfortable with fraction arithmetic. The basic fraction operations are summarized in Table 1-1.

TABLE 1-1

Property	Example(s)
Multiplying Fractions $\frac{a}{b} \cdot \frac{c}{d} = \frac{ac}{bd}$ Multiply the numerators and denominators.	• $\frac{5}{4} \cdot \frac{1}{3} = \frac{5}{12}$ • $9(\frac{3}{4}) = \frac{9}{1} \cdot \frac{3}{4} = \frac{27}{4}$
Dividing Fractions $\frac{a}{b} \div \frac{c}{d} = \frac{a}{b} \cdot \frac{d}{c}$ Change to a multiplication problem by inverting ("flipping") the second fraction.	• $\frac{1}{2} \div \frac{1}{3} = \frac{1}{2} \cdot \frac{3}{1} = \frac{3}{2}$ • $\frac{3}{4} \div \frac{2}{5} = \frac{3}{4} \cdot \frac{5}{2} = \frac{15}{8}$
Adding Fractions (with like denominators) $\frac{a}{b} + \frac{c}{b} = \frac{a+c}{b}$ When the denominators are the same, add/subtract the numerators.	• $\frac{1}{5} + \frac{2}{5} = \frac{1+2}{5} = \frac{3}{5}$ • $\frac{6}{7} - \frac{2}{7} = \frac{6-2}{7} = \frac{4}{7}$
Adding Fractions (with unlike denominators) $\frac{a}{b} + \frac{c}{d} = \frac{ad+bc}{bd}$ When denominators are not alike, rewrite each fraction so that they have a common denominator.	• $\frac{1}{4} + \frac{3}{5} = \frac{1(5)+4(3)}{4(5)}$ $= \frac{5+12}{20} = \frac{17}{20}$
Simplifying Fractions $\frac{ab}{cb} = \frac{a}{c}$ If the numerator and denominator have a common factor, it can be divided out (also called *canceled*).	• $\frac{24}{36} = \frac{2(12)}{3(12)} = \frac{2}{3} \cdot \frac{12}{12}$ $= \frac{2}{3}(1) = \frac{2}{3}$

Simplifying Rational Expressions

A rational expression (or any kind of fraction) is in *lowest terms* if the numerator and denominator have no common factors (other than 1, of course). We simplify (or *reduce*) a fraction by dividing the numerator and denominator by their common factor(s). This process is also called *canceling.*

EXAMPLE 1-8

Simplify the rational expression.

- $\dfrac{6x^2y}{15xy}$

The first step in simplifying a rational expression is to factor its numerator and denominator, and then we divide out any common factor.

$$\frac{6x^2y}{15xy} = \frac{(3xy)(2x)}{(3xy)(5)} = \frac{2x}{5}$$

- $\dfrac{x^2+2x-8}{x^2-3x+2} = \dfrac{(x+4)(x-2)}{(x-1)(x-2)} = \dfrac{x+4}{x-1}$
- $\dfrac{x^2-4}{x^2+5x+6} = \dfrac{(x-2)(x+2)}{(x+3)(x+2)} = \dfrac{x-2}{x+3}$

Still Struggling

Only *factors* can be canceled, not *terms*. For example, $\frac{x^2}{x+4}$ is already in lowest terms.

Multiplying Rational Expressions

We multiply rational expressions by multiplying their numerators and denominators. If any numerator has a factor in common with any denominator, we can cancel it before multiplying.

EXAMPLE 1-9

Find the product.

- $\dfrac{15xy^2}{14}\cdot\dfrac{8}{3xy}$

We begin by writing the numerators and denominators so that the common factors are obvious. This step is not necessary.

2 and 3*xy* are common factors

$$\frac{15xy^2}{14}\cdot\frac{8}{3xy}=\frac{(3xy)(5y)\cdot(4)(2)}{(2)(7)\cdot 3xy}=\frac{5y(4)}{7}=\frac{20y}{7}$$

- $\dfrac{2x^2+3x-2}{x^2+2x-15}\cdot\dfrac{x^2+9x+20}{3x^2+10x-8}$

We must factor the numerators and denominators so that we can tell if there are any common factors.

$$\frac{2x^2+3x-2}{x^2+2x-15}\cdot\frac{x^2+9x+20}{3x^2+10x-8}=\frac{(2x-1)(x+2)}{(x+5)(x-3)}\cdot\frac{(x+4)(x+5)}{(x+4)(3x-2)}$$

$$=\frac{(2x-1)(x+2)}{(x-3)(3x-2)}\quad x+4 \text{ and } x+5 \text{ are canceled.}$$

$$=\frac{2x^2+3x-2}{3x^2-11x+6}\quad \text{Use the FOIL method.}$$

Adding Rational Expressions

We cannot add rational expressions until they have the same denominator. Once we have found a common denominator, we rewrite the fractions so that they have a common denominator and then add their numerators.

EXAMPLE 1-10

Perform the addition (or subtraction).

- $\dfrac{1}{2x}+\dfrac{2}{3x}-\dfrac{x}{6}$

We want the smallest denominator, called the *least common denominator* (LCD), that is divisible by each of $2x$, $3x$, and 6. This would be $6x$.

Multiply the first fraction by 3/3	Multiply the second fraction by 2/2	Multiply the third fraction by x/x
$6x = 2x(3)$	$6x = 3x(2)$	$6x = 6(x)$

$$\frac{1}{2x} + \frac{2}{3x} - \frac{x}{6} = \frac{1}{2x} \cdot \frac{3}{3} + \frac{2}{3x} \cdot \frac{2}{2} - \frac{x}{6} \cdot \frac{x}{x}$$

$$= \frac{3}{6x} + \frac{4}{6x} - \frac{x^2}{6x} = \frac{3 + 4 - x^2}{6x} = \frac{-x^2 + 7}{6x}$$

- $\dfrac{x+1}{x^2 - 16} + \dfrac{5}{x^2 + x - 12}$

We begin by factoring each denominator and then finding the LCD.

The LCD is $(x-4)(x+4)(x-3)$

$$\frac{x+1}{x^2 - 16} + \frac{5}{x^2 + x - 12} = \frac{x+1}{(x-4)(x+4)} + \frac{5}{(x-3)(x+4)}$$

$$= \frac{(x+1)(x-3)}{(x-4)(x+4)(x-3)} + \frac{5(x-4)}{(x-4)(x+4)(x-3)}$$

Use the FOIL method on the numerator

$$= \frac{x^2 - 2x - 3}{(x-4)(x+4)(x-3)} + \frac{5x - 20}{(x-4)(x+4)(x-3)}$$

Add the numerators; Combine like terms

$$= \frac{x^2 - 2x - 3 + 5x - 20}{(x-4)(x+4)(x-3)} = \frac{x^2 + 3x - 23}{(x-4)(x+4)(x-3)}$$

PRACTICE

Simplify the rational expression in Problems 1-3.

1. $\dfrac{30x^2y^2z}{18x^2yz}$ 2. $\dfrac{6 - 3x}{2 - x}$ 3. $\dfrac{3x^2 - 10x - 8}{x^2 + 2x - 24}$

Multiply the rational expression in Problems 4-6. Leave the denominator in factored form.

4. $\dfrac{2x}{5y} \cdot \dfrac{15y^2}{8}$ 5. $\dfrac{1}{4x^2 - 25} \cdot \dfrac{2x^2 + 7x + 5}{x + 1}$

6. $\dfrac{x^2+10x+25}{4x^2+5x-6}\cdot\dfrac{4x-3}{x^2+6x+5}$

Perform the addition and/or subtraction of the rational expression in Problems 7-9. Leave the denominator in factored form.

7. $\dfrac{1}{3x^2}+\dfrac{2}{15y}-\dfrac{3}{xy}$

8. $\dfrac{1}{x^2-x}+\dfrac{1}{x^3+x^2}+\dfrac{1}{x^2-1}$

9. $\dfrac{2}{4x^2+3x-1}-\dfrac{1}{16x^2-1}+\dfrac{4}{x^2+3x+2}$

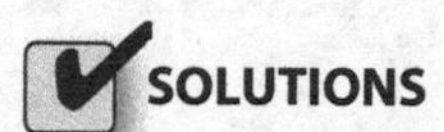

SOLUTIONS

1. $\dfrac{30x^2y^2z}{18x^2yz}=\dfrac{6x^2yz(5y)}{6x^2yz(3)}=\dfrac{5y}{3}$

2. $\dfrac{6-3x}{2-x}=\dfrac{3(2-x)}{1(2-x)}=\dfrac{3}{1}=3$

3. $\dfrac{3x^2-10x-8}{x^2+2x-24}=\dfrac{(3x+2)(x-4)}{(x+6)(x-4)}=\dfrac{3x+2}{x+6}$

4. $\dfrac{2x}{5y}\cdot\dfrac{15y^2}{8}=\dfrac{2x}{5y}\cdot\dfrac{5y(3y)}{2(4)}=\dfrac{3xy}{4}$

5. $\dfrac{1}{4x^2-25}\cdot\dfrac{2x^2+7x+5}{x+1}=\dfrac{1}{(2x-5)(2x+5)}$

$$\times\frac{(2x+5)(x+1)}{x+1}=\frac{1}{2x-5}$$

6. $\dfrac{x^2+10x+25}{4x^2+5x-6}\cdot\dfrac{4x-3}{x^2+6x+5}=\dfrac{(x+5)(x+5)}{(4x-3)(x+2)}\cdot\dfrac{4x-3}{(x+5)(x+1)}$

$$=\frac{x+5}{(x+2)(x+1)}$$

The LCD is $15x^2y$

7. $\dfrac{1}{3x^2}+\dfrac{2}{15y}-\dfrac{3}{xy}=\dfrac{1}{3x^2}\cdot\dfrac{5y}{5y}+\dfrac{2}{15y}\cdot\dfrac{x^2}{x^2}-\dfrac{3}{xy}\cdot\dfrac{15x}{15x}$

$$=\frac{5y}{15x^2y}+\frac{2x^2}{15x^2y}-\frac{45x}{15x^2y}=\frac{5y+2x^2-45x}{15x^2y}$$

8. $\dfrac{1}{x^2-x}+\dfrac{1}{x^3+x^2}+\dfrac{1}{x^2-1}$

Factor the denominators

$$=\frac{1}{x(x-1)}+\frac{1}{x^2(x+1)}+\frac{1}{(x-1)(x+1)}$$

The LCD is $x^2(x-1)(x+1)$

$$=\frac{1}{x(x-1)}\cdot\frac{x(x+1)}{x(x+1)}+\frac{1}{x^2(x+1)}\cdot\frac{x-1}{x-1}+\frac{1}{(x-1)(x+1)}\cdot\frac{x^2}{x^2}$$

Add the numerators

$$=\frac{x^2+x+x-1+x^2}{x^2(x-1)(x+1)}=\frac{2x^2+2x-1}{x^2(x-1)(x+1)}$$

9. $\dfrac{2}{4x^2+3x-1}-\dfrac{1}{16x^2-1}+\dfrac{4}{x^2+3x+2}$

Factor the denominators

$$=\frac{2}{(4x-1)(x+1)}-\frac{1}{(4x-1)(4x+1)}+\frac{4}{(x+2)(x+1)}$$

The LCD is $(4x-1)(4x+1)(x+1)(x+2)$

$$=\frac{2}{(4x-1)(x+1)}\cdot\frac{(4x+1)(x+2)}{(4x+1)(x+2)}$$

$$-\frac{1}{(4x-1)(4x+1)}\cdot\frac{(x+1)(x+2)}{(x+1)(x+2)}$$

$$+\frac{4}{(x+2)(x+1)}\cdot\frac{(4x-1)(4x+1)}{(4x-1)(4x+1)}$$

$$=\frac{2[(4x+1)(x+2)]-[(x+1)(x+2)]+4[(4x-1)(4x+1)]}{(4x-1)(4x+1)(x+1)(x+2)}$$

Use the FOIL method

$$=\frac{2(4x^2+9x+2)-(x^2+3x+2)+4(16x^2-1)}{(4x-1)(4x+1)(x+1)(x+2)}$$

Use the Distributive Property

$$=\frac{8x^2+18x+4-x^2-3x-2+64x^2-4}{(4x-1)(4x+1)(x+1)(x+2)}$$

Combine like terms

$$=\frac{71x^2+15x-2}{(4x-1)(4x+1)(x+1)(x+2)}$$

TABLE 1-2 Basic Exponent Properties

Let a and b be nonzero real numbers and let m and n be any real numbers.

Property		Example
E.1	$a^m a^n = a^{m+n}$	$y^2 y^4 = y^{2+4} = y^6$
E.2	$\frac{a^m}{a^n} = a^{m-n}$	$\frac{x^5}{x^2} = x^{5-2} = x^3$
E.3	$\frac{1}{a} = a^{-1}$	$\frac{1}{4} = 4^{-1}$
E.4	$\frac{1}{a^n} = a^{-n}$	$\frac{1}{t^2} = t^{-2}$
E.5	$a^0 = 1$	$5^0 = 1$
E.6	$(a^m)^n = a^{mn}$	$(x^2)^5 = x^{(2)(5)} = x^{10}$
E.7	$(ab)^n = a^n b^n$	$(3x)^2 = 3^2 x^2 = 9x^2$
E.8	$(\frac{a}{b})^n = \frac{a^n}{b^n}$	$(\frac{x}{y})^4 = \frac{x^4}{y^4}$

Exponents and Roots

On occasion, we will work with exponents and radicals (such as square roots), so we should review their basic properties. Exponent properties are summarized in Table 1-2, and radical properties are summarized in Table 1-3.

We use exponent properties to rewrite expressions, sometimes using more than one property on the same expression.

EXAMPLE 1-11

Use exponent properties to rewrite the expression. Do not leave negative exponents in the answer.

- $3(2x^2)^4 = 3(2^4 \cdot x^{(2)(4)}) = 3(16)(x^8) = 48x^8$

TABLE 1-3 Basic Radical Properties

Let a and b be real numbers and let m and n be positive integers.
If n is even, a cannot be negative. The root symbol $\sqrt{\ }$ is called a *radical.*

Property		Property	
R.1	$\sqrt[n]{a} = b$ if $b^n = a$	R.2	$\sqrt[n]{a^n} = (\sqrt[n]{a})^n$
R.3	$\sqrt[n]{a^n} = a$	R.4	$\sqrt[n]{ab} = \sqrt[n]{a}\sqrt[n]{a}$
R.5	$\sqrt[n]{\frac{a}{b}} = \frac{\sqrt[n]{a}}{\sqrt[n]{b}}$, $b \neq 0$	R.6	$\sqrt[n]{a} = a^{1/n}$
R.7	$\sqrt[n]{a^m} = a^{m/n}$		

- $5x^3 \cdot 2x^{-3} = 5 \cdot 2 \cdot x^{3-3} = 10x^0 = 10(1) = 10$
- $8(x^{-1})^2 = 8(x^{(-1)(2)}) = 8x^{-2} = 8\left(\frac{1}{x^2}\right) = \frac{8}{x^2}$
- $\left(\frac{2x}{3y}\right)^2 = \frac{(2x)^2}{(3y)^2} = \frac{2^2x^2}{3^2y^2} = \frac{4x^2}{9y^2}$
- $\frac{6x^4}{(2x^3)^2} = \frac{6x^4}{2^2x^{(3)(2)}} = \frac{6x^4}{4x^6} = \frac{3x^4}{2x^6} = \frac{3}{2}x^{4-6} = \frac{3}{2}x^{-2} = \frac{3}{2} \cdot \frac{1}{x^2} = \frac{3}{2x^2}$

Exponents and radicals are closely related. For example, $\sqrt[3]{8} = 2$ because $2^3 = 8$: we say that 2 is the cube root of 8. We will use radicals (also called *roots*) and exponents in solving certain kinds of equations and in our work with logarithms and exponents in Chap. 12. The basic radical properties are summarized in Table 1-3.

EXAMPLE 1-12

Use radical properties to rewrite the expression. When necessary assume that the variable represents a positive number.

- $\sqrt{4x^2y} \overset{\text{Property R.4}}{=} \sqrt{4}\sqrt{x^2}\sqrt{y} \overset{\text{Property R.3}}{=} 2x\sqrt{y}$
- $\sqrt{\frac{16x}{9y^2}} \overset{\text{Property R.5}}{=} \frac{\sqrt{16x}}{\sqrt{9y^2}} \overset{\text{Property R.4}}{=} \frac{\sqrt{16}\sqrt{x}}{\sqrt{9}\sqrt{y^2}} \overset{\text{Property R.3}}{=} \frac{4\sqrt{x}}{3y}$
- $\sqrt{36x^6} \overset{\text{Property R.4 and E.6}}{=} \sqrt{36}\sqrt{(x^3)^2} \overset{\text{Property R.3}}{=} 6x^3$

Later, when we use the quadratic formula, we will have quantities that involve a square root. Usually, these quantities need to be simplified. Of course, an expression such as $\sqrt{36}$ is easy to simplify, but other quantities need a little more work. If a quantity has a perfect square as a factor, then that factor needs to come out of the square root. For example, 12 has 4, a perfect square, as a factor, so $\sqrt{12}$ needs to be simplified. We can do so with Property R.3 and Property R.4.

$$\sqrt{12} = \sqrt{4 \cdot 3} \overset{\text{Property R.4}}{=} \sqrt{4}\sqrt{3} \overset{\text{Property R.3}}{=} 2\sqrt{3}$$

Sometimes, we must do this simplification in combination with reducing a fraction. For example, $\frac{8+\sqrt{12}}{2}$ can be simplified as a fraction *after* simplifying $\sqrt{12}$.

$$\frac{8+\sqrt{12}}{2} = \overset{\text{Everything is divisible by 2}}{\frac{8+2\sqrt{3}}{2}} = \overset{\text{Factor the numerator}}{\frac{2(4+\sqrt{3})}{2}} \overset{\text{Cancel}}{=} 4+\sqrt{3}$$

EXAMPLE 1-13

Simplify the square root and, if necessary, the fraction.

- $\sqrt{18} = \sqrt{9(2)} = \sqrt{9}\sqrt{2} = 3\sqrt{2}$
- $\frac{6-\sqrt{24}}{8} = \frac{6-\sqrt{4(6)}}{8} = \overset{\text{Pull out the square}}{\frac{6-2\sqrt{6}}{8}}$

  $= \overset{\text{Everything divisible by 2}}{\frac{2(3-\sqrt{6})}{8}} = \overset{\text{Cancel}}{\frac{3-\sqrt{6}}{4}}$

Radical Properties R.6 and R.7 allow us to replace root symbols with fraction exponents and vice versa.

EXAMPLE 1-14

Rewrite the expression using Radical Properties R.6 and R.7.

- $\sqrt{y} = y^{1/2}$
- $\sqrt{x^3} = x^{3/2}$
- $\sqrt[3]{6y} = (6y)^{1/3}$
- $19^{1/2} = \sqrt{19}$
- $7^{2/3} = \sqrt[3]{7^2} = \sqrt[3]{49}$

PRACTICE

Use the Radical Properties to rewrite the expression in Problems 1-6. Assume variables are positive where necessary.

1. $\sqrt{\frac{16}{25}}$
2. $\sqrt{100x^2y^4}$
3. $\sqrt{\frac{81x^2}{y^2}}$
4. $\sqrt{27}$
5. $\sqrt{20}$
6. $\sqrt{63x^2}$

Simplify the fraction in Problems 7-9.

7. $\dfrac{6+\sqrt{18}}{9}$ 8. $\dfrac{10-\sqrt{28}}{4}$ 9. $\dfrac{12+\sqrt{80}}{4}$

Rewrite the expression in Problems 10-14 using Radical Properties R.6 and R.7.

10. $10^{1/3}$ 11. $3^{3/4}$ 12. $\sqrt{2x}$ 13. $\sqrt[3]{14}$ 14. $\sqrt[5]{x^4}$

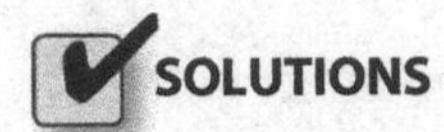

SOLUTIONS

1. $\sqrt{\dfrac{16}{25}} = \dfrac{\sqrt{16}}{\sqrt{25}} = \dfrac{4}{5}$

2. $\sqrt{100x^2y^4} = \sqrt{10^2x^2(y^2)^2} = 10xy^2$

3. $\sqrt{\dfrac{81x^2}{y^2}} = \dfrac{\sqrt{81}\sqrt{x^2}}{\sqrt{y^2}} = \dfrac{9x}{y}$

4. $\sqrt{27} = \sqrt{9(3)} = \sqrt{9}\sqrt{3} = 3\sqrt{3}$

5. $\sqrt{20} = \sqrt{4(5)} = \sqrt{4}\sqrt{5} = 2\sqrt{5}$

6. $\sqrt{63x^2} = \sqrt{9(7)x^2} = \sqrt{9}\sqrt{7}\sqrt{x^2} = 3x\sqrt{7}$

7. $\dfrac{6+\sqrt{18}}{9} = \dfrac{6+\sqrt{9(2)}}{9} = \dfrac{6+3\sqrt{2}}{9} = \dfrac{3(2+\sqrt{2})}{9} = \dfrac{2+\sqrt{2}}{3}$

8. $\dfrac{10-\sqrt{28}}{4} = \dfrac{10-\sqrt{4(7)}}{4} = \dfrac{10-2\sqrt{7}}{4}$

$= \dfrac{2(5-\sqrt{7})}{4} = \dfrac{5-\sqrt{7}}{2}$

9. $\dfrac{12+\sqrt{80}}{4} = \dfrac{12+\sqrt{16(5)}}{4} = \dfrac{12+4\sqrt{5}}{4}$

$= \dfrac{4(3+\sqrt{5})}{4} = 3+\sqrt{5}$

10. $10^{1/3} = \sqrt[3]{10}$ 11. $3^{3/4} = \sqrt[4]{3^3} = \sqrt[4]{27}$ 12. $\sqrt{2x} = (2x)^{1/2}$

13. $\sqrt[3]{14} = 14^{1/3}$ 14. $\sqrt[5]{x^4} = x^{4/5}$

Summary

In this chapter, we learned how to

- *Use the Distributive Property*. This property allows us to write expressions in expanded form or factored form. If every term in an expression is divisible by the same quantity, this quantity can be factored: $ab + ac = a(b + c)$.
- *Use the FOIL method*. The letters in "FOIL" help us to keep track of four products necessary to use the Distributive Property on expressions of the form $(a + b)(c + d)$. We also learned how to factor expressions of the form $x^2 + 3x + 2$.
- *Simplify rational expressions*. A rational expression is simplified if the only factor in common with the numerator and denominator is 1. To simplify the rational expression, we factor the numerator and denominator and divide out (cancel) any common factors.
- *Multipy and divide rational expressions*. We find the product of two rational expressions by multiplying their numerators and multiplying their denominators. We cancel any common factors before multiplying. We divide one rational expression by another by inverting (flipping) the second fraction and then multiplying them.
- *Work with exponent and radical properties*. These are summarized in Table 1-4. We assume that a and b are nonzero real numbers, and if a root is even, then we assume that a and b are not negative. Finally, we assume that m and n are positive integers.

TABLE 1-4

Exponent Property	Radical Property
E.1 $a^m a^n = a^{m+n}$	R.1 $\sqrt[n]{a} = b$ if $b^n = a$
E.2 $\frac{a^m}{a^n} = a^{m-n}$	R.2 $\sqrt[n]{a^n} = (\sqrt[n]{a})^n$
E.3 $\frac{1}{a} = a^{-1}$, $a \neq 0$	R.3 $\sqrt[n]{a^n} = a$
E.4 $\frac{1}{a^n} = a^{-n}$, $a \neq 0$	R.4 $\sqrt[n]{ab} = \sqrt[n]{a}\sqrt[n]{b}$
E.5 $a^0 = 1$	R.5 $\sqrt[n]{\frac{a}{b}} = \frac{\sqrt[n]{a}}{\sqrt[n]{b}}$, $b \neq 0$
E.6 $(a^m)^n = a^{mn}$	R.6 $\sqrt[n]{a} = a^{1/n}$
E.7 $(ab)^n = a^n b^n$	R.7 $\sqrt[n]{a^m} = a^{m/n}$
E.8 $\left(\frac{a}{b}\right)^n = \frac{a^n}{b^n}$	

QUIZ

Assume that $x \geq 0$ and $y \geq 0$, if they appear under an even root.

1. $8xy(3x + 5y - xy + 2) =$

A. $24x^2y + 40xy^2 + 8x^2y^2 + 16xy$
B. $56x^2y^2 + 16xy$
C. $24x^2y + 40xy^2 - 8x^2y^2 + 16xy$
D. $24x^2y + 40xy - 8x^2y^2 - 16xy$

2. $-4y(3 - 2y - 10x) =$

A. $-12y + 8y^2 - 40xy$
B. $-12y + 8y^2 + 40xy$
C. $-12y - 8y^2 - 40xy$
D. $-12y - 8y^2 + 40xy$

3. $4y^2 + 8y - 12xy =$

A. $4y(y + 2 - 3x)$
B. $4y(y + 2y - 3xy)$
C. $2y(2y^2 + 4 - 3x)$
D. $4y^2(1 + 2 - 3x)$

4. $(2x - 5)(x - 3) =$

A. $2x^2 + x + 15$
B. $2x^2 - 11x + 15$
C. $2x^2 + 15$
D. $2x^2 - 15$

5. $(3x + 1)^2 =$

A. $9x^2 + 1$
B. $3x^2 + 1$
C. $3x^2 + 9x + 1$
D. $9x^2 + 6x + 1$

6. $49x^2 - \frac{1}{4} =$

A. $(7x - \frac{1}{2})^2$
B. $(7x - \frac{1}{2})(7x + \frac{1}{2})$
C. $(49x - \frac{1}{2})(x - \frac{1}{2})$
D. $(49x - \frac{1}{2})(x + \frac{1}{2})$

7. $\frac{1}{x^2 - 4} + \frac{2}{x^2 + 4x + 4} =$

A. $\frac{x^2 + 6x + 2}{(x-2)(x+2)^2}$
B. $\frac{3x - 2}{(x-2)(x+2)^2}$
C. $\frac{3}{(x-2)(x+2)^2}$
D. $\frac{2}{(x-2)(x+2)^2}$

8. $\frac{x^2 + 4x}{x^2 - 2x - 3} \cdot \frac{x - 3}{x^2 + 9x + 20} =$

A. $\frac{-x}{x^2 + 6x + 5}$
B. $\frac{x - 4}{x^2 + 6x + 5}$
C. $\frac{x}{x^2 + 6x + 5}$
D. $\frac{4}{x^2 + 6x + 5}$

9. $\sqrt{32x^6} =$

A. $4x^2\sqrt{2x}$ B. $16x^3\sqrt{2}$ C. $16x^3$ D. $4x^3\sqrt{2}$

10. $\frac{(6x)^2(2y)^3}{30x^3y^2} =$

A. $\frac{2y}{5x}$ B. $\frac{8y}{x}$ C. $\frac{48}{5x^2y}$ D. $\frac{48y}{5x}$

11. $\sqrt{\frac{32x^2}{25y^2}} =$

A. $\frac{4x\sqrt{2}}{5y}$ B. $\frac{16x}{5y}$ C. $\frac{8x\sqrt{2}}{5y}$ D. $\frac{2x\sqrt{8}}{5y}$

12. $(5x)^{2/3} =$

A. $\sqrt[3]{25x^2}$ B. $\sqrt[3]{5x^2}$ C. $\sqrt{125x^3}$ D. $\sqrt{5x^3}$

13. $\sqrt[5]{21y^2} =$

A. $21y^{2/5}$ B. $(21y^2)^{1/5}$ C. $(21y)^{2/5}$ D. $(21y)^{1/5}$

14. $\frac{18+\sqrt{27}}{6} =$

A. $\frac{3+3\sqrt{3}}{2}$ B. $\frac{1+\sqrt{3}}{2}$ C. $\frac{6+\sqrt{3}}{2}$ D. $\frac{2+\sqrt{3}}{2}$

chapter 2

Linear Equations and Inequalities

Throughout any college algebra course (and many other mathematics courses), students solve equations and inequalities, many of them *linear*. An equation is a linear equation if the power on the variable(s) is 1. In this chapter, we begin with basic linear equations. From there, we will move onto linear inequalities (solved using almost the same strategy) and to equations that lead to linear equations after a few steps. Finally, we will work with absolute value equations and inequalities.

CHAPTER OBJECTIVES

In this chapter, you will

- Solve linear equations
- Solve linear inequalities
- Solve equations that lead to linear equations
- Solve compound inequalities
- Solve absolute value equations and inequalities

Basic Linear Equations

The strategy for solving linear equations (outlined in Table 2-1) can be adapted to solving other types of equations, one of which we will see later in Chap. 2.

EXAMPLE 2-1

Solve the equation.

- **$4x - 9 = 1$**

$$4x - 9 = 1$$

$$4x = 10 \qquad \text{Step 2: Add 9 (Step 1 is not necessary.)}$$

$$\frac{4x}{4} = \frac{10}{4} \qquad \text{Step 3: Divide by 4.}$$

$$x = \frac{5}{2} \qquad \text{Simplify.}$$

- **$2(5x - 3) + 1 = 4(x + 1)$**

$$2(5x - 3) + 1 = 4(x + 1)$$

$$10x - 6 + 1 = 4x + 4 \qquad \text{Step 1: Use the Distributive Property.}$$

$$10x - 5 = 4x + 4 \qquad \text{Simplify the left side.}$$

TABLE 2-1

	Solving Basic Linear Equations	Example
1.	If necessary, simplify each side of the equation; use the Distributive Property and combine like terms.	$5x - 3 = 3(x + 1) - 2$ $5x - 3 = 3x + 3 - 2$ $5x - 3 = 3x + 1$
2.	Collect terms with the variable on one side of the equation and terms without a variable on the other side.	$5x - 3 = 3x + 1$ $-3x + 3 - 3x + 3$ $2x = 4$
3.	Divide each side of the equation by the variable's coefficient (the number in front of the variable), and simplify.	$\frac{2x}{2} = \frac{4}{2}$ $x = 2$

$6x = 9$ **Step 2: Subtract 4*x*, add 5.**

$\frac{6x}{6} = \frac{9}{6}$ **Step 3: Divide by *x*'s coefficient.**

$x = \frac{3}{2}$ **Simplify.**

If fractions appear in an equation, we can eliminate them and be left with a simpler equation. After we find the least common denominator (LCD), we multiply each side of the equation by the LCD and simplify.

EXAMPLE 2-2

Solve the equation.

- $\frac{3}{4}(x+2) = 6$

$\frac{3}{4}(x+2) = 6$ **The LCD is 4.**

$\left[4\left(\frac{3}{4}\right)\right](x+2) = 4(6)$ **Multiply each side by the LCD.**

$3(x+2) = 24$ **Now, proceed as before.**

$3x + 6 = 24$ **Step 1**

$3x = 18$ **Step 2**

$x = 6$ **Step 3**

- $\frac{5}{6}x + 2 = -\frac{1}{15}x + 1$

The smallest number divisible by both 6 and 15 is 30, so we begin by multiplying each side by 30.

$30\left(\frac{5}{6}x + 2\right) = 30\left(-\frac{1}{15}x + 1\right)$ **Multiplying by the LCD eliminates the fractions.**

$25x + 60 = -2x + 30$ **Distribute 30: $30\left(\frac{5}{6}\right) = 25$; $30\left(-\frac{1}{15}\right) = -2$.**

$$27x = -30 \qquad \text{Add } 2x\text{, subtract 60.}$$

$$x = -\frac{30}{27} = -\frac{10}{9} \qquad \text{Divide by 27 and simplify.}$$

PRACTICE

Solve the equation.

1. $7(2x - 3) = 5(x + 1) - 10$
2. $\frac{3}{5}(x - 4) = 6$
3. $\frac{1}{4}(2x + 9) = \frac{2}{3}(x + 4)$

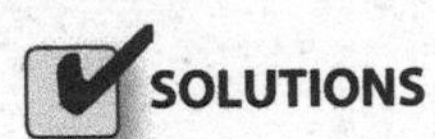

SOLUTIONS

1. Begin with the Distributive Property on the left and the right sides.

$$7(2x - 3) = 5(x + 1) - 10$$

$$14x - 21 = 5x + 5 - 10$$

$$14x - 21 = 5x - 5$$

$$9x = 16$$

$$x = \frac{16}{9}$$

2. The LCD is 5.

$$\left[5\left(\frac{3}{5}\right)\right](x - 4) = 5(6)$$

$$3(x - 4) = 30$$

$$3x - 12 = 30$$

$$3x = 42$$

$$x = \frac{42}{3} = 14$$

3. The LCD is 12.

$$\left[12\left(\frac{1}{4}\right)\right](2x+9)=\left[12\left(\frac{2}{3}\right)\right](x+4)$$

$$3(2x+9)=8(x+4)$$

$$6x+27=8x+32$$

$$-2x=5$$

$$x=-\frac{5}{2}$$

Equations Leading to Linear Equations

In a linear equation, all the powers on the variable are 1; there are no variables in a denominator and no variable under a radical.

Not linear $x^2-4x-5=0$ Not linear $3\sqrt{x}-8=1$ Not linear $\frac{1}{x-3}=\frac{4}{5}$

Some nonlinear equations become linear equations after only a few steps. On occasion, the solutions we find are *extraneous*. This means that the solution to the linear equation is not a solution to the original equation. We will know that a solution is extraneaous if substituting it in the original equation causes something "illegal" such as division by 0.

We begin with equations involving x in the denominator. We will treat these the same as regular linear equations that have fractions in them: we identify the LCD (which will have x in it) and then multiply each side of the equation by the LCD.

EXAMPLE 2-3

- $\frac{1}{x}-\frac{1}{x+1}=\frac{1}{4x}$

The LCD is $4x(x+1)$, so we multiply each side of the equation by this quantity and then simplify.

$$4x(x+1)\left(\frac{1}{x}-\frac{1}{x+1}\right)=4x(x+1)\left(\frac{1}{4x}\right)$$

$$4x(x+1)\left(\frac{1}{x}\right) - 4x(x+1)\left(\frac{1}{x+1}\right) = x+1$$ Distribute the LCD.

$$4(x+1) - 4x = x+1$$ This is now a linear equation.

$$4x + 4 - 4x = x+1$$

$$4 = x+1$$

$$3 = x$$

Since $x = 3$ does not cause a 0 in any denominator, the solution is $x = 3$.

- $\dfrac{x+1}{x-4} = \dfrac{x-1}{x+3}$

The LCD is $(x-4)(x+3)$.

$$(x-4)(x+3)\left(\frac{x+1}{x-4}\right) = (x-4)(x+3)\left(\frac{x-1}{x+3}\right)$$

$$(x+3)(x+1) = (x-4)(x-1)$$ The denominators cancel.

$$x^2 + 4x + 3 = x^2 - 5x + 4$$ Use the FOIL method.

$$4x + 3 = -5x + 4$$ The x^2s cancel.

$$9x = 1$$

$$x = \frac{1}{9}$$

- $\dfrac{x^2-1}{x+1} = -2$

Notice that the fraction on the left can be simplified because the numerator, $x^2 - 1$, factors.

$$\frac{x^2-1}{x+1} = -2$$

$$\frac{(x-1)(x+1)}{x+1} = -2$$ Cancel.

$$x - 1 = -2$$

$$x = -1$$

This solution is extraneous because we cannot allow $x = -1$ in the original equation. The equation, then, has no solution.

Some equations involving radicals also lead to linear equations. Once we isolate the radical on one side of the equation, we raise each side of the equation to a power that eliminates the radical. For example, squaring each side of $\sqrt{x} = 3$ gives us $x = 3^2 = 9$. Remember, though, if the root is even (such as a square root), then the number under the radical cannot be negative.

EXAMPLE 2-4

- $4\sqrt{x} + 1 = 9$

We isolate $\sqrt{x}$ by subtracting 1 from each side and then dividing by 4.

$$4\sqrt{x} + 1 = 9$$

$$4\sqrt{x} = 8$$

$$\sqrt{x} = \frac{8}{4} = 2$$

$$\left(\sqrt{x}\right)^2 = 2^2 \qquad \textbf{Square each side.}$$

$$x = 4$$

- $\frac{1}{2}\sqrt[3]{4x} = 3$

$$\sqrt[3]{4x} = 6 \qquad \textbf{Multiply both sides by 2 to isolate the radical.}$$

$$\left(\sqrt[3]{4x}\right)^3 = 6^3 \qquad \textbf{Raise each side by 3 to eliminate the root.}$$

$$4x = 216$$

$$x = \frac{216}{4} = 54$$

- $\sqrt{x} = -6$

The square root of a real number is never negative, so the equation has no (real) solution. (Later, we will work with *complex numbers*, which can have negative square roots.)

- $\sqrt{3x+1} = 4$

The radical is already isolated, so we begin by squaring each side.

$$\left(\sqrt{3x+1}\right)^2 = 4^2$$

$$3x+1 = 16$$

$$3x = 15$$

$$x = 5$$

PRACTICE

Solve the equation.

1. $\dfrac{4}{2x-1} = 8$ 2. $\dfrac{3}{x-2} + \dfrac{1}{x+4} = \dfrac{5}{x-2}$ 3. $\dfrac{2x-9}{x-1} = \dfrac{2x+5}{x+1}$

4. $\dfrac{x^2+3x-10}{x-2} = 7$ **(Hint: The fraction can be simplified.)**

5. $4\sqrt{x} = 12$ 6. $8\sqrt{x} - 1 = 15$ 7. $2\sqrt{x-3} = 1$

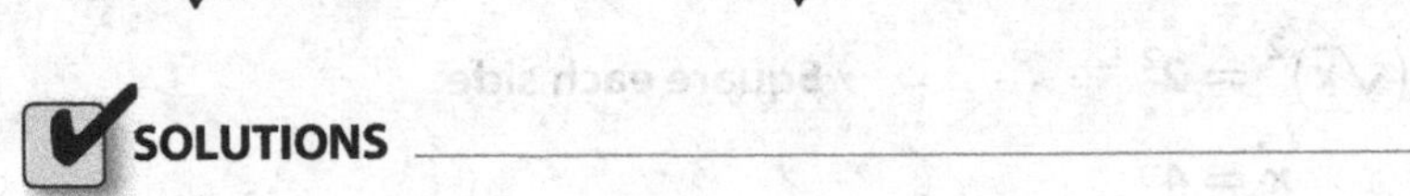

SOLUTIONS

1. **We begin by multiplying each side of the equation by $2x - 1$.**

$$(2x-1)\left(\frac{4}{2x-1}\right) = (2x-1)8$$

$$4 = 16x - 8 \qquad \text{Simplify on the left, distribute 8 on the right.}$$

$$12 = 16x$$

$$\frac{12}{16} = x \qquad \text{The solution is } x = \frac{12}{16} = \frac{3}{4}.$$

2. **The LCD is $(x-2)(x+4)$. We begin by multiplying each side by the LCD.**

$$(x-2)(x+4)\left(\frac{3}{x-2} + \frac{1}{x+4}\right) = (x-2)(x+4)\left(\frac{5}{x-2}\right)$$

$$(x-2)(x+4)\left(\frac{3}{x-2}\right)+(x-2)(x+4)\left(\frac{1}{x+4}\right)=(x+4)5$$

$$(x+4)3+x-2=(x+4)5$$

$$3x+12+x-2=5x+20$$

$$4x+10=5x+20$$

$$-10=x$$

3. The LCD is $(x-1)(x+1)$, so we begin by multiplying each side by this quantity.

$$(x-1)(x+1)\left(\frac{2x-9}{x-1}\right)=(x-1)(x+1)\left(\frac{2x+5}{x+1}\right)$$

$$(x+1)(2x-9)=(x-1)(2x+5)$$ Use the FOIL method to expand each side.

$$2x^2-7x-9=2x^2+3x-5$$ $2x^2$ on each side cancels.

$$-7x-9=3x-5$$

$$-10x=4$$

$$x=-\frac{4}{10}=-\frac{2}{5}$$

4. The numerator factors: $x^2+3x-10=(x+5)(x-2)$.

$$\frac{(x+5)(x-2)}{x-2}=7$$ Simplify the fraction.

$$x+5=7$$

$$x=2$$ There is no solution because $x=2$ is extraneous.

5. Divide each side by 4 to isolate the radical.

$$4\sqrt{x}=12$$

$$\sqrt{x}=\frac{12}{4}=3$$ Divide each side by 4.

$$\left(\sqrt{x}\right)^2=3^2$$ Square each side.

$$x=9$$

6. **Isolate the radical by adding 1 and then dividing each side by 8.**

$$8\sqrt{x} - 1 = 15$$

$$8\sqrt{x} = 16$$

$$\sqrt{x} = \frac{16}{8} = 2$$

$$x = 4 \qquad \text{Square each side.}$$

7. **Divide each side by 2 before squaring each side.**

$$\frac{2\sqrt{x-3}}{2} = \frac{1}{2}$$

$$\left(\sqrt{x-3}\right)^2 = \left(\frac{1}{2}\right)^2$$

$$x - 3 = \frac{1}{4}$$

$$x = 3 + \frac{1}{4} = \frac{13}{4}$$

Absolute Value Equations

The *absolute value* of a number is its distance from 0. Because distances are not negative, the absolute value of a number is never negative. The symbol for the absolute value is a pair of absolute value bars, "| |." Because −3 is 3 units away from 0 on the number line, $|-3| = 3$. A number written without absolute values bars gives both the distance from 0 as well as the direction. For example, −3 is 3 units to the *left* of 0 and 3 is 3 units to the *right* of 0, but $|-3| = 3$ simply means 3 units *away* from 0. Because 0 is no distance from 0, $|0| = 0$.

EXAMPLE 2-5

- $|100| = 100$
- $|-83| = 83$
- $\left|-\frac{5}{2}\right| = \frac{5}{2}$
- $|5 - 11| = 6$
- $|10 - 1| = 9$
- $|68 - 90| = 22$

PRACTICE

1. $|-6.75| =$
2. $|8| =$
3. $|-4| =$
4. $|8 - 19| =$
5. $|13 - 25| =$
6. $|0| =$

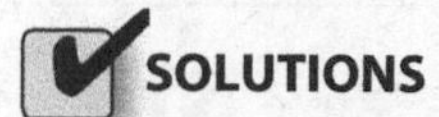

SOLUTIONS

1. $|-6.75| = 6.75$
2. $|8| = 8$
3. $|-4| = 4$
4. $|8 - 19| = 11$
5. $|13 - 25| = 12$
6. $|0| = 0$

The equation "$|x| = 5$" is really the question, "What numbers are 5 units away from 0?" Two numbers are 5 units from 0, 5 and -5, so there are two solutions, $x = 5$ and $x = -5$. Absolute value equations often have two solutions. In general, we solve an equation of the type "|Expression| = positive number" by solving the two equations: "Expression = negative number" and "Expression = positive number." Equations such as $|x| = -6$ have no solution because no number has a negative distance from 0.

EXAMPLE 2-6

Solve the equation.

- $|x| = 16$

The solutions are $x = 16, -16$.

- $|x + 3| = 5$

$$x + 3 = -5 \qquad\qquad x + 3 = 5$$

$$x = -8 \qquad\qquad x = 2$$

The solutions are $x = -8$ and $x = 2$.

- $|6 - 8x| = 0$

Because 0 and -0 are the same number, there is only one equation to solve.

$$6 - 8x = 0$$

$$-8x = -6$$

$$x = \frac{-6}{-8} = \frac{3}{4}$$

Sometimes an absolute value expression is part of a more complex equation. We need to isolate the absolute value expression on one side of the equation, and then we can solve it as before.

EXAMPLE 2-7

Solve the equation.

- $2|x-4|+7=13$

$$2|x-4|+7=13$$

$$-7 \quad -7$$

$$2|x-4|=6$$

$$\frac{2|x-4|}{2}=\frac{6}{2}$$

$$|x-4|=3$$

$$x-4=3 \qquad\qquad x-4=-3$$

$$x=7 \qquad\qquad x=1$$

- $\frac{1}{3}|2x+8|-10=-2$

$$\frac{1}{3}|2x+8|-10=-2$$

$$+10 \quad +10$$

$$\frac{1}{3}|2x+8|=8$$

$$3\left(\frac{1}{3}|2x+8|\right)=3(8)$$

$$|2x+8|=24$$

$$2x+8=24 \qquad\qquad 2x+8=-24$$

$$2x=16 \qquad\qquad 2x=-32$$

$$x=8 \qquad\qquad x=-16$$

PRACTICE

Solve the equation.

1. $|-5x+1| = 6$
2. $\left|\frac{3}{4}x - 8\right| = 1$
3. $\left|x + \frac{1}{2}\right| = \frac{2}{3}$
4. $-|3x+4| = -2$
5. $|2x-9| = 0$
6. $4|5x-2| + 3 = 11$
7. $3 - 2|x-9| = 1$
8. $\frac{2}{5}|4x-3| - 9 = -1$

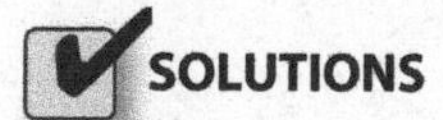

SOLUTIONS

1.

$$-5x + 1 = 6 \qquad -5x + 1 = -6$$

$$-5x = 5 \qquad -5x = -7$$

$$x = -1 \qquad x = \frac{7}{5}$$

2.

$$\frac{3}{4}x - 8 = 1 \qquad \frac{3}{4}x - 8 = -1$$

$$\frac{3}{4}x = 9 \qquad \frac{3}{4}x = 7$$

$$x = \frac{4}{3} \cdot 9 = 12 \qquad x = \frac{4}{3} \cdot 7 = \frac{28}{3}$$

3.

$$x + \frac{1}{2} = \frac{2}{3} \qquad x + \frac{1}{2} = -\frac{2}{3}$$

$$x = -\frac{1}{2} + \frac{2}{3} \qquad x = -\frac{1}{2} - \frac{2}{3}$$

$$x = -\frac{3}{6} + \frac{4}{6} = \frac{1}{6} \qquad x = -\frac{3}{6} - \frac{4}{6} = -\frac{7}{6}$$

4. $-|3x+4| = -2$ becomes $|3x+4| = 2$

$$3x + 4 = 2 \qquad 3x + 4 = -2$$

$$3x = -2 \qquad 3x = -6$$

$$x = -\frac{2}{3} \qquad x = -2$$

5.
$$2x - 9 = 0$$
$$2x = 9$$
$$x = \frac{9}{2}$$

6.
$$4|5x - 2| + 3 = 11$$
$$\quad -3 \quad -3$$
$$4|5x - 2| = 8$$
$$\frac{4|5x - 2|}{4} = \frac{8}{4}$$
$$|5x - 2| = 2$$

$$5x - 2 = 2 \qquad 5x - 2 = -2$$
$$5x = 4 \qquad 5x = 0$$
$$x = \frac{4}{5} \qquad x = \frac{0}{5} = 0$$

7.
$$3 - 2|x - 9| = 1$$
$$-3 \quad -3$$
$$-2|x - 9| = -2$$
$$\frac{-2|x - 9|}{-2} = \frac{-2}{-2}$$
$$|x - 9| = 1$$

$$x - 9 = 1 \qquad x - 9 = -1$$
$$x = 10 \qquad x = 8$$

8.
$$\frac{2}{5}|4x - 3| - 9 = -1$$
$$+9 \quad +9$$
$$\frac{2}{5}|4x - 3| = 8$$

$$\frac{5}{2}\cdot\frac{2}{5}|4x-3| = \frac{5}{2}\cdot 8$$

$$|4x-3| = 20$$

$$4x-3 = 20 \qquad\qquad 4x-3=-20$$

$$4x = 23 \qquad\qquad 4x = -17$$

$$x = \frac{23}{4} \qquad\qquad x = -\frac{17}{4}$$

Linear Inequalities

The strategy we used to solve linear equations is almost the same as the strategy that we will use to solve linear inequalities, with one important change. When multiplying or dividing both sides of a linear inequality by a negative number, we *must* reverse the inequality symbol. For example, if $-2x < 8$, then $\frac{-2x}{-2} > \frac{8}{-2}$. Because the solution to most inequalities is an interval of numbers, we will give our answers using interval notation. These rules are outlined in Table 2-2.

EXAMPLE 2-8

Solve the inequality and write the solution in interval notation.

- $\frac{3}{4}x - 1 \geq 5$

We begin by multiplying each side of the inequality by 4.

$$4\left(\frac{3}{4}x-1\right) \geq 4(5)$$

$$3x - 4 \geq 20$$

$$3x \geq 24$$

$$x \geq 8 \qquad \textbf{The solution is } [8, \infty).$$

- $-3(x-4)+2 < 8$

$$-3(x-4)+2 < 8 \qquad \textbf{Begin by distributing } -3.$$

$$-3x+12+2 < 8 \qquad \textbf{Combine like terms.}$$

TABLE 2-2

Inequality	Number Line	Interval
$x < a$	a	$(-\infty, a)$
$x \leq a$	a	$(-\infty, a]$
$x > a$	a	(a, ∞)
$x \geq a$	a	$[a, \infty)$
$a < x < b$	a b	(a, b)
$a \leq x \leq b$	a b	$[a, b]$
$x < a$ or $x > b$	a b	$(-\infty, a) \cup (b, \infty)$
$x \leq a$ or $x \geq b$	a b	$(-\infty, a] \cup [b, \infty)$
All x	All real numbers	$(-\infty, \infty)$

$$-3x + 14 < 8$$

$$-3x < -6$$ **When dividing by −3, reverse the inequality.**

$$\frac{-3x}{-3} > \frac{-6}{-3}$$

$$x > 2$$ **The solution is $(2, \infty)$.**

Compound Inequalities

A *compound* inequality represents two inequalities. For example, the compound inequality $10 < x + 4 < 12$ represents the inequalities $10 < x + 4$ and $x + 4 < 12$. The solution to a compound inequality is where the solution to

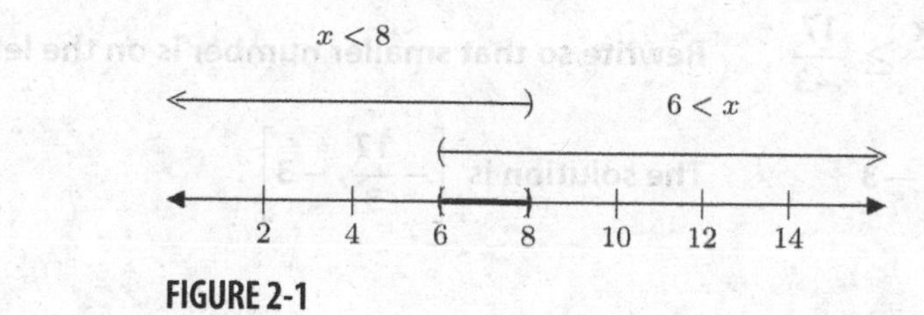

FIGURE 2-1

one inequality overlaps the solution to the other inequality. The solution to each inequality is plotted in Fig. 2-1. These solutions overlap between $x = 6$ and $x = 8$, so the solution to the compound inequality is (6, 8). (In the formal language of mathematics, we say that the solution to $10 < x + 4 < 12$ is where the solution to $10 < x + 4$ *intersects* the solution to $x + 4 < 12$.)

Our strategy for solving compound linear inequalities is the same as the one we used for linear inequalities except that we have three "sides" instead of two.

EXAMPLE 2-9

Solve the inequality, and write the solution in interval notation.

- $-3 < 4x + 9 < 7$

We isolate x by subtracting 9 from all three quantities and then dividing by 4.

$$-3 < 4x + 9 < 7$$
$$-9 \quad -9 \quad -9$$
$$-12 < 4x < -2$$
$$\frac{-12}{4} < \frac{4x}{4} < \frac{-2}{4}$$
$$-3 < x < -\frac{1}{2}$$

The solution is $\left(-3, -\frac{1}{2}\right)$.

- $5 \le \frac{1-3x}{2} \le 9$

We begin by clearing the fraction.

$$2(5) \le 2\left(\frac{1-3x}{2}\right) \le 2(9)$$ **Distribute 2.**

$10 \le 1 - 3x \le 18$ Subtract 1 from each quantity.

$9 \le -3x \le 17$ When dividing by −3, reverse the inequalities.

$\frac{9}{-3} \ge \frac{-3x}{-3} \ge \frac{17}{-3}$ Rewrite so that smaller number is on the left.

$-\frac{17}{3} \le x \le -3$ The solution is $\left[-\frac{17}{3}, -3\right]$.

Still Struggling

An inequality of the form "Larger No. < x < Smaller No." has no solution because x cannot be in two places at once on the number line.

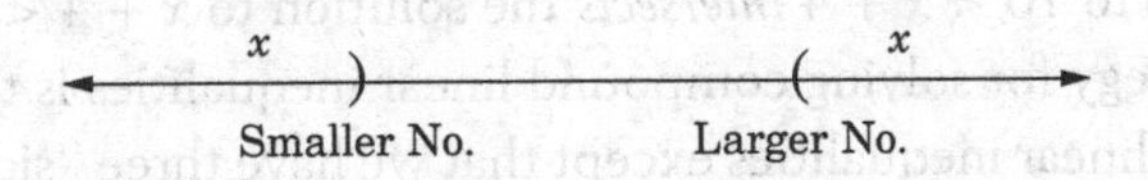

PRACTICE

Solve the inequality, and write the solution in interval notation.

1. $8x - 3 \le 21$
2. $-\frac{3}{4}x + 1 > 7$
3. $-2 \le \frac{3x+2}{6} \le 4$
4. $6 < 1 - 5x < 11$

SOLUTIONS

1.
$$8x - 3 \le 21$$
$$8x \le 24$$
$$x \le 3$$
The solution is $(-\infty, 3]$.

2.
$$-\frac{3}{4}x + 1 > 7$$
$$-\frac{3}{4}x > 6$$
$$x < \left(-\frac{4}{3}\right)(6) = -8$$
The solution is $(-\infty, -8)$.

3. $$-2 \leq \frac{3x+2}{6} \leq 4$$

$$6(-2) \leq 6\left(\frac{3x+2}{6}\right) \leq 6(4)$$

$$-12 \leq 3x+2 \leq 24$$

$$-14 \leq 3x \leq 22$$

$$-\frac{14}{3} \leq x \leq \frac{22}{3}$$ The solution is $\left[-\frac{14}{3}, \frac{22}{3}\right]$.

4. $$6 < 1 - 5x < 11$$

$$5 < -5x < 10$$

$$\frac{5}{-5} > \frac{-5x}{-5} > \frac{10}{-5}$$ Rewrite so that -2 is on the left.

$$-2 < x < -1$$ The solution is $(-2, -1)$.

Absolute Value Inequalities

The inequality "$|x| < 4$" is, in mathematical symbols, the question, "What real numbers are closer to 0 than 4 is?" A look at the number line in Fig. 2-2 might help with this question.

From the number line we can see that the numbers between -4 and 4 have an absolute value less than 4. The solution to $|x| < 4$ is the interval $(-4, 4)$. In inequality notation, the solution is $-4 < x < 4$.

Similarly, the solution to the inequality $|x| > 3$ is all numbers farther from 0 than 3 is. See Fig. 2-3.

The solution is all numbers smaller than -3 or larger than 3. In interval notation, the solution is $(-\infty, -3) \cup (3, \infty)$. The "$\cup$" symbol means "or." In inequality notation, the solution is $x < -3$ *or* $x > 3$. The sentence

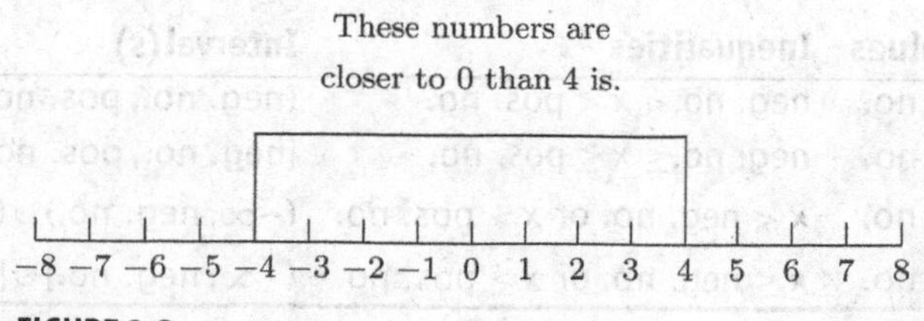

FIGURE 2-2

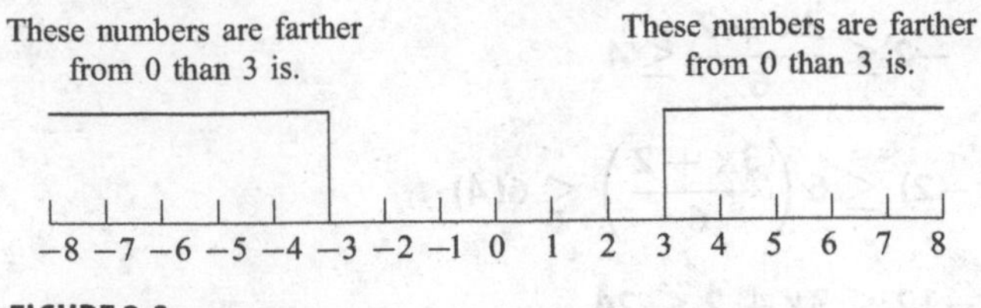

FIGURE 2-3

"$3 < x < -3$" has no meaning because no number x is *both* larger than 3 and smaller than -3.

We will use the strategy outlined in Table 2-3 for solving absolute value inequalities.

EXAMPLE 2-10

Solve the inequality, and write the solution in both inequality and interval notation.

Absolute Values	Inequalities	Interval(s)
• $\|x\| < 1$	$-1 < x < 1$	$(-1, 1)$
• $\|x\| > 16$	$x < -16$ or $x > 16$	$(-\infty, -16) \cup (16, \infty)$
• $\|x\| \geq 3$	$x \leq -3$ or $x \geq 3$	$(-\infty, -3] \cup [3, \infty)$
• $\|x\| \leq 5$	$-5 \leq x \leq 5$	$[-5, 5]$

Some absolute value inequalities, like absolute value equations, have no solution: $|x| < -6$. Because absolute values are not negative, no number has an absolute value smaller than -6. If we switch the inequality sign, $|x| > -6$, then we have an inequality for which *every* real number is a solution.

TABLE 2-3

Absolute Values	Inequalities	Interval(s)
• $\|x\| <$ pos. no.	neg. no. $< x <$ pos. no.	(neg. no., pos. no.)
• $\|x\| \leq$ pos. no.	neg. no. $\leq x \leq$ pos. no.	[neg. no., pos. no.]
• $\|x\| >$ pos. no.	$x <$ neg. no. or $x >$ pos. no.	$(-\infty$, neg. no.$) \cup ($pos. no., $\infty)$
• $\|x\| \geq$ pos. no.	$x \leq$ neg. no. or $x \geq$ pos. no.	$(-\infty$, neg. no.$] \cup [$pos. no., $\infty)$

PRACTICE

Solve the inequality, and write the solution in both inequality and interval notation.

1. $|x| > 12$ **2.** $|x| \leq 9$ **3.** $|x| < 10$ **4.** $|x| \geq 25$

SOLUTIONS

1. $|x| > 12$ $x < -12$ or $x > 12$ $(-\infty, -12) \cup (12, \infty)$
2. $|x| \leq 9$ $-9 \leq x \leq 9$ $[-9, 9]$
3. $|x| < 10$ $-10 < x < 10$ $(-10, 10)$
4. $|x| \geq 25$ $x \leq -25$ or $x \geq 25$ $(-\infty, -25] \cup [25, \infty)$

For some absolute value inequalities, finding the inequalities is only the first step toward finding the solution.

EXAMPLE 2-11

Solve the inequality, and give the solution in inequality and interval notation.

- $|4x - 5| \geq 9$

From the number line in Fig. 2-4, we see that either $4x - 5 \leq -9$ or $4x - 5 \geq 9$. These are the inequalities we need to solve.

$$4x - 5 \leq -9 \qquad\qquad 4x - 5 \geq 9$$

$$4x \leq -4 \qquad\qquad 4x \geq 14$$

$$x \leq -1 \qquad\qquad x \geq \frac{14}{4} = \frac{7}{2}$$

The solution in interval notation is $(-\infty, -1] \cup \left[\frac{7}{2}, \infty\right)$.

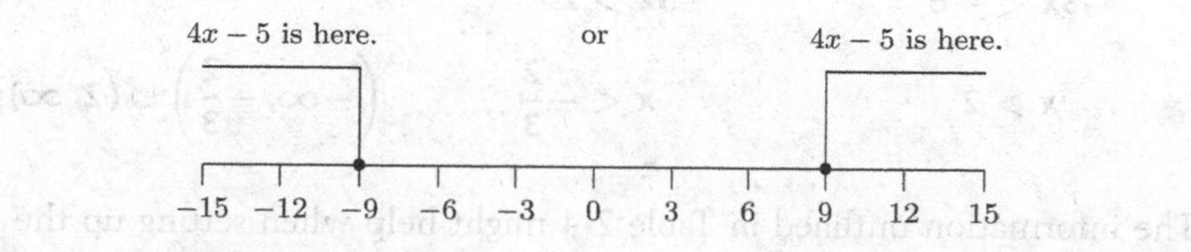

FIGURE 2-4

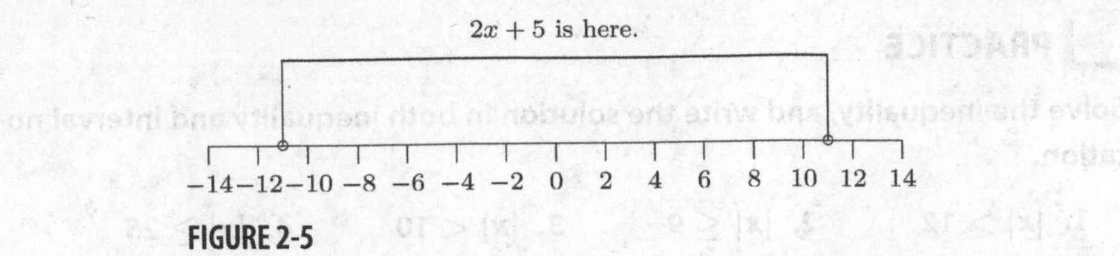

FIGURE 2-5

- $|2x + 5| < 11$

From the number line in Fig. 2-5, we see that $2x + 5$ is between -11 and 11. This means that $-11 < 2x + 5 < 11$.

$$-11 < 2x + 5 < 11$$

$$-16 < 2x < 6$$

$$-8 < x < 3 \qquad (-8, 3)$$

- $|9 - 3x| < 12$

$$-12 < 9 - 3x < 12$$

$$-21 < -3x < 3$$

$$\frac{-21}{-3} > \frac{-3}{-3}x > \frac{3}{-3} \qquad \text{Reverse the signs at this step.}$$

$$7 > x > -1 \quad \text{or} \quad -1 < x < 7 \qquad (-1, 7)$$

- $\left|\frac{2 - 3x}{4}\right| > 1$

$$\frac{2-3x}{4} < -1 \qquad \frac{2-3x}{4} > 1$$

$$2 - 3x < -4 \qquad 2 - 3x > 4$$

$$-3x < -6 \qquad -3x > 2$$

$$x > 2 \qquad x < -\frac{2}{3} \qquad \left(-\infty, -\frac{2}{3}\right) \cup (2, \infty)$$

The information outlined in Table 2-4 might help when setting up the inequalities for an absolute value inequality.

TABLE 2-4

Absolute Value Inequalities	Solve These Inequalities
$\lvert \text{Expression} \rvert > \text{pos. no.}$	$\text{Expression} < \text{neg. no.}$ or $\text{Expression} > \text{pos. no.}$
$\lvert \text{Expression} \rvert \geq \text{pos. no.}$	$\text{Expression} \leq \text{neg. no.}$ or $\text{Expression} \geq \text{pos. no.}$
$\lvert \text{Expression} \rvert < \text{pos. no.}$	$\text{neg. no.} < \text{Expression} < \text{pos. no.}$
$\lvert \text{Expression} \rvert \leq \text{pos. no.}$	$\text{neg. no.} \leq \text{Expression} \leq \text{pos. no.}$

PRACTICE

Solve the inequality, and write the solution in both inequality and interval notation.

1. $|3x+4| < 5$
2. $|x-2| > 4$
3. $|6-2x| \leq 4$
4. $\left|\frac{x-4}{3}\right| < 2$
5. $|8-3x| \geq 5$

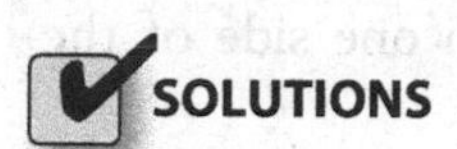

SOLUTIONS

1.
$$-5 < 3x+4 < 5$$
$$-9 < 3x < 1$$
$$-3 < x < \frac{1}{3} \qquad \left(-3, \frac{1}{3}\right)$$

2.
$$x-2 < -4 \qquad x-2 > 4$$
$$x < -2 \qquad x > 6 \qquad (-\infty, -2) \cup (6, \infty)$$

3.
$$-4 \leq 6-2x \leq 4$$
$$-10 \leq -2x \leq -2$$
$$\frac{-10}{-2} \geq \frac{-2}{-2}x \geq \frac{-2}{-2}$$
$$5 \geq x \geq 1 \qquad 1 \leq x \leq 5 \qquad [1, 5]$$

4. $$-2 < \frac{x-4}{3} < 2$$

$$3(-2) < 3\left(\frac{x-4}{3}\right) < 3(2)$$

$$-6 < x-4 < 6$$

$$-2 < x < 10 \qquad (-2, 10)$$

5. $8-3x \le -5$ $\qquad$ $8-3x \ge 5$

$-3x \le -13$ $\qquad$ $-3x \ge -3$

$\frac{-3}{-3}x \ge \frac{-13}{-3}$ $\qquad$ $\frac{-3}{-3}x \le \frac{-3}{-3}$

$x \ge \frac{13}{3}$ $\qquad$ $x \le 1$ $\qquad$ $(-\infty, 1] \cup \left[\frac{13}{3}, \infty\right)$

Sometimes absolute value expressions are part of more complicated inequalities. As before, we isolate the absolute value expression on one side of the inequality and then solve the inequality.

EXAMPLE 2-12

Solve the inequality, and write the solution in both interval and inequality notation.

- $3|x+4| - 7 \ge 5$

$$3|x+4| - 7 \ge 5$$

$3|x+4| \ge 12$ $\qquad$ **Add 7 to each side.**

$|x+4| \ge 4$ $\qquad$ **Divide each side by 3.**

$x+4 \le -4$ $\qquad$ $x+4 \ge 4$

$x \le -8$ $\qquad$ $x \ge 0$ $\qquad$ $(-\infty, -8] \cup [0, \infty)$

- $8 - 4\left|\frac{1}{2}x - 1\right| \ge 3$

$$8 - 4\left|\frac{1}{2}x - 1\right| \geq 3$$

$$-4\left|\frac{1}{2}x - 1\right| \geq -5 \quad \text{Subtract 8 from each side.}$$

$$\frac{-4\left|\frac{1}{2}x - 1\right|}{-4} \leq \frac{-5}{-4} \quad \text{Divide each side by } -4\text{; reverse the inequality.}$$

$$\left|\frac{1}{2}x - 1\right| \leq \frac{5}{4}$$

$$-\frac{5}{4} \leq \frac{1}{2}x - 1 \leq \frac{5}{4}$$

$$-\frac{1}{4} \leq \frac{1}{2}x \leq \frac{9}{4} \quad \text{Add } 1 = \frac{4}{4}.$$

$$2\left(-\frac{1}{4}\right) \leq 2\left(\frac{1}{2}x\right) \leq 2\left(\frac{9}{4}\right) \quad \text{Clear the fraction by multiplying by 2.}$$

$$-\frac{1}{2} \leq x \leq \frac{9}{2} \qquad \left[-\frac{1}{2}, \frac{9}{2}\right]$$

PRACTICE

Solve the inequality, and write the solution in both interval and inequality notation.

1. $\frac{1}{3}|x + 4| - 5 < 2$
2. $4 + 2|3x + 5| \geq 6$
3. $9 + |6 - 2x| \leq 11$
4. $-3|7x + 4| - 4 < -13$

SOLUTIONS

1.

$$\frac{1}{3}|x + 4| - 5 < 2$$

$$\frac{1}{3}|x + 4| < 7$$

$$3\left(\frac{1}{3}|x + 4|\right) < 3(7)$$

$$|x + 4| < 21$$

$$-21 < x + 4 < 21$$

$$-25 < x < 17 \qquad (-25, 17)$$

2.

$$4 + 2|3x+5| \geq 6$$
$$2|3x+5| \geq 2$$
$$\frac{2|3x+5|}{2} \geq \frac{2}{2}$$
$$|3x+5| \geq 1$$

$$3x+5 \leq -1 \qquad 3x+5 \geq 1$$
$$3x \leq -6 \qquad 3x \geq -4$$
$$x \leq -2 \qquad x \geq -\frac{4}{3} \qquad (-\infty, -2] \cup \left[-\frac{4}{3}, \infty\right)$$

3.

$$9 + |6-2x| \leq 11$$
$$|6-2x| \leq 2$$

$$-2 \leq 6-2x \leq 2$$
$$-8 \leq -2x \leq -4$$
$$\frac{-8}{-2} \geq \frac{-2}{-2}x \geq \frac{-4}{-2}$$
$$4 \geq x \geq 2 \quad \text{or} \quad 2 \leq x \leq 4 \quad [2, 4]$$

4.

$$-3|7x+4| - 4 < -13$$
$$-3|7x+4| < -9$$
$$\frac{-3|7x+4|}{-3} > \frac{-9}{-3}$$
$$|7x+4| > 3$$

$$7x+4 < -3 \qquad 7x+4 > 3$$
$$7x < -7 \qquad 7x > -1$$
$$x < -1 \qquad x > -\frac{1}{7} \qquad (-\infty, -1) \cup \left(-\frac{1}{7}, \infty\right)$$

Summary

In this chapter, we learned how to

- *Solve linear equations*. We solve a linear equation by simplifying each side of the equation, collecting variable terms on one side of the equation and terms without a variable in them on the other side, and dividing each side of the equation by the coefficient of the variable. If the equation contains any fractions, we can eliminate the fraction(s) by multiplyng each side of the equation by the least common denominator (LCD).
- *Solve linear inequalities*. We solve linear inequalities with the same strategy except that when multiplying or dividing each side of the inequality by a negative number, we must reverse the inequality symbol. Normally, we give solutions to inequalities using interval notation.
- *Solve equations leading to linear equations*. If an equation contains a rational expression (a fraction containing a variable), it might become a linear equation after multiplying each side by the LCD. If the LCD contains a variable, then multiplying each side by it might give us an extraneous solution. An extraneous solution is a solution to the linear equation but not to the original equation. If an equation contains a radical (that has a variable under the root symbol), then we isolate the radical expression and raise each side of the equation by a power that eliminates the root symbol. This also has the possibility of giving us an extraneous solution.
- *Solve a compound inequality*. A compound inequality actually represents two separate inequalities: "Smaller number < expression < larger number." This compound inequality represents the separate inequalities: "Smaller number < expression *and* Expression < larger number." We solve these inequalities with the same strategy, except that there are three quantities instead of two.
- *Solve an absolute value equation*. The absolute value of a number is its distance from 0. The absolute value of a number or expression is denoted with absolute value bars, | |. We solve most absolute value equations of the form "|Expression| = Positive number" by setting the expression inside the absolute value bars equal to the number or its negative. This gives us two separate equations to solve.

- *Solve absolute value inequalities*. Once the absolute value expression is isolated on one side of the inequality, we determine which two inequalities to solve. Refer to Table 2-3 for these inequalities.

QUIZ

Solve the equation or inequality.

1. $\frac{3}{5}(x+3) = \frac{1}{6}$

   A. $x = -\frac{13}{6}$ B. $x = -\frac{49}{18}$ C. $x = -\frac{85}{30}$ D. $x = \frac{1}{9}$

2. $\frac{x+4}{x-3} = \frac{x-3}{x+2}$

   A. $x = -\frac{5}{12}$ B. $x = -\frac{17}{6}$ C. $x = \frac{1}{12}$ D. $x = -\frac{1}{6}$

3. $\sqrt{4x - 11} = 3$

   A. $x = \frac{5}{4}$ B. $x = 4$ C. $x = \frac{25}{24}$ D. $x = 5$

4. $|6x + 5| = 4$

   A. $x = \frac{1}{6}, -\frac{2}{3}$ B. $x = -\frac{1}{6}, -\frac{3}{2}$ C. $x = -\frac{1}{2}, \frac{2}{3}$ D. $x = -\frac{1}{2}, -\frac{1}{6}$

5. $3|x + 4| + 5 = 2$

   A. $x = -5, -3$ B. $x = 5, -3$ C. $x = -5, 3$ D. There is no solution.

6. $\frac{1}{3x-1} + 4 = 9$

   A. $x = \frac{3}{5}$ B. $x = \frac{2}{5}$ C. $x = \frac{2}{3}$ D. There is no solution.

7. $6\sqrt{x} + 4 = 7$

   A. $x = \frac{1}{2}$ B. $x = \frac{1}{9}$ C. $x = \frac{1}{4}$ D. There is no solution.

8. $\frac{4x+3}{x+2} = \frac{4x+1}{x-6}$

   A. $x = \frac{4}{3}$ B. $x = -\frac{2}{3}$ C. $x = -\frac{1}{5}$ D. $x = \frac{3}{5}$

9. $8x + 6 < -10$

   A. $(-\infty, -2)$ B. $(-\infty, -\frac{1}{2})$ C. $(-2, \infty)$ D. $(-\frac{1}{2}, \infty)$

10. $|8x + 1| > 5$

A. $(-\infty, \frac{1}{2}) \cup (-\frac{3}{4}, \infty)$ B. $(\frac{1}{2}, -\frac{3}{4})$ C. $(-\frac{3}{4}, \frac{1}{2})$ D. $(-\infty, -\frac{3}{4}) \cup (\frac{1}{2}, \infty)$

11. $|-\frac{1}{2}x + 5| \leq 3$

A. $[2, 4]$ B. $[4, 16]$ C. $[-2, -1]$ D. There is no solution.

12. $-1 < \frac{6x+3}{2} < 5$

A. $(-\frac{1}{6}, \frac{1}{6})$ B. $(\frac{1}{6}, \frac{7}{6})$ C. $(-\frac{5}{6}, \frac{7}{6})$ D. $(-\frac{1}{6}, \frac{7}{6})$

13. $|9x + 4| \leq 1$

A. $(-\infty, -\frac{5}{9}] \cup [-\frac{1}{3}, \infty)$ B. $[-\frac{5}{9}, -\frac{1}{3}]$
C. $[-\frac{1}{3}, -\frac{5}{9}]$ D. There is no solution.

14. $|3x + 2| > 8$

A. $(-\frac{10}{3}, \infty) \cup (2, \infty)$ B. $(-\frac{10}{3}, 2)$ C. $(2, -\frac{10}{3})$ D. $(-\infty, -\frac{10}{3}) \cup (2, \infty)$

chapter 3

Quadratic Equations

We now work with *quadratic equations*, another important family of equations. A quadratic equation is an equation that can be written in the form $ax^2 + bx + c = 0$, where $a \neq 0$. We will solve quadratic equations throughout this book. In this chapter, we learn three approaches: factoring, the quadratic formula, and completing the square. Some equations are more easily solved using one method over the other. We will also use completing the square in our work with circles (Chap. 4), parabolas (Chap. 5), and with quadratic functions (Chap. 8).

CHAPTER OBJECTIVES

In this chapter, you will

- Solve a quadratic equation by factoring
- Solve a quadratic equation using the quadratic formula
- Solve a quadratic equation by completing the square

Solving Quadratic Equations by Factoring

We begin with factoring, which relies on the fact that if $ab = 0$, at least one of a or b is 0 (i.e., if the product of two or more numbers is 0, then at least one of them is 0). The strategy we will use is outlined in Table 3-1.

EXAMPLE 3-1

Solve the equation.

- $x^2 - 6x + 8 = 0$

Once we have factored $x^2 - 6x + 8$, we set each factor equal to 0 and then solve for x.

$$x^2 - 6x + 8 = 0$$

$$(x - 2)(x - 4) = 0$$

$x - 2 = 0 \qquad x - 4 = 0$

$x = 2 \qquad x = 4$ **The solutions are $x = 2$ and $x = 4$.**

- $x^2 - 16 = 0$

As $x^2 - 16$ is the difference of two squares, we use the factoring formula $a^2 - b^2 = (a - b)(a + b)$.

$$x^2 - 16 = 0$$

$$(x - 4)(x + 4) = 0$$

$x - 4 = 0 \qquad x + 4 = 0$

$x = 4 \qquad x = -4$ **The solutions are $x = 4$ and $x = -4$.**

TABLE 3-1 Solving a Quadratic Equation by Factoring

1.	Make sure one side of the equation is 0.
2.	Factor the nonzero side.
3.	Set each factor equal to 0, and then solve the equation.

- $2x^2 - 7x = 30$

The factoring method only works when one side of the equation is 0, so we begin by subtracting 30 from each side.

$$2x^2 - 7x - 30 = 0$$

$$(2x + 5)(x - 6) = 0$$

$2x + 5 = 0 \qquad x - 6 = 0$

$x = -\frac{5}{2} \qquad x = 6 \qquad$ **The solutions are** $x = -\frac{5}{2}$ **and** $x = 6$.

Still Struggling

In order to use the factoring method to solve a quadratic equation, we *must* have 0 on one side.

The Quadratic Formula

The factoring method for solving quadratic equations works great as long as we can easily factor the nonzero side of the equation. If we cannot easily factor the nonzero side, we can use the quadratic formula.

$$\text{If } ax^2 + bx + c = 0, \text{ for } a \neq 0 \text{ then } x = \frac{-b \pm \sqrt{b^2 - 4ac}}{2a}$$

The disadvantage of this method is that it requires a lot of arithmetic.

EXAMPLE 3-2

Solve the equation.

- $x^2 - 6x - 2 = 0$

The left side of this equation does not factor easily, so we use the quadratic formula. After identifying *a*, *b*, and *c*, we substitute them in the formula and then perform the arithmetic: $a = 1$, $b = -6$, **and** $c = -2$.

$$\sqrt{44} = \sqrt{4}\cdot\sqrt{11} = 2\sqrt{11}$$

$$x = \frac{-(-6) \pm \sqrt{(-6)^2 - 4(1)(-2)}}{2(1)} = \frac{6 \pm \sqrt{44}}{2} = \frac{6 \pm 2\sqrt{11}}{2}$$

Factor 2 in the numerator and cancel.

$$= \frac{2(3 \pm \sqrt{11})}{2} = 3 \pm \sqrt{11}$$

- $3x^2 + 8x = -4$

We must have a 0 on one side before we identify *a*, *b*, and *c*.

$$\overset{a}{3}x^2 + \overset{b}{8}x + \overset{c}{4} = 0$$

$$x = \frac{-8 \pm \sqrt{8^2 - 4(3)(4)}}{2(3)} = \frac{-8 \pm \sqrt{16}}{6} = \frac{-8 \pm 4}{6} = \frac{-8+4}{6}, \frac{-8-4}{6}$$

$$= \frac{-4}{6}, \frac{-12}{6} = -\frac{2}{3}, -2$$

Still Struggling

When using the quadratic formula, *all* of $-b \pm \sqrt{b^2 - 4ac}$ is over $2a$.

PRACTICE

Use the factoring method to solve the equation in Problems 1-5. Use the quadratic formula to solve the equation in Problems 6-8.

1. $x^2 + 5x - 6 = 0$
2. $x^2 - 7x + 12 = 0$
3. $6x^2 - x - 2 = 0$
4. $x^2 - 25 = 0$
5. $6x^2 - 2 = -11x$
6. $x^2 - 8x + 3 = 0$
7. $5x^2 + 2x - 1 = 0$
8. $3x^2 - x = 4$

SOLUTIONS

1. $x^2 + 5x - 6 = 0$

$$(x+6)(x-1) = 0$$

$$x + 6 = 0 \qquad x - 1 = 0$$

$$x = -6 \qquad x = 1$$

2. $x^2 - 7x + 12 = 0$

$$(x-3)(x-4) = 0$$

$$x - 3 = 0 \qquad x - 4 = 0$$

$$x = 3 \qquad x = 4$$

3. $6x^2 - x - 2 = 0$

$$(3x-2)(2x+1) = 0$$

$$3x - 2 = 0 \qquad 2x + 1 = 0$$

$$x = \frac{2}{3} \qquad x = -\frac{1}{2}$$

4. $x^2 - 25 = 0$

$$(x-5)(x+5) = 0$$

$$x - 5 = 0 \qquad x + 5 = 0$$

$$x = 5 \qquad x = -5$$

5. $6x^2 - 2 = -11x$: rewrite as $6x^2 + 11x - 2 = 0$

$$(6x-1)(x+2) = 0$$

$$6x - 1 = 0 \qquad x + 2 = 0$$

$$x = \frac{1}{6} \qquad x = -2$$

6. $x^2 - 8x + 3 = 0$: $a = 1, b = -8, c = 3$

$$x = \frac{-(-8) \pm \sqrt{(-8)^2 - 4(1)(3)}}{2(1)} = \frac{8 \pm \sqrt{52}}{2} = \frac{8 \pm 2\sqrt{13}}{2} \quad \left(\sqrt{52} = \sqrt{4 \cdot 13}\right)$$

$$= \frac{\overset{\text{Factor 2}}{2(4 \pm \sqrt{13})}}{2} = 4 \pm \sqrt{13}$$

7. $5x^2 + 2x - 1 = 0$: $a = 5, b = 2, c = -1$

$$x = \frac{-2 \pm \sqrt{2^2 - 4(5)(-1)}}{2(5)} = \frac{-2 \pm \sqrt{24}}{10} = \frac{-2 \pm 2\sqrt{6}}{10}$$

$$= \frac{2(-1 \pm \sqrt{6})}{10} = \frac{-1 \pm \sqrt{6}}{5}$$

8. $3x^2 - x = 4$; rewrite as $3x^2 - x - 4 = 0$

$$x = \frac{-(-1) \pm \sqrt{(-1)^2 - 4(3)(-4)}}{2(3)} = \frac{1 \pm \sqrt{49}}{6} = \frac{1 \pm 7}{6}$$

$$= \frac{1+7}{6}, \frac{1-7}{6} = \frac{8}{6}, \frac{-6}{6} = \frac{4}{3}, -1$$

Completing the Square

Our last method for solving quadratic equations is *completing the square*. We will also use this method later when sketching the graphs of circles and parabolas. The goal of completing the square is to write the equation in the form "$(x+a)^2 = k$" or "$(x-a)^2 = k$," where a and k are some numbers. We begin with the mechanics of "completing the square."

Let us expand $(x+a)^2$ and $(x-a)^2$ with the FOIL method to see what a perfect square looks like.

$$(x+a)^2 = (x+a)(x+a) \qquad (x-a)^2 = (x-a)(x-a)$$

$$= x^2 + 2ax + a^2 \qquad = x^2 - 2ax + a^2$$

The constant term is a^2, and the coefficient of x is $2a$ or $-2a$. This means that in a perfect square, the constant term is the square of half of x's coefficient.

$$\left(\frac{2a}{2}\right)^2 = a^2 \text{ and } \left(\frac{-2a}{2}\right)^2 = a^2$$

EXAMPLE 3-3

- $(x+3)^2 = x^2 + 6x + 9$ **Half of 6 is 3 and 3^2 is 9.**
- $(x-5)^2 = x^2 - 10x + 25$ **Half of −10 is −5 and $(-5)^2$ is 25.**
- $\left(x - \frac{1}{2}\right)^2 = x^2 - x + \frac{1}{4}$ **Half of −1 is $-\frac{1}{2}$ and $\left(-\frac{1}{2}\right)^2$ is $\frac{1}{4}$.**

The first step in most "completing the square" problems is deciding what number we add to $x^2 + bx$ in order to "complete the square." We find this number by dividing b by 2 and then squaring the result.

EXAMPLE 3-4

Fill in the blank with the number that completes the square.

	Fill in the blank with this.	This is a perfect square.
• $x^2 + 12x + \_\_\_$	$\frac{12}{2} = 6$ and $6^2 = \boxed{36}$	$x^2 + 12x + \underline{36}$
• $x^2 - 4x + \_\_\_$	$-\frac{4}{2} = -2$ and $(-2)^2 = \boxed{4}$	$x^2 - 4x + \underline{4}$
• $x^2 + \frac{1}{3}x + \_\_\_$	$\frac{1/3}{2} = \frac{1}{6}$ and $\left(\frac{1}{6}\right)^2 = \boxed{\frac{1}{36}}$	$x^2 + \frac{1}{3}x + \underline{\frac{1}{36}}$

PRACTICE

Fill in the blank with the number that completes the square.

1. $x^2 + 18x + \_\_\_$
2. $x^2 + 14x + \_\_\_$
3. $x^2 - 22x + \_\_\_$
4. $x^2 + 30x + \_\_\_$
5. $x^2 - 7x + \_\_\_$
6. $x^2 + \frac{1}{4}x + \_\_\_$

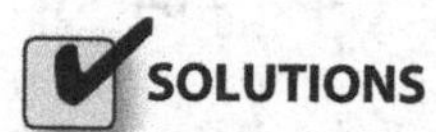

SOLUTIONS

1. $x^2 + 18x + \left(\frac{18}{2}\right)^2 = x^2 + 18x + \underline{81}$
2. $x^2 + 14x + \left(\frac{14}{2}\right)^2 = x^2 + 14x + \underline{49}$
3. $x^2 - 22x + \left(\frac{-22}{2}\right)^2 = x^2 - 22x + \underline{121}$
4. $x^2 + 30x + \left(\frac{30}{2}\right)^2 = x^2 + 30x + \underline{225}$
5. $x^2 - 7x + \left(-\frac{7}{2}\right)^2 = x^2 - 7x + \underline{\frac{49}{4}}$
6. $x^2 + \frac{1}{4}x + \left(\frac{1/4}{2}\right)^2 = x^2 + \frac{1}{4}x + \underline{\frac{1}{64}}$

The last step in completing the square is to rewrite the expression as a perfect square. First, we write $(x - \_\_\_)^2$ if the first sign is a minus sign or $(x + \_\_\_)^2$ if the first sign is a plus sign. We then fill in the blank with half of x's coefficient, which is also the square root of the constant term.

EXAMPLE 3-5

- $x^2 + 12x + 36 = (x + \_\_\_)^2 = (x+6)^2$ $\quad \frac{12}{2} = 6$ and $\sqrt{36} = 6$
- $x^2 - 4x + 4 = (x - \_\_\_)^2 = (x-2)^2$ $\quad \frac{4}{2} = 2$ and $\sqrt{4} = 2$
- $x^2 + \frac{1}{3}x + \frac{1}{36} = (x + \_\_\_)^2 = \left(x + \frac{1}{6}\right)^2$ $\quad \frac{1/3}{2} = \frac{1}{6}$ and $\sqrt{\frac{1}{36}} = \frac{1}{6}$

PRACTICE

Write the expression as a squared expression. These are the same problems used in the previous set of Practice problems.

1. $x^2 + 18x + 81$
2. $x^2 + 14x + 49$
3. $x^2 - 22x + 121$
4. $x^2 + 30x + 225$
5. $x^2 - 7x + \frac{49}{4}$
6. $x^2 + \frac{1}{4}x + \frac{1}{64}$

SOLUTIONS

1. $x^2 + 18x + 81 = (x+9)^2$
2. $x^2 + 14x + 49 = (x+7)^2$
3. $x^2 - 22x + 121 = (x-11)^2$
4. $x^2 + 30x + 225 = (x+15)^2$
5. $x^2 - 7x + \frac{49}{4} = \left(x - \frac{7}{2}\right)^2$
6. $x^2 + \frac{1}{4}x + \frac{1}{64} = \left(x + \frac{1}{8}\right)^2$

To solve a quadratic equation in the form $(x-a)^2 = k$ or $(x+a)^2 = k$, we take the square root of each side of the equation, using a "±" symbol to find both solutions (usually, quadratic equations have two solutions). This process is called *extracting roots*.

$$(x-a)^2 = k \qquad\qquad (x+a)^2 = k$$

$$x - a = \pm\sqrt{k} \qquad\qquad x + a = \pm\sqrt{k}$$

$$x = a \pm \sqrt{k} \qquad\qquad x = -a \pm \sqrt{k}$$

EXAMPLE 3-6

Solve the equation.

- $(x-1)^2 = 9$

$$x-1 = \pm\sqrt{9} = \pm 3$$
$$x = 1 \pm 3 = 1+3,\ 1-3 = 4,\ -2$$

- $\left(x+\frac{1}{2}\right)^2 = 5$

$$x+\frac{1}{2} = \pm\sqrt{5}$$
$$x = -\frac{1}{2} \pm \sqrt{5}$$

- $(x-6)^2 = 0$

$$x-6 = \pm\sqrt{0} = 0$$
$$x = 6$$

PRACTICE

Solve the equation.

1. $(x-2)^2 = 4$
2. $(x-4)^2 = 9$
3. $(x+5)^2 = 10$
4. $\left(x+\frac{1}{3}\right)^2 = 1$

SOLUTIONS

1. $(x-2)^2 = 4$

$$x-2 = \pm\sqrt{4} = \pm 2$$
$$x = 2 \pm 2 = 2+2,\ 2-2 = 4,\ 0$$

2. $(x-4)^2 = 9$

$$x-4 = \pm\sqrt{9} = \pm 3$$
$$x = 4 \pm 3 = 4+3,\ 4-3 = 7,\ 1$$

3. $(x+5)^2 = 10$

$$x+5 = \pm\sqrt{10}$$
$$x = -5 \pm \sqrt{10}$$

4. $\left(x+\frac{1}{3}\right)^2=1$

$$x+\frac{1}{3}=\pm\sqrt{1}\pm 1$$

$$x=-\frac{1}{3}\pm 1=-\frac{1}{3}+1,\ -\frac{1}{3}-1=-\frac{1}{3}+\frac{3}{3},\ -\frac{1}{3}-\frac{3}{3}=\frac{2}{3},\ -\frac{4}{3}$$

We are now ready to solve quadratic equations by completing the square. The strategy is outlined in Table 3-2.

EXAMPLE 3-7

Solve the quadratic equation by completing the square.

- $x^2+6x-7=0$

$x^2+6x=7$	Step 1
$x^2+6x+9=9+7=16$	Step 3
$(x+3)^2=16$	Step 4
$x+3=\pm 4$	Step 5
$x=-3\pm 4=-3+4,\ -3-4=1,\ -7$	Step 6

- $2x^2-2x-24=0$

$2x^2-2x=24$	Step 1
$x^2-x=12$	Step 2: Divide each side by 2.

TABLE 3-2 Strategy for Solving Quadratic Equations by Completing the Square
If $ax^2+bx+c=0$ and $a\neq 0$,
1. Move the constant term to the other side of the equation.
2. Divide each side of the equation by a, if a is not 1.
3. Find the constant term that completes the square on the left side of the equation. Add this number to each side of the equation.
4. Rewrite the left-hand side of the equation as a perfect square.
5. Take the square root of each side of the equation. Remember to use a $\pm$ symbol.
6. Solve for x and simplify the right-hand side, if necessary.

$$x^2 - x + \frac{1}{4} = 12 + \frac{1}{4} \qquad \textbf{Step 3}$$

$$= \frac{48}{4} + \frac{1}{4} = \frac{49}{4} \qquad \textbf{Simplify}$$

$$\left(x - \frac{1}{2}\right)^2 = \frac{49}{4} \qquad \textbf{Step 4}$$

$$x - \frac{1}{2} = \pm\sqrt{\frac{49}{4}} = \pm\frac{7}{2} \qquad \textbf{Step 5}$$

$$x = \frac{1}{2} \pm \frac{7}{2} = \frac{1}{2} + \frac{7}{2},\ \frac{1}{2} - \frac{7}{2} = 4, -3 \qquad \textbf{Step 6}$$

You might be interested to know that the quadratic formula comes from solving the equation $ax^2 + bx + c = 0$ by completing the square.

Still Struggling

Do not complete the square on $ax^2 + bx = -c$ when a is not 1. Divide by a and then complete the square.

PRACTICE

Solve for x by completing the square.

1. $x^2 - 10x + 24 = 0$
2. $x^2 + 6x + 5 = 0$
3. $x^2 + 5x + 6 = 0$
4. $x^2 - 3x = 4$
5. $4x^2 + 11x = -6$
6. $x^2 + 7x + 2 = 0$

SOLUTIONS

1.

$$x^2 - 10x + 24 = 0$$

$$x^2 - 10x = -24$$

$$x^2 - 10x + 25 = -24 + 25$$

$$(x - 5)^2 = 1$$

$$x - 5 = \pm\sqrt{1} = \pm 1$$

$$x = 5 \pm 1 = 6, 4$$

2.

$$x^2 + 6x + 5 = 0$$

$$x^2 + 6x = -5$$

$$x^2+6x+9=-5+9=4$$
$$(x+3)^2=4$$
$$x+3=\pm\sqrt{4}=\pm 2$$
$$x=-3\pm 2=-1,\ -5$$

3.
$$x^2+5x+6=0$$
$$x^2+5x=-6$$
$$x^2+5x+\left(\frac{5}{2}\right)^2=-6+\left(\frac{5}{2}\right)^2=-6+\frac{25}{4}=\frac{-24}{4}+\frac{25}{4}=\frac{1}{4}$$
$$\left(x+\frac{5}{2}\right)^2=\frac{1}{4}$$
$$x+\frac{5}{2}=\pm\sqrt{\frac{1}{4}}=\pm\frac{1}{2}$$
$$x=-\frac{5}{2}\pm\frac{1}{2}=-\frac{5}{2}+\frac{1}{2},\ -\frac{5}{2}-\frac{1}{2}=-\frac{4}{2},\ -\frac{6}{2}$$
$$=-2,\ -3$$

4.
$$x^2-3x=4$$
$$x^2-3x+\left(-\frac{3}{2}\right)^2=4+\left(-\frac{3}{2}\right)^2=4+\frac{9}{4}=\frac{16}{4}+\frac{9}{4}=\frac{25}{4}$$
$$\left(x-\frac{3}{2}\right)^2=\frac{25}{4}$$
$$x-\frac{3}{2}=\pm\sqrt{\frac{25}{4}}=\pm\frac{5}{2}$$
$$x=\frac{3}{2}\pm\frac{5}{2}=\frac{3}{2}+\frac{5}{2},\ \frac{3}{2}-\frac{5}{2}=4,\ -1$$

5.
$$4x^2+11x=-6$$
$$\frac{4x^2}{4}+\frac{11x}{4}=\frac{-6}{4}=-\frac{3}{2}$$
$$x^2+\frac{11}{4}x=-\frac{3}{2}$$
$$x^2+\frac{11}{4}x+\left(\frac{11/4}{2}\right)^2=-\frac{3}{2}+\left(\frac{11}{8}\right)^2=-\frac{3}{2}+\frac{121}{64}$$
$$=-\frac{96}{64}+\frac{121}{64}=\frac{25}{64}$$

$$\left(x+\frac{11}{8}\right)^2=\frac{25}{64}$$

$$x+\frac{11}{8}=\pm\sqrt{\frac{25}{64}}=\pm\frac{5}{8}$$

$$x=-\frac{11}{8}\pm\frac{5}{8}=-\frac{11}{8}+\frac{5}{8},\ -\frac{11}{8}-\frac{5}{8}$$

$$=-\frac{6}{8},\ -\frac{16}{8}=-\frac{3}{4},\ -2$$

6. $$x^2+7x+2=0$$

$$x^2+7x=-2$$

$$x^2+7x+\left(\frac{7}{2}\right)^2=-2+\left(\frac{7}{2}\right)^2=-2+\frac{49}{4}=-\frac{8}{4}+\frac{49}{4}=\frac{41}{4}$$

$$\left(x+\frac{7}{2}\right)^2=\frac{41}{4}$$

$$x+\frac{7}{2}=\pm\sqrt{\frac{41}{4}}=\pm\frac{\sqrt{41}}{\sqrt{4}}=\pm\frac{\sqrt{41}}{2}$$

$$x=-\frac{7}{2}\pm\frac{\sqrt{41}}{2}$$

Summary

In this chapter, we learned how to

- *Solve a quadratic equation by factoring.* The factoring method is based on the fact that $ab=0$ implies that at least one of a or b is 0. In order for this method to work, one side of the equation must be 0. We factor the nonzero side of the equation, set each factor equal to 0, and solve for x. This method works best when the nonzero side of the equation is easy to factor.
- *Solve a quadratic equation with the quadratic formula.* For an equation of the form $ax^2+bx+c=0$, the solution(s) to the equation comes from the formula $x=\frac{-b\pm\sqrt{b^2-4ac}}{2a}$.
- *Solve a quadratic equation by completing the square.* This strategy is based on the fact that $x^2=k$ implies $x=\pm\sqrt{k}$. After writing the equation in the form $x^2+bx=c$, we "complete the square" by adding $(\frac{b}{2})^2$ to each side of the equation. This makes the left side of the equation $(x+\frac{b}{2})^2$. We then take the square root of each side and solve for x.

QUIZ

1. **Use the factoring method to solve for x: $x^2 - 2x = 24$.**

   A. $x = 2, -12$ B. $x = -12, 2$ C. $x = -6, 4$ D. $x = -4, 6$

2. **Fill in the blank so that the quantity is a perfect square: $x^2 - \frac{3}{4}x +$ ______.**

   A. $\frac{9}{8}$ B. $\frac{3}{2}$ C. $\frac{9}{64}$ D. $\frac{9}{16}$

3. **Solve for x by taking the square root of each side: $(x - 5)^2 = 36$.**

   A. $x = 31, 41$ B. $x = -31, 19$ C. $x = -19, 31$ D. $x = -1, 11$

4. **Use the quadratic formula to solve the equation: $4x^2 - 2x - 1 = 0$.**

   A. $x = \frac{1 \pm \sqrt{5}}{4}$ B. $x = 1 \pm \frac{\sqrt{5}}{4}$ C. $x = \frac{1 \pm 2\sqrt{5}}{4}$ D. $x = \frac{1}{4} \pm 2\sqrt{5}$

5. **Fill in the blank: $(x +$ ___ $)^2 = x^2 + \frac{4}{5}x + \frac{4}{25}$.**

   A. $\frac{4}{25}$ B. $\frac{2}{25}$ C. $\frac{4}{5}$ D. $\frac{2}{5}$

6. **Fill in the blank so that the quantity is a perfect square: $x^2 + 9x +$ ____.**

   A. $\frac{9}{2}$ B. $\frac{81}{4}$ C. $\frac{9}{4}$ D. $\frac{81}{2}$

7. **Use the factoring method to solve for x: $x^2 - 6x + 5 = 0$.**

   A. $x = 1, 5$ B. $x = -1, -5$ C. $x = 2, 3$ D. $x = -2, -3$

8. **Solve the equation using any method: $6x^2 - x = 1$.**

   A. $x = -\frac{1}{3}, 2$ B. $x = -\frac{1}{2}, 3$ C. $x = -\frac{1}{3}, \frac{1}{2}$ D. $x = -\frac{1}{3}, \frac{1}{3}$

9. **The equation $4x^2 - 16x + 7 = 0$ is equivalent to**

   A. $(x-2)^2 = \frac{9}{4}$ B. $(x+4)^2 = \frac{57}{4}$ C. $(x-4)^2 = \frac{57}{4}$ D. $(x-2)^2 = 57$

chapter 4

The xy-Coordinate Plane

After learning how to plot points, we learn how to use the distance and midpoint formulas. We are then introduced to sketching the graph of an equation which is a major topic in college algebra. After learning how to plot points, we learn how to find clues from the equation of a circle to plot its graph with only a little work (a major topic throughout this book). If the equation is not in the form where we can readily identify the circle's center and radius, we learn how to use completing the square to find this information.

CHAPTER OBJECTIVES

In this chapter, you will

- Plot points in the *xy*-plane
- Use the distance formula to calculate the distance between two points
- Use the midpoint formula to find the point that is halfway between two points
- Work with equations of circles
- Sketch a circle from its equation

Plotting Points

The xy-coordinate plane (or *plane*) is formed by two number lines. The vertical number line is called the y-axis, and the horizontal number line is called the x-axis. The number lines cross at 0. This point is called the *origin*. Points on the plane can be located and identified by *coordinates*: (x, y). The first number is called the *x-coordinate*. This number describes how far left or right to move from the origin to locate the point. A negative number tells us that we move to the left, and a positive number tells us that we move to the right. The second number in the ordered pair (x, y) is called the *y-coordinate*. This number describes how far up or down to move from the origin to locate the point. A negative number tells us that we move down, and a postive number tells us that we move up.

(+right, +up) (−left,+up) (+right,−down) (−left,−down)

EXAMPLE 4-1

- **(4, 1) Right 4, up 1**
- **(−1, −5) Left 1, down 5**
- **(0, 2) No horizontal movement, up 2**
- **(−2, 5) Left 2, up 5**
- **(5, −3) Right 5, down 3**
- **(−3, 0) Left 3, no vertical movement**

These points are plotted in Fig. 4-1.

PRACTICE

Describe the horizontal and vertical movements and locate the point on the plane.

1. (4, 5) **2. (−1, −3)** **3. (2, 2)**
4. (−6, 4) **5. (0, 3)** **6. (−4, 0)**

SOLUTIONS

The solutions are plotted in Fig. 4-2.

1. (4, 5) Right 4, up 5 **2. (−1, −3) Left 1, down 3**
3. (2, 2) Right 2, up 2 **4. (−6, 4) Left 6, up 4**

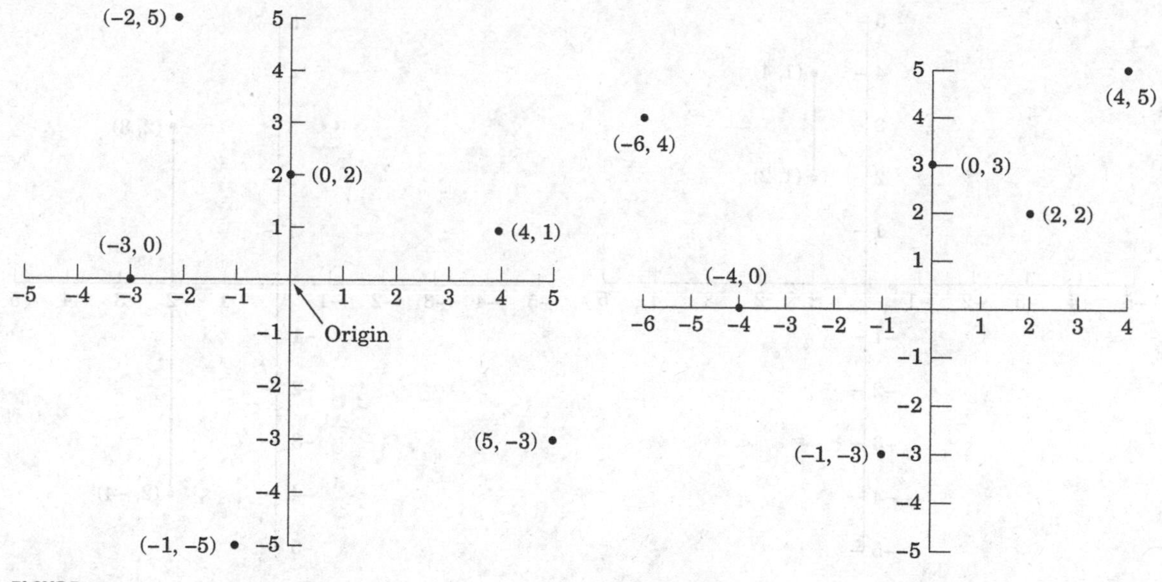

FIGURE 4-1

FIGURE 4-2

5. (0, 3) No horizontal movement, up 3
6. (−4, 0) Left 4, no vertical movement

The Distance Between Two Points

At times we need to find the distance between two points. If the points are on the same vertical line (the x-coordinates are the same), the distance between the points is the *absolute value* of the difference between the y-coordinates. If the points are on the same horizontal line (the y-coordinates are the same), the distance between the points is the absolute value of the difference between the x-coordinates.

EXAMPLE 4-2

- **The distance between (1, 4) and (1, 2) is $|4-2| = |2| = 2$. As you can see in Fig. 4-3, these points are on the same vertical line.**
- **The distance between (2, 3) and (2, −4) is $|-4-3| = |-7| = 7$. See Fig. 4-4.**

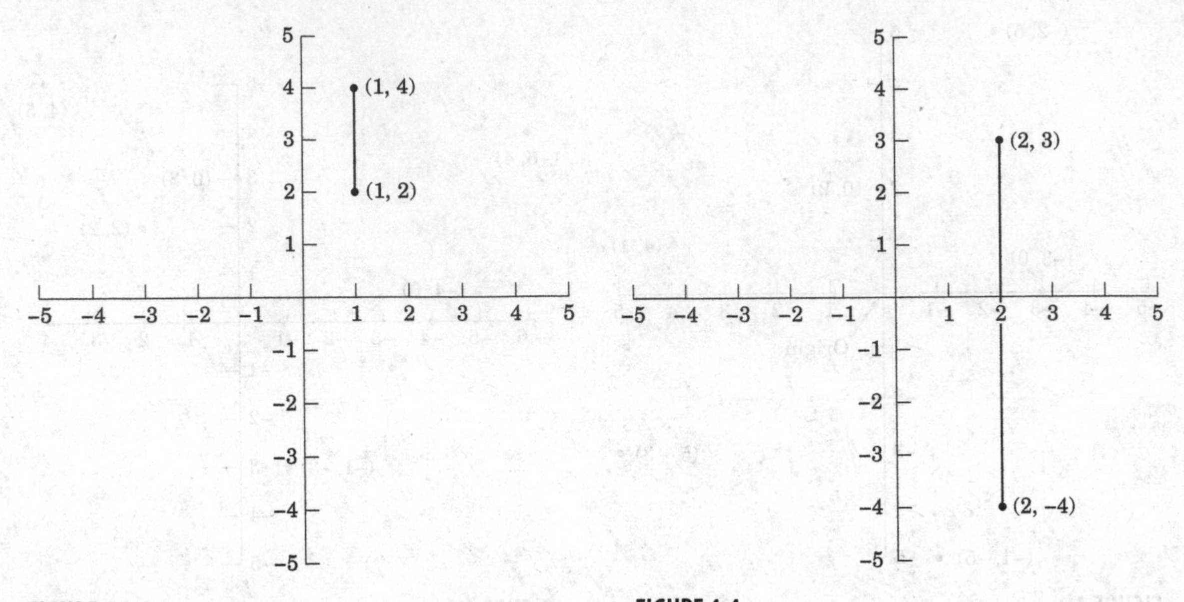

FIGURE 4-3

FIGURE 4-4

- **The distance between $(-5, 3)$ and $(-1, 3)$ is $|-5-(-1)| = |-5+1| = |-4| = 4$. See Fig. 4-5.**

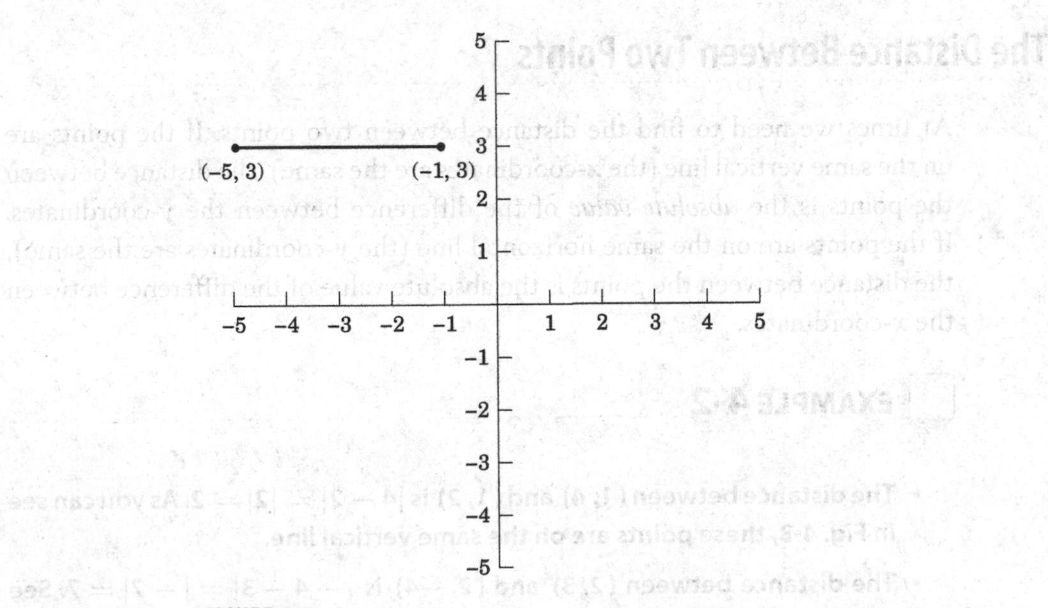

FIGURE 4-5

PRACTICE

Find the distance between the two points.

1. **(5, 6) and (1, 6)**
2. **(4, 3) and (4, 8)**
3. **(−1, 2) and (−1, 9)**
4. **(6, 0) and (2, 0)**

SOLUTIONS

1. $|5-1| = |4| = 4$
2. $|3-8| = |-5| = 5$
3. $|2-9| = |-7| = 7$
4. $|6-2| = |4| = 4$

In the rest of the chapter, we will work with formulas involving two points. We call one of them Point 1 with coordinates (x_1, y_1) and the other, Point 2, with coordintates (x_2, y_2). Suppose we have two points that are not on the same vertical line or on the same horizontal line. By using the Pythagorean theorem in a clever way, we can find the distance between *any* two points, as in Fig. 4-6.

We draw a vertical line through one of the points and a horizontal line through the other. The point where these lines cross has the x-coordinate of one of the points and the y-coordinate of the other. See Fig. 4-7.

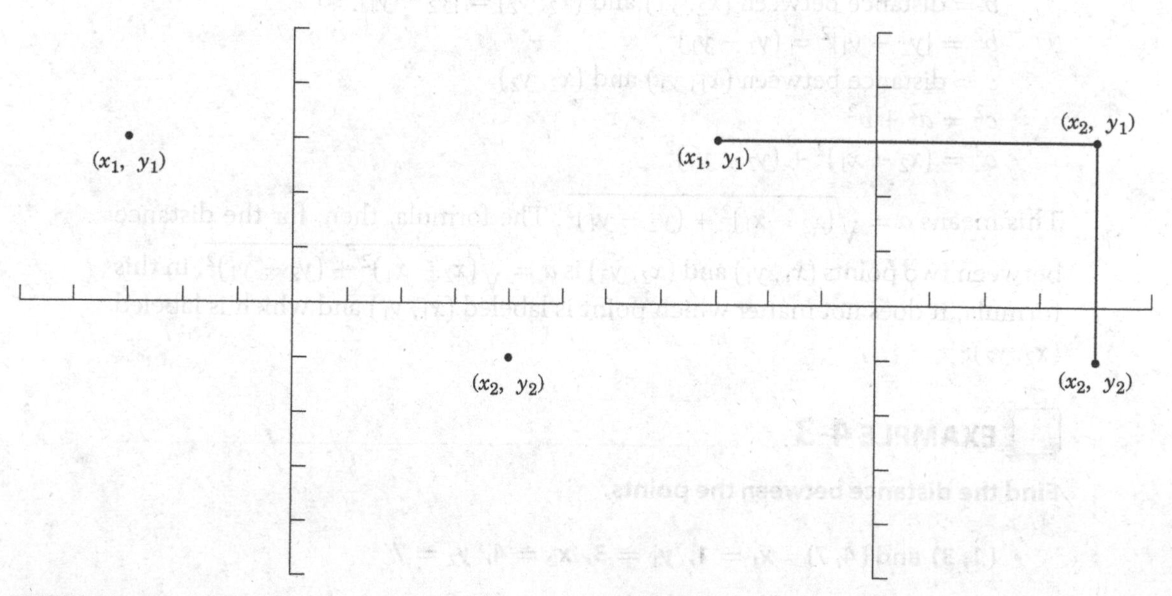

FIGURE 4-6 **FIGURE 4-7**

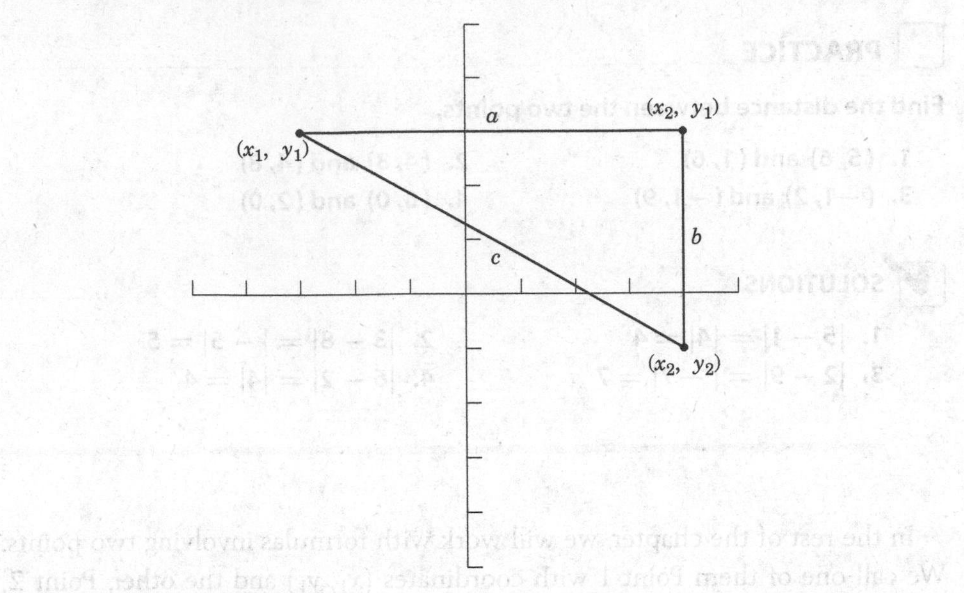

FIGURE 4-8

The three points form a right triangle. The length of the hypotenuse of this triangle is the distance between (x_1, y_1) and (x_2, y_2). We can find this length using the Pythagorean theorem: $a^2 + b^2 = c^2$. See Fig. 4-8.

$a =$ distance between (x_1, y_1) and $(x_2, y_1) = |x_2 - x_1|$, so
$a^2 = |x_2 - x_1|^2 = (x_2 - x_1)^2$
$b =$ distance between (x_2, y_1) and $(x_2, y_2) = |y_2 - y_1|$, so
$b^2 = |y_2 - y_1|^2 = (y_2 - y_1)^2$
$c =$ distance between (x_1, y_1) and (x_2, y_2)
$c^2 = a^2 + b^2$
$c^2 = (x_2 - x_1)^2 + (y_2 - y_1)^2$

This means $c = \sqrt{(x_2 - x_1)^2 + (y_2 - y_1)^2}$. The formula, then, for the distance between two points (x_1, y_1) and (x_2, y_2) is $d = \sqrt{(x_2 - x_1)^2 + (y_2 - y_1)^2}$. In this formula, it does not matter which point is labeled (x_1, y_1) and which is labeled (x_2, y_2).

EXAMPLE 4-3

Find the distance between the points.

- **(1, 3) and (4, 7)** $x_1 = 1,\ y_1 = 3,\ x_2 = 4,\ y_2 = 7$

$$d = \sqrt{(4-1)^2 + (7-3)^2} = \sqrt{3^2 + 4^2} = \sqrt{25} = 5$$

- **(−2, 5) and (1, 4)** $x_1 = -2,\ y_1 = 5,\ x_2 = 1,\ y_2 = 4$

$$\sqrt{(1-(-2))^2+(4-5)^2} = \sqrt{(1+2)^2+(4-5)^2}$$
$$= \sqrt{3^2+(-1)^2} = \sqrt{10}$$

- **(0, −7) and (−2, 4)** $x_1 = 0,\ y_1 = -7,\ x_2 = -2,\ y_2 = 4$

$$d = \sqrt{(-2-0)^2+(4-(-7))^2} = \sqrt{(-2)^2+(4+7)^2}$$
$$= \sqrt{125} = 5\sqrt{5}$$

This formula even works for two points on the same horizontal or vertical line. For example, we know the distance between (3, 8) and (3, 6) is 2. Let's see what happens in the formula.

$$d = \sqrt{(3-3)^2+(6-8)^2} = \sqrt{0^2+(-2)^2} = \sqrt{4} = 2$$

PRACTICE

Find the distance between the points.

1. (−1, 4), (3, 3)
2. (6, −4), (−2, −5)
3. (0, 8), (2, 1)
4. (7, −3), (5, −3)

SOLUTIONS

1. $d = \sqrt{(3-(-1))^2+(3-4)^2} = \sqrt{(3+1)^2+(-1)^2} = \sqrt{17}$
2. $d = \sqrt{(-2-6)^2+(-5-(-4))^2} = \sqrt{(-8)^2+(-5+4)^2}$
   $= \sqrt{65}$
3. $d = \sqrt{(2-0)^2+(1-8)^2} = \sqrt{2^2+(-7)^2} = \sqrt{53}$
4. $d = \sqrt{(5-7)^2+(-3-(-3))^2} = \sqrt{(-2)^2+(-3+3)^2} = 2$

Sometimes we are asked to show that groups of points form shapes such as squares or right triangles. We can use the distance formula to show that the lengths of the sides of a square are equal or that the lengths of the sides of a right triangle follow the Pythagorean Theorem.

EXAMPLE 4-4

- Show that the points $(-2, \frac{13}{2})$, $(1, 2)$, and $(4, 4)$ are the vertices of a right triangle.

To use the distance formula on this problem, we show that if we square and then add the lengths of the two legs (the sides that are not the hypotenuse), this should equal the square of the hypotenuse: $a^2 + b^2 = c^2$. While not really necessary, we plot the points to see which two sides are the legs and which side is the hypotenuse. The triangle is shown in Fig. 4-9.

From the graph in Fig. 4-9, we see that a is the distance from $(-2, \frac{13}{2})$ to $(1, 2)$; b is the distance from $(1, 2)$ to $(4, 4)$; and c is the distance from $(-2, \frac{13}{2})$ to $(4, 4)$.

$$a = \sqrt{(1-(-2))^2 + \left(2 - \frac{13}{2}\right)^2} = \sqrt{9 + \frac{81}{4}} = \sqrt{\frac{117}{4}}, \text{ so } a^2$$

$$= \left(\sqrt{\frac{117}{4}}\right)^2 = \frac{117}{4}$$

$$b = \sqrt{(4-1)^2 + (4-2)^2} = \sqrt{9+4} = \sqrt{13}, \text{ so } b^2 = \left(\sqrt{13}\right)^2 = 13$$

$$c = \sqrt{(4-(-2))^2 + \left(4 - \frac{13}{2}\right)^2} = \sqrt{36 + \frac{25}{4}} = \sqrt{\frac{169}{4}}, \text{ so } c^2$$

$$= \left(\sqrt{\frac{169}{4}}\right)^2 = \frac{169}{4}$$

Is it true that $a^2 + b^2 = c^2$?

$$a^2 + b^2 = \frac{117}{4} + 13 = \frac{117}{4} + \frac{52}{4} = \frac{169}{4} = c^2$$

Because $a^2 + b^2 = c^2$ is a true statement, $(-2, \frac{13}{2})$, $(1, 2)$, and $(4, 4)$ are the vertices of a right triangle.

- Show that $(2, 5)$, $(6, 3)$, and $(2, 1)$ are the vertices of an isosceles triangle.

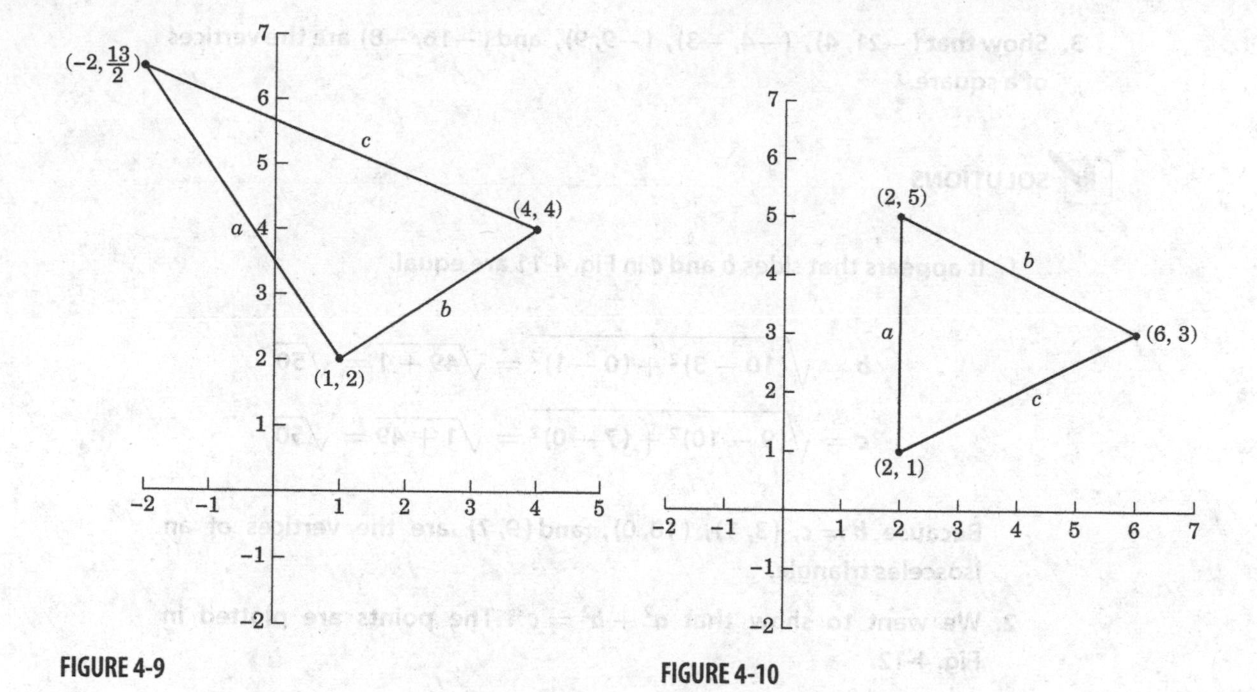

FIGURE 4-9

FIGURE 4-10

Two sides of an isosceles triangle have the same length. If we plot the points, we should be able to tell which sides have equal length. These points are plotted in Fig. 4-10.

It appears that sides b and c have equal length.

$b =$ the distance between (2, 5) and (6, 3), and $c =$ the distance between (2, 1) and (6, 3).

$$b = \sqrt{(6-2)^2 + (3-5)^2} = \sqrt{16+4} = \sqrt{20}$$

$$c = \sqrt{(6-2)^2 + (3-1)^2} = \sqrt{16+4} = \sqrt{20}$$

Because $b = c$, the points (2, 5), (6, 3), and (2, 1) are the vertices of an isosceles triangle.

PRACTICE

1. Show that (3, 1), (10, 0), and (9, 7) are the vertices of an isosceles triangle.
2. Show that (−3, 2), (4, 0), and (−1, 9) are the vertices of a right triangle.

3. Show that $(-21, 4)$, $(-4, -3)$, $(-9, 9)$, and $(-16, -8)$ are the vertices of a square.

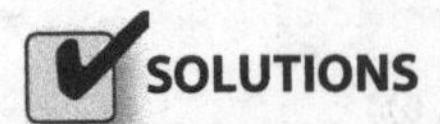

SOLUTIONS

1. It appears that sides *b* and *c* in Fig. 4-11 are equal.

$$b = \sqrt{(10-3)^2 + (0-1)^2} = \sqrt{49+1} = \sqrt{50}$$

$$c = \sqrt{(9-10)^2 + (7-0)^2} = \sqrt{1+49} = \sqrt{50}$$

Because $b = c$, $(3, 1)$, $(10, 0)$, and $(9, 7)$ are the vertices of an isosceles triangle.

2. We want to show that $a^2 + b^2 = c^2$. The points are plotted in Fig. 4-12.

$$a = \sqrt{(-1-(-3))^2 + (9-2)^2} = \sqrt{(-1+3)^2 + 7^2}$$
$$= \sqrt{4+49} = \sqrt{53}$$

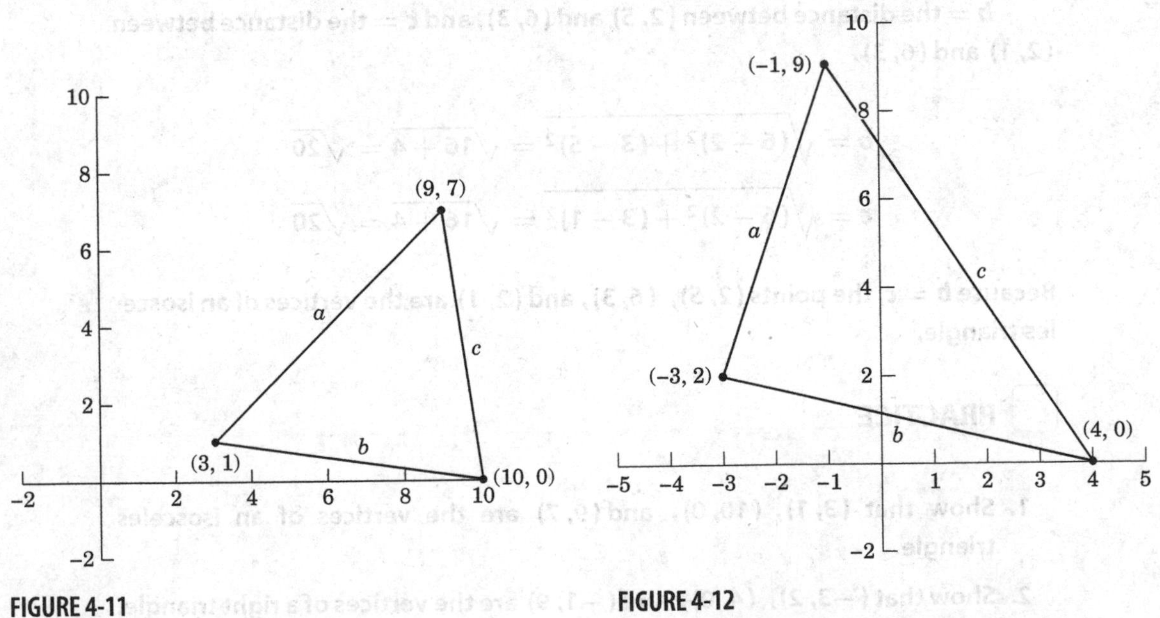

FIGURE 4-11

FIGURE 4-12

$$b = \sqrt{(-3-4)^2 + (2-0)^2} = \sqrt{(-7)^2 + 2^2} = \sqrt{49+4}$$
$$= \sqrt{53}$$
$$c = \sqrt{(-1-4)^2 + (9-0)^2} = \sqrt{(-5)^2 + 9^2} = \sqrt{25+81}$$
$$= \sqrt{106}$$
$$a^2 + b^2 = \left(\sqrt{53}\right)^2 + \left(\sqrt{53}\right)^2 = 106 = \left(\sqrt{106}\right)^2 = c^2$$

This means that $(-3, 2)$, $(4, 0)$, and $(-1, 9)$ are the vertices of a right triangle.

3. **We want to show that $a = b = c = d$. See Fig. 4-13.**

$$a = \sqrt{(-16-(-21))^2 + (-8-4)^2}$$
$$= \sqrt{(-16+21)^2 + (-12)^2} = \sqrt{25+144} = 13$$
$$b = \sqrt{(-4-(-16))^2 + (-3-(-8))^2}$$
$$= \sqrt{(-4+16)^2 + (-3+8)^2} = \sqrt{144+25} = 13$$

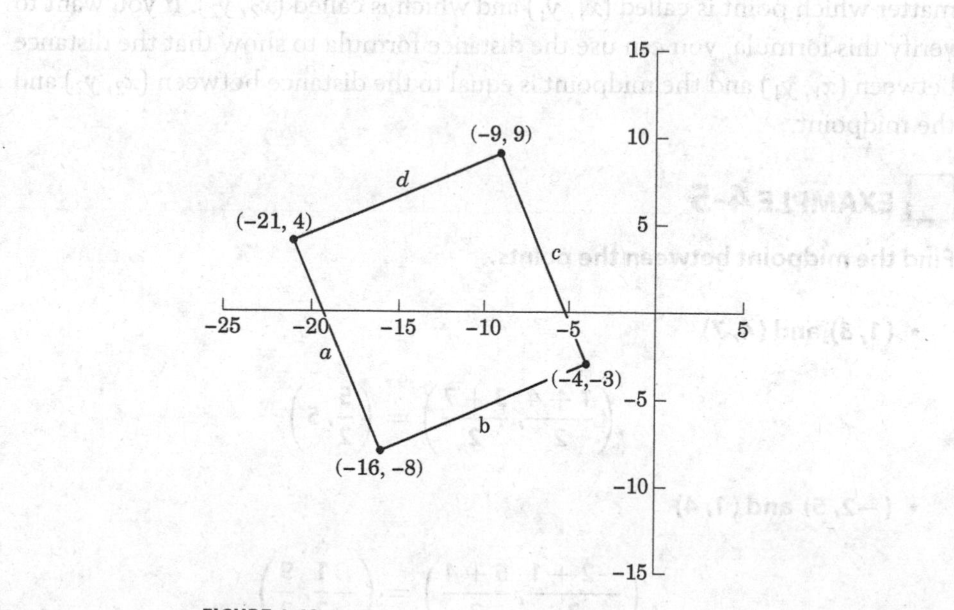

FIGURE 4-13

$$c = \sqrt{(-9-(-4))^2 + (9-(-3))^2}$$

$$= \sqrt{(-9+4)^2 + (9+3)^2} = \sqrt{25+144} = 13$$

$$d = \sqrt{(-21-(-9))^2 + (4-9)^2}$$

$$= \sqrt{(-21+9)^2 + (-5)^2} = \sqrt{144+25} = 13$$

Because $a = b = c = d$, $(-21, 4)$, $(-4, -3)$, $(-9, 9)$, and $(-16, -8)$ are the vertices of a square.

The Midpoint Formula

The *midpoint* between two points (x_1, y_1) and (x_2, y_2) is the point that is halfway between them. We find it with the midpoint formula:

$$\left(\frac{x_1+x_2}{2}, \frac{y_1+y_2}{2}\right)$$

This formula is easy to remember if we think of finding the *average* of the x-values and the *average* of the y-values. As with the distance formula, it does not matter which point is called (x_1, y_1) and which is called (x_2, y_2). If you want to verify this formula, you can use the distance formula to show that the distance between (x_1, y_1) and the midpoint is equal to the distance between (x_2, y_2) and the midpoint.

EXAMPLE 4-5

Find the midpoint between the points.

- **$(1, 3)$ and $(4, 7)$**

$$\left(\frac{1+4}{2}, \frac{3+7}{2}\right) = \left(\frac{5}{2}, 5\right)$$

- **$(-2, 5)$ and $(1, 4)$**

$$\left(\frac{-2+1}{2}, \frac{5+4}{2}\right) = \left(-\frac{1}{2}, \frac{9}{2}\right)$$

- **(0, −7) and (−2, 4)**

$$\left(\frac{0+(-2)}{2}, \frac{-7+4}{2}\right) = \left(-1, -\frac{3}{2}\right)$$

PRACTICE

Find the midpoint between the points.

1. (3, 5) and (1, 2)
2. (6, −4) and (−2, −5)
3. (0, 8) and (2, 1)
4. (7, −3) and (5, −3)

SOLUTIONS

1. $\left(\frac{3+1}{2}, \frac{5+2}{2}\right) = \left(2, \frac{7}{2}\right)$
2. $\left(\frac{6+(-2)}{2}, \frac{-4+(-5)}{2}\right) = \left(2, -\frac{9}{2}\right)$
3. $\left(\frac{0+2}{2}, \frac{8+1}{2}\right) = \left(1, \frac{9}{2}\right)$
4. $\left(\frac{7+5}{2}, \frac{-3+(-3)}{2}\right) = (6, -3)$

Circles

An equation with two variables can be *graphed* on the *xy*-plane. Think of a graph as a "picture" of all solutions to the equation. Every point on the graph is a solution to the equation, and every solution to the equation is represented by a point on the graph.

For example, in the equation $x + y = 5$, *any* pair of numbers whose sum is 5 is represented by a point on the graph of the equation. Some of those pairs of numbers are (0, 5), (1, 4), (3, 2), (4, 1), (5, 0), (6, −1), (7, −2), (−4, 9), (10, −5), $(4\frac{1}{2}, \frac{1}{2})$, and $(-\frac{1}{2}, 5\frac{1}{2})$. These points are plotted in Fig. 4-14.

All of these points lie on a line. For this reason, equations such as $x + y = 5$ are called *linear equations*. If we were to draw a line through these points, the sum of the coordinates for every point on the line is 5.

Graphing equations is a major part of college algebra (and many other math courses). We will learn how to choose points to plot so that we get a good idea of what an equation's graph looks like with a minimum amount of work. Let

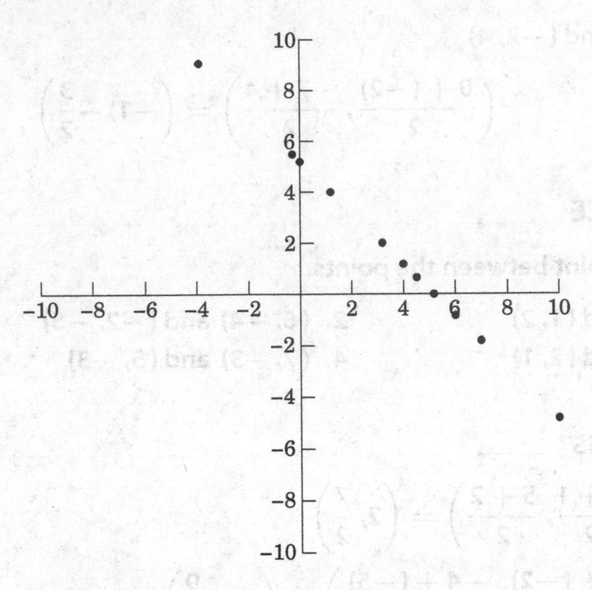

FIGURE 4-14

us begin with circles. Every point on a circle is at the same distance from the center of the circle. That distance is called the *radius*. This fact along with the distance formula allows us to discover a formula for the circle in the xy-plane. Call the center of the circle (h, k). That is, the x-coordinate of the circle is h and the y-coordinate is k. Call the radius of the circle r. A point (x, y) is on the circle if its distance from (h, k) is r. When we put this information in the distance formula, it becomes

$$\sqrt{(x-h)^2+(y-k)^2}=r$$

If we square both sides of this equation, we have

$$(x-h)^2+(y-k)^2=r^2$$

This is the formula for a circle in the xy-plane with radius r and center (h, k). When we are given the center and radius of a circle, we only need to put these three numbers, h, k, r, in the formula to find an equation for that circle.

EXAMPLE 4-6

Find an equation for the circle with the given radius and center.

- **Center (1, 4) , radius 3**

Here $h = 1$, $k = 4$, $r = 3$, $r^2 = 9$ and $(x - h)^2 + (y - k)^2 = r^2$ becomes $(x - 1)^2 + (y - 4)^2 = 9$.

- Center (0, 9), radius 4

With $h = 0$, $k = 9$, $r = 4$, $r^2 = 16$, we have $(x - 0)^2 + (y - 9)^2 = 16$. We simplify $(x - 0)^2$ to x^2, giving us $x^2 + (y - 9)^2 = 16$.

- Center (3, 2), radius $\frac{1}{2}$

With $h = 3$, $k = 2$, $r = \frac{1}{2}$, $r^2 = \frac{1}{4}$, we have $(x - 3)^2 + (y - 2)^2 = \frac{1}{4}$.

- Center (−2, 1), radius $r = 6$

With $h = -2$, $k = 1$, $r = 6$, $r^2 = 36$, we have $(x - (-2))^2 + (y - 1)^2 = 36$. Simplifying $(x - (-2))^2$ as $(x + 2)^2$, gives us $(x + 2)^2 + (y - 1)^2 = 36$.

PRACTICE

Find an equation for the circle with the given radius and center.

1. Center (4, 1), radius 7
2. Center (3, 6), radius 1
3. Center (0, 2), radius 3
4. Center (0, 0), radius 5
5. Center (−5, −2), radius 8

SOLUTIONS

1. $h = 4$, $k = 1$, $r = 7$, $r^2 = 49$: $(x - 4)^2 + (y - 1)^2 = 49$
2. $h = 3$, $k = 6$, $r = 1$, $r^2 = 1$: $(x - 3)^2 + (y - 6)^2 = 1$
3. $h = 0$, $k = 2$, $r = 3$, $r^2 = 9$

$$(x - 0)^2 + (y - 2)^2 = 9$$
$$x^2 + (y - 2)^2 = 9$$

4. $h = 0$, $k = 0$, $r = 5$, $r^2 = 25$

$$(x - 0)^2 + (y - 0)^2 = 25$$
$$x^2 + y^2 = 25$$

5. $h = -5,\ k = -2,\ r = 8,\ r^2 = 64$

$$(x-(-5))^2 + (y-(-2))^2 = 64$$
$$(x+5)^2 + (y+2)^2 = 64$$

When the equation of a circle is in the form $(x-h)^2 + (y-k)^2 = r^2$, we have a good idea of what its graph looks like. If h, k, and r are integers, we can even graph the circle with practically no work. We mark the center and go up, down, left, and right r units to plot four points on the circle. Next we draw a circle through these four points. We then erase the mark for the center because the center is not really a point on the circle.

EXAMPLE 4-7

- $(x-2)^2 + (y+1)^2 = 4$

The center is at $(2, -1)$ and the radius is 2. The center and four points are plotted in Fig. 4-15. The circle is given in Fig. 4-16.

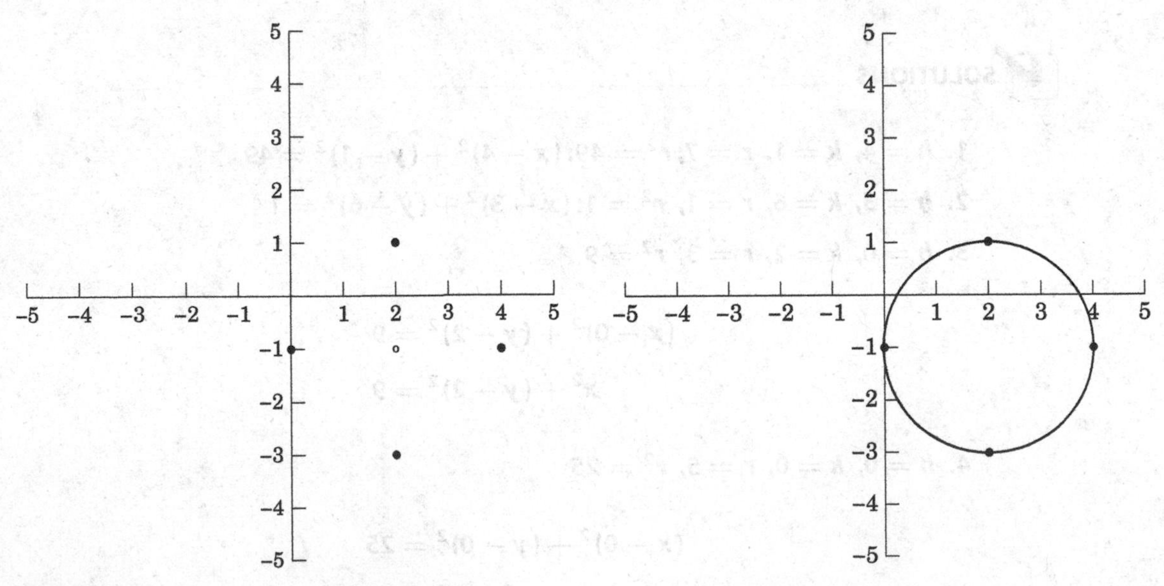

FIGURE 4-15

FIGURE 4-16

PRACTICE

Identify the center and radius of the circle and sketch its graph.

1. $(x-3)^2+(y-2)^2=9$
2. $(x+1)^2+(y+3)^2=4$
3. $(x-4)^2+y^2=1$
4. $x^2+y^2=16$

SOLUTIONS

1. **See Fig. 4-17; center: (3, 2); radius: 3**
2. **See Fig. 4-18; center: (−1, −3); radius: 2**
3. **See Fig. 4-19; center: (4, 0); radius: 1**
4. **See Fig. 4-20; center: (0, 0); radius: 4**

We can find an equation of the circle without directly knowing its center and radius. When given the endpoints of a diameter (a line segment that stretches the full width of a circle), we can find the center of the circle by finding the midpoint of the diameter. Once we know (h, k), we can use the coordinates of one of the points for x and y in the equation $(x-h)^2+(y-k)^2=r^2$ to find r^2.

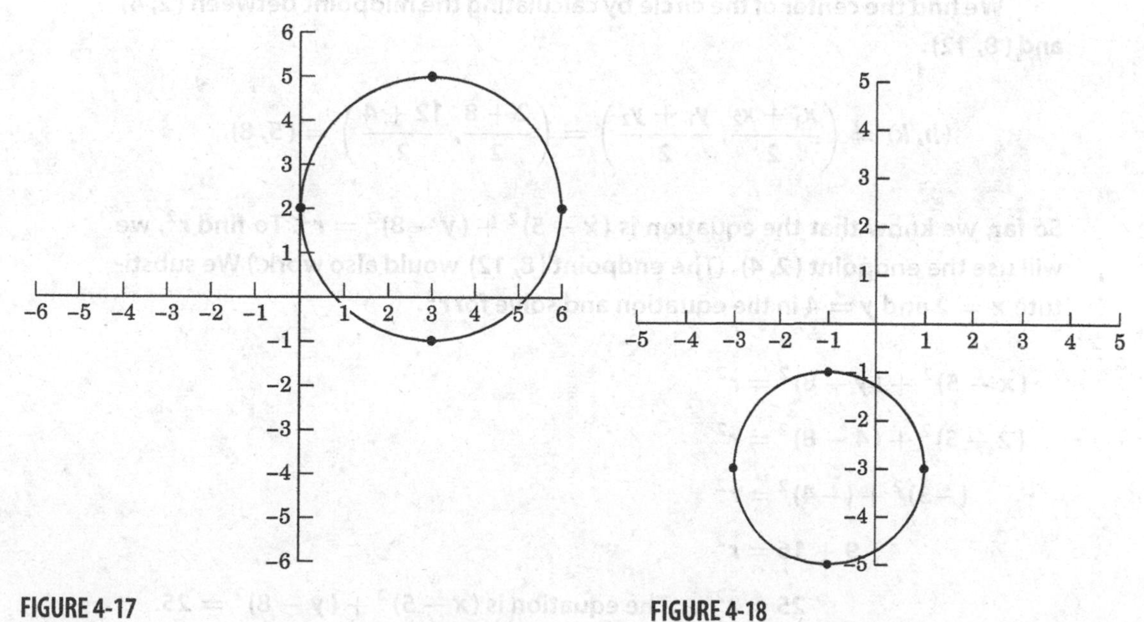

FIGURE 4-17

FIGURE 4-18

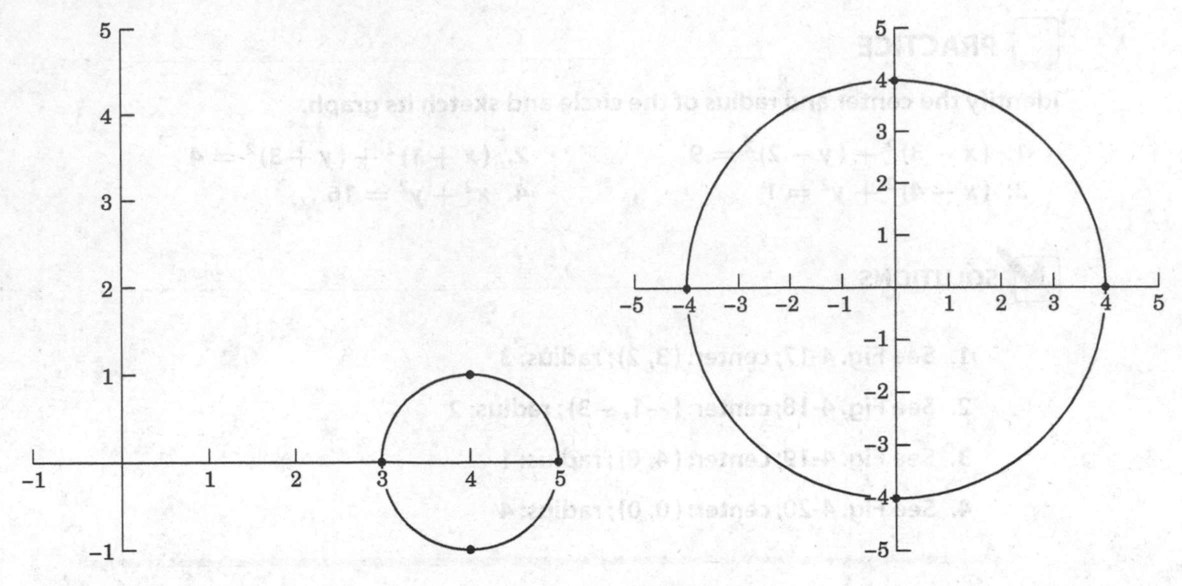

FIGURE 4-19

FIGURE 4-20

EXAMPLE 4-8

- **The endpoints of a diameter of a circle are (2, 4) and (8, 12). Find an equation for the circle.**

We find the center of the circle by calculating the midpoint between (2, 4) and (8, 12).

$$(h, k) = \left(\frac{x_1 + x_2}{2}, \frac{y_1 + y_2}{2}\right) = \left(\frac{2+8}{2}, \frac{12+4}{2}\right) = (5, 8)$$

So far, we know that the equation is $(x-5)^2 + (y-8)^2 = r^2$. To find r^2, we will use the endpoint (2, 4). (The endpoint (8, 12) would also work.) We substitute $x = 2$ and $y = 4$ in the equation and solve for r^2.

$$(x-5)^2 + (y-8)^2 = r^2$$

$$(2-5)^2 + (4-8)^2 = r^2$$

$$(-3)^2 + (-4)^2 = r^2$$

$$9 + 16 = r^2$$

$$25 = r^2$$ The equation is $(x-5)^2 + (y-8)^2 = 25$.

- **A circle has center $(-4, 3)$ and the point $(2, 11)$ is on the circle. Find an equation for the circle.**

Because $(-4, 3)$ is the center, we already know the equation is $(x+4)^2 + (y-3)^2 = r^2$. Let $x = 2$ and $y = 11$ in this equation to find r^2.

$$(2+4)^2 + (11-3)^2 = r^2$$

$$6^2 + 8^2 = r^2$$

$$100 = r^2 \quad \text{The equation is } (x+4)^2 + (y-3)^2 = 100.$$

Use the information given about the circle to find its equation.

1. **A diameter of the circle has endpoints $(2, 1)$ and $(-4, 9)$.**
2. **A diameter of the circle has endpoints $(0, 4)$ and $(-12, 9)$.**
3. **The center of a circle has coordinates $(1, 8)$ and the point $(13, 13)$ is on the circle.**
4. **The center of a circle is $(5, 0)$ and $(-5, 6)$ is on the circle.**

1. **The midpoint is the center of the circle.**

$$(h, k) = \left(\frac{2+(-4)}{2}, \frac{1+9}{2}\right) = (-1, 5)$$

So far, we know the equation is $(x+1)^2 + (y-5)^2 = r^2$. We could use either $(2, 1)$ or $(-4, 9)$ in the equation to find r^2. Here, we use $(2, 1)$.

$$(2+1)^2 + (1-5)^2 = r^2$$

$$9 + 16 = r^2$$

$$25 = r^2 \quad \text{The equation is } (x+1)^2 + (y-5)^2 = 25.$$

2. The midpoint is the center of the circle.

$$(h,k)=\left(\frac{0+(-12)}{2},\frac{4+9}{2}\right)=\left(-6,\frac{13}{2}\right)$$

So far, we know the equation is $(x+6)^2+(y-\frac{13}{2})^2=r^2$. We use $(0,4)$ to find r^2.

$$(0+6)^2+\left(4-\frac{13}{2}\right)=r^2$$

$$6^2+\left(-\frac{5}{2}\right)^2=r^2$$

$$36+\frac{25}{4}=r^2$$

$$\frac{169}{4}=r^2$$

The equation is $(x+6)^2+\left(y-\frac{13}{2}\right)^2=\frac{169}{4}$.

3. The center of the circle is $(1,8)$. This means that the circle equation begins as $(x-1)^2+(y-8)^2=r^2$. We use $(13,13)$ to find r^2.

$$(13-1)^2+(13-8)^2=r^2$$

$$144+25=r^2$$

$$169=r^2$$

The equation is $(x-1)^2+(y-8)^2=169$.

4. Because the center is $(5,0)$, we know the equation begins as $(x-5)^2+y^2=r^2$. We use $(-5,6)$ to find r^2.

$$(-5-5)^2+6^2=r^2$$

$$100+36=r^2$$

$$136=r^2$$

The equation is $(x-5)+y^2=136$.

Equations of circles are not always written in the form $(x-h)^2+(y-k)^2=r^2$. For example, the equation $(x-2)^2+(y+3)^2=16$ might be written in its expanded form.

$$(x-2)^2+(y+3)^2=16$$
$$(x-2)(x-2)+(y+3)(y+3)=16$$
$$x^2-4x+4+y^2+6y+9=16 \quad \text{After using the FOIL method}$$
$$x^2+y^2-4x+6y-3=0$$

In this section, we are given equations like the one above and use "completing the square" to rewrite them in the form $(x-h)^2+(y-k)^2=r^2$.

EXAMPLE 4-9

Identify the center and radius of the circle.

- $\mathbf{x^2+y^2-12x+4y+36=0}$

For the first step, we move the constant term (the number without a variable) to the right side of the equation, writing the left side with the *x*-terms together and the *y*-terms together.

$$\mathbf{x^2-12x+y^2+4y=-36}$$

Next, we complete the square for the *x*-terms and the *y*-terms and then we add both numbers to each side of the equation.

$$\mathbf{x^2-12x+36+y^2+4y+4=-36+36+4}$$

In the last step, we write the left side of the equation as the sum of two perfect squares. For $\mathbf{x^2-12x+36=(x-\_\_)^2}$, we have $\mathbf{\sqrt{36}=6}$. For $\mathbf{y^2+4y+4=(y+\_\_)^2}$, we have $\mathbf{\sqrt{4}=2}$.

$$\mathbf{(x-6)^2+(y+2)^2=4}$$

Now we see that this circle has center (6, −2) and radius 2.

- $\mathbf{x^2+y^2-8x-4y=-11}$

$$x^2 - 8x + \_\_ + y^2 - 4y + \_\_ = -11 + \_\_ + \_\_$$

Fill in the blanks with these numbers

$$\left(\frac{8}{2}\right)^2 = 16 \text{ and } \left(\frac{4}{2}\right)^2 = 4$$

$$x^2 - 8x + \underline{16} + y^2 - 4y + \underline{4} = -11 + \underline{16} + \underline{4}$$

For $x^2 - 8x + 16 = (x - \_\_)^2$, we use $\sqrt{16} = 4$. For $y^2 - 4y + 4 = (y - \_\_)^2$, we use $\sqrt{4} = 2$. The equation is $(x - 4)^2 + (y - 2)^2 = 9$, so the center is $(4, 2)$ and the radius is 3.

- $x^2 + y^2 - 2y - 14 = 0$

Because x^2 already is a perfect square, we only need to complete the square on the y-terms.

$$x^2 + y^2 - 2y + \_\_ = 14 + \_\_ \qquad \left(\frac{2}{2}\right)^2 = 1$$

$$x^2 + y^2 - 2y + 1 = 14 + 1$$

For $y^2 - 2y + 1 = (y - \_\_)^2$, we use $\sqrt{1} = 1$.

$$x^2 + (y - 1)^2 = 15$$ **The center is $(0, 1)$, and the radius is $\sqrt{15}$.**

PRACTICE

Identify the center and radius of the circle.

1. $x^2 + y^2 - 14x - 10y + 68 = 0$
2. $x^2 + y^2 + 4x - 8y + 11 = 0$
3. $x^2 + y^2 - 12x = -21$

SOLUTIONS

1. $x^2 - 14x + \_\_ + y^2 - 10y + \_\_ = -68 + \_\_ + \_\_$

$$x^2 - 14x + 49 + y^2 - 10y + 25 = -68 + 49 + 25$$

$$(x - 7)^2 + (y - 5)^2 = 6$$ **The center is $(7, 5)$ and the radius is $\sqrt{6}$.**

2. $x^2 + 4x + \_\_ + y^2 - 8y + \_\_ = -11 + \_\_ + \_\_$

$$x^2 + 4x + 4 + y^2 - 8y + 16 = -11 + 4 + 16$$

$$(x + 2)^2 + (y - 4)^2 = 9$$

The center is $(-2, 4)$ and the radius is 3.

3. $x^2 - 12x + \_\_ + y^2 = -21 + \_\_$

$$x^2 - 12x + 36 + y^2 = -21 + 36$$

$$(x - 6)^2 + y^2 = 15$$

The center is $(6, 0)$ and the radius is $\sqrt{15}$.

Sometimes the coefficient of x^2 and y^2 is not 1. In this case, we must divide both sides of the equation by this number *before* completing the square. It is worth mentioning that in equations of circles, x^2 and y^2 always have the same coefficient. If the coefficients are different, the graph of the equation is not a circle.

EXAMPLE 4-10

- $3x^2 + 3y^2 - 30x - 12y + 84 = 0$

$$3x^2 + 3y^2 - 30x - 12y + 84 = 0$$

$$\frac{1}{3}\left(3x^2 + 3y^2 - 30x - 12y\right) = \frac{1}{3}(-84)$$

$$x^2 + y^2 - 10x - 4y = -28$$

$$x^2 - 10x + \_\_ + y^2 - 4y + \_\_ = -28 + \_\_ + \_\_$$

$$x^2 - 10x + 25 + y^2 - 4y + 4 = -28 + 25 + 4$$

$$(x - 5)^2 + (y - 2)^2 = 1$$

Summary

In this chapter, we learned how to

- *Plot points in the xy-plane*. Every point in the xy plane can be labeled with an ordered pair (x, y). The x-coordinate describes the horizontal direction and distance to move from the origin. Similarly, the y-coordinate describes the vertical direction and distance to move from the origin.
- *Use the distance formula*. The distance between the points (x_1, y_1) and (x_2, y_2) is calculated with the formula $d = \sqrt{(x_2 - x_1)^2 + (y_2 - y_1)^2}$. We used this formula to determine whether or not points in the plane form certain shapes (squares, right triangles, etc.).
- *Use the midpoint formula*. The point halfway between the points (x_1, y_1) and (x_2, y_2) is $(\frac{x_1 + x_2}{2}, \frac{y_1 + y_2}{2})$.
- *Sketch the graph of a circle*. The graph of an equation of the form $(x - h)^2 + (y - k)^2 = r^2$ is a circle with center (h, k) and radius r. After plotting a point for the center, we move up, down, left, and right r units and then draw a circle through these four points.
- *Complete the square to rewrite an equation of a circle*. If an equation of a circle is not in the form $(x - h)^2 + (y - k)^2 = r^2$, we can complete the square to rewrite the equation in this form.

QUIZ

1. **Find the distance between the points $\left(-\frac{2}{3}, 1\right)$ and $\left(\frac{4}{3}, 5\right)$.**

   A. $2\sqrt{5}$ B. $\frac{2\sqrt{37}}{3}$ C. $\sqrt{41}$ D. $\frac{2\sqrt{82}}{3}$

2. **Find the midpoint between the points $\left(-\frac{2}{3}, 1\right)$ and $\left(\frac{4}{3}, 5\right)$.**

   A. $(-1, 2)$ B. $\left(\frac{1}{3}, 3\right)$ C. $(-1, 3)$ D. $\left(-\frac{1}{3}, -2\right)$

3. **True or false: The points $(0, 0)$, $(6, 2)$, $(8, -4)$, and $(2, -6)$ are the vertices of a square.**

   A. True B. False C. It is impossible to tell.

4. **What is the center and radius of the circle $x^2 + (y+4)^2 = 9$?**

   A. Center: $(0, 4)$; radius: 3 B. Center: $(0, 4)$; radius: 9
   C. Center: $(0, -4)$; radius: 3 D. Center: $(0, -4)$; radius: 9

5. **What is the center and radius of the circle $x^2 - 4x + y^2 + 8y + 16 = 0$?**

   A. Center: $(2, -8)$; radius: 4 B. Center: $(2, -4)$; radius: 4
   C. Center: $(2, -8)$; radius: 2 D. Center: $(2, -4)$; radius: 2

6. **Find an equation for the circle with center $\left(\frac{1}{2}, -1\right)$ and radius 4.**

   A. $(x + \frac{1}{2})^2 + (y - 1)^2 = 16$ B. $(x - \frac{1}{2})^2 + (y + 1)^2 = 4$
   C. $(x + \frac{1}{2})^2 + (y - 1)^2 = 4$ D. $(x - \frac{1}{2})^2 + (y + 1)^2 = 16$

7. **Find an equation for the circle with center $(5, 3)$ and whose graph contains the point $(-3, 9)$.**

   A. $x^2 - 10x + y^2 - 6y + 56 = 0$ B. $x^2 - 10x + y^2 - 6y - 56 = 0$
   C. $x^2 - 10x + y^2 - 6y + 76 = 0$ D. $x^2 - 10x + y^2 - 6y - 66 = 0$

8. **Find an equation for the circle in Fig. 4-21.**

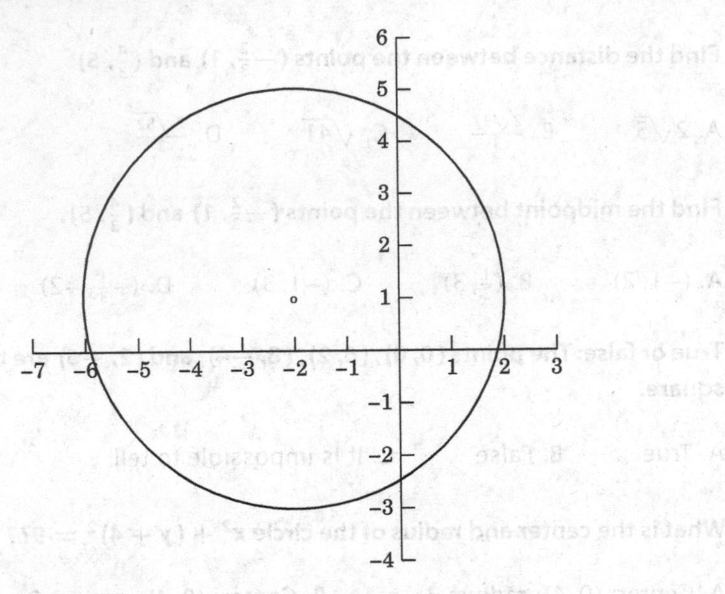

FIGURE 4-21

A. $(x+2)^2+(y-1)^2=2$ B. $(x+2)^2+(y-1)^2=16$
C. $(x-2)^2+(y+1)^2=2$ D. $(x-2)^2+(y+1)^2=16$

9. **Find the center and radius of the circle whose equation is $x^2-x+y^2+2y+\frac{1}{4}=0$.**

A. Center: $(\frac{1}{4},-1)$; radius: $\frac{1}{2}$ B. Center: $(-\frac{1}{4},-1)$; radius: 1
C. Center: $(\frac{1}{2},-1)$; radius: 1 D. Center: $(-\frac{1}{2},1)$; radius: $\frac{1}{2}$

10. **Find an equation for the circle with a diameter whose endpoints are (14, −5) and (2, 11).**

A. $(x-6)^2+(y-4)^2=256$ B. $(x-6)^2+(y-4)^2=16$
C. $(x-8)^2+(y-3)^2=100$ D. $(x-8)^2+(y-3)^2=20$

chapter 5

Lines and Parabolas

In this chapter, we continue our work with equations and their graphs. In Chap. 4, we learned how to find the center and radius of a circle from its equation. Here, we learn how to use similar information from an equation to graph lines and parabolas. We also use linear equations to solve applied problems. In Chap. 8, we will continue our work with the graph of a parabola, using what we have learned from this chapter.

CHAPTER OBJECTIVES

In this chapter, you will

- Sketch the graph of a line from an equation
- Locate the intercepts for a graph algebraically
- Calculate the slope of a line
- Find an equation for a line from two points on its graph
- Determine whether or not two lines are perpendicular or parallel
- Use linear equations to solve applied problems
- Sketch the graph of a parabola from its equation
- Algebraically find the vertex for the graph of a parabola

Introduction to Lines

The graph of an equation of the form $Ax + By = C$ is a line. An equation that can be written in this form is called a *linear equation*. We only need two points to sketch a line. It does not matter which two points, but we choose points that are easy to plot. We can pick two x-values at random. We then substitute them in the equation to compute the y-values.

EXAMPLE 5-1

Find the y-values for the given x-values and use the two points to sketch the line.

- $2x + 3y = 6$

We can choose *any* two numbers for x. Here we use $x = 0$ and $x = 6$.

$$\begin{aligned} 2(0) + 3y &= 6 & \qquad 2(6) + 3y &= 6 \\ 3y &= 6 & 12 + 3y &= 6 \\ y &= 2 & 3y &= -6 \\ & & y &= -2 \end{aligned}$$

After plotting the points $(0, 2)$ and $(6, -2)$, we draw a line through them. The line is shown in Fig. 5-1.

- $4x - y = 7$

We use $x = 0$ and $x = 2$.

$$\begin{aligned} 4(0) - y &= 7 & \qquad 4(2) - y &= 7 \\ -y &= 7 & 8 - y &= 7 \\ y &= -7 & -y &= -1, \text{ so } y = 1 \end{aligned}$$

The points $(0, -7)$ and $(2, 1)$ and the line are plotted in Fig. 5-2.

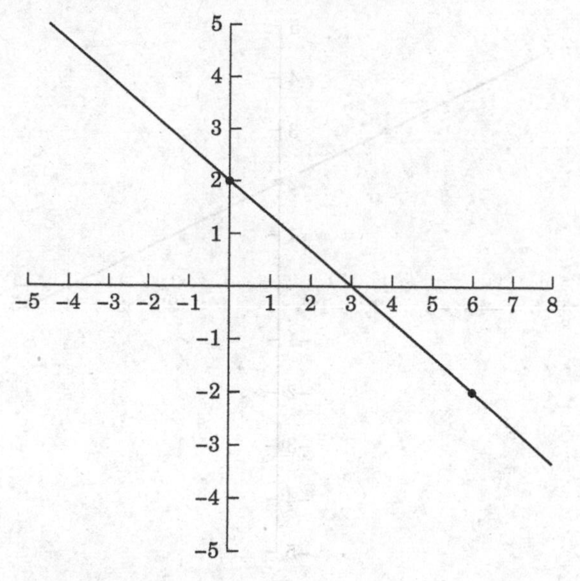

FIGURE 5-1

FIGURE 5-2

PRACTICE

Find the *y*-values for the given *x*-values and use the points to sketch the line.

1. $-3x + y = 5, \; x = 0$ and $x = -1$
2. $2x + 4y = 8, \; x = 0$ and $x = -2$
3. $x - 4y = 12, \; x = 0$ and $x = 4$
4. $-3x + 4y = -6, \; x = 0$ and $x = 4$

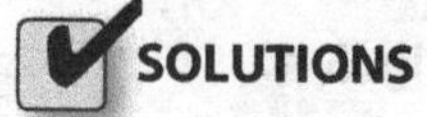

SOLUTIONS

1. We substitute $x = 0$ and $x = -1$ in $-3x + y = 5$ to find y.

$$-3(0) + y = 5 \qquad\qquad -3(-1) + y = 5$$

$$y = 5 \qquad\qquad 3 + y = 5$$

$$y = 2$$

We plot $(0, 5)$ and $(-1, 2)$. The graph is shown in Fig. 5-3.

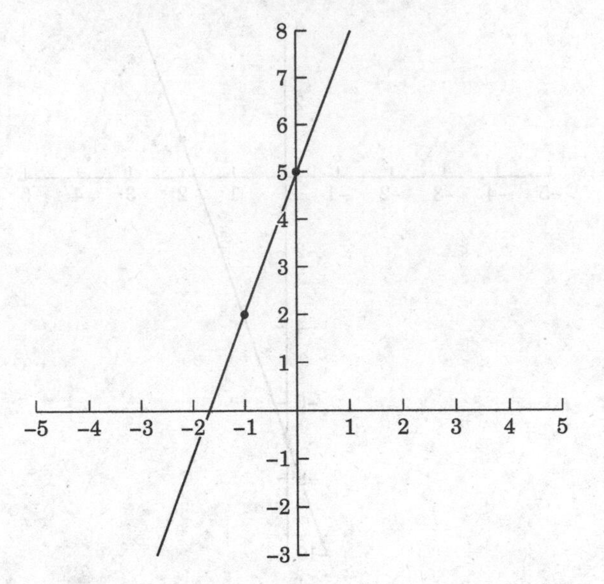

FIGURE 5-3

FIGURE 5-4

2. We substitute $x = 0$ and $x = -2$ in $2x + 4y = 8$ to find y.

$$2(0) + 4y = 8 \qquad 2(-2) + 4y = 8$$
$$4y = 8 \qquad -4 + 4y = 8$$
$$y = 2 \qquad 4y = 12$$
$$y = 3$$

We plot $(0, 2)$ and $(-2, 3)$. The graph is shown in Fig. 5-4.

3. We substitute $x = 0$ and $x = 4$ in $x - 4y = 12$ to find y.

$$0 - 4y = 12 \qquad 4 - 4y = 12$$
$$y = -3 \qquad -4y = 8$$
$$y = -2$$

We plot $(0, -3)$ and $(4, -2)$. The graph is shown in Fig. 5-5.

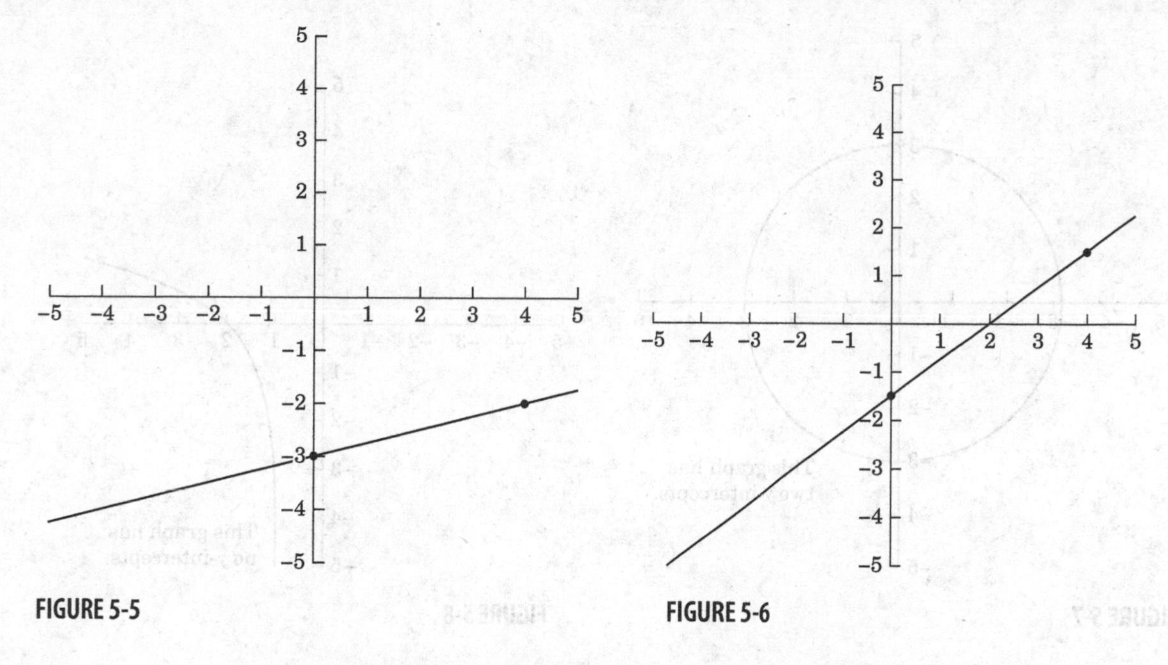

FIGURE 5-5

FIGURE 5-6

4. Let $x = 0$ and $x = 4$ in $-3x + 4y = -6$ to find y.

$$-3(0) + 4y = -6 \qquad -3(4) + 4y = -6$$

$$4y = -6 \qquad -12 + 4y = -6$$

$$y = \frac{-6}{4} \qquad 4y = -6 + 12 = 6$$

$$y = -\frac{3}{2} \qquad y = \frac{6}{4} = \frac{3}{2}$$

Now, we plot $(0, -\frac{3}{2})$ and $(4, \frac{3}{2})$. The graph is shown in Fig. 5-6.

Intercepts

You might have noticed that we plotted a point for $x = 0$ in all the previous Examples and Practice problems. This point was chosen for two reasons: one, computing y is easier if $x = 0$; and two, it is an important point in its own right. A point on a graph whose x-coordinate is 0 is called a *y-intercept*. This is where the graph crosses or touches the y-axis. Many of the graphs in this book have exactly one y-intercept. Some graphs have more than one y-intercept and some have none. See the graphs in Figs. 5-7 and 5-8 for examples.

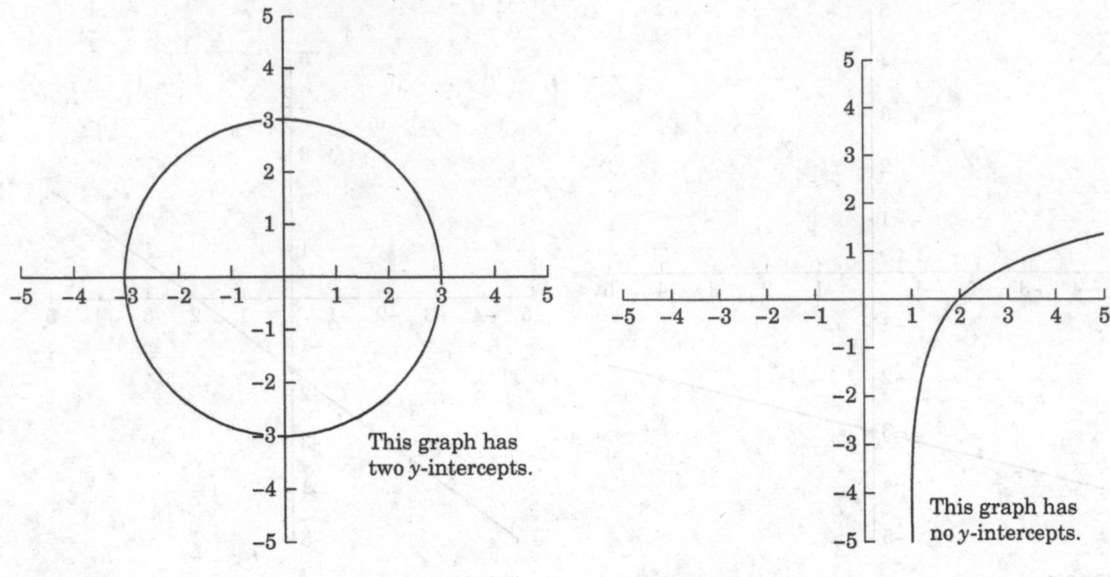

FIGURE 5-7

FIGURE 5-8

The y-coordinate is 0 for points on the graph that cross (or touch) the x-axis. This point is called an *x-intercept*. Some of the graphs in this book have exactly one x-intercept, some have more than one, and still others do not have any. See the graphs shown in Figs. 5-9–5-11 for examples.

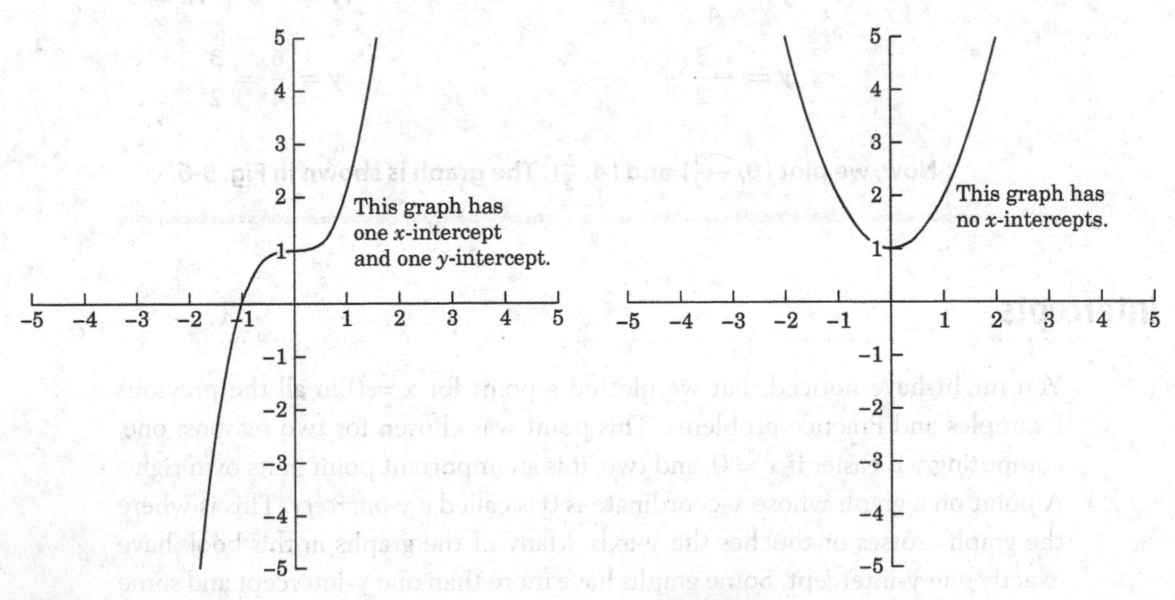

FIGURE 5-9

FIGURE 5-10

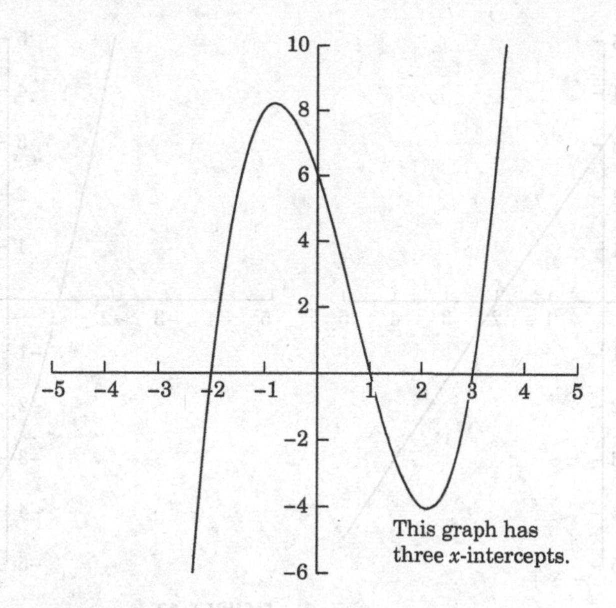

FIGURE 5-11

Rather than say $(a, 0)$ is an x-intercept, we say for short, a is an x-intercept. The x-intercept in Fig. 5-9 is -1 and the y-intercept is 1. The x-intercepts in Fig. 5-11 are -2, 1, and 3. The y-intercept is 6.

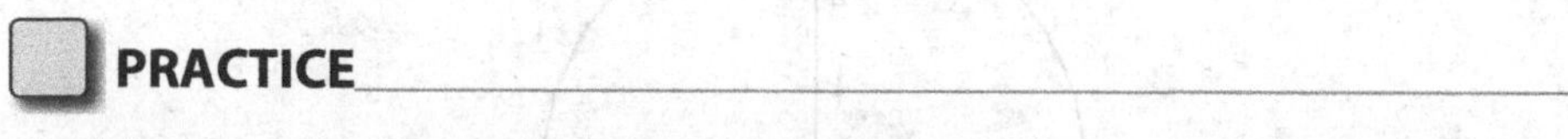

PRACTICE

Find the *x*- and *y*-intercepts on the graphs.

1. **The intercepts for the graph shown in Fig. 5-12 are __________.**
2. **The intercepts for the graph shown in Fig. 5-13 are __________.**
3. **The intercepts for the graph shown in Fig. 5-14 are __________.**

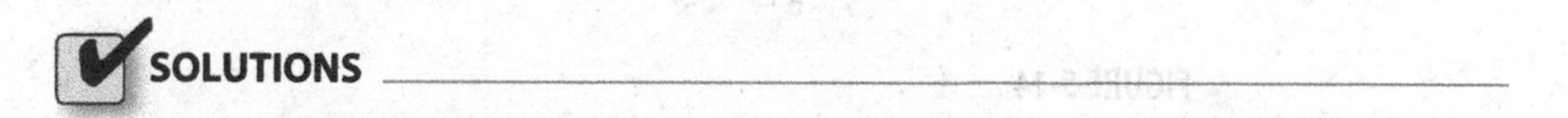

SOLUTIONS

1. **The *x*-intercept is 2, and the *y*-intercept is 3.**
2. **The *x*-intercepts are −1 and 3, and the *y*-intercept is −3.**
3. **The *x*-intercepts are −3 and 3, and the *y*-intercepts are −3 and 3.**

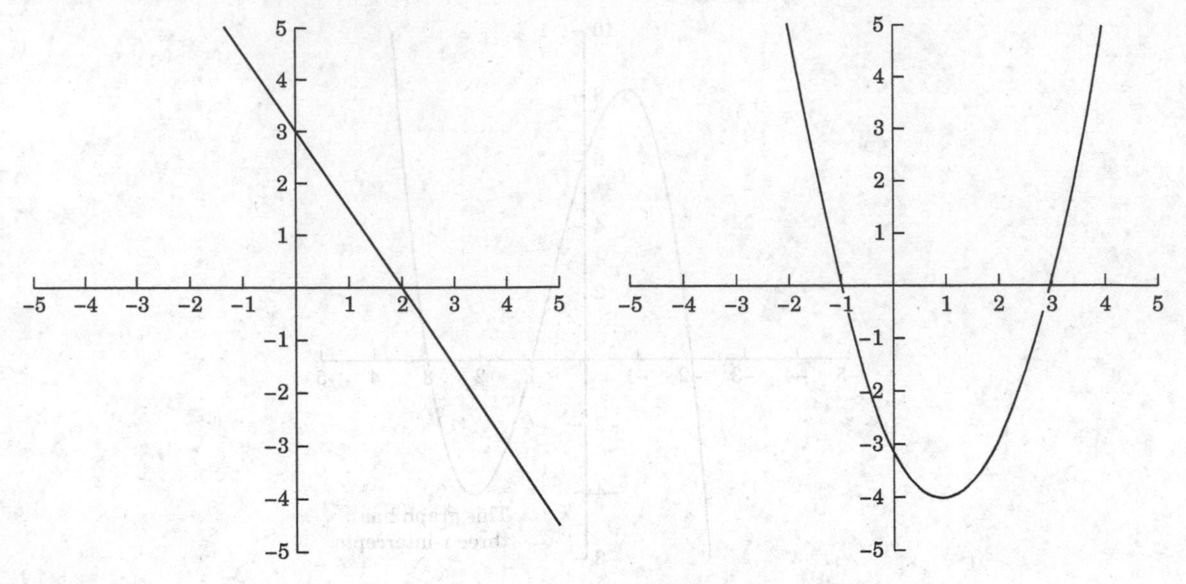

FIGURE 5-12

FIGURE 5-13

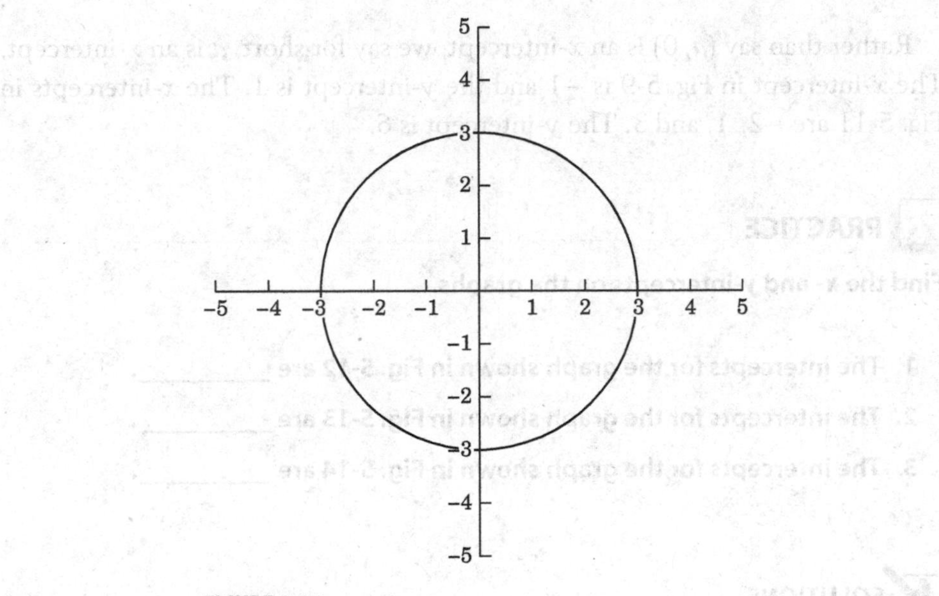

FIGURE 5-14

We can find intercepts for a graph without looking at it. We find the intercepts algebraically (if an equation has intercepts) by substituting 0 for one of the variables and solving for the other variable. To find the x-intercept, let $y = 0$ and solve for x. To find the y-intercept, let $x = 0$ and solve for y.

EXAMPLE 5-2

Find the intercepts algebraically.

- $3x + 2y = 6$

$$y = 0 \qquad\qquad x = 0$$

$$3x + 2(0) = 6 \qquad\qquad 3(0) + 2y = 6$$

$$3x = 6 \qquad\qquad 2y = 6$$

$$x = 2 \qquad\qquad y = 3$$

The x-intercept is 2, and the y-intercept is 3. The graph for this equation is in Fig. 5-12.

- $y = x^2 - x - 2$

$$y = 0 \qquad\qquad x = 0$$

$$x^2 - x - 2 = 0 \qquad\qquad y = 0^2 - 0 - 2$$

$$(x - 2)(x + 1) = 0 \qquad\qquad y = -2$$

$$x - 2 = 0 \text{ and } x + 1 = 0$$

$$x = 2 \text{ and } x = -1$$

The x-intercepts are 2 and -1, and the y-intercept is -2.

- $x^2 + y^2 = 16$

$$y = 0 \qquad\qquad x = 0$$

$$x^2 + 0^2 = 16 \qquad\qquad 0^2 + y^2 = 16$$

$$x^2 = 16 \qquad\qquad y^2 = 16$$

$$x = \pm 4 \qquad\qquad y = \pm 4$$

The x-intercepts are -4 and 4, and the y-intercepts are -4 and 4.

- $y = \dfrac{x+8}{x-2}$

The only way a fraction can be zero is if the numerator is zero. Here the numerator is $x + 8$, so we solve $x + 8 = 0$ to find the x-intercept.

$$y = 0 \qquad\qquad x = 0$$

$$x + 8 = 0 \qquad\qquad y = \frac{0+8}{0-2}$$

$$x = -8 \qquad\qquad y = -4$$

The x-intercept is -8, and the y-intercept is -4.

PRACTICE

Find the x- and y-intercepts algebraically.

1. $x - 2y = 4$ 2. $y = 3x - 12$ 3. $y = x^2 + 3x - 4$ 4. $y = \dfrac{x+6}{x+12}$

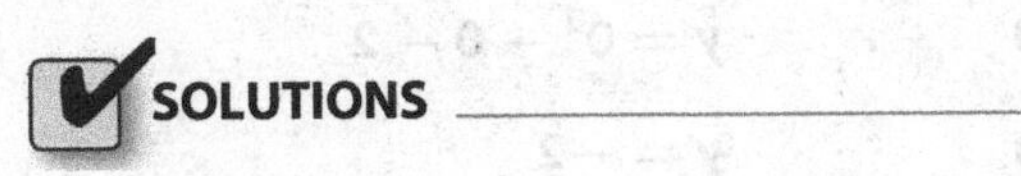

SOLUTIONS

1. $x - 2y = 4$

$$x - 2(0) = 4 \qquad\qquad 0 - 2y = 4$$

$$x = 4 \qquad\qquad -2y = 4$$

$$y = -2$$

The x-intercept is 4, and the y-intercept is -2.

2. $y = 3x - 12$

$$3x - 12 = 0 \qquad\qquad y = 3(0) - 12$$

$$3x = 12 \qquad\qquad y = -12$$

$$x = 4$$

The x-intercept is 4, and the y-intercept is -12.

3. $y = x^2 + 3x - 4$

$$x^2 + 3x - 4 = 0 \qquad y = 0^2 + 3(0) - 4$$

$$(x + 4)(x - 1) = 0 \qquad y = -4$$

$$x + 4 = 0 \text{ and } x - 1 = 0$$

$$x = -4 \text{ and } x = 1$$

The x-intercepts are -4 and 1, and the y-intercept is -4.

4. $y = \dfrac{x + 6}{x + 12}$

When finding the x-intercept, we only need to solve $x + 6 = 0$ because the only way a fraction can be zero is if its numerator is zero.

$$x + 6 = 0 \qquad y = \frac{0 + 6}{0 + 12}$$

$$x = -6 \qquad y = \frac{6}{12} = \frac{1}{2}$$

The x-intercept is -6, and the y-intercept is $\frac{1}{2}$.

We can tell whether or not a graph has intercepts by looking at it. What happens if we *do not* have the graph? A graph does not have an x-intercept if when letting $y = 0$ in its equation does not give us a real number solution. A graph does not have a y-intercept if when letting $x = 0$ in its equation does not give us a real number solution.

EXAMPLE 5-3

The graph of $y = x^2 + 4$ in Fig. 5-15 does not have any x-intercepts. Let's see what happens if we try to find the x-intercepts algebraically.

$$x^2 + 4 = 0$$

$$x^2 = -4$$

$$x = \pm\sqrt{-4}$$

Because $\sqrt{-4}$ is not a real number, the equation $x^2 + 4 = 0$ does not have a real solution.

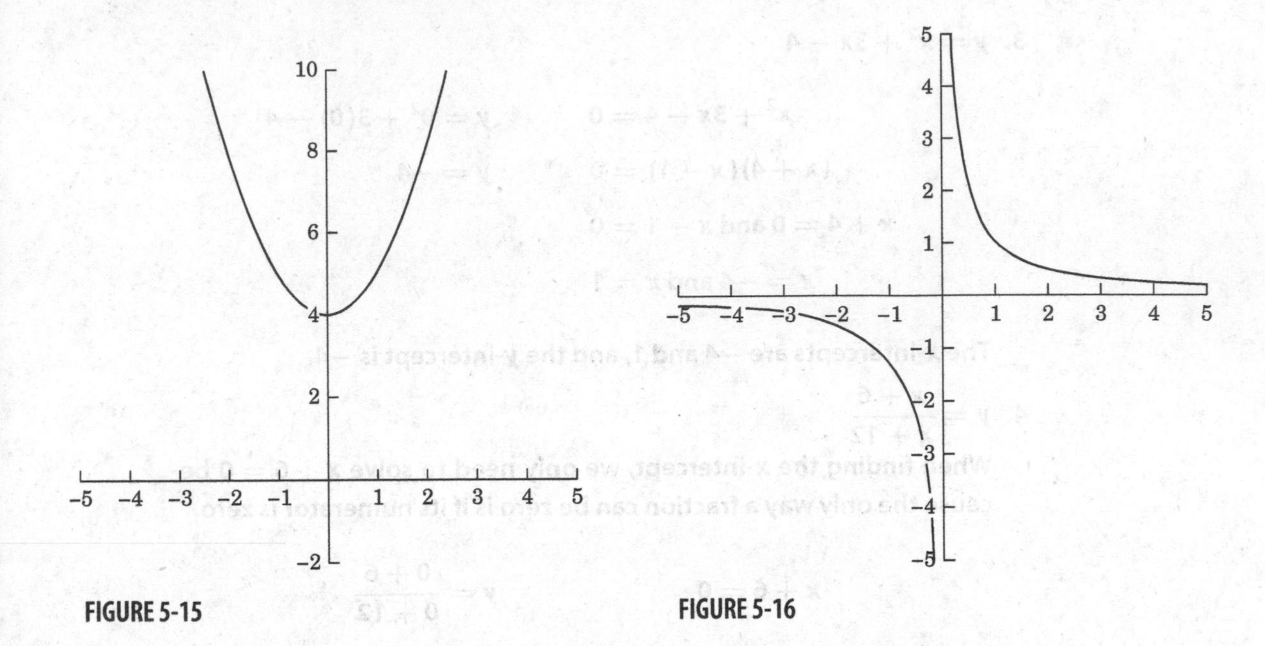

FIGURE 5-15

FIGURE 5-16

The graph of $y = \frac{1}{x}$ in Fig. 5-16 does not have any intercepts at all.

$$y = 0 \qquad x = 0$$

$$\frac{1}{x} = 0 \qquad \frac{1}{0} = y$$

As was mentioned earlier, a fraction can equal zero only if the numerator is zero. The equation $\frac{1}{x} = 0$ has no solution because the fraction is zero but the numerator, 1, is never zero. This shows us that the graph of $y = \frac{1}{x}$ has no *x*-intercept. The equation $\frac{1}{0} = y$ has no solution because $\frac{1}{0}$ is not a number. This shows us that the graph of $y = \frac{1}{x}$ has no *y*-intercept either.

The Slope of a Line

Another important part of a line is its *slope*. The slope is a measure of a line's tilt. Some lines have a steep slope and others have a more gradual slope. A line that tilts upward has a different slope from a line that tilts downward.

A line has a steep slope if a small horizontal change results in a large vertical change. The line in Fig. 5-17 has a steeper slope than the line in Fig. 5-18.

A line has a more gradual slope if a large horizontal change results in a small vertical change. The line in Fig. 5-18 has a more gradual slope.

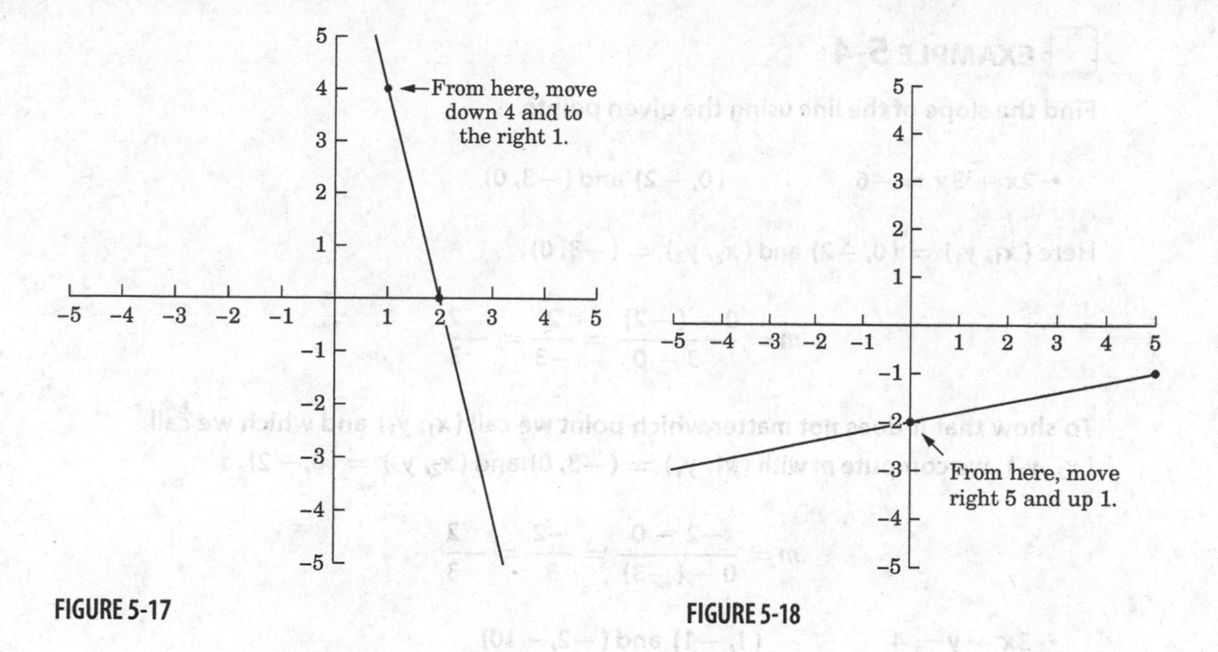

FIGURE 5-17

FIGURE 5-18

The slope of a line is measured by a number. This number is a quotient (a fraction) where the vertical change is divided by the horizontal change. In Fig. 5-17, to move from one point to the other, we have a vertical change of down 4 and a horiztonal change right 1. This means that the slope of the line $4x + y = 8$ is $\frac{-4}{1}$. In Fig. 5-18, to move from one point to the other, we moved up 1 and to the right 5. The slope to the line $x - 5y = 10$ is $\frac{1}{5}$.

One of the nice things about the slope of a line is that it does not matter which two points we use—the quotient of the vertical change to the horizontal change is the same. Suppose we use two other points on the line $4x + y = 8$. If we moved from $(-4, 24)$ to $(2, 0)$, then we would go down 24 (a change of -24) and to the right 6. This quotient is $\frac{-24}{6} = \frac{-4}{1}$, the same as with the two points in Fig. 5-17. This idea leads us to the slope formula. This formula is important and is worth memorizing.

If (x_1, y_1) and (x_2, y_2) are two points on a line, then the slope, m, of the line is the number

$$m = \frac{y_2 - y_1}{x_2 - x_1} = \frac{\text{vertical change}}{\text{horizontal change}}$$

EXAMPLE 5-4

Find the slope of the line using the given points.

- $2x + 3y = -6$ $(0, -2)$ and $(-3, 0)$

Here $(x_1, y_1) = (0, -2)$ and $(x_2, y_2) = (-3, 0)$.

$$m = \frac{0-(-2)}{-3-0} = \frac{2}{-3} = -\frac{2}{3}$$

To show that it does not matter which point we call (x_1, y_1) and which we call (x_2, y_2), we compute m with $(x_1, y_1) = (-3, 0)$ and $(x_2, y_2) = (0, -2)$.

$$m = \frac{-2-0}{0-(-3)} = \frac{-2}{3} = -\frac{2}{3}$$

- $3x - y = 4$ $(1, -1)$ and $(-2, -10)$

$$m = \frac{-10-(-1)}{-2-1} = \frac{-9}{-3} = \frac{3}{1} = 3$$

- $x - 2y = -2$ $(4, 3)$ and $(-2, 0)$

$$m = \frac{0-3}{-2-4} = \frac{-3}{-6} = \frac{1}{2}$$

PRACTICE

Find the slope of the line using the given points.

1. $2x + 3y = -12$; $(0, -4)$ and $(3, -6)$
2. $2x - y = 1$; $(0, -1)$ and $(1, 1)$
3. $x - y = 4$; $(3, -1)$ and $(2, -2)$
4. $x + 2y = 6$; $(2, 2)$ and $(-4, 5)$
5. $3x - 5y = 10$; $(10, 4)$ and $(5, 1)$

SOLUTIONS

1. $m = \dfrac{-6-(-4)}{3-0} = \dfrac{-2}{3} = -\dfrac{2}{3}$
2. $\dfrac{1-(-1)}{1-0} = \dfrac{2}{1} = 2$
3. $m = \dfrac{-2-(-1)}{2-3} = \dfrac{-1}{-1} = 1$
4. $m = \dfrac{5-2}{-4-2} = \dfrac{3}{-6} = -\dfrac{1}{2}$
5. $m = \dfrac{1-4}{5-10} = \dfrac{-3}{-5} = \dfrac{3}{5}$

Horizontal and Vertical Lines

The y-values of a horizontal line are the same number. The equation for a horizontal line is in the form "$y =$ number." For example, the line in Fig. 5-19 is the graph of $y = 3$.

What would the slope of a horizontal line be? No matter which two points we choose, their y-values are the same. This means that y_1 and y_2 are equal, so $y_2 - y_1 = 0$.

$$m = \frac{y_2 - y_1}{x_2 - x_1} = \frac{0}{x_2 - x_1} = 0$$

The slope of *any* horizontal line is 0.

The x-values of a vertical line are the same number. The equation for a vertical line is in the form "$x =$ number." The equation for the vertical line in Fig. 5-20 is $x = 5$.

Because all of the x-values on a vertical line are the same, x_2 and x_1 are the same. This means that the denominator of the slope of a vertical line is 0, so the slope is undefined.

$$m = \frac{y_2 - y_1}{x_2 - x_1} = \frac{y_2 - y_1}{0}$$

In addition to saying that the slope of a vertical line is undefined, we also say it does not exist. To say that the slope of a line does not exist is *not* the same

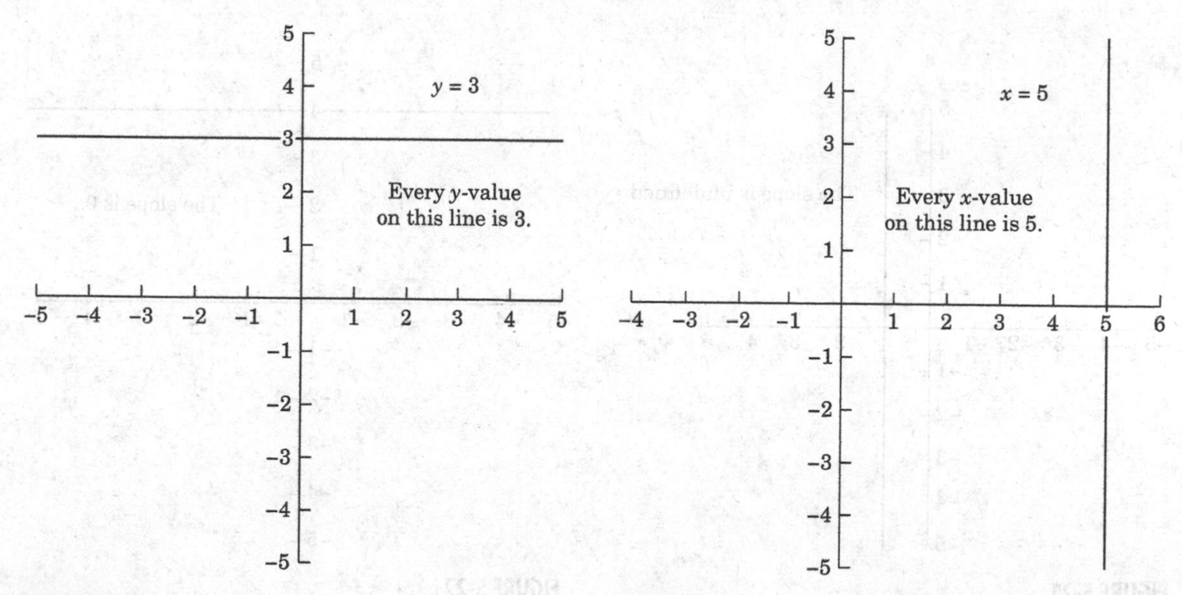

FIGURE 5-19

FIGURE 5-20

as saying that the slope is 0. The slope of a horizontal line is 0; the slope of a vertical line does not exist.

Still Struggling

Be careful to keep the coordinates of each point aligned in the slope formula. A common error is to switch the order of the coordinates in the numerator or denominator.

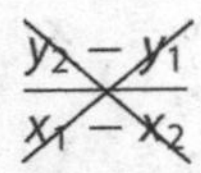

PRACTICE

Graph each line. State whether the slope is 0 or does not exist.

1. $x = 1$
2. $y = 4$
3. $x = -\frac{3}{2}$

SOLUTIONS

1. See Fig. 5-21.
2. See Fig. 5-22.
3. See Fig. 5-23.

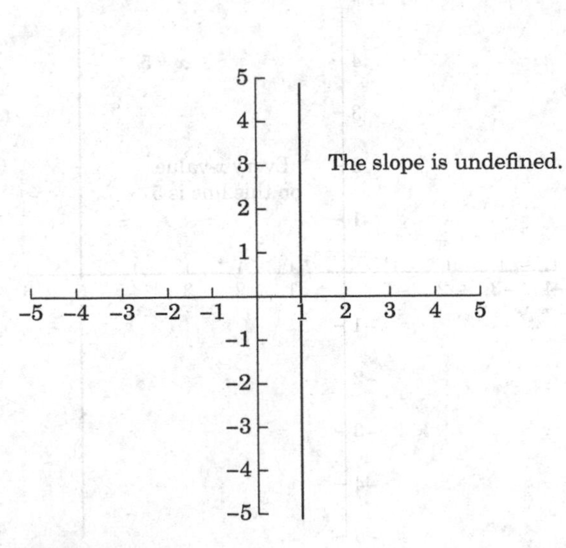
The slope is undefined.

FIGURE 5-21

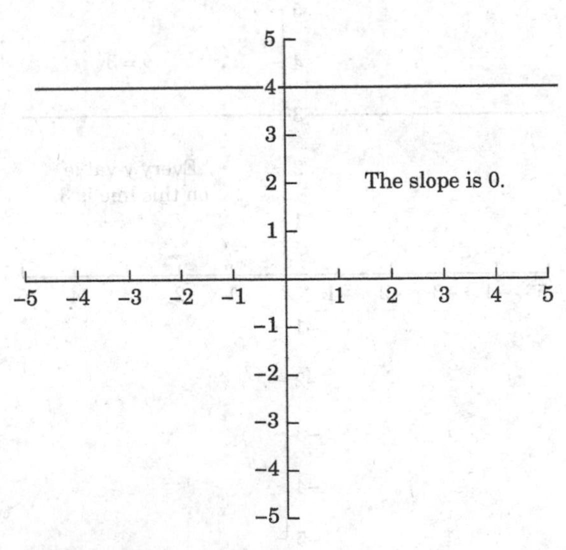
The slope is 0.

FIGURE 5-22

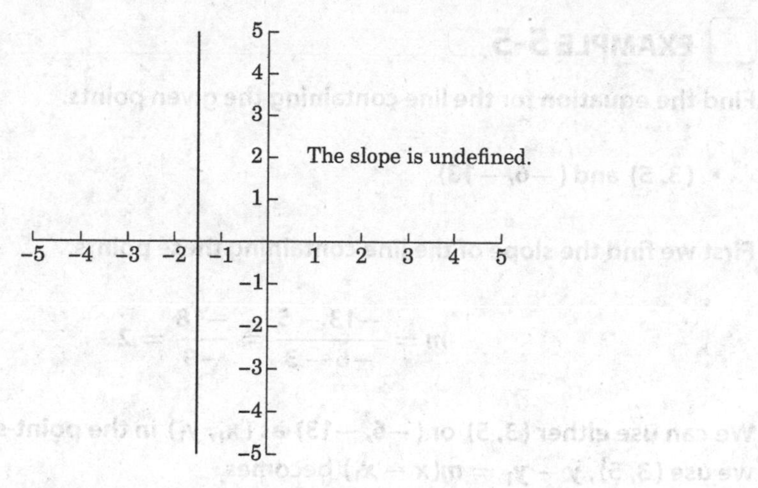

FIGURE 5-23

Finding an Equation for a Line

Two points on a line not only allow us to graph the line and calculate its slope, they also give us enough information to find an equation for the line. First, we use the slope formula to find the slope of the line. Second, we use the slope and one of the points in the *point-slope formula*.

$$y - y_1 = m(x - x_1)$$

This formula comes directly from the slope formula. All that was done to the slope formula was to replace (x_2, y_2) with (x, y) and to clear the fraction.

$$m = \frac{y - y_1}{x - x_1}$$

$$(x - x_1)m = (x - x_1) \cdot \frac{y - y_1}{x - x_1}$$

$$(x - x_1)m = y - y_1$$

$$y - y_1 = m(x - x_1)$$

In the following examples and Practice problems, we write linear equations in the general form: $Ax + By = C$, where A, B, and C are integers and A is not negative.

EXAMPLE 5-5

Find the equation for the line containing the given points.

- **(3, 5) and (−6, −13)**

First we find the slope of the line containing these points.

$$m = \frac{-13-5}{-6-3} = \frac{-18}{-9} = 2$$

We can use either (3, 5) or (−6, −13) as (x_1, y_1) in the point-slope formula. If we use (3, 5), $y - y_1 = m(x - x_1)$ becomes

$$y - 5 = 2(x - 3)$$

Now we write this equation in the general form.

$$y - 5 = 2(x - 3) \quad \text{Distribute 2.}$$

$$y - 5 = 2x - 6 \quad \text{We need } x \text{ and } y \text{ on the same side.}$$

$$-2x + y = -1 \quad \text{Collect variables on the left, constants on the right.}$$

$$-(-2x + y) = -(-1) \quad A \text{ needs to be positive.}$$

$$2x - y = 1$$

To show that it does not matter which point we choose for (x_1, y_1), we now find this equation using the point (−6, −13).

$$y - (-13) = 2(x - (-6))$$

$$y + 13 = 2x + 12$$

$$-2x + y = -1$$

$$2x - y = 1$$

- **(−4, 0) and (0, 4) (These are the intercepts.)**

$$m = \frac{4-0}{0-(-4)} = \frac{4}{4} = 1$$

We use $(-4, 0)$ as (x_1, y_1).

$$y - 0 = 1(x - (-4))$$

$$y = x + 4$$

$$-x + y = 4$$

$$x - y = -4$$

- $(-3, -2)$ and $(6, 1)$

$$m = \frac{1 - (-2)}{6 - (-3)} = \frac{3}{9} = \frac{1}{3}$$

We use $(-3, -2)$ as (x_1, y_1).

$$y - (-2) = \frac{1}{3}(x - (-3))$$

$$y + 2 = \frac{1}{3}(x + 3)$$

$$3(y + 2) = \left(\frac{1}{3} \cdot 3\right)(x + 3) \quad \text{Multiply by the LCD.}$$

$$3y + 6 = x + 3$$

$$-x + 3y = -3$$

$$x - 3y = 3$$

- $(4, 6)$ and $(-3, 6)$

The y-values are the same, making this a horizontal line. The equation for this line is $y = 6$—no work is necessary. The method used above still works on horizontal lines, though.

$$m = \frac{6 - 6}{-3 - 4} = \frac{0}{-7} = 0$$

$$y - 6 = 0(x - 4)$$

$$y - 6 = 0 \quad \text{or } y = 6$$

- $(-2, 0)$ and $(-2, 5)$

This line is a vertical line because the x-values are the same. The equation for this line is $x = -2$. No work is necessary (or even possible).

PRACTICE

Find an equation for the line containing the given points. Write the equation in the general form ($Ax + By = C$, where A, B, and C are integers and A is not negative) or in the form $x =$ number or $y =$ number.

1. $(1, 2)$ and $(5, -2)$
2. $(2, -7)$ and $(-1, 5)$
3. $(4, -6)$ and $(4, 2)$
4. $(5, -1)$ and $(-10, -10)$
5. $(1, 5)$ and $(4, 2)$
6. $(4, 8)$ and $(-1, 8)$
7. $\left(-1, \frac{3}{2}\right)$ and $\left(2, -\frac{3}{2}\right)$

SOLUTIONS

1. $m = \dfrac{-2-2}{5-1} = \dfrac{-4}{4} = -1$

$$y - 2 = -1(x - 1)$$
$$y - 2 = -x + 1$$
$$x + y = 3$$

2. $m = \dfrac{5-(-7)}{-1-2} = \dfrac{12}{-3} = -4$

$$y - (-7) = -4(x - 2)$$
$$y + 7 = -4x + 8$$
$$4x + y = 1$$

3. The x-values are the same, making this a vertical line. The equation is $x = 4$.

4. $m = \dfrac{-10-(-1)}{-10-5} = \dfrac{-9}{-15} = \dfrac{3}{5}$

$$y - (-1) = \frac{3}{5}(x - 5)$$

$$5(y + 1) = \left(5 \cdot \frac{3}{5}\right)(x - 5) \qquad \textbf{Multiply by the LCD.}$$

$$5y + 5 = 3(x - 5)$$

$$5y + 5 = 3x - 15$$

$$-3x + 5y = -20$$

$$3x - 5y = 20$$

5. $m = \dfrac{2 - 5}{4 - 1} = \dfrac{-3}{3} = -1$

$$y - 5 = -1(x - 1)$$

$$y - 5 = -x + 1$$

$$x + y = 6$$

6. Because the y-values are the same, this line is horizontal. The equation is $y = 8$.

7. $m = \dfrac{-\frac{3}{2} - \frac{3}{2}}{2 - (-1)} = \dfrac{-\frac{6}{2}}{3} = \dfrac{-3}{3} = -1$

$$y - \frac{3}{2} = -1(x - (-1))$$

$$y - \frac{3}{2} = -(x + 1)$$

$$y - \frac{3}{2} = -x - 1$$

$$x + y = \frac{3}{2} - 1$$

$$x + y = \frac{1}{2}$$

$$2(x + y) = 2 \cdot \frac{1}{2} \qquad \textbf{We want } C \textbf{ to be an integer.}$$

$$2x + 2y = 1$$

The Slope-Intercept Form of a Line

Now we are ready to learn a new form of the line. Remember that when a circle is in the form $(x-h)^2+(y-k)^2=r^2$, we know the circle's center and radius. The form of the line that gives the same kind of information is called the *slope-intercept* form of the line. When an equation is in this form, we immediately know the line's slope and its y-intercept.

To discover this form, we now examine a Practice problem from an earlier Practice set. Two points on the line were $(0,-4)$ (this is the y-intercept) and $(3,-6)$. The slope of the line is $-\frac{2}{3}$ and the general form of the equation is $2x+3y=-12$. Let us solve this equation for y (this means to isolate y on one side of the equation).

$$2x+3y=-12$$

$$3y=-2x-12$$

$$\frac{3y}{3}=\frac{-2x}{3}-\frac{12}{3}$$

$$y=-\frac{2}{3}x-4$$

The coefficient of x is $-\frac{2}{3}$, which is the slope; and the constant term is -4, the y-intercept. This happens *every* time a linear equation is solved for y. This is why $y=mx+b$ is called the slope-intercept form of the line. Because a vertical line has no y term and no slope, there is no slope-intercept form for a vertical line.

EXAMPLE 5-6

- **$y=3x+4$ The slope is 3. The y-intercept is 4.**
- **$y=x-2$ The slope is 1. They y-intercept is -2.**
- **$y=-\frac{1}{2}$ This equation could be rewritten as $y=0x-\frac{1}{2}$. The slope is 0. The y-intercept is $-\frac{1}{2}$.**
- **$y=x$ This equation could be rewritten as $y=1x+0$. The slope is 1. They y-intercept is 0.**

PRACTICE

Identify the slope and y-intercept for the line.

1. $y = -2x + 6$
2. $y = \frac{3}{4}x - 5$
3. $y = x + \frac{2}{3}$
4. $y = 4x$
5. $y = 10$

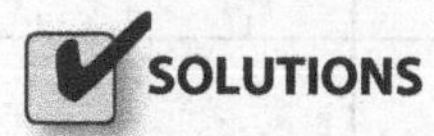

SOLUTIONS

1. The slope is -2, and the y-intercept is 6.
2. The slope is $\frac{3}{4}$, and the y-intercept is -5.
3. The slope is 1, and the y-intercept is $\frac{2}{3}$.
4. The equation can be rewritten as $y = 4x + 0$. The slope is 4, and the y-intercept is 0.
5. The equation can be rewritten as $y = 0x + 10$. The slope is 0, and the y-intercept is 10.

Graphing the Line Using the Slope and y-Intercept

We can sketch the graph of a line using the slope and any point on the line. In particular, we can graph a line using the slope and y-intercept. Remember what information the slope is giving: the vertical change over the horizontal change. We begin by plotting the y-intercept, and then we use the slope to find another point on the line. Finally, we draw a line through these two points.

EXAMPLE 5-7

- $y = \frac{2}{3}x + 1$

Because the y-intercept is 1, we begin by plotting $(0, 1)$ (see Fig. 5-24).

Because the slope is $\frac{2}{3}$, we locate another point from $(0, 1)$ by moving up 2 and then to the right 3. See Fig. 5-25 for the second point and Fig. 5-26 for the line.

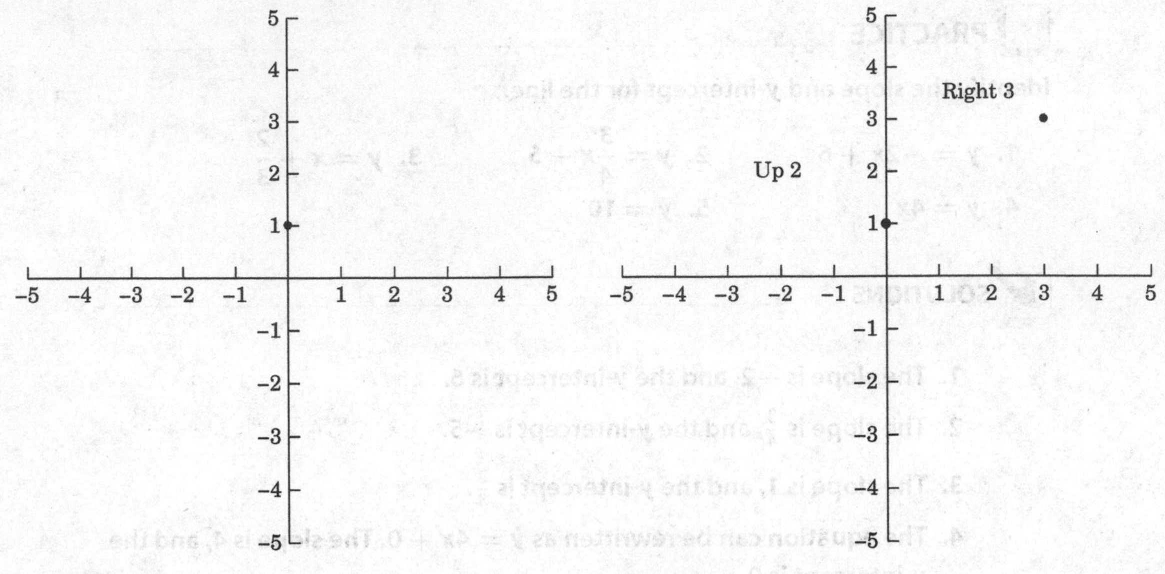

FIGURE 5-24

FIGURE 5-25

- $y = \frac{3}{5}x - 2$

First, we plot (0, −2). Next, we move up 3 units and to the right 5 units. This line is graphed in Fig. 5-27.

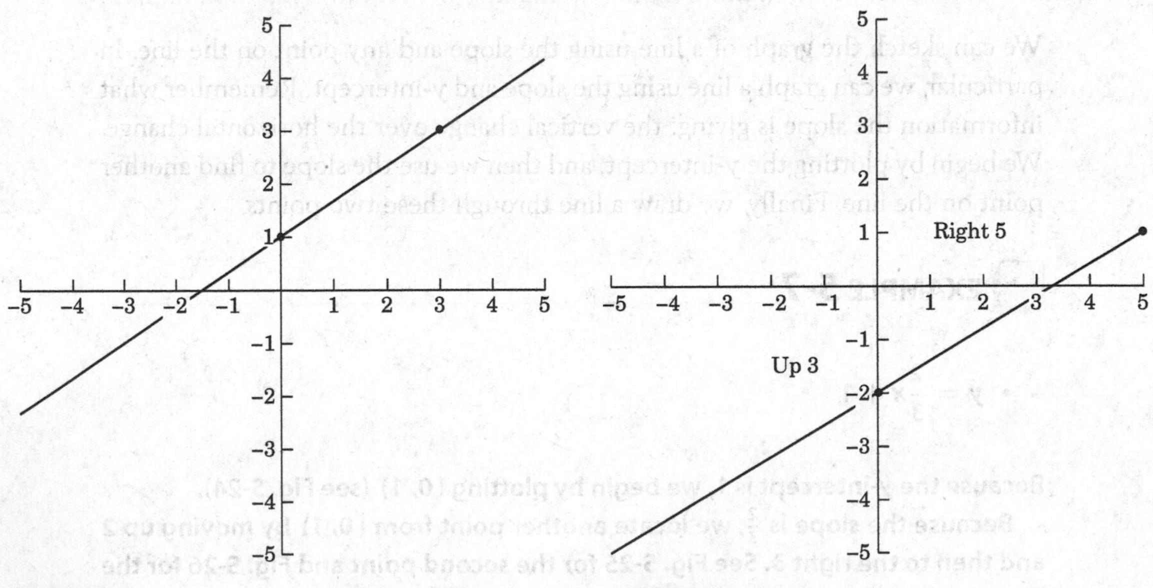

FIGURE 5-26

FIGURE 5-27

- $y = 3x + 1 = \frac{3}{1}x + 1$

We plot (0, 1) and then move up 3 units and to the right 1 unit. This line is graphed in Fig. 5-28.

- $y = -\frac{5}{2}x = -\frac{5}{2}x + 0 = \frac{-5}{2}x + 0 = \frac{5}{-2}x + 0$

We plot (0, 0). We either move down 5 units then to the right 2 units or move up 5 units then to the left 2 units. The line is graphed in Fig. 5-29.

- $y = 2 = 0x + 2$

We plot (0, 2). Go to the left or right any distance; do not move up or down. See Fig. 5-30.

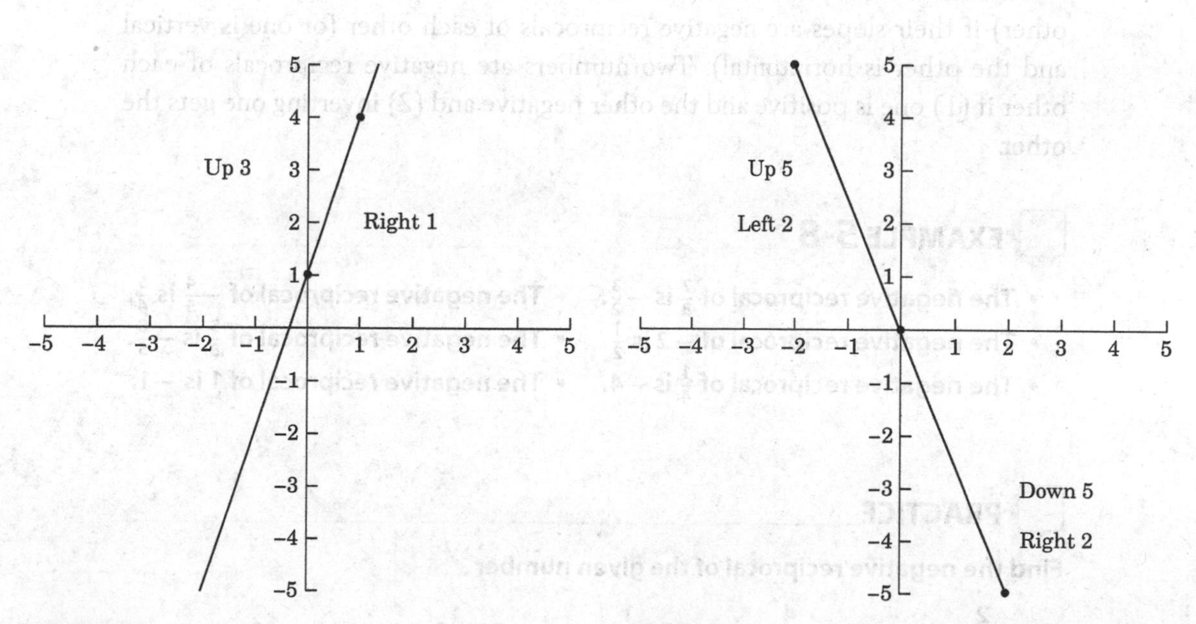

FIGURE 5-28

FIGURE 5-29

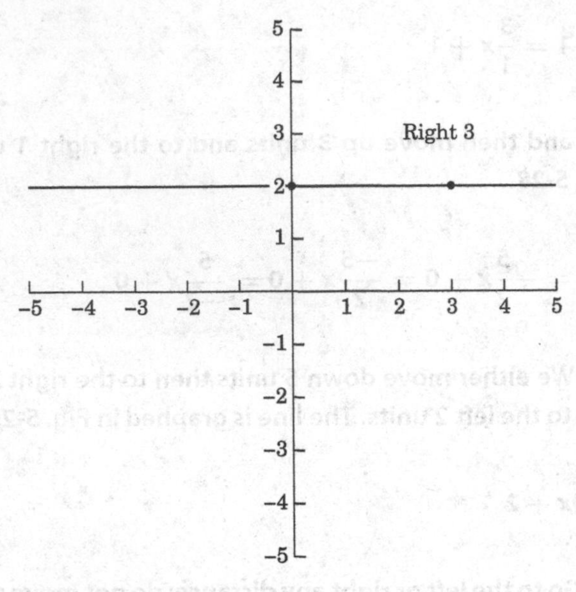

FIGURE 5-30

Parallel and Perpendicular Lines

Two lines are parallel if they have the same slope or if each slope is undefined.

Two lines are perpendicular (i.e., they form a 90° angle where they cross each other) if their slopes are negative reciprocals of each other (or one is vertical and the other is horiztontal). Two numbers are negative reciprocals of each other if (1) one is positive and the other negative and (2) inverting one gets the other.

EXAMPLE 5-8

- **The negative reciprocal of $\frac{2}{3}$ is $-\frac{3}{2}$.**
- **The negative reciprocal of -2 is $\frac{1}{2}$.**
- **The negative reciprocal of $\frac{1}{4}$ is -4.**
- **The negative reciprocal of $-\frac{4}{5}$ is $\frac{5}{4}$.**
- **The negative reciprocal of $\frac{5}{8}$ is $-\frac{8}{5}$.**
- **The negative reciprocal of 1 is -1.**

PRACTICE

Find the negative reciprocal of the given number

1. $\frac{2}{7}$ **2.** $-\frac{4}{3}$ **3.** $\frac{1}{5}$ **4.** -3 **5.** -1

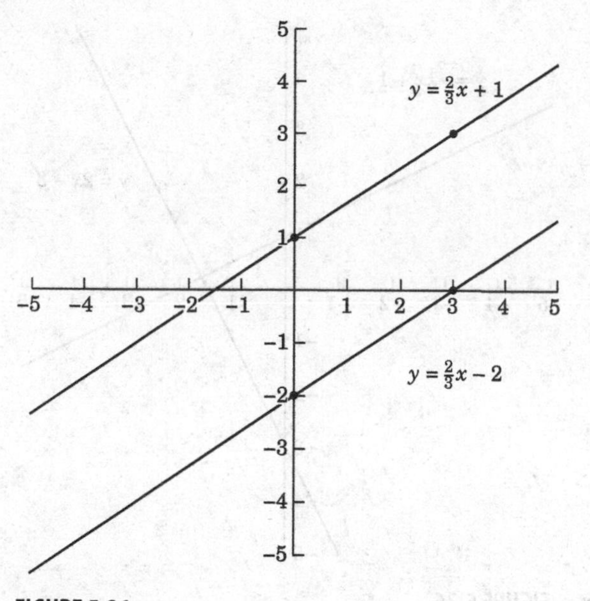

FIGURE 5-31

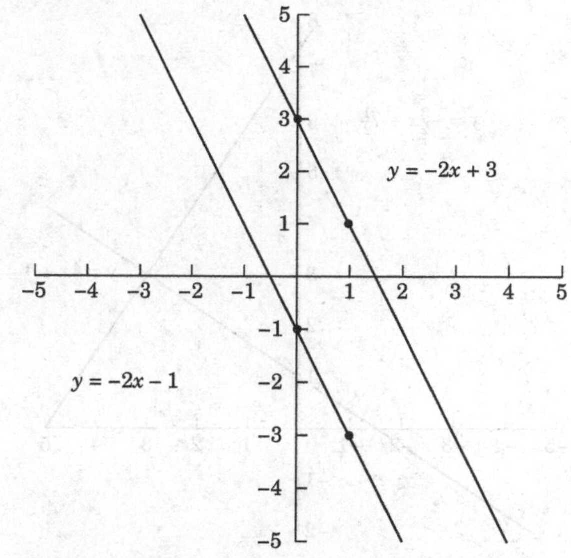

FIGURE 5-32

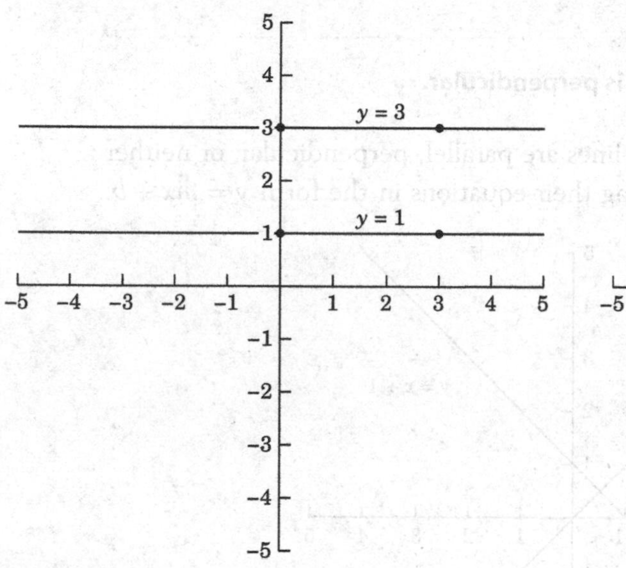

FIGURE 5-33

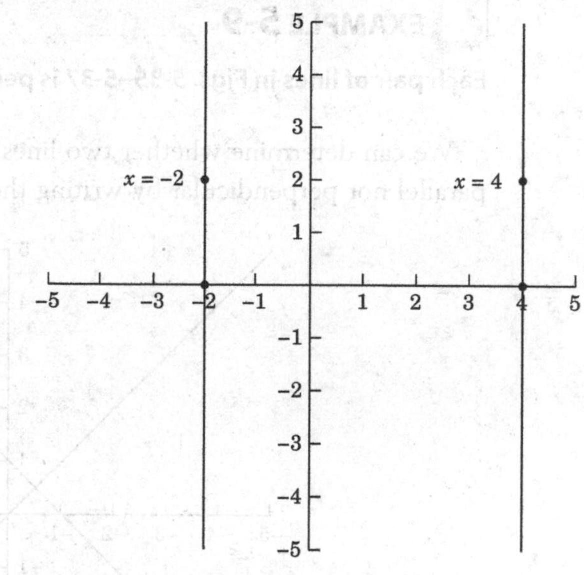

FIGURE 5-34

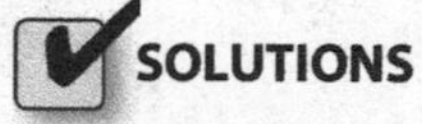

SOLUTIONS

1. $-\frac{7}{2}$ 2. $\frac{3}{4}$ 3. -5 4. $\frac{1}{3}$ 5. 1

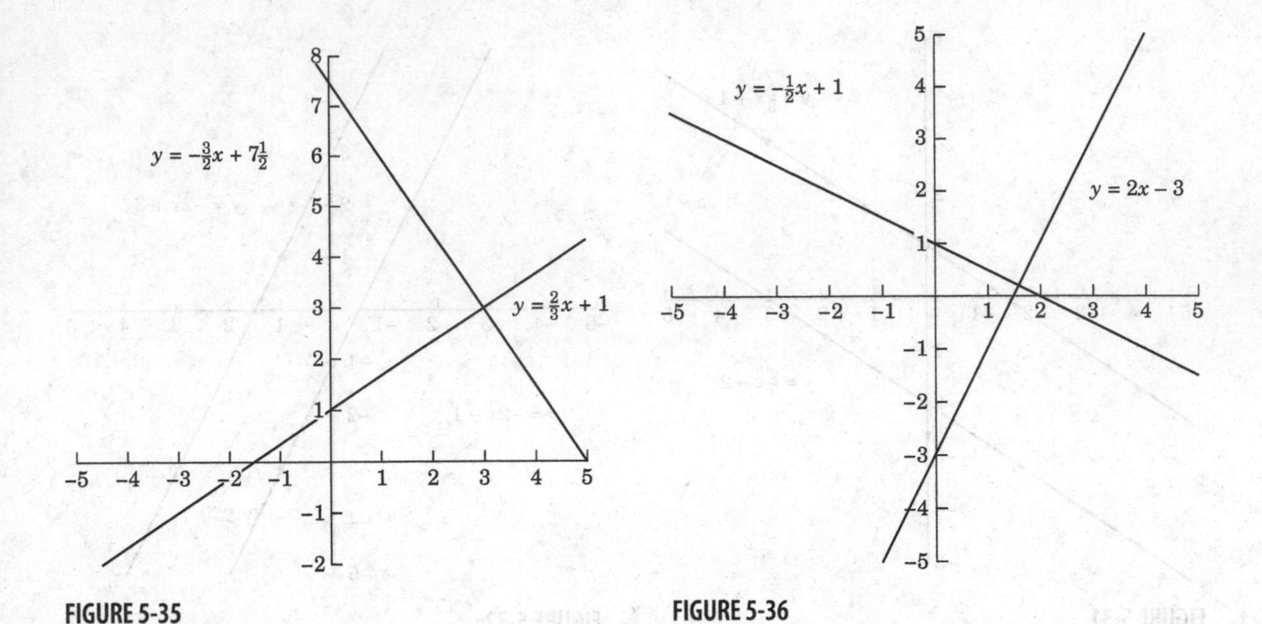

FIGURE 5-35

FIGURE 5-36

EXAMPLE 5-9

Each pair of lines in Figs. 5-35–5-37 is perpendicular.

We can determine whether two lines are parallel, perpendicular, or neither parallel nor perpendicular by writing their equations in the form $y = mx + b$.

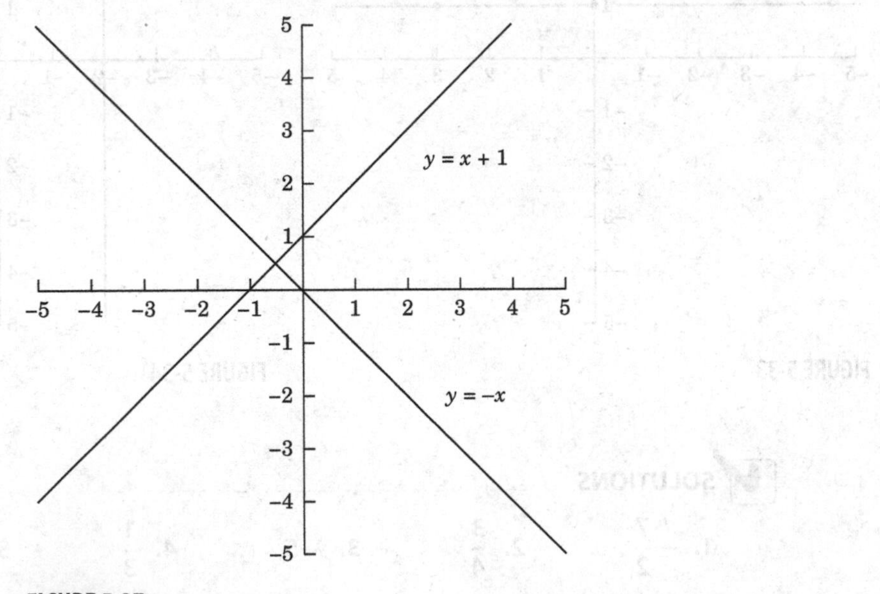

FIGURE 5-37

If m is the same for each line (or both are vertical), the lines are parallel. If one m is the negative reciprocal of the other (or one is vertical and the other horizontal), the lines are perpendicular. Otherwise, the lines are neither parallel nor perpendicular.

EXAMPLE 5-10

Determine if the lines are parallel, perpendicular, or neither.

- $2x - y = 4 \qquad 2x - y = -5$

First, we solve for y in each equation and then compare their slopes.

$$2x - y = 4 \qquad\qquad 2x - y = -5$$

$$-y = -2x + 4 \qquad\qquad -y = -2x - 5$$

$$y = 2x - 4 \qquad\qquad y = 2x + 5$$

Each slope is 2 so the lines are parallel.

- $x - 3y = -6 \qquad 3x + y = 2$

$$x - 3y = -6 \qquad\qquad 3x + y = 2$$

$$-3y = -x - 6 \qquad\qquad y = -3x + 2$$

$$y = \frac{-x}{-3} - \frac{6}{-3}$$

$$y = \frac{1}{3}x + 2$$

The slopes are negative reciprocals of each other, so the lines are perpendicular.

- $3x - 4y = 4 \qquad -4x + 3y = 9$

$$3x - 4y = 4 \qquad\qquad -4x + 3y = 9$$

$$-4y = -3x + 4 \qquad\qquad 3y = 4x + 9$$

$$y = \frac{3}{4}x - 1 \qquad\qquad y = \frac{4}{3}x + 3$$

The slopes are not equal and they are not negative reciprocals, so the lines are neither parallel nor perpendicular.

- $x = 4$ $\qquad\qquad\qquad\qquad$ $y = -2$

The first line is vertical and the second line is horiztonal. These lines are perpendicular.

PRACTICE

Determine if the lines are parallel, perpendicular, or neither.

1. $3x - 8y = 8$ and $3x - 8y = -16$
2. $3x - 4y = -8$ and $4x - 3y = 3$
3. $3x - 5y = -10$ and $5x + 3y = -3$
4. $y = 6$ and $y = -4$
5. $3x - 3y = 2$ and $4x + 4y = 1$
6. $2x - y = 4$ and $6x - 3y = 9$
7. $x = 1$ and $y = 1$

SOLUTIONS

1. **Parallel**

$$3x - 8y = 8 \qquad\qquad 3x - 8y = -16$$

$$-8y = -3x + 8 \qquad\qquad -8y = -3x - 16$$

$$y = \frac{3}{8}x - 1 \qquad\qquad y = \frac{3}{8}x + 2$$

2. **Neither**

$$3x - 4y = -8 \qquad\qquad 4x - 3y = 3$$

$$-4y = -3x - 8 \qquad\qquad -3y = -4x + 3$$

$$y = \frac{3}{4}x + 2 \qquad\qquad y = \frac{4}{3}x - 1$$

3. **Perpendicular**

$$3x - 5y = -10 \qquad\qquad 5x + 3y = -3$$

$$-5y = -3x - 10 \qquad\qquad 3y = -5x - 3$$

$$y = \frac{3}{5}x + 2 \qquad\qquad y = -\frac{5}{3}x - 1$$

4. **Both lines are horizontal, so they are parallel.**

5. Perpendicular (−1 and 1 are negative reciprocals)

$$3x - 3y = 2 \qquad 4x + 4y = 1$$

$$-3y = -3x + 2 \qquad 4y = -4x + 1$$

$$y = 1x - \frac{2}{3} \qquad y = -1x + \frac{1}{4}$$

$$y = x - \frac{2}{3} \qquad y = -x + \frac{1}{4}$$

6. Parallel

$$2x - y = 4 \qquad 6x - 3y = 9$$

$$-y = -2x + 4 \qquad -3y = -6x + 9$$

$$y = 2x - 4 \qquad y = 2x - 3$$

7. One line is vertical and the other is horizontal, so these lines are perpendicular.

We can use a faster method for finding an equation for a line when we know the slope and a point. We can substitute the slope and the coordinates of the point we know into the slope-intercept form ($y = mx + b$). The only unknown would then be b.

EXAMPLE 5-11

Find the slope-intercept form of the line containing the given point and having the given slope.

- **The slope is 4 and the point (1, −2) is on the line.**

Because $m = 4$, $y = mx + b$ becomes $y = 4x + b$. We can find b by letting $x = 1$ and $y = -2$ in $y = 4x + b$.

$$-2 = 4(1) + b$$

$$-2 = 4 + b$$

$$-6 = b$$

The line containing $(1, -2)$ with slope 4 is $y = 4x - 6$.

- The slope is 2 and the x-intercept is 3.

To say that the x-intercept is 3 is the same as saying $(3, 0)$ is a point on the line.

$$y = 2x + b$$
$$0 = 2(3) + b$$
$$-6 = b$$
$$y = 2x - 6$$

PRACTICE

Find the slope-intercept form of the line containing the given point and having the given slope.

1. The slope is $\frac{1}{2}$ and $(2, 5)$ is on the line.
2. The slope is $-\frac{4}{9}$ and $(18, -8)$ is on the line.
3. The slope is -5 and the x-intercept is 1.
4. The slope is -2 and the y-intercept is 6.

SOLUTIONS

1. $m = \frac{1}{2}$, $x = 2$, $y = 5$

$$y = mx + b$$
$$5 = \frac{1}{2}(2) + b$$
$$5 = 1 + b$$
$$4 = b$$

The line is $y = \frac{1}{2}x + 4$.

2. $m = -\frac{4}{9}, \ x = 18, \ y = -8$

$$y = mx + b$$

$$-8 = -\frac{4}{9}(18) + b$$

$$-8 = -8 + b$$

$$0 = b \qquad \text{The line is } y = -\frac{4}{9}x.$$

3. $m = -5$

To say that the x-intercept is 1 is another way of saying $(1, 0)$ is on the line, so $x = 1, \ y = 0$.

$$y = mx + b$$

$$0 = -5(1) + b$$

$$5 = b \qquad \text{The line is } y = -5x + 5.$$

4. The y-intercept is 6 and $m = -2$. There is nothing we need to do but write the equation: $y = -2x + 6$.

Linear Applications

The relationship between many pairs of variables can be described by a linear equation. These variables are called *linearly related*. For example, if one is paid \$12 per hour, the daily pay (before deductions) would be described by the equation $p = 12h$, where p represents the daily pay, and h represents the number of hours worked for the day. The slope of this line is 12 and the p-intercept (this is like the y-intercept) is 0. We begin our work with linear applications by answering such questions as, "If you were paid \$60, how many hours did you work?" Later we will use two pairs of numbers to find a linear equation that models the situation.

EXAMPLE 5-12

- **An electric company bills y dollars for x kilowatt-hours used each month. The equation for each family's electric bill is $y = 0.16x + 25$. If the Tobias family's electric bill for 1 month was \$153, how much electricity was used?**

The information given in the problem is $y = 153$ for $y = 0.16x + 25$. We substitute $y = 153$ in the equation and solve for x, the number of kilowatt-hours used.

$$y = 0.16x + 25$$

$$153 = 0.16x + 25$$

$$128 = 0.16x$$

$$\frac{128}{0.16} = x$$

$$800 = x$$

The Tobias family used 800 kilowatt-hours of electricity.

- **The relationship between degrees Celsius and Fahrenheit is $F = \frac{9}{5}C + 32$.**

(a) If the temperature in Fahrenheit is 86 degrees, what is the temperature in Celsius?

(b) If the temperature is −20 degrees Celsius, what is the temperature on the Fahrenheit scale?

(a) We substitute $F = 86$ in $F = \frac{9}{5}C + 32$ and solve for C.

$$F = \frac{9}{5}C + 32$$

$$86 = \frac{9}{5}C + 32$$

$$54 = \frac{9}{5}C$$

$$\frac{5}{9} \cdot 54 = C$$

$$30 = C$$

The temperature is 30 degrees Celsius.

(b) We substitute $C = -20$ and compute F.

$$F = \frac{9}{5}(-20) + 32$$

$$F = -36 + 32 = -4$$

The temperature is -4 degrees Fahrenheit.

- For the years 2000–2009, enrollment at a small college is approximated by the equation $y = 75x + 1100$, where y represents the number of students enrolled and x represents the number of years after 2000. Use the equation to approximate enrollment for the years 2000, 2006, and 2009. In what year was enrollment about 1475?

Because x represents the number of years after 2000, $x = 0$ is the year 2000; $x = 6$ is the year 2006; and $x = 9$ is the year 2009. (Because the equation is only good for the years 2000–2009, the only values of x we can use are $x = 0, 1, 2, \ldots, 9$.) We want to find y for $x = 0$, $x = 6$, $x = 9$.

When $x = 0$, $y = 75(0) + 1100$. Enrollment for 2000 was about 1100.

When $x = 6$, $y = 75(6) + 1100 = 1550$. Enrollment for 2006 was about 1550.

When $x = 9$, $y = 75(9) + 1100 = 1775$. Enrollment for 2009 was about 1775.

For the question "In what year was enrollment about 1475?" we let $y = 1475$ and solve for x.

$$1475 = 75x + 1100$$

$$375 = 75x$$

$$5 = x$$

Enrollment was about 1475 in the year $2000 + 5 = 2005$.

PRACTICE

1. Whitney is paid a salary plus a commission on sales. Her salary is given by the equation $y = 0.08x + 25{,}000$, where y is her annual salary and x is her annual sales level.
   (a) If Whitney's annual sales level was \$190,000, what was her annual salary?
   (b) If Whitney's annual salary was \$35,080, what was her annual sales level?

2. The relationship between degrees Celsius and degrees Fahrenheit is given by the equation $C = \frac{5}{9}(F - 32)$.
   (a) What is the temperature on the Celsius scale when it is 113 degrees Fahrenheit?
   (b) What is the temperature on the Fahrenheit scale when it is 35 degrees Celsius?

3. A package delivery company added vans to its fleet at one of its centers between the years 2005 and 2012. The number of vans in the center's fleet is given by the equation $y = 10x + 90$, where y is the number of vans and x is the number of years after 2005.
   (a) How many vans were in the fleet for the years 2005, 2009, and 2012?
   (b) In what year did the center have 110 vans?

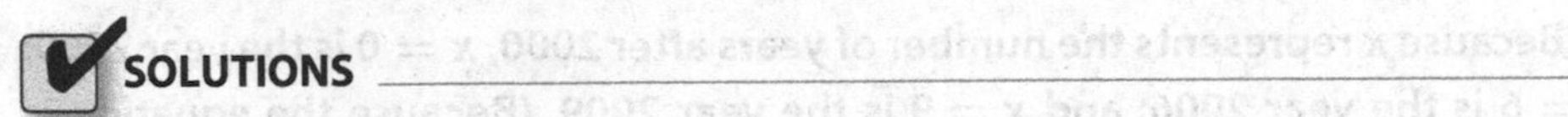

SOLUTIONS

1. (a) Whitney's annual sales level was \$190,000. Let $x = 190{,}000$ in the equation $y = 0.08x + 25{,}000$ and compute y.

$$y = 0.08(190{,}000) + 25{,}000$$

$$y = 15{,}200 + 25{,}000 = 40{,}200.$$ Her annual salary was \$40,200.

   (b) Her annual salary was \$35,080. Let $y = 35{,}080$ in the equation and solve for x.

$$35{,}080 = 0.08x + 25{,}000$$

$$10{,}080 = 0.08x$$

$$\frac{10{,}080}{0.08} = x$$

$$126{,}000 = x$$ Her annual sales level was \$126,000.

2. (a) We substitute $F = 113$ in $C = \frac{5}{9}(F - 32)$.

$$C = \frac{5}{9}(113 - 32)$$

$$C = \frac{5}{9}(81) = 45$$ The temperature is 45 degrees Celsius.

(b) We substitute $C = 35$ in the equation and solve for F.

$$35 = \frac{5}{9}(F - 32)$$

$$9(35) = \left(9 \cdot \frac{5}{9}\right)(F - 32) \quad \text{Clear the fraction.}$$

$$315 = 5(F - 32)$$

$$315 = 5F - 160$$

$$475 = 5F$$

$$95 = F \quad \text{The temperature is 95 degrees Fahrenheit.}$$

3. (a) The year 2005 is 0 years after 2005, so $x = 0$. We substitute $x = 0$ in the equation $y = 10x + 90$ and compute y.

$$y = 10(0) + 90$$

$$y = 90 \quad \text{The center had 90 vans in its fleet in the year 2005.}$$

The year 2009 is 4 years after 2005, so $x = 4$.

$$y = 10(4) + 90$$

$$y = 40 + 90 = 130 \quad \text{The center had 130 vans in its fleet in the year 2009.}$$

The year 2012 is 7 years after 2005, so $x = 7$.

$$y = 10(7) + 90$$

$$y = 70 + 90 = 160 \quad \text{The center had 160 vans in its fleet in the year 2012.}$$

(b) Let $y = 110$ in the equation and solve for x.

$$110 = 10x + 90$$

$$20 = 10x$$

$$2 = x \quad \text{There were 110 vans in the fleet when } x = 2 \text{ in the year 2007.}$$

In the last problems in this section, we are given enough information to find a linear equation. In the first problem set, we are given enough information to find two points on the line. In the second problem set, we are given enough information to find a point and the slope.

EXAMPLE 5-13

- **A company pays its entry-level sales representatives a commission that is a percentage of their monthly sales plus a certain base salary. This month, the sales representative from City A earned $5000 from sales of $35,000. The sales representative from City B earned $5300 on sales of $37,500. What percentage of monthly sales does the company pay in commission? What is its base salary?**

Salaries that are based on commission (with or without a base salary) are based on a linear equation. If y is the salary, m is the commission percentage, and x is the sales level, then the equation is $y = mx$ (without a base salary) and $y = mx + b$ (with base salary b). What do the ordered pairs (x, y) mean for this problem? The x-coordinate is a representative's sales level, and the y-coordinate is his pay amount. With this in mind, we can view the sentence, "The sales representative from City A earned $5000 from sales of $35,000" as the ordered pair $(35{,}000, 5000)$ on the line $y = mx + b$. The other sales representative's pay amount of $5300 on sales of $37,500 becomes the point $(37{,}500, 5300)$. Now that we have two points on a line, we can find the equation for the line containing the points.

$$m = \frac{5300 - 5000}{37{,}500 - 35{,}000} = \frac{300}{2500} = 0.12$$

Using $m = 0.12$ and $(35{,}000, 5000)$ in $y - y_1 = m(x - x_1)$, we have

$$y - 5000 = 0.12(x - 35{,}000)$$

$$y - 5000 = 0.12x - 4200$$

$$y = 0.12x + 800$$

The commission rate is 12% of sales and the monthly base salary is $800.

- **The manager of a grocery store notices that sales of bananas are proportionate to sales of milk. On one Friday, 400 pounds of bananas and 1700 gallons of milk are sold. On the following Friday, 360 pounds of bananas and 1540 gallons of milk are sold. Find an equation that gives the number of gallons of milk sold in terms of the number of pounds of bananas sold.**

We now find a linear equation in the form $y = mx + b$. For some problems, it does not matter which quantity x represents and which y represents. In this problem, it does matter because of the sentence, "Find an equation that gives the number of gallons of milk sold in terms of the number of pounds of bananas sold." The equation $y = mx + b$ gives y in terms of x. This means that y needs to represent the number of gallons of milk and x the number of pounds of bananas. The ordered pair is (bananas, milk). Our points, then, are (360, 1540) and (400, 1700).

$$m = \frac{1700 - 1540}{400 - 360} = \frac{160}{40} = 4$$

Using $m = 4$ and (400, 1700) in $y - y_1 = m(x - x_1)$, we have

$$y - 1700 = 4(x - 400)$$
$$y - 1700 = 4x - 1600$$
$$y = 4x + 100$$

What does the slope mean in these two problems? In the first equation, $y = 0.12x + 800$, the slope tells us how a sales representative's pay increases for each one-dollar increase in sales.

$$m = \frac{\$0.12}{\$1.00} = \frac{\text{increase in pay}}{\text{increase in sales}}$$

In the second equation, $y = 4x + 100$, the slope tells us that each pound of bananas sold results in four gallons of milk sold.

$$m = 4 = \frac{4}{1} = \frac{\text{4 gallons of milk}}{\text{1 pound of bananas}}$$

PRACTICE

1. Anthony, a marketing director, notices that the sales level for a certain product and amount spent on television advertising are linearly related. When \$6000 is spent on television advertising, sales for the product are \$255,000, and when \$8000 is spent on television advertising, sales for the product are \$305,000. Find an equation that gives the sales level for the product in terms of amount spent on television advertising.

2. **Show that the formula $C = \frac{5}{9}(F - 32)$ gives the degrees Celsius in terms of degrees Fahrenheit. Use the fact that water freezes at 0°C and 32°F and boils at 100°C and 212°F.**
3. **A car rental company charges a daily fee plus a mileage fee. A businesswoman's bill for one day was $42.55 after driving 55 miles. The bill for the next day was $36.40 after driving 40 miles. How much did it cost for each mile? What was the daily fee?**

SOLUTIONS

1. Because we want the sales level in terms of the amount spent on television advertising, we let y represent the sales level and x represent the amount spent on advertising. Our points are (6000, 255,000) and (8000, 305,000).

$$m = \frac{305{,}000 - 255{,}000}{8000 - 6000} = \frac{50{,}000}{2000} = 25$$

$$y - 255{,}000 = 25(x - 6000)$$

$$y - 255{,}000 = 25x - 150{,}000$$

$$y = 25x + 105{,}000$$

(Every dollar in advertising results in $25 in sales.)

2. We treat C like y and F like x. Our points have the form (degrees Fahrenheit, degrees Celsius): (32, 0) and (212, 100).

$$m = \frac{100 - 0}{212 - 32} = \frac{100}{180} = \frac{5}{9}$$

$$C - 0 = \frac{5}{9}(F - 32)$$

$$C = \frac{5}{9}(F - 32)$$

3. In the equation $y = mx + b$, we let x represent the number of miles driven and y represent the daily cost. The ordered pair (x, y) is (miles, cost). The points are (55, 42.55) and (40, 36.40).

$$m = \frac{36.40 - 42.55}{40 - 55} = \frac{-6.15}{-15} = 0.41$$

$$y - 36.40 = 0.41(x - 40)$$
$$y - 36.40 = 0.41x - 16.40$$
$$y = 0.41x + 20$$

The daily fee is \$20 and each mile costs \$0.41.

In these last problems, we are given one pair of numbers (or a pair is implied), which is a point on the line, and information on the rate of change. The rate of change is the slope. Instead of using the point-slope form of a line, we will use the *y*-intercept form, $y = mx + b$.

EXAMPLE 5-14

- **A utility company charges $12\frac{1}{2}$ cents per kilowatt-hour for electricity plus a monthly base charge. Find an equation that gives the monthly costs in terms of the number of kilowatt-hours of electricity used if the bill for one month for a certain family is \$161.25 for 1050 kilowatt-hours of electricity used.**

In the equation $y = mx + b$ (where y is the cost and x is the number of kilowatt-hours used), the slope is the cost per kilowatt-hour of electricity used. This means that $m = 0.125$ and a point is $(1050, 161.25)$. For $y = mx + b$, we have $y = 161.25$, $m = 0.125$, and $x = 1050$.

$$161.25 = 0.125(1050) + b$$

$$30 = b \qquad \text{The equation is } y = 0.125x + 30.$$

- **A recipe calls for 2 cups of biscuit mix and $\frac{2}{3}$ cups of milk. Find a linear equation that gives the amount of milk in terms of the amount of biscuit mix.**

Because we need to give the milk in terms of the biscuit mix, we let y represent the number of cups of milk and x represent the number of cups of biscuit mix. The ordered pair (x, y) is (biscuit mix, milk). Also, the slope is $\frac{\text{change in } y}{\text{change in } x}$ which is $\frac{\text{change in milk}}{\text{change in mix}}$. We use the fact that if we increase the number of cups of biscuit mix by 2 cups, we need to increase the number of cups of milk by $\frac{2}{3}$, giving us a slope of

$$\frac{\frac{2}{3}}{2} = \frac{2}{3} \div 2 = \frac{2}{3} \cdot \frac{1}{2} = \frac{1}{3}$$

So far we have $y = \frac{1}{3}x + b$. We need more information to find b. Although another point is not explicitly given, we can figure one out—when no biscuit mix is used, no milk is used. In other words, $(0, 0)$ is a point on the line. This means that b, the y-intercept, is 0. The equation is $y = \frac{1}{3}x$.

- **The dosage for a certain cattle drug is 4.5 cc per 100 pounds of body weight. Find an equation that gives the amount of the drug in terms of a cow's weight.**

We want the amount of the drug in terms of a cow's weight, so we let y represent the number of cc's of the drug and x represent a cow's weight in pounds. What is the slope of our line?

$$m = \frac{\text{change in drug amount}}{\text{change in weight}} = \frac{4.5}{100} = 0.045$$

This means that 0.045 cc's of the drug are needed for each pound of a cow's weight. Again, we assume $(0, 0)$ is a point on the line, so $b = 0$. The equation is $y = 0.045x$.

PRACTICE

1. The manufacturing cost of each unit of a product is \$1.75. The total cost to produce 20,000 units one week was \$41,000. Find an equation that gives the total cost in terms of the number of units produced.
2. The manager of a movie theater believes that for every 200 tickets sold, 15 buckets of popcorn are sold. Find an equation that gives the amount of buckets of popcorn sold in terms of the number of tickets sold.
3. An office manager notices that the copier in his office uses one container of toner for every 25 reams of paper. Find an equation that gives the amount of toner used in terms of the amount of paper used.
4. A garden hose is used to fill a tall rectangular tank. The water level rises six inches every 20 minutes. If the water level was already eight inches before the water was turned on, find an equation that gives the water level in terms of the time the hose is used (before the tank overflows, that is).

SOLUTIONS

1. We let y represent the total cost and x the number of units produced. This means that $(20{,}000, 41{,}000)$ is a point on the line. Each unit costs $1.75 to produce, so the slope is 1.75. We let $x = 20{,}000$, $y = 41{,}000$, and $m = 1.75$ in $y = mx + b$.

$$41{,}000 = 1.75(20{,}000) + b$$
$$6000 = b$$

The equation is $y = 1.75x + 6000$. ($6000 represents *fixed costs*, costs such as rent, loan payments, salaries, etc.)

2. We let y represent the number of buckets of popcorn sold and x represent the number of tickets sold. The slope is

$$m = \frac{\text{change in popcorn sales}}{\text{change in tickets sold}} = \frac{15}{200} = 0.075$$

We assume 0 buckets of popcorn are sold when 0 tickets are sold, so $(0, 0)$ is on the line and b, the y-intercept, is 0. The equation is $y = 0.075x$.

3. We let y represent the number of toner containers used and x the number of reams of paper used. The slope is

$$m = \frac{\text{change in toner used}}{\text{change in paper used}} = \frac{1}{25}$$

The point $(0, 0)$ is on the graph, so b, the y-intercept, is 0. The equation is $y = \frac{1}{25}x$.

4. We let y represent the water level in inches and x the time in minutes that the hose is used. When the time is 0 minutes, the water level is 8 inches, giving us the point $(0, 8)$. This means that b, the y-intercept, is 8. The slope is

$$m = \frac{\text{change in water level}}{\text{change in time}} = \frac{6 \text{ inches}}{20 \text{ minutes}} = 0.3$$

The equation is $y = 0.3x + 8$.

Parabolas

The graph of any quadratic equation ($y = ax^2 + bx + c$) looks like one of the graphs in Fig. 5-38.

These graphs are called *parabolas*. Parabolas occur frequently in nature. To see a parabola, toss a small object up and watch its path—it follows a parabola.

The graph of every parabola has a *vertex*, the point where the graph turns around. For a parabola that opens up, the vertex is the lowest point. The vertex is the highest point for a graph that opens down. When a quadratic equation is in the form $y = a(x - h)^2 + k$, the vertex is the point (h, k).

EXAMPLE 5-15

Identify a and the vertex.

- $y = 2(x - 4)^2 + 5$

Here, $a = 2$, $h = 4$, $k = 5$, so the vertex is $(4, 5)$.

- $y = (x + 2)^2 - 1$

Here, $a = 1$, $h = -2$, $k = -1$, so the vertex is $(-2, -1)$.

- $y = -3(x - 1)^2 + 2$

Here, $a = -3$, $h = 1$, $k = 2$, so the vertex is $(1, 2)$.

- $y = 2x^2 - 4 = 2(x - 0)^2 - 4$

Here, $a = 2$, $h = 0$, $k = -4$, so the vertex is $(0, -4)$.

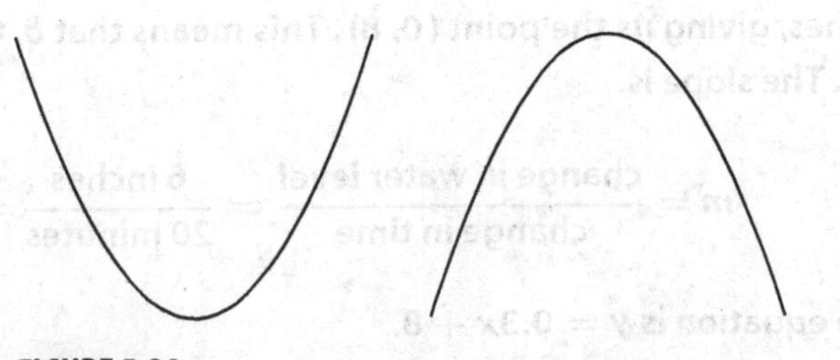

FIGURE 5-38

- $y = (x - 8)^2 = (x - 8)^2 + 0$

Here, $a = 1$, $h = 8$, $k = 0$, so the vertex is $(8, 0)$.

PRACTICE

Identify *a* and the vertex.

1. $y = -(x-2)^2 + 4$
2. $y = 10(x+1)^2 - 2$
3. $y = \frac{1}{2}(x+5)^2 + 4$
4. $y = (x+6)^2$
5. $y = -x^2$

SOLUTIONS

1. The vertex is $(2, 4)$, and $a = -1$.
2. The vertex is $(-1, -2)$, and $a = 10$.
3. The vertex is $(-5, 4)$, and $a = \frac{1}{2}$.
4. The vertex is $(-6, 0)$, and $a = 1$.
5. The vertex is $(0, 0)$, and $a = -1$.

Sketching the Graph of a Parabola

When graphing parabolas, we begin with the vertex. We then plot two points to the left and to the right of the vertex. One pair of points should be fairly close to the vertex to show the curving around the vertex. Another pair should be further away to show how steep the ends are. What do "fairly close" and "a little further away" mean? There is no standard answer. For some parabolas, one unit is "close," but for others, one unit is "far away." It all depends on a. A good rule of thumb is to plot two points a units to the left and to the right of the vertex and two other points that are $2a$ units to the left and right of the vertex. The sign on a is also important. When a is positive, the parabola opens up. When a is negative, the parabola opens down (see Fig. 5-39).

When sketching the graph of a parabola, we begin the T-table with five x-values (see Table 5-1).

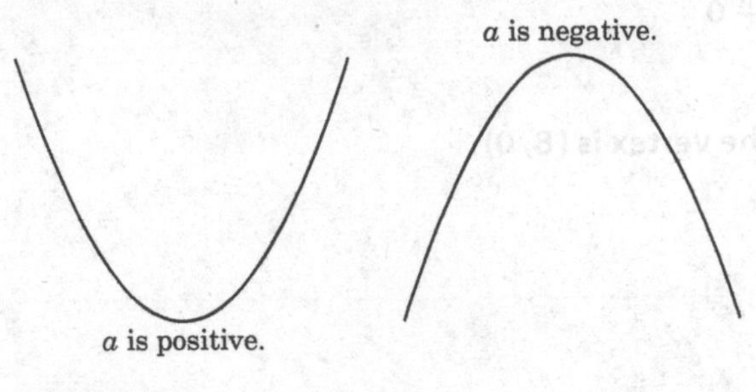

FIGURE 5-39

TABLE 5-1	
x	y
$h-2a$	
$h-a$	
h	k vertex
$h+a$	
$h+2a$	

Because parabolas are symmetric (the left half is a mirror image of the right half), the y-values for $h-a$ and $h+a$ are the same; and the y-values for $h-2a$ and $y+2a$ are the same. This fact can save us a little work.

EXAMPLE 5-16

Sketch the graph of the parabola.

- $y = 2(x-4)^2 - 8$

$$a = 2,\ h = 4,\ k = -8$$

See Table 5-2 and Fig. 5-40.

TABLE 5-2	
x-values	**y-values**
$h-2a = 4-2(2) = 0$	$y = 2(0-4)^2 - 8 = 24$
$h-a = 4-2 = 2$	$y = 2(2-4)^2 - 8 = 0$
$h+a = 4+2 = 6$	$y = 2(6-4)^2 - 8 = 0$
$h+2a = 4+2(2) = 8$	$y = 2(8-4)^2 - 8 = 24$

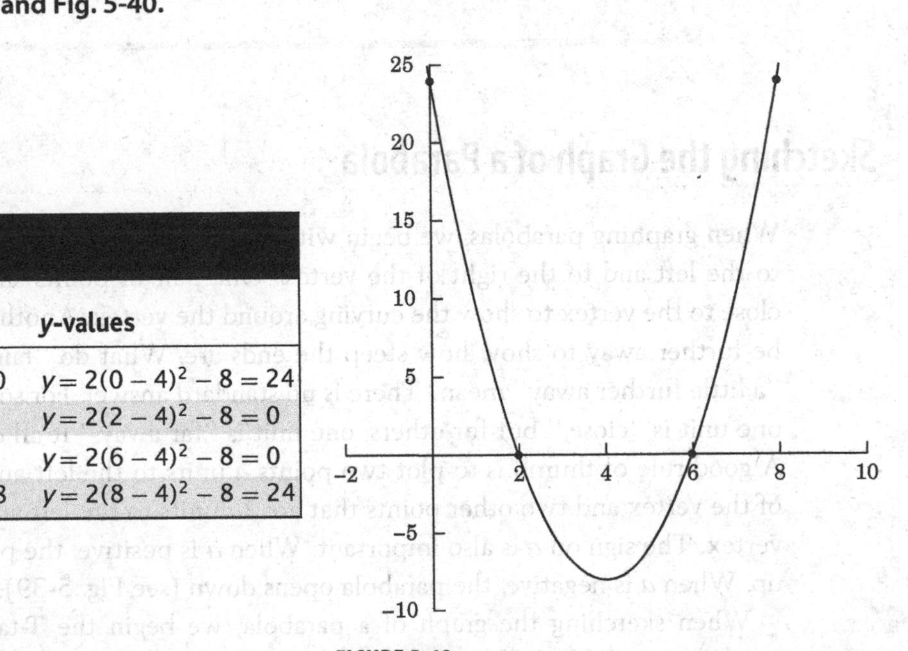

FIGURE 5-40

TABLE 5-3

x-values	*y*-values
$h-2a=1-2(-2)=5$	$y=-2(5-1)^2+5=-27$
$h-a=1-(-2)=3$	$y=-2(3-1)^2+5=-3$
$h+a=1+(-2)=-1$	$y=-2(-1-1)^2+5=-3$
$h+2a=1+2(-2)=-3$	$y=-2(-3-1)^2+5=-27$

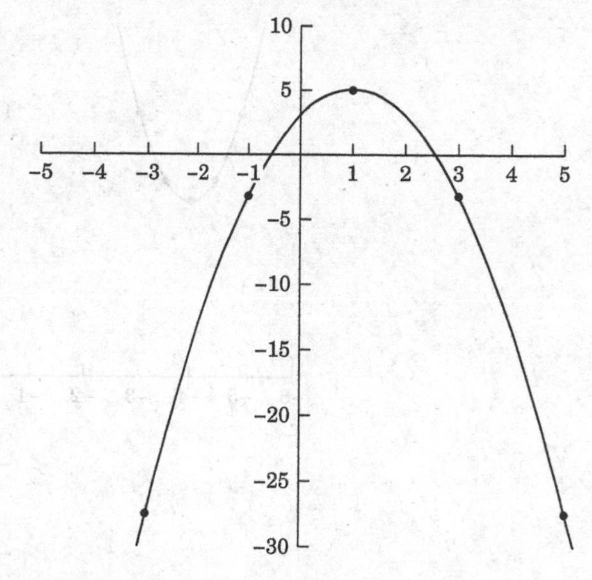

FIGURE 5-41

- $\boldsymbol{y=-2(x-1)^2+5}$

$$\boldsymbol{a=-2,\ h=1,\ k=5}$$

See Table 5-3 and Fig. 5-41.

- $\boldsymbol{y=\frac{1}{2}(x+4)^2+1}$

$$\boldsymbol{a=\frac{1}{2},\ h=-4,\ k=1}$$

See Table 5-4 and Fig. 5-42.

TABLE 5-4

x-values	*y*-values
$h-2a=-4-2\left(\frac{1}{2}\right)=-5$	$y=\frac{1}{2}(-5+4)^2+1=1\frac{1}{2}$
$h-a=-4-\frac{1}{2}=-4\frac{1}{2}$	$y=\frac{1}{2}\left(-4\frac{1}{2}+4\right)^2+1=1\frac{1}{8}$
$h+a=-4+\frac{1}{2}=-3\frac{1}{2}$	$y=\frac{1}{2}\left(-3\frac{1}{2}+4\right)^2+1=1\frac{1}{8}$
$h+2a=-4+2\left(\frac{1}{2}\right)=-3$	$y=\frac{1}{2}(-3+4)^2+1=1\frac{1}{2}$

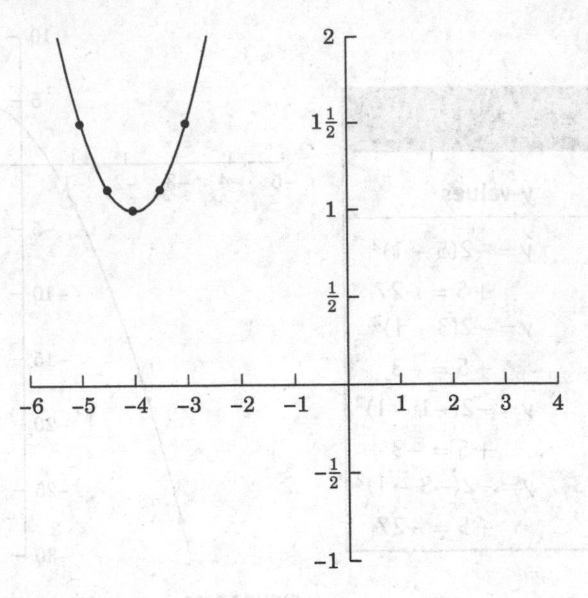

FIGURE 5-42

It might seem that $y = a(x - h)^2 + k$ should either be $y = a(x + h)^2 + k$ or $y = a(x - h)^2 - k$. The reason that signs in front of h and k are different is that k (the y-coordinate of the vertex) is on the same side as x in the formula. If k were on the same side as y, then the signs on h and k would be the same: $y - k = a(x - h)^2$.

PRACTICE

Sketch the graph of the parabola.

1. $y = -(x - 2)^2 + 4$
2. $y = (x - 3)^2 - 1$
3. $y = \frac{1}{2}(x + 5)^2 + 4$
4. $y = (x + 6)^2$
5. $y = -x^2$ (Square x and then take the negative, e.g., $-3^2 = -9$.)

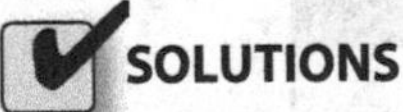

SOLUTIONS

1. $a = -1$, $h = 2$, $k = 4$; see Table 5-5 and Fig. 5-43.
2. $a = 1$, $h = 3$, $k = -1$; see Table 5-6 and Fig. 5-44.
3. $a = \frac{1}{2}$, $h = -5$, $k = 4$; see Table 5-7 and Fig. 5-45.
4. $a = 1$, $h = -6$, $k = 0$; see Table 5-8 and Fig. 5-46.
5. $a = -1$, $h = 0$, $k = 0$; see Table 5-9 and Fig. 5-47.

TABLE 5-5	
x	*y*
4	0
3	3
2	4
1	3
0	0

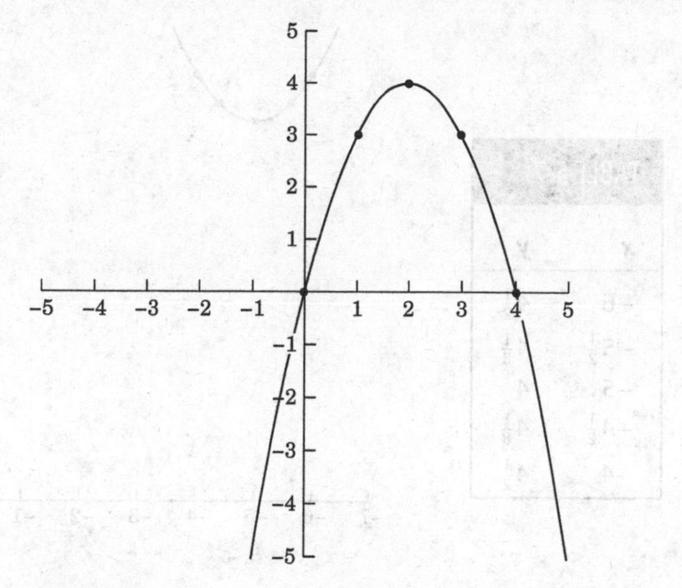

FIGURE 5-43

TABLE 5-6	
x	*y*
1	3
2	0
3	−1
4	0
5	3

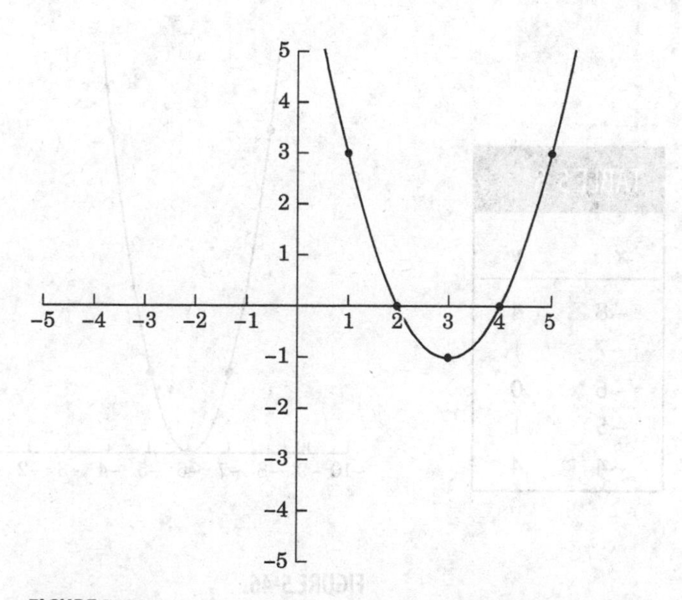

FIGURE 5-44

TABLE 5-7

x	y
-6	$4\frac{1}{2}$
$-5\frac{1}{2}$	$4\frac{1}{8}$
-5	4
$-4\frac{1}{2}$	$4\frac{1}{8}$
-4	$4\frac{1}{2}$

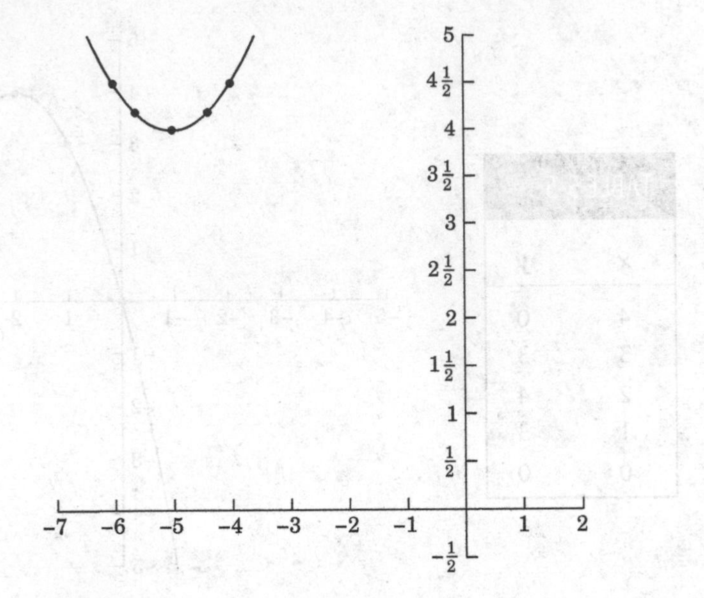

FIGURE 5-45

TABLE 5-8

x	y
-8	4
-7	1
-6	0
-5	1
-4	4

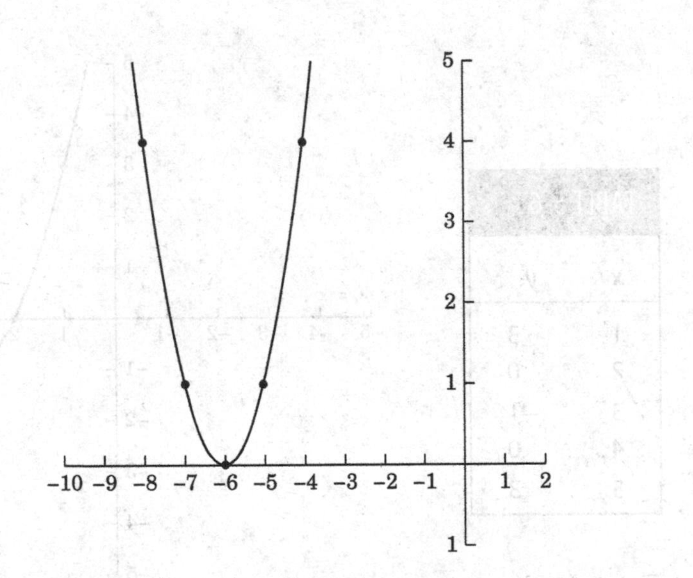

FIGURE 5-46

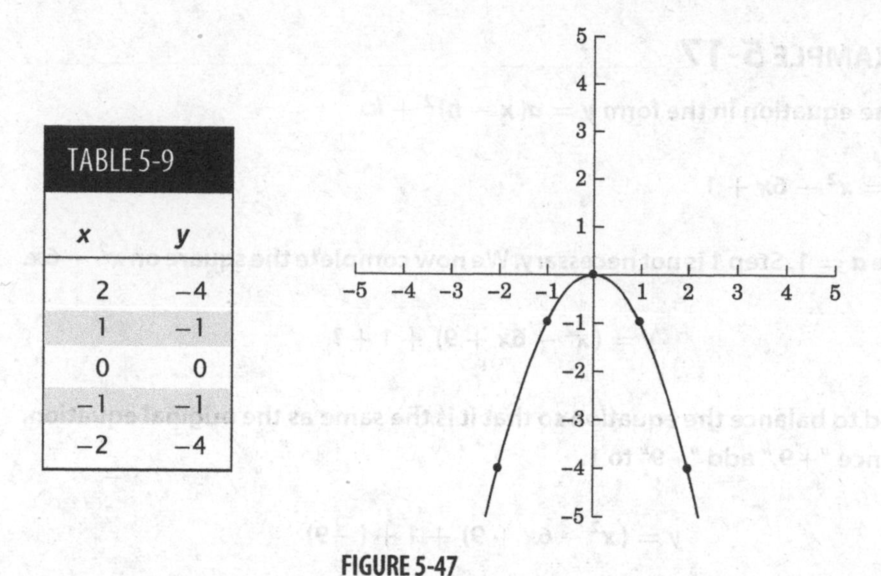

TABLE 5-9

x	y
2	−4
1	−1
0	0
−1	−1
−2	−4

FIGURE 5-47

Locating the Vertex by Completing the Square

We normally write quadratic equations in the form $y = ax^2 + bx + c$ (called the *general* form), not $y = a(x - h)^2 + k$ (called the *standard* form). As with circle equations, to write an equation in the first form into one that is in the second form we complete the square. Completing the square on these equations can be tricky. The strategy is outlined in Table 5-10.

TABLE 5-10 Completing the Square to Write an Equation in the Form $y = a(x - h)^2 + k$

Method	Example
$y = ax^2 + bx + c$	$y = 2x^2 - 8x + 1$
1. Factor a from the x^2 and x terms.	$y = 2(x^2 - 4x) + 1$
2. Complete the square on the x^2 and x terms.	$y = 2(x^2 - 4x + 4) + 1 + ?$
3. Compute the constant that must be added to c to "balance" the equation.	$y = 2(x^2 - 4x + 4) + 1 - 2(4)$ $y = 2(x^2 - 4x + 4) - 7$
4. Write the expression in the parentheses as a perfect square.	$y = 2(x - 2)^2 - 7$

EXAMPLE 5-17

Write the equation in the form $y = a(x - h)^2 + k$.

- $y = x^2 - 6x + 1$

Because $a = 1$, Step 1 is not necessary. We now complete the square on $x^2 - 6x$.

$$y = (x^2 - 6x + 9) + 1 + ?$$

We need to balance the equation so that it is the same as the original equation. To balance "+9," add "−9" to 1.

$$y = (x^2 - 6x + 9) + 1 + (-9)$$
$$y = (x^2 - 6x + 9) - 8$$
$$y = (x - 3)^2 - 8$$

- $y = 2x^2 + 16x - 1$

We factor 2 from $2x^2 + 16x$. We must do this step *before* completing the square.

$$y = 2(x^2 + 8x) - 1$$
$$y = 2(x^2 + 8x + 16) - 1 + ?$$

Adding "−16" to −1 might seem to be the next step. This would not result in an equivalent equation. We simplify $y = 2(x^2 + 8x + 16) - 1$ to see what effect adding 16 in the parentheses has on the equation.

$$y = 2(x^2 + 8x + 16) - 1 = 2x^2 + 16x + 32 - 1$$

By writing "+16" in the parentheses, we are really adding $2(16) = 32$. To balance the equation, we need to add "−32" to −1.

$$y = 2(x^2 + 8x + 16) - 1 - 32$$
$$y = 2(x^2 + 8x + 16) - 33$$
$$y = 2(x + 4)^2 - 33$$

- $y = -3x^2 + 12x + 2$

First we factor -3 from $-3x^2 + 12x$.

$$\frac{-3x^2}{-3} = x^2 \qquad \frac{12x}{-3} = -4x \qquad y = -3(x^2 - 4x) + 2$$

We are ready to complete the square.

$$y = -3(x^2 - 4x + 4) + 2 + ?$$

By writing "+4" in the parentheses, we are really adding $-3(4) = -12$. We balance this by adding 12 to 2.

$$y = -3(x^2 - 4x + 4) + 2 + 12$$
$$y = -3(x^2 - 4x + 4) + 14$$
$$y = -3(x - 2)^2 + 14$$

Still Struggling

Be sure to divide the x^2 and x terms by the coefficient of x^2 *before* completing the square.

PRACTICE

Write the quadratic equations in the form $y = a(x - h)^2 + k$.

1. $y = x^2 - 10x + 6$
2. $y = -x^2 - 4x + 3$
3. $y = 5x^2 + 10x + 6$
4. $y = 3x^2 + 12x - 4$
5. $y = -2x^2 + 4x + 3$

SOLUTIONS

1.
$$y = x^2 - 10x + 6$$
$$y = (x^2 - 10x + 25) + 6 + (-25)$$
$$y = (x^2 - 10x + 25) - 19$$
$$y = (x - 5)^2 - 19$$

2.
$$y = -x^2 - 4x + 3$$
$$y = -(x^2 + 4x) + 3$$
$$y = -(x^2 + 4x + 4) + 3 + ?$$

By writing "+4" in the parentheses, we are adding $-(4) = -4$. We balance this by adding 4 to 3.

$$y = -(x^2 + 4x + 4) + 3 + 4$$
$$y = -(x^2 + 4x + 4) + 7$$
$$y = -(x + 2)^2 + 7$$

3.
$$y = 5x^2 + 10x + 6$$
$$y = 5(x^2 + 2x) + 6$$
$$y = 5(x^2 + 2x + 1) + 6 + ?$$

By writing "+1" in the parentheses, we are adding $5(1) = 5$. We balance this by adding -5 to 6.

$$y = 5(x^2 + 2x + 1) + 6 + (-5)$$
$$y = 5(x^2 + 2x + 1) + 1$$
$$y = 5(x + 1)^2 + 1$$

4.
$$y = 3x^2 + 12x - 4$$
$$y = 3(x^2 + 4x) - 4$$
$$y = 3(x^2 + 4x + 4) - 4 + ?$$

By writing "+4" inside the parentheses, we added 3(4) = 12. We balance this by adding −12 to −4.

$$y = 3(x^2 + 4x + 4) - 4 + (-12)$$

$$y = 3(x^2 + 4x + 4) - 16$$

$$y = 3(x + 2)^2 - 16$$

5.

$$y = -2x^2 + 4x + 3$$

$$y = -2(x^2 - 2x) + 3$$

$$y = -2(x^2 - 2x + 1) + 3 + ?$$

By writing "+1" inside the parentheses, we are adding −2(1) = −2. We balance this by adding 2 to 3.

$$y = -2(x^2 - 2x + 1) + 3 + 2$$

$$y = -2(x^2 - 2x + 1) + 5$$

$$y = -2(x - 1)^2 + 5$$

We can use a shortcut to find the vertex of a parabola without having to write the equation in the form $y = a(x - h)^2 + k$. The shortcut involves a formula for h: $h = \frac{-b}{2a}$. We find k by letting $x = h$ in the equation. The shortcut for h comes from completing the square on $y = ax^2 + bx + c$.

EXAMPLE 5-18

Find the vertex using the formula for h: $h = \frac{-b}{2a}$.

- $y = x^2 + 6x + 4$

$a = 1$ $\quad$ $b = 6$ $\quad$ $h = \dfrac{-b}{2a} = \dfrac{-6}{2(1)} = -3$

$k = (-3)^2 + 6(-3) + 4 = -5$ $\quad$ The vertex is $(-3, -5)$.

- $y = 2x^2 - 12x - 7$

$$a = 2 \qquad b = -12 \qquad h = \frac{-b}{2a} = \frac{-(-12)}{2(2)} = 3$$

$$k = 2(3)^2 - 12(3) - 7 = -25 \qquad \text{The vertex is } (3, -25).$$

- $y = -x^2 + 2x - 4$

$$a = -1, \qquad b = 2 \qquad h = \frac{-b}{2a} = \frac{-2}{2(-1)} = 1$$

$$k = -(1)^2 + 2(1) - 4 = -3 \qquad \text{The vertex is } (1, -3).$$

PRACTICE

Find the vertex using the formula for h: $h = \frac{-b}{2a}$.

1. $y = x^2 + 6x + 5$
2. $y = \frac{1}{2}x^2 - 3x + 4$
3. $y = 4x^2 - 6x + 8$
4. $y = -x^2 - 5x + 3$

SOLUTIONS

1. $a = 1$, $b = 6$

$$h = \frac{-b}{2a} = \frac{-6}{2(1)} = -3$$

$$k = (-3)^2 + 6(-3) + 5 = -4 \qquad \text{The vertex is } (-3, -4).$$

2. $a = \frac{1}{2}$, $b = -3$

$$h = \frac{-b}{2a} = \frac{-(-3)}{2 \cdot \frac{1}{2}} = \frac{3}{1} = 3$$

$$k = \frac{1}{2}(3)^2 - 3(3) + 4 = -\frac{1}{2} \qquad \text{The vertex is } \left(3, -\frac{1}{2}\right).$$

3. $a = 4,\ b = -6$

$$h = \frac{-b}{2a} = \frac{-(-6)}{2(4)} = \frac{3}{4}$$

$$k = 4\left(\frac{3}{4}\right)^2 - 6\left(\frac{3}{4}\right) + 8$$

$$= 4\left(\frac{9}{16}\right) - \frac{9}{2} + 8 = \frac{23}{4} \quad \text{The vertex is } \left(\frac{3}{4}, \frac{23}{4}\right).$$

4. $a = -1,\ b = -5$

$$h = \frac{-b}{2a} = \frac{-(-5)}{2(-1)} = -\frac{5}{2}$$

$$k = -\left(-\frac{5}{2}\right)^2 - 5\left(-\frac{5}{2}\right) + 3$$

$$= -\left(\frac{25}{4}\right) + \frac{25}{2} + 3 = \frac{37}{4} \quad \text{The vertex is } \left(-\frac{5}{2}, \frac{37}{4}\right).$$

Summary

In this chapter, we learned how to

- *Algebraically locate the intercepts for a graph.* The intercepts for a graph (if it has any) are points where the graph crosses (or simply touches) an axis. We find the x-intercept (if it has any) by setting $y = 0$ and solving for x. Similarly, we find the y-intercept (if it has any) by setting $x = 0$ and solving for y.
- *Calculate the slope of a line.* We can calculate the slope of a line by substituting the coordinates of two points on the line, (x_1, y_1) and (x_2, y_2), in the slope formula: $m = \frac{y_2 - y_1}{x_2 - x_1}$. The slope of a horizontal line is 0, and the slope of a vertical line is not defined.
- *Sketch the graph of a line.* We can sketch the graph of a line either by plotting two points or by plotting one point and using the slope, to locate another point. If we choose to sketch the line by plotting two points, we can randomly choose two x-values (or two y-values), and plug these values

into the equation to find the other coordinate. Once we have two points plotted, we draw a line through them.

- *Find the slope-intercept form of a line.* The slope-intercept form of a line is $y = mx + b$. In this form, the y-intercept is b, and the slope is m. We can sketch the line based on this information alone: we plot the y-intercept and use the slope to locate a second point.
- *Use two points on the line to find an equation for the line.* We learned two strategies for finding an equation for the line. Once we know the slope (which we might have to calculate), we put the slope and the coordinates of one point, (x_1, y_1), into the point-slope form of a line: $y - y_1 = m(x - x_1)$. We either solve this equation for y or algebraically rewrite it in the form $Ax + By = C$, where A, B, and C are integers and A is not negative. The equation for a horizontal line is $y =$ number, and the equation for a vertical line is $x =$ number. The second strategy we used was to substitute the coordinates of the point and m in the formula $y = mx + b$ to find b.
- *Determine whether or not a pair of lines is parallel or perpendicular.* Two lines are parallel if their slopes are equal (or if they are both vertical lines). Two lines are perpendicular if their slopes are negative reciprocals of each other (or if one is vertical and the other, horizontal).
- *Solve linear applications.* If a pair of variables (such as pay and number of hours worked) is linearly related, we can find an equation that represents the relationship. Once we have an equation, we can answer any number of questions about the variables.
- *Sketch the graph of a parabola.* The graph of an equation of the form $y = a(x - h)^2 + k$ is a parabola with vertex (h, k). If a is positive, the graph opens up, and if a is negative, it opens down. We sketch the graph by plotting the vertex and two points to its right and two points to its left. In this chapter, we plotted points for $h - 2a$, $h - a$, $h + a$, and $h + 2a$.
- *Use completing the square to find the vertex for a parabola.* If an equation is in the form $y = ax^2 + bx + c$, we can locate the vertex in one of two ways: we complete the square to rewrite the equation in the form $y = a(x - h)^2 + k$ or we use a formula for finding h. The formula for finding h is $h = \frac{-b}{2a}$. Once we have h, we substitute h for x in the equation to find k.

QUIZ

1. **Find the slope of the line containing the points $(-4, \frac{1}{3})$ and $(-1, -\frac{2}{3})$.**

   A. $-\frac{4}{25}$ B. $\frac{4}{19}$ C. $-\frac{1}{3}$ D. $-\frac{4}{9}$

2. **What is the vertex for the graph of $y = 3(x-4)^2 + 10$?**

   A. (4, 10) B. (12, 10) C. (−3, 10) D. (−12, 10)

3. **What is the slope and *y*-intercept for the graph of $y = \frac{3}{5}x + 2$?**

   A. The slope is $-\frac{3}{5}$, and the *y*-intercept is 2.
   B. The slope is $\frac{3}{5}$, and the *y*-intercept is 2.
   C. The slope is $-\frac{3}{5}$, and the *y*-intercept is −2.
   D. The slope is $\frac{3}{5}$, and the *y*-intercept is $-\frac{10}{3}$.

4. **Rewrite the equation $y = 3x^2 + 6x - 1$ in the form $y = a(x-h)^2 + k$.**

   A. $y = 3(x+2)^2 - 9$ B. $y = 3(x+1)^2 - 2$
   C. $y = 3(x+2)^2 + 8$ D. $y = 3(x+1)^2 - 4$

5. **Are the lines $5x - 4y = 10$ and $4x - 5y = 3$ parallel, perpendicular, or neither?**

   A. Parallel B. Perpendicular C. Neither

6. **What are the intercepts for the graph of $y = x^2 - 9x + 20$?**

   A. The *x*-intercepts are 4, 5, and the *y*-intercept is 20.
   B. The *x*-intercepts are −4, −5, and the *y*-intercept is 20.
   C. The *x*-intercepts are −4, 5, and the *y*-intercept is 20.
   D. There are no *x*-intercepts, and the *y*-intercept is 20.

7. **For $y = 2(x^2 + 10x + \_\_\_) + 6 + \_\_\_$, what number(s) should be put in the blanks to write the equation in the form $y = a(x-h)^2 + k$?**

   A. Write 25 in the first blank and −25 in the second.
   B. Write 25 in the first blank and 10 in the second.
   C. Write 25 in the first blank and −10 in the second.
   D. Write 25 in the first blank and −50 in the second.

8. The slope of a certain line is $-\frac{5}{8}$. What is the slope of a line that is perpendicular to it?

A. $\frac{5}{8}$ B. $\frac{8}{5}$ C. $-\frac{5}{8}$ D. $-\frac{8}{5}$

9. Find an equation for the line containing the points (5, 6) and (−3, 2).

A. $2x - y = 4$ B. $x + 2y = 17$ C. $2x + y = -4$ D. $x - 2y = -7$

10. What is the vertex for the graph of $y = 5x^2 - 10x - 4$?

A. (1, −9) B. (−1, 11) C. (2, −4) D. (−2, 36)

11. The equation $y = 0.15x + 25$ is the formula for computing the monthly bill for electricity for the Gardner family, where x is the number of kilowatt-hours used, and y the amount of the bill. How many kilowatt-hours were used if the monthly bill is $126.25?

A. 665 B. 675 C. 685 D. 695

12. The graph in Fig. 5-48 is the graph of which equation?

A. $4x - 3y = -6$ B. $3x + 4y = 8$ C. $4x + 3y = 6$ D. $3x - 4y = -8$

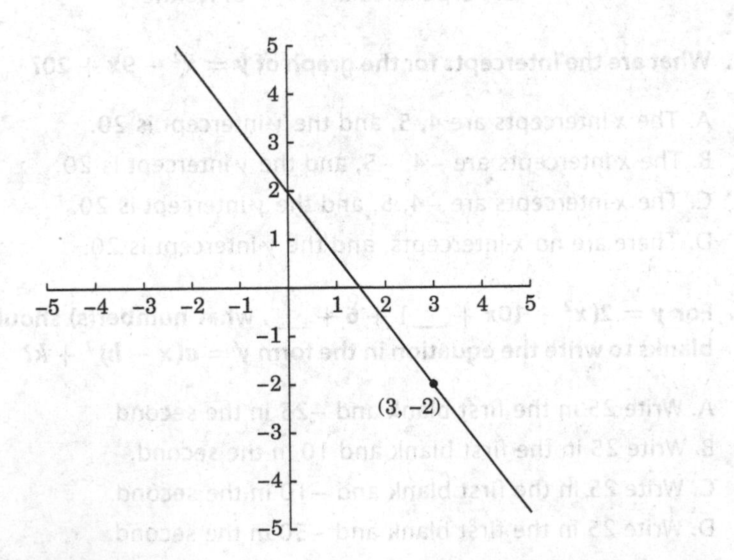

FIGURE 5-48

13. Which of the following lines is perpendicular to the line $y = -\frac{1}{3}$?

A. $x = -\frac{1}{3}$ B. $y = -3$ C. $y = 3$

D. None of these lines is perpendicular to the line $y = -\frac{1}{3}$.

14. A biscuit mix calls for $\frac{1}{2}$ cup of milk for $1\frac{1}{2}$ cups of mix. Find an equation that gives the amount of milk in terms of the amount of mix.

A. $y = 3x$ B. $y = \frac{2}{3}x$ C. $y = \frac{1}{3}x$ D. $y = \frac{3}{2}x$

15. The graph in Fig. 5-49 is the graph of which equation?

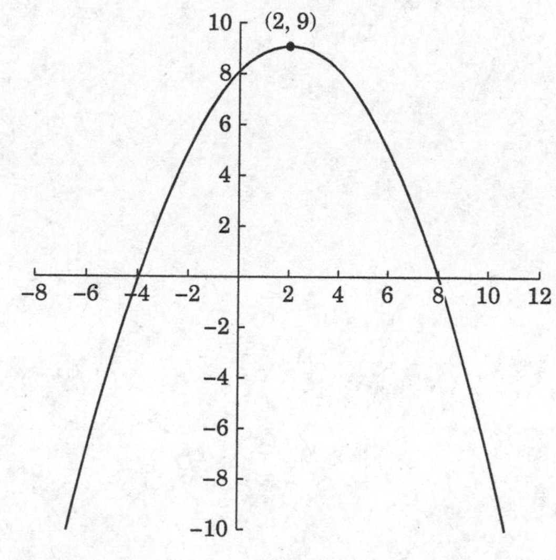

FIGURE 5-49

A. $y = -\frac{1}{2}(x-2)^2 + 9$ B. $y = -\frac{1}{2}(x-1)^2 + 9$

C. $y = -\frac{1}{4}(x-1)^2 + 9$ D. $y = -\frac{1}{4}(x-2)^2 + 9$

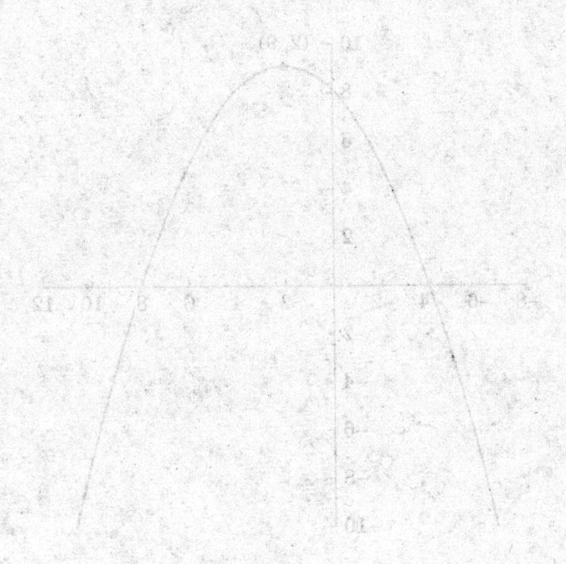

chapter 6

Nonlinear Inequalities

We will solve nonlinear inqualities in this chapter and use the skills that we learn here in Chap. 7. You might recall from Chap. 2 that we solve linear inequalities using the same basic strategy we use for solving linear equations, except that if we multiplied or divided each side of the inequality by a negative number, we reversed the inequality. A nonlinear inequality cannot be solved with the same method that we used to solve nonlinear equations (e.g., a quadratic equation). Consider the inequality $\frac{1}{x+1} < 2$. We cannot "clear the fraction" by multiplying each side of the inequality by $x + 1$. Why not? For some values of x, $x + 1$ is positive (and we would not have to reverse the inequality), but for other values of x, $x + 1$ is negative, which would require that we reverse the inequality. We must use a different approach. So that this approach makes sense, we begin by examining the graph of the equation related to the inequality.

CHAPTER OBJECTIVES

In this chapter, you will

- Determine algebraically where a graph is above or below the x-axis
- Solve nonlinear inequalities both graphically and algebraically

Solving Nonlinear Inequalities Graphically

On occasion, we need to know where a graph is above or below the x-axis. For what x-values is this graph in Fig. 6-1 above the x-axis? Below the x-axis?

The graph is above the x-axis to the left of $x = -3$ and between $x = 0$ and $x = 5$ (see the solid part of the graph in Fig. 6-2). The graph is below the x-axis between $x = -3$ and $x = 0$ and to the right of $x = 5$ (see the dashed part of the graph in Fig. 6-2).

When answering questions about graphs, we usually write the answer using interval notation. For example, to represent "to the left of $x = -3$," we write $(\infty, -3)$. To represent "between $x = 0$ and $x = 5$," we write $(0, 5)$.

EXAMPLE 6-1

Determine where the following graphs are above the x-axis and where they are below the x-axis.

- **The graph in Fig. 6-3 is above the x-axis on $(-\infty, -\frac{1}{2})$ (to the left of $x = -\frac{1}{2}$) and on $(3, \infty)$ (to the right of $x = 3$). The graph is below the x-axis on $(-\frac{1}{2}, 3)$ (between $x = -\frac{1}{2}$ and $x = 3$).**
- **The graph in Fig. 6-4 is above the x-axis on $(2, \infty)$ (to the right of $x = 2$) and below the x-axis on $(-\infty, 2)$ (to the left of $x = 2$).**

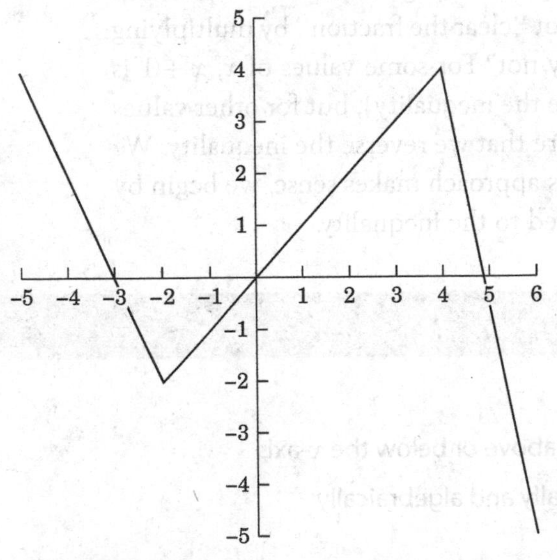

FIGURE 6-1

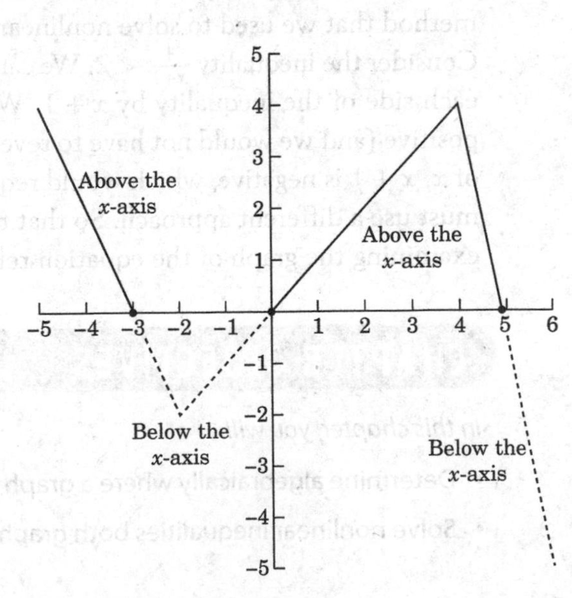

FIGURE 6-2

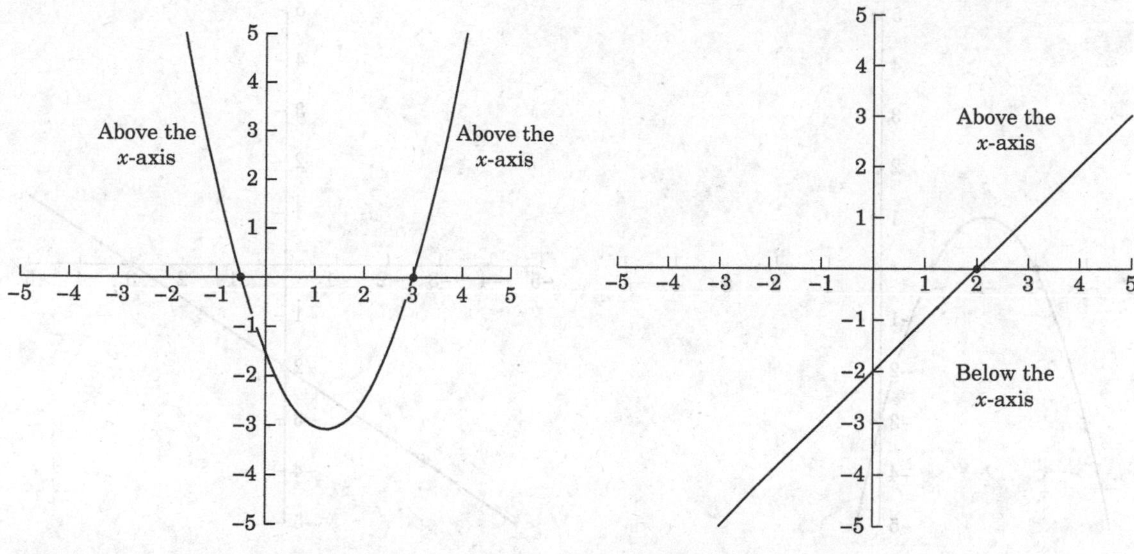

FIGURE 6-3

FIGURE 6-4

- **The graph in Fig. 6-5 is above the x-axis on $(-3, 1)$ (between $x = -3$ and $x = 1$) and on $(4, \infty)$ (to the right of $x = 4$). The graph is below the x-axis on $(-\infty, -3)$ (to the left of $x = -3$) and on $(1, 4)$ (between $x = 1$ and $x = 4$).**
- **The graph in Fig. 6-6 is never above the x-axis. The graph is below the x-axis on $(-\infty, \infty)$ (this is interval notation for "all real numbers").**

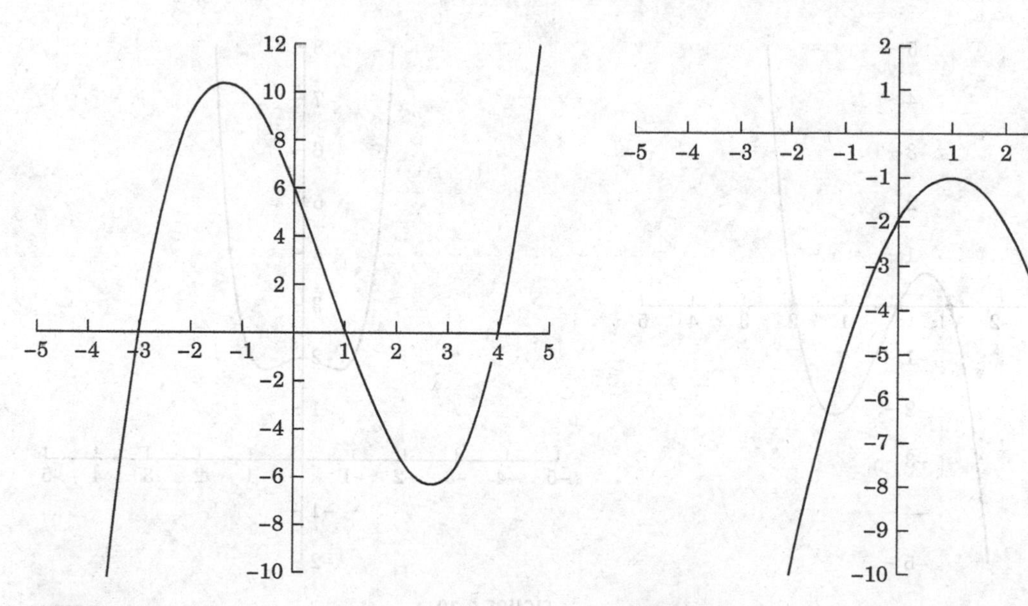

FIGURE 6-5

FIGURE 6-6

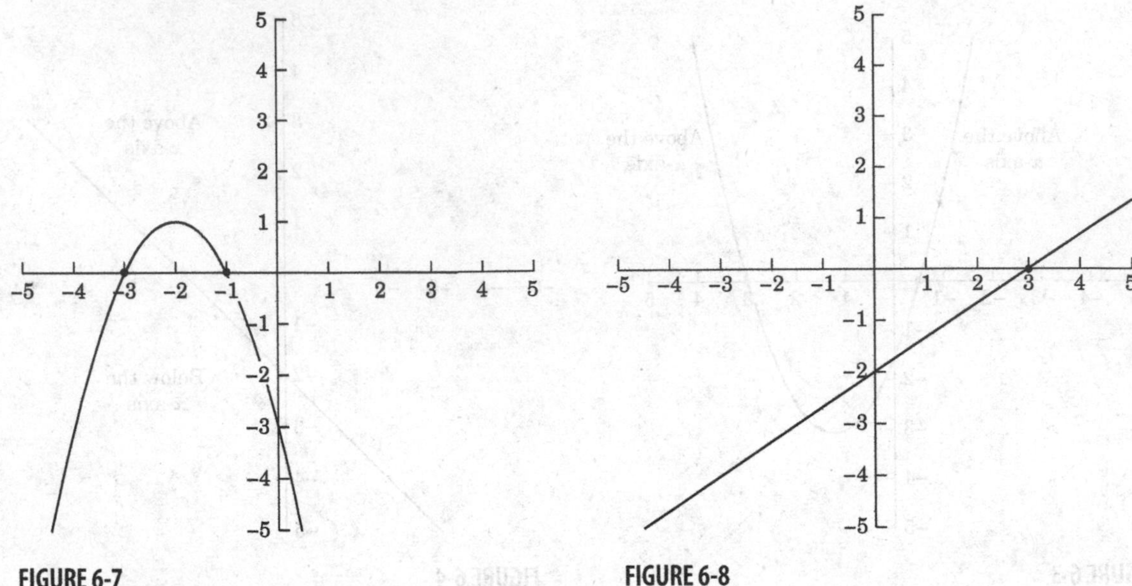

FIGURE 6-7

FIGURE 6-8

PRACTICE

Determine where the following graphs are above the *x*-axis and where they are below the *x*-axis.

1. See Fig. 6-7.
2. See Fig. 6-8.
3. See Fig. 6-9.
4. See Fig. 6-10.

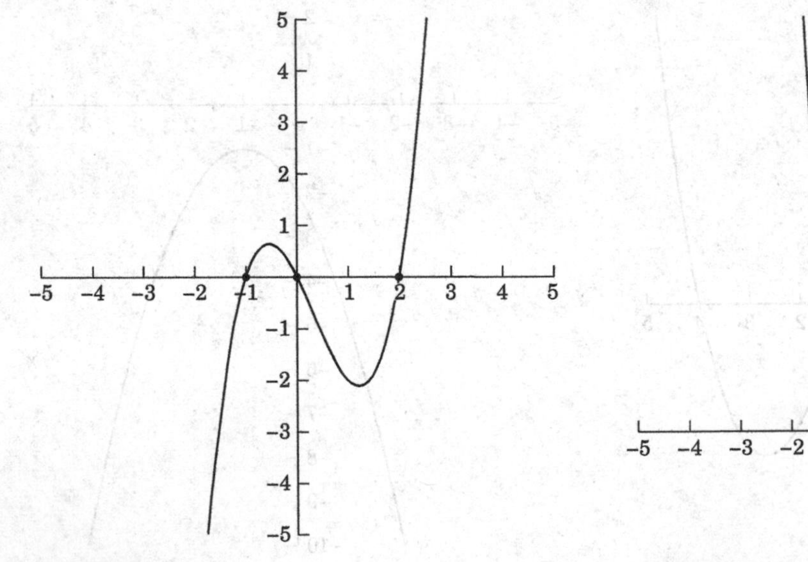

FIGURE 6-9

FIGURE 6-10

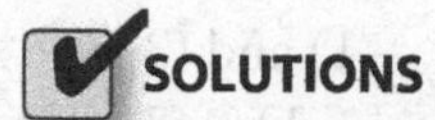

SOLUTIONS

1. **The graph is above the x-axis on $(-3, -1)$. The graph is below the x-axis on $(-\infty, -3)$ and $(-1, \infty)$.**
2. **The graph is above the x-axis on $(3, \infty)$ and below the x-axis on $(-\infty, 3)$.**
3. **The graph is above the x-axis on $(-1, 0)$ and $(2, \infty)$. The graph is below the x-axis on $(-\infty, -1)$ and $(0, 2)$.**
4. **The graph is above the x-axis on $(-\infty, \infty)$ (everywhere). The graph is never below the x-axis.**

Solving Nonlinear Inequalities

In order for the algebraic method for solving nonlinear inequalities to make sense, let us solve the inequality $x^2 - 2x - 3 > 0$ using a graph. This inequality is asking the question, "For what values of x are the y-values for $y = x^2 - 2x - 3$ positive?" The graph of $y = x^2 - 2x - 3$ (which we will call the *related graph*) is in Fig. 6-11.

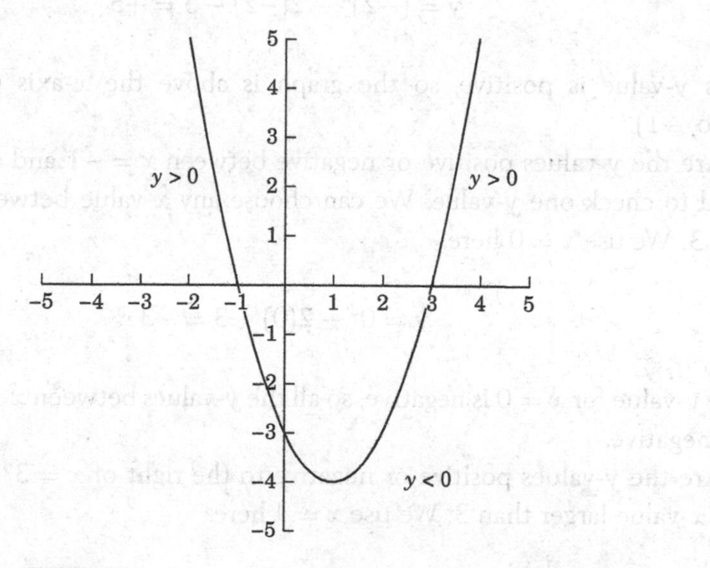

FIGURE 6-11

We see from the graph that $x^2 - 2x - 3$ is positive for $(-\infty, -1)$ and $(3, \infty)$. According to the graph, the solution to $x^2 - 2x - 3 > 0$ is $(-\infty, -1) \cup (3, \infty)$. (The union symbol, "$\cup$", is set notation for the word "or.")

Notice that the y-values for the graph of $y = x^2 - 2x - 3$ change between positive and negative *only* on either side of an x-intercept. This is why our method for solving a nonlinear inequality begins with using algebra to locate any x-intercepts. After we have located all the x-intercepts, we choose a test value in each interval. Once we have substituted these test values in the equation, we determine where the y-values are positive and where they are negative.

Let us now use this strategy for solving the original inequality. We find the x-intercepts for $y = x^2 - 2x - 3$ by setting $y = 0$ and solving for x.

$$0 = x^2 - 2x - 3$$

$$0 = (x - 3)(x + 1)$$

$$x - 3 = 0 \qquad\qquad x + 1 = 0$$

$$x = 3 \qquad\qquad x = -1$$

The two x-intercepts are -1 and 3. Are the y-values to the left of $x = -1$ positive or negative? We can answer this question by choosing *any* x-value smaller than -1. We use $x = -2$ here. Is the y-value for $x = -2$ positive or negative?

$$y = (-2)^2 - 2(-2) - 3 = +5$$

This y-value is positive, so the graph is above the x-axis on the interval $(-\infty, -1)$.

Are the y-values positive or negative between $x = -1$ and $x = 3$? We only need to check one y-value. We can choose any x-value between $x = -1$ and $x = 3$. We use $x = 0$ here.

$$y = 0^2 - 2(0) - 3 = -3$$

The y-value for $x = 0$ is negative, so all the y-values between $x = -1$ and $x = 3$ are negative.

Are the y-values positive or negative to the right of $x = 3$? We can choose any x-value larger than 3. We use $x = 4$ here.

$$y = 4^2 - 2(4) - 3 = +5$$

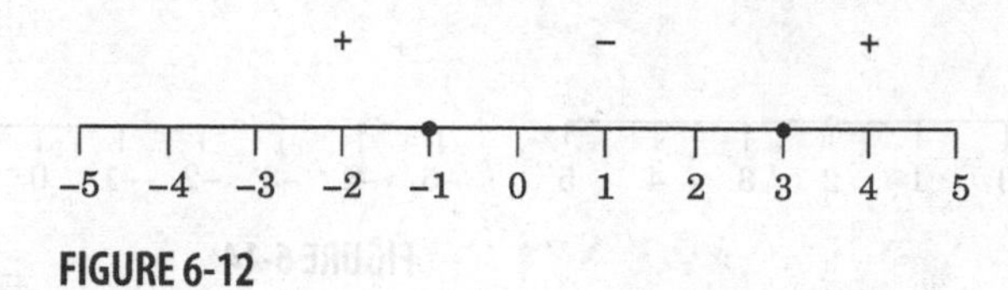

FIGURE 6-12

The y-value for $x = 4$ is positive, so the graph is above the x-axis on the interval $(3, \infty)$.

Values of x for which the graph is above the x-axis are to the left of $x = -1$ or to the right of $x = 3$, so we have confirmed that the solution for the inequality $x^2 - 2x - 3 > 0$ is $(-\infty, -1) \cup (3, \infty)$.

Sign Graphs

Using a *sign graph* helps us to keep track of x-intervals having positive y-values and those having negative y-values. First we draw the number line. Next we compute the x-intercepts and then mark them on the sign graph. After determining where y-values are positive (using the test points), we write a plus sign over this interval or these intervals. We write a minus sign over the interval or intervals having negative y-values. The sign graph for the inequality $x^2 - 2x - 3 > 0$ is shown in Fig. 6-12.

The steps that we used to solve the inequality $x^2 - 2x - 3 > 0$ are outlined in Table 6-1.

TABLE 6-1 Solving Certain* Nonlinear Inequalities

1.	If necessary, rewrite the inequality with 0 on one side.
2.	Find any x-intercepts of the related graph.
3.	Mark the x-intercepts on a sign graph.
4.	Choose an x-value in each interval to test whether the y-value is positive or negative.
5.	Mark each interval with a plus sign or a minus sign, depending on whether the y-value for the interval is positive or negative.
6.	Look at the inequality to decide if the solution is interval(s) marked with the plus or minus sign.
7.	Write the solution using interval notation.

*By "Certain Nonlinear Inequalities," we mean that the related graph has no breaks in it. Later, we will learn how to extend this strategy to include inequalities whose related graphs might have a break in them.

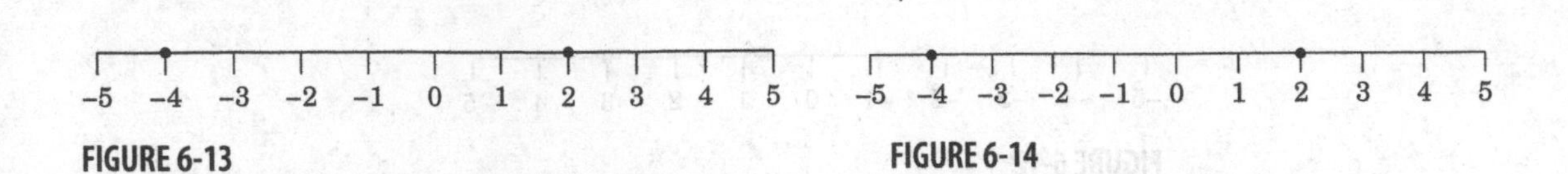

FIGURE 6-13 FIGURE 6-14

EXAMPLE 6-2

Solve the inequality and write the solution using interval notation.

- $x^2 + 2x - 8 < 0$

Step 1 is not necessary because one side of the inequality is already 0. We find the x-intercept(s) for the graph of $y = x^2 + 2x - 8$ by setting y equal to 0 and solving for x.

$$0 = x^2 + 2x - 8$$

$$0 = (x + 4)(x - 2)$$

$$x + 4 = 0 \qquad x - 2 = 0$$

$$x = -4 \qquad x = 2$$

Next we mark the x-intercepts on the sign graph (see Fig. 6-13).

We use $x = -5$ to represent the interval to the left of $x = -4$; $x = 0$ for the interval between $x = -4$ and $x = 2$; and $x = 3$ for the interval to the right of $x = 2$.

Letting $x = -5$ in the equation, $y = (-5)^2 + 2(-5) - 8 = +7$. We record a plus sign to the left of -4 on the sign graph (see Fig. 6-14). (Note: we can use either the expanded form $x^2 + 2x - 8$ or the factored form $(x + 4)(x - 2)$ when determining the sign of the test points.)

Letting $x = 0$, $y = 0^2 + 2(0) - 8 = -8$. We record a minus sign between $x = -4$ and $x = 2$ (see Fig. 6-15).

Letting $x = 3$, $y = 3^2 + 2(3) - 8 = +7$. We record a plus sign to the right of $x = 2$ (see Fig. 6-16).

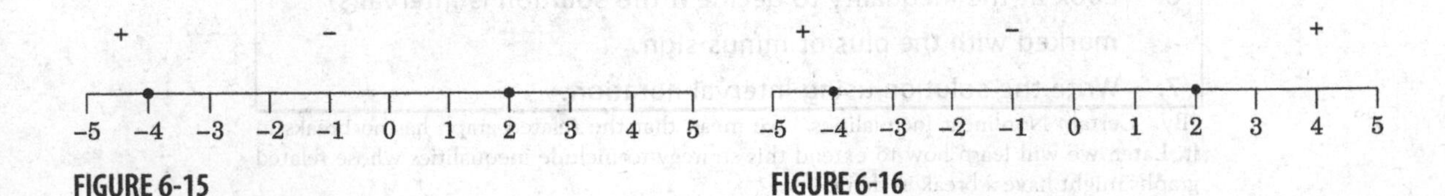

FIGURE 6-15 FIGURE 6-16

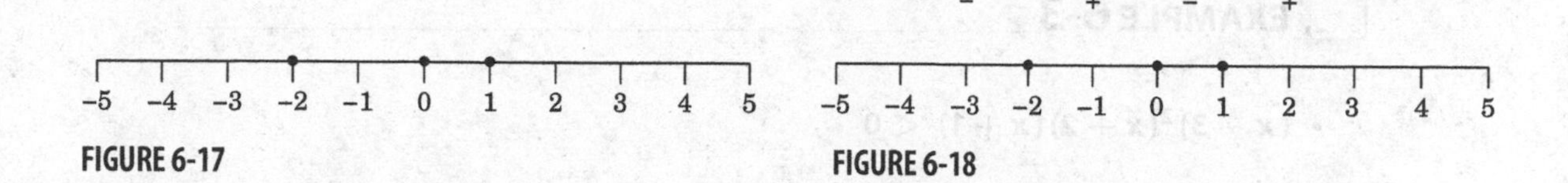

FIGURE 6-17

FIGURE 6-18

The inequality reads "< 0," which means we want the interval marked with a minus sign. The solution, then, is the interval between $x = -4$ and $x = 2$: $(-4, 2)$.

- **$x^3 + x^2 - 2x \geq 0$**

We find the x intercepts by factoring $y = x^3 + x^2 - 2x$ and setting each factor equal to 0.

$$x^3 + x^2 - 2x = x(x^2 + x - 2) = x(x + 2)(x - 1)$$

$$x = 0 \qquad x + 2 = 0 \qquad x - 1 = 0$$

$$x = -2 \qquad x = 1$$

Now we can mark the x-intercepts on the graph (see Fig. 6-17).

We use $x = -3$ for the interval to the left of $x = -2$; $x = -1$ for the interval between $x = -2$ and $x = 0$; $x = 0.5$ for the interval between $x = 0$ and $x = 1$; and $x = 2$ for the interval to the right of $x = 1$.

Letting $x = -3$, $y = (-3)^3 + (-3)^2 - 2(-3) = -12$. We record a minus sign to the left of $x = -2$.

Letting $x = -1$, $y = (-1)^3 + (-1)^2 - 2(-1) = +2$. We record a plus sign between $x = -2$ and $x = 0$.

Letting $x = 0.5$, $y = (0.5)^3 + (0.5)^2 - 2(0.5) = -0.625$. We record a minus sign between $x = 0$ and $x = 1$.

Letting $x = 2$, $y = 2^3 + 2^2 - 2(2) = +8$. We record a plus sign to the right of $x = 1$ (See Fig. 6-18).

The inequality is "≥ 0," which means we want the positive intervals. The solution is $[-2, 0] \cup [1, \infty)$.

It seems that the signs always alternate between plus and minus signs, but the signs do not alternate on the sign graph for one of the problems in the next example.

EXAMPLE 6-3

- $(x-3)^2(x+2)(x+1) < 0$

$$(x-3)^2 = 0 \qquad x+2=0 \qquad x+1=0$$
$$x-3=0 \qquad x=-2 \qquad x=-1$$
$$x=3$$

For $x=-3$, $y=(-3-3)^2(-3+2)(-3+1) = +72$. We record a plus sign to the left of $x=-2$.

For $x=-1.5$, $y=(-1.5-3)^2(-1.5+2)(-1.5+1) = -5.0625$. We record a minus sign between $x=-1$ and $x=-2$.

For $x=0$, $y=(0-3)^2(0+2)(0+1) = +18$. We record a plus sign between $x=-1$ and $x=3$.

For $x=4$, $y=(4-3)^2(4+2)(4+1) = +30$. We record a plus sign to the right of $x=3$ (see Fig. 6-19).

The inequality is "< 0," which means we want the interval marked with a minus sign. The solution is $(-2,-1)$.

- $x^2+5x+9<3$

We must subtract 3 from each side of the inequality so that 0 is on one side.

$$x^2+5x+6<0$$
$$x^2+5x+6=(x+2)(x+3)$$
$$x+2=0 \qquad x+3=0$$
$$x=-2 \qquad x=-3$$

For $x=-4$, $y=(-4+2)(-4+3)=+2$.

For $x=-2.5$, $y=(-2.5+2)(-2.5+3)=-0.25$.

For $x=0$, $y=(0+2)(0+3)=+6$. See Fig. 6-20.

The solution is $(-3,-2)$.

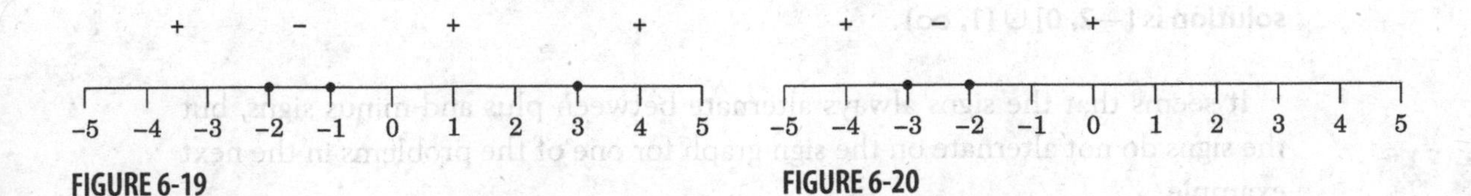

FIGURE 6-19

FIGURE 6-20

- $x^2 + 1 > 0$

The equation $x^2 + 1 = 0$ has no (real) solution, so the graph of $y = x^2 + 1$ has no x-intercept. This means that either all y-values are positive or they are all negative. We need to check only one y-value, so we can test any x-value. We use $x = 0$.

For $x = 0, y = 0^2 + 1 = +1$. Because this y-value is positive, *all* y-values are positive. The solution is $(-\infty, \infty)$.

PRACTICE

Solve the inequality and write the solution using interval notation.

1. $x^2 + 3x - 4 < 0$ 2. $-x^2 - x + 6 \geq -14$ 3. $x^4 - 13x^2 + 36 \geq 0$

Hint: $x^4 - 13x^2 + 36 = (x^2 - 4)(x^2 - 9) = (x - 2)(x + 2)(x - 3)(x + 3)$

4. $x^2 + 9 \leq 0$

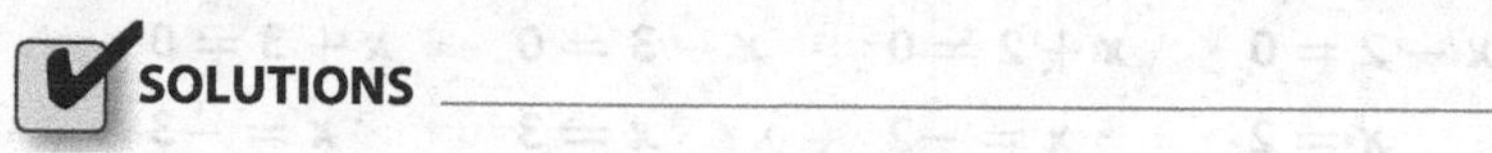

SOLUTIONS

1. $x^2 + 3x - 4 = (x + 4)(x - 1)$

$$x + 4 = 0 \qquad\qquad x - 1 = 0$$

$$x = -4 \qquad\qquad x = 1$$

For $x = -5, y = (-5 + 4)(-5 - 1) = 6$.
For $x = 0, y = (0 + 4)(0 - 1) = -4$.
For $x = 2, y = (2 + 4)(2 - 1) = 6$.
See Fig. 6-21.
The solution is $(-4, 1)$.

2. First we must add 14 to both sides of the inequality to get $-x^2 - x + 20 \geq 0$. Factoring the nonzero side gives us $-x^2 - x + 20 = -(x^2 + x - 20) = -(x + 5)(x - 4)$.

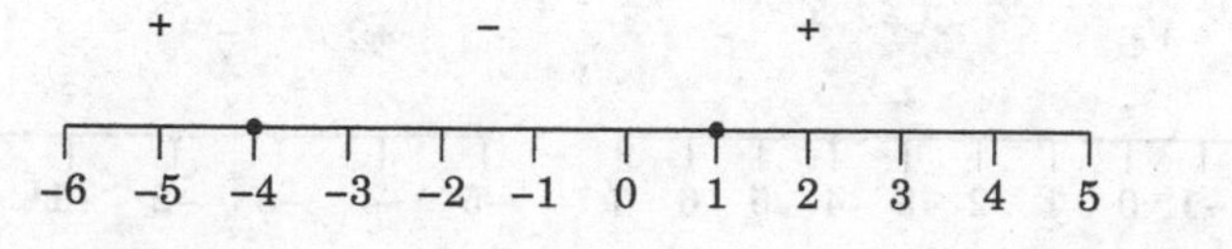

FIGURE 6-21

$$-(x+5)=0 \qquad x-4=0$$

$$-x-5=0 \qquad x=4$$

$$-x=5$$

$$x=-5$$

For $x=-6$, $y=-(-6+5)(-6-4)=-10$.
For $x=0$, $y=-(0+5)(0-4)=20$.
For $x=5$, $y=-(5+5)(5-4)=-10$ (see Fig. 6-22).
The solution is $[-5, 4]$.

3. $x^4-13x^2+36=(x^2-4)(x^2-9)=(x-2)(x+2)(x-3)(x+3)$

$$x-2=0 \qquad x+2=0 \qquad x-3=0 \qquad x+3=0$$

$$x=2 \qquad x=-2 \qquad x=3 \qquad x=-3$$

For $x=-4$, $y=(-4-2)(-4+2)(-4-3)(-4+3)=+84$.
For $x=-2.5$, $y=(-2.5-2)(-2.5+2)(-2.5-3)(-2.5+3)=-6.1875$.
For $x=0$, $y=(0-2)(0+2)(0-3)(0+3)=+36$.
For $x=2.5$, $y=(2.5-2)(2.5+2)(2.5-3)(2.5+3)=-6.1875$.
For $x=4$, $y=(4-2)(4+2)(4-3)(4+3)=+84$ (see Fig. 6-23).
The solution is $(-\infty, -3] \cup [-2, 2] \cup [3, \infty)$.

4. The equation $x^2+9=0$ has no (real) solution. This means that the graph of $y=x^2+9$ has no x-intercepts, so either all y-values are positive or they are all negative. We need to check only one y-value. Let $x=0$: $0^2+9=9$.

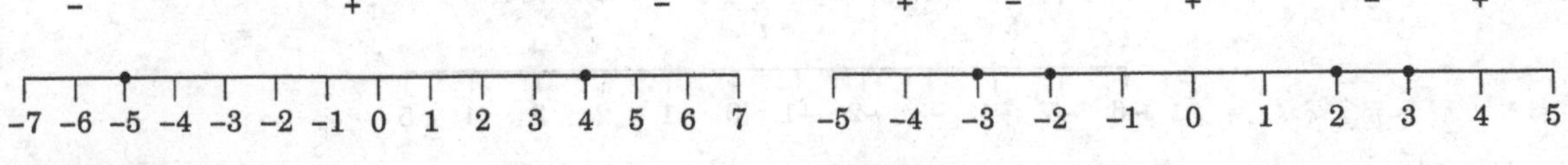

FIGURE 6-22

FIGURE 6-23

This y-value is positive, so all y-values are positive. Because the inequality is "≤ 0," we want negative y-values. The inequality has no solution.

Rational Inequalities

If an equation has a variable in a denominator, then its graph likely comes in separate pieces. For example, part of the graph of $y = \frac{1}{x}$ (shown in Fig. 6-24) lies to the left of the y-axis and the other part lies to the right of the y-axis.

The graph of such equations has a break at every value of x that causes a zero in its denominator. A break in the graph of these equations acts like an x-intercept—the y-values can change from positive to negative (or from negative to positive) on either side of the break.

EXAMPLE 6-4

The graph in Fig. 6-25 is the graph of the equation $y = \frac{x+2}{x-1}$.

This graph has both an x-intercept (at $x = -2$) and a break (at $x = 1$). The y-values are positive to the left of the x-intercept and to the right of the break. The y-values are negative between the x-intercept and the break.

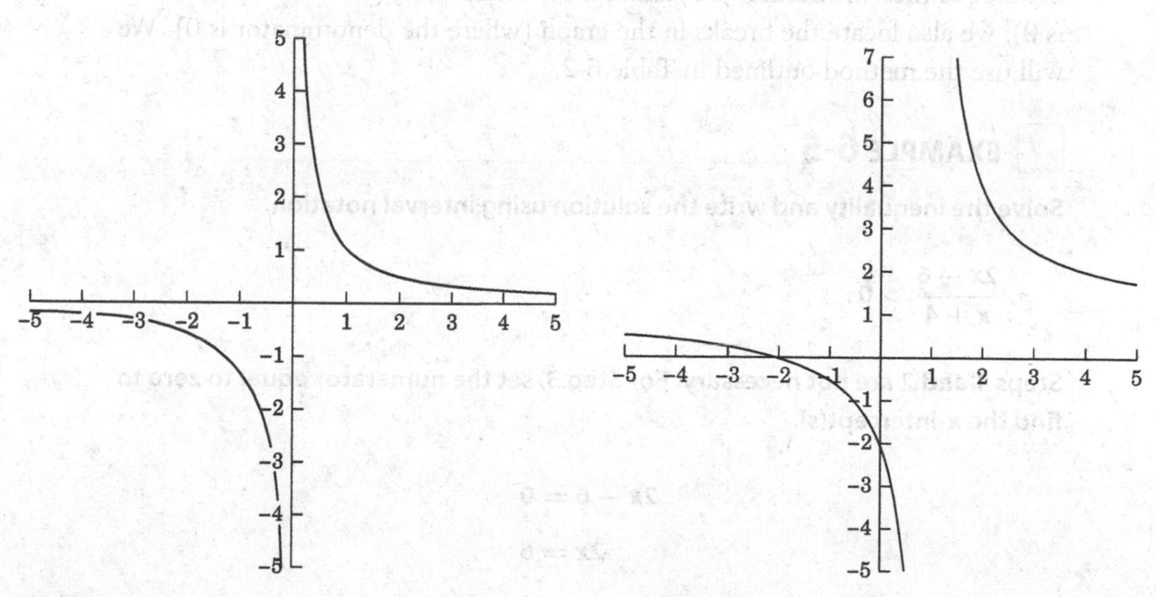

FIGURE 6-24

FIGURE 6-25

TABLE 6-2 Solving Rational Inequalities

1.	Rewrite the inequality with 0 on one side.
2.	Rewrite the nonzero side of the inequality as a single fraction.
3.	Find the x-intercept(s) of the related graph by setting the numerator equal to 0 and solving for x. If the numerator does not have a variable, the graph has no x-intercept. Mark the sign graph with any x-intercept.
4.	Find the break(s) in the graph by setting the denominator equal to 0 and solving for x. Mark this x-value or these values on the sign graph. If there is no x-value that makes the denominator equal to 0, then there is no break in the graph.
5.	Test an x-value in each interval on the sign graph. If the y-value is positive, record a plus sign over the interval. If the y-value is negative, record a minus sign over the interval.
6.	Look at the inequality to determine whether the solution is the interval(s) marked with a "plus" sign or a "minus" sign. Be careful to exclude from the solution any x-value that causes a 0 in a denominator.

Solving inequalities with variables in a denominator is much like solving earlier inequalities. In addition to locating the x-intercept(s) (where the numerator is 0), we also locate the breaks in the graph (where the denominator is 0). We will use the method outlined in Table 6-2.

EXAMPLE 6-5

Solve the inequality and write the solution using interval notation.

- $\dfrac{2x-6}{x+4} > 0$

Steps 1 and 2 are not necessary. For Step 3, set the numerator equal to zero to find the x-intercept(s).

$$2x - 6 = 0$$

$$2x = 6$$

$$x = 3$$

We mark $x = 3$ on the sign graph.

For Step 4, we set the denominator equal to zero to find any break in the graph.

$$x + 4 = 0$$

$$x = -4$$

We mark $x = -4$ on the sign graph. For Step 5, we test an x-value smaller than 4 (here we use $x = -5$), between -4 and 3 (here we use $x = 0$), and larger than 3 (we use $x = 4$ here).

$$x = -5 \qquad y = \frac{2(-5) - 6}{-5 + 4} = 16$$

$$x = 0 \qquad y = \frac{2(0) - 6}{0 + 4} = -\frac{3}{2}$$

$$x = 4 \qquad y = \frac{2(4) - 6}{4 + 4} = \frac{1}{4}$$

See Fig. 6-26.

The solution is $(-\infty, -4) \cup (3, \infty)$.

- $\dfrac{5x + 8}{x^2 + 1} < 0$

We set the numerator equal to zero to find the x-intercept(s).

$$5x + 8 = 0$$

$$5x = -8$$

$$x = -\frac{8}{5}$$

We mark $x = -\frac{8}{5}$ on the sign graph.

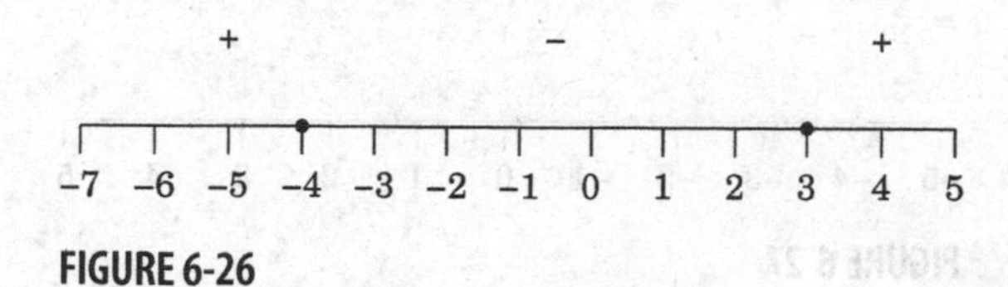

FIGURE 6-26

We set the denominator equal to zero to find any break in the graph of $y = \frac{5x+8}{x^2+1}$.

$$x^2 + 1 = 0$$

This equation has no (real) solution, so there are no breaks in the graph.

We test $x = -2$ for the point to the left of $x = -\frac{8}{5}$ and $x = 0$ for the point to the right of $x = -\frac{8}{5}$.

$$x = -2 \qquad y = \frac{5(-2) + 8}{(-2)^2 + 1} = -\frac{2}{5}$$

$$x = 0 \qquad y = \frac{5(0) + 8}{0^2 + 1} = 8 \qquad \text{See Fig. 6-27.}$$

The solution is $(-\infty, -\frac{8}{5})$.

- $\frac{3 - 2x}{x + 4} > 2$

One side of the inequality must be 0, so we subtract 2 from each side.

$$\frac{3 - 2x}{x + 4} - 2 > 0$$

Now we can write the left side as a single fraction.

$$\frac{3 - 2x}{x + 4} - 2 = \frac{3 - 2x}{x + 4} - 2\left(\frac{x + 4}{x + 4}\right) = \frac{3 - 2x}{x + 4} - \frac{2(x + 4)}{x + 4}$$

$$= \frac{3 - 2x - 2(x + 4)}{x + 4} = \frac{3 - 2x - 2x - 8}{x + 4} = \frac{-4x - 5}{x + 4}$$

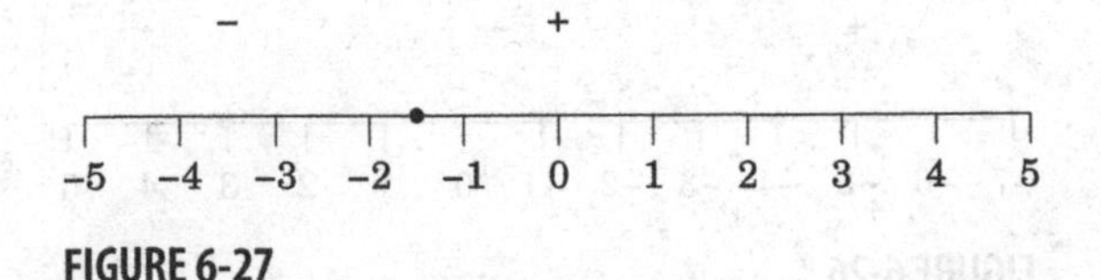

FIGURE 6-27

The inequality can be rewritten so that the left side is a single fraction.

$$\frac{-4x-5}{x+4} > 0$$

$$-4x-5=0 \qquad\qquad x+4=0$$

$$-4x=5 \qquad\qquad x=-4$$

$$x=-\frac{5}{4}$$

We mark $x=-\frac{5}{4}$ and $x=-4$ on the sign graph.

We test $x=-5$ for a point to the left of $x=-4$, $x=-2$ for the point between $x=-4$ and $x=-\frac{5}{4}$, and $x=0$ for the point to the right of $x=-\frac{5}{4}$.

$$x=-5 \qquad y=\frac{-4(-5)-5}{-5+4}=-15$$

$$x=-2 \qquad y=\frac{-4(-2)-5}{-2+4}=\frac{3}{2}$$

$$x=0 \qquad y=\frac{-4(0)-5}{0+4}=-\frac{5}{4}$$

See Fig. 6-28.

The solution is $(-4, -1\frac{1}{4})$.

- $\dfrac{x^2+6x+8}{x-3} \le 0$

$$x^2+6x+8=0$$

$$(x+2)(x+4)=0$$

$$x+2=0 \qquad\qquad x+4=0$$

$$x=-2 \qquad\qquad x=-4$$

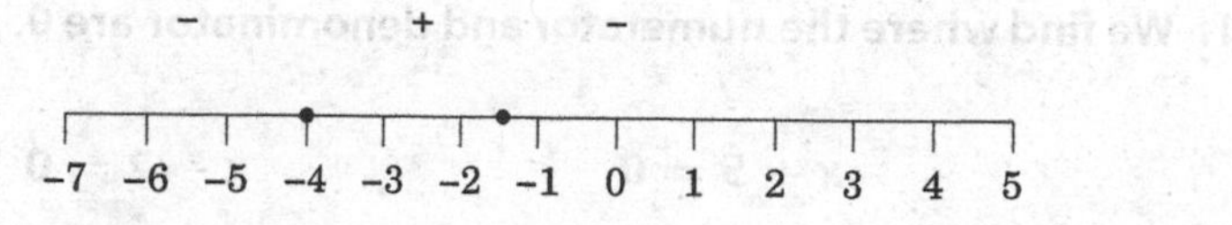

FIGURE 6-28

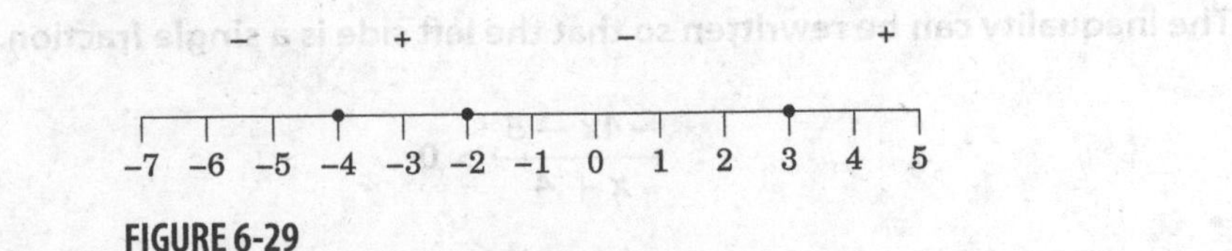

FIGURE 6-29

From $x - 3 = 0$, we also have $x = 3$. We mark $x = -2$, $x = -4$, and $x = 3$ on the sign graph. Next, we test an x-value smaller than -4 (here we use $x = -5$), between -2 and -4 (we use $x = -3$ here), between -2 and 3 (here we use $x = 0$), and larger than 3 (we use $x = 4$ here).

$$x = -5 \qquad y = \frac{(-5)^2 + 6(-5) + 8}{-5 - 3} = -\frac{3}{8}$$

$$x = -3 \qquad y = \frac{(-3)^2 + 6(-3) + 8}{-3 - 3} = \frac{1}{6}$$

$$x = 0 \qquad y = \frac{0^2 + 6(0) + 8}{0 - 3} = -\frac{8}{3}$$

$$x = 4 \qquad y = \frac{4^2 + 6(4) + 8}{4 - 3} = 48 \qquad \text{See Fig. 6-29.}$$

The solution is *not* $(-\infty, -4] \cup [-2, 3]$. By using a bracket around 3, we are implying that $x = 3$ is a solution, but $x = 3$ leads to a zero in the denominator, so it cannot be a solution. We use a parenthesis around 3 to indicate that $x = 3$ is not part of the solution. The solution is $(-\infty, -4] \cup [-2, 3)$.

PRACTICE

Solve the inequality and write the solution using interval notation.

1. $\dfrac{x + 5}{x - 2} > 0$ 2. $\dfrac{2x - 8}{x^2 + 3x + 2} < 0$ 3. $\dfrac{3x + 2}{x - 4} \le 1$

SOLUTIONS

1. We find where the numerator and denominator are 0.

$$x + 5 = 0 \qquad\qquad x - 2 = 0$$

$$x = -5 \qquad\qquad x = 2$$

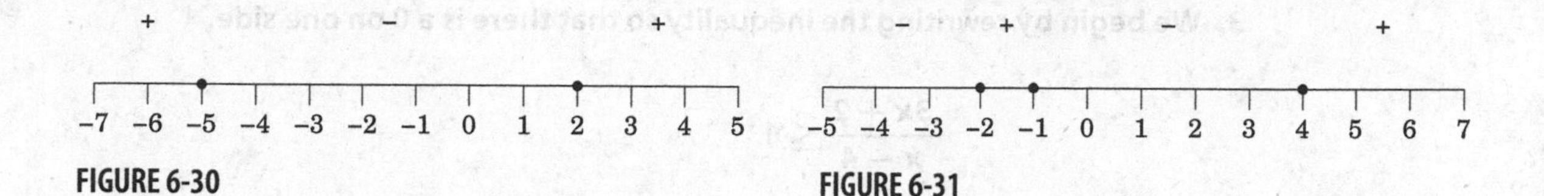

FIGURE 6-30

FIGURE 6-31

$$x = -6 \qquad y = \frac{-6+5}{-6-2} = \frac{1}{8}$$

$$x = 0 \qquad y = \frac{0+5}{0-2} = -\frac{5}{2}$$

$$x = 3 \qquad y = \frac{3+5}{3-2} = 8$$

See Fig. 6-30.

The solution is $(-\infty, -5) \cup (2, \infty)$.

2. Setting the numerator equal to 0 gives us $x = 4$ $(2x - 8 = 0)$. Now, we find where the denominator is 0.

$$x^2 + 3x + 2 = 0$$

$$(x+2)(x+1) = 0$$

$$x + 2 = 0 \qquad x + 1 = 0$$

$$x = -2 \qquad x = -1$$

$$x = -3 \qquad y = \frac{2(-3) - 8}{(-3)^2 + 3(-3) + 2} = -7$$

$$x = -1.5 \qquad y = \frac{2(-1.5) - 8}{(-1.5)^2 + 3(1.5) + 2} = 44$$

$$x = 0 \qquad y = \frac{2(0) - 8}{0^2 + 3(0) + 2} = -4$$

$$x = 5 \qquad y = \frac{2(5) - 8}{5^2 + 3(5) + 2} = \frac{1}{21}$$

See Fig. 6-31.

The solution is $(-\infty, -2) \cup (-1, 4)$.

3. We begin by rewriting the inequality so that there is a 0 on one side.

$$\frac{3x+2}{x-4} \leq 1$$

$$\frac{3x+2}{x-4} - 1 \leq 0$$

$$\frac{3x+2}{x-4} - 1\left(\frac{x-4}{x-4}\right) \leq 0 \quad \text{Write as a single fraction.}$$

$$\frac{3x+2-(x-4)}{x-4} \leq 0$$

$$\frac{3x+2-x+4}{x-4} \leq 0$$

$$\frac{2x+6}{x-4} \leq 0 \quad \text{Find where the numerator and denominator are 0.}$$

$$2x+6=0 \qquad\qquad x-4=0$$

$$x=-3 \qquad\qquad x=4$$

$$x=-4 \qquad \frac{2(-4)+6}{-4-4} = \frac{1}{4}$$

$$x=0 \qquad \frac{2(0)+6}{0-4} = -\frac{3}{2}$$

$$x=5 \qquad \frac{2(5)+6}{5-4} = 16 \qquad \text{See Fig. 6-32.}$$

The solution is $[-3, 4)$. The solution is not $[-3, 4]$ because $x = 4$ leads to a zero in the denominator.

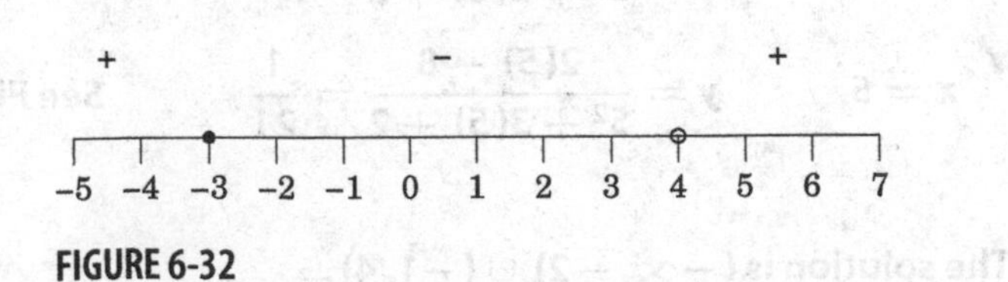

FIGURE 6-32

Summary

In this chapter, we learned how to

- *Examine a graph to determine where the graph is above the x-axis and where it is below it.*
- *Solve a nonlinear inequality graphically*. When asked to solve a nonlinear inequality, once in the form "Expression > 0," for example, we examine the graph of $y =$ Expression. If the inequality is "> 0" or "≥ 0," we determine where the graph is *above* the x-axis. If the inequality is "< 0" or "≤ 0," we determine where the graph is below the x-axis.
- *Solve a nonlinear inequality algebraically*. From our work with graphs, we noticed that if the graph comes in one piece, then the only place it can change from going above the x-axis to below it (or vice versa) is on either side of an x-intercept. For this reason, we begin to solve a nonlinear inequality by finding the x-intercepts for its related graph and investigating where the graph lies above/below the x-axis. We did so with the aid of a sign graph. We record the x-intercepts on the sign graph, select a test value in each interval and record whether the y-values in each interval are positive or negative.
- *Solve a rational inequality algebraically*. If an inequality involves a fraction containing an x in its denominator, then the graph has a break where the denominator is 0. The graph of the related equation can change from being above the x-axis to below it (and vice versa) on either side of an x-intercept and on either side of a break in the graph. For this reason, we must investigate where the denominator of any fraction is 0 and mark these values on the sign graph. When writing the solution, we must be careful not to include any x-value that causes a 0 in the denominator.

QUIZ

Solve the inequality.

1. $x^2 - 2x - 24 < 0$

A. $(-\infty, -4) \cup (6, \infty)$ B. $(-4, 6)$ C. $(-\infty, -6) \cup (4, \infty)$ D. $(-6, 4)$

2. $2x^2 + x \geq 6$

A. $(-\infty, -2] \cup [\frac{3}{2}, \infty)$ B. $[-2, \frac{3}{2}]$ C. $(-\infty, -\frac{3}{2}] \cup [2, \infty)$ D. $[-\frac{3}{2}, 2]$

3. $x^2 + 81 < 0$

A. $(-\infty, -9) \cup (9, \infty)$ B. $(-9, 9)$ C. $(-\infty, \infty)$ D. There is no solution.

4. $\frac{x+6}{x-4} > 0$

A. $(-\infty, -6) \cup (4, \infty)$ B. $(-6, 4)$
C. $(-6, \infty) \cup (4, \infty)$ D. There is no solution.

5. $\frac{x^2-25}{x+1} \leq 0$

A. $(-\infty, -5] \cup (-1, 5]$ B. $(-\infty, -5) \cup (1, 5)$
C. $[-5, -1) \cup [5, \infty)$ D. $(-\infty, -5) \cup (5, \infty)$

6. $(x-9)^2(x+4) > 0$

A. $(-4, \infty)$ B. $(-4, 9) \cup (9, \infty)$ C. $(-\infty, -4)$ D. $(-\infty, -4) \cup (9, \infty)$

7. $\frac{2x+6}{3x-4} < 5$

A. $(-\infty, \frac{4}{3}) \cup (2, \infty)$ B. $(-\infty, -3) \cup (\frac{4}{3}, \infty)$ C. $(-3, \frac{4}{3})$ D. $(-\infty, \frac{4}{3})$

8. $-x^2 - x + 6 > 0$

A. $(-2, 3)$ B. $(-\infty, -2) \cup (3, \infty)$ C. $(-3, 2)$ D. $(-\infty, -3) \cup (2, \infty)$

chapter 7

Functions

A function is a special type of relationship between variables where the value of one variable *depends* on the value of one or more other variables. Functions occur all around us. For example, a person's weight *depends* on many variables—age, sex, height, food intake, activity level, and so on. An hourly worker's paycheck *depends* on the number of hours worked. In this chapter, we learn the definition of a function, how to tell whether or not an equation is a function, how to evaluate functions, and how to find their domain and range. These will be important topics throughout the remainder of this book (except for Chap. 11) as we work with several important families of functions.

CHAPTER OBJECTIVES

In this chapter, you will learn how to

- Determine whether or not an equation gives y as a function of x
- Evaluate a function both algebraically and graphically
- Find the domain and range of a function graphically
- Find the domain of a function algebraically
- Use the graph of a function to determine where the function is increasing and/or decreasing
- Sketch the graph of a piecewise function
- Calculate the difference quotient for a given function

Introduction to Functions

Officially, a function is a relation between two sets, A and B, where every element in A is assigned exactly one element in B. What this means for x and y is that for *every* x-value, there is *exactly* one y-value.

We will work with functions that can be represented by an equation in x and y. For example, the linear equation $y = mx + b$ is actually a linear function, and the quadratic equation $y = ax^2 + bx + c$ is actually a quadratic function. No matter what we put in for x, there is exactly one y-value for that particular x for these equations. We call x the *independent* variable and y the *dependent* variable.

What kind of equations are not functions? An equation is not a function if there is any x-value that has more than one y-value. For example, in the equation $x^2 + y^2 = 9$, y is not a function of x. If we let $x = 0$, then we have $y^2 = 9$, so $y = 3$ or $y = -3$. This means $x = 0$ has *two* y-values, 3 and -3.

When asked to determine whether or not an equation "gives y as a function of x," we can do one of two things. If y is not a function of x, all we need to do is to find one x-value that has two y-values (as we did above with $x = 0$). If we suspect that y is a function of x, we solve the equation for y. Once we have y isolated in the equation, we then decide if there can be any x-value that has more than one y-value.

Determine if y is a function of x.

- $y^3 + x^2 - 3x = 7$

We begin by solving for y.

$$y^3 = -x^2 + 3x + 7$$

$$y = \sqrt[3]{-x^2 + 3x + 7} \quad \textbf{(We only need +/− for even roots.)}$$

Each x-value has only one y-value, so y is a function of x.

- $(x + 1)^2 + y^2 = 9$

We solve this equation for *y*.

$$(x+1)^2 + y^2 = 9 \qquad \text{Subtract } (x+1)^2 \text{ from each side.}$$

$$y^2 = 9 - (x+1)^2 \qquad \text{Take the square root of each side.}$$

$$y = \pm\sqrt{9-(x+1)^2}$$

Most *x*-values have *two* *y*-values, $y = \sqrt{9-(x+1)^2}$ and $y = -\sqrt{9-(x+1)^2}$. This means that *y* is not a function of *x*.

PRACTICE

Determine if *y* is a function of *x*.

1. $x^2 + (y-3)^2 = 16$ 2. $x^2 - 2y = 4$ 3. $|y| = x$

SOLUTIONS

1. **Here we solve for *y* and will observe that there are two *y*-values for some values of *x*. We could actually avoid this work by finding a single value of *x* having two *y*-values, for example, $x = 0$ has two *y*-values: $y = 7$ and $y = -1$. Either way, we see that *y* is not a function of *x*.**

$$x^2 + (y-3)^2 = 16$$

$$(y-3)^2 = 16 - x^2$$

$$y - 3 = \pm\sqrt{16-x^2}$$

$$y = 3 \pm \sqrt{16-x^2}$$

2. ***y* is a function of *x* because we can solve for *y* (without using a $\pm$ symbol):**

$$x^2 - 2y = 4$$

$$-2y = 4 - x^2$$

$$y = \frac{4-x^2}{-2}$$

3. In the equation $|y| = x$, y is not a function of x because every nonnegative x-value has two y-values. For example, if $x = 3$, $|y| = 3$ has the solutions $y = 3$ and $y = -3$.

Evaluating Functions

Functions are often given letter names. The most common name is "$f(x)$." The notation "$f(x)$" means "the function f evaluated at x." We pronounce "$f(x)$" as "f at x" or "f of x." Usually y and $f(x)$ are the same. For example, instead of writing $y = 2x + 1$, we write $f(x) = 2x + 1$.

Evaluating a function at a quantity means to substitute the quantity for x. If the function is $f(x) = 2x + 1$, then to evaluate the function at 3 means to find the y-value for $x = 3$. To "find f at 3" or "evaluate $f(3)$" means to let $x = 3$ in the equation.

EXAMPLE 7-2

- **Find $f(1)$ and $f(0)$ for $f(x) = 3x^2 + 4$.**

We substitute the number in the parentheses for x.

$$f(1) = 3(1)^2 + 4 = 3(1) + 4 = 7 \qquad f(0) = 3(0)^2 + 4 = 3(0) + 4 = 4$$

- **Find $f(0)$, $f(10)$, and $f(1)$ for $f(x) = \frac{6x+5}{x^2+2}$.**

$$f(0) = \frac{6(0) + 5}{(0)^2 + 2} = \frac{5}{2} \qquad f(10) = \frac{6(10) + 5}{10^2 + 2} = \frac{60 + 5}{100 + 2} = \frac{65}{102}$$

$$f(1) = \frac{6(1) + 5}{1^2 + 2} = \frac{6 + 5}{1 + 2} = \frac{11}{3}$$

- **Find $g(0)$, $g(1)$, and $g(-6)$ for $g(t) = \sqrt{3 - t}$. (Treat the variable t like the variable x.)**

$$g(0) = \sqrt{3 - 0} = \sqrt{3} \qquad g(1) = \sqrt{3 - 1} = \sqrt{2}$$

$$g(-6) = \sqrt{3 - (-6)} = \sqrt{9} = 3$$

Functions that have no variable (other than y, $f(x)$, or $g(t)$, etc.) are called *constant functions*. The y-values do not change. No matter what x is, the y-value (or the function's value) stays the same.

EXAMPLE 7-3

- **Evaluate $f(x) = 10$ for $x = 3, x = -8$, and $x = \pi$.**

No matter what x is, $f(x) = 10$.

$$f(3) = 10 \qquad f(-8) = 10 \qquad f(\pi) = 10$$

PRACTICE

1. **Find $f(4)$, $f(-6)$, and $f(0)$ for $f(x) = 3x - 2$.**
2. **Find $f(-10)$, $f(6)$, and $f(\pi^{23})$ for $f(x) = \sqrt{17}$.**
3. **Find $h(0), h(5)$, and $h(-2)$ for $h(t) = (2t + 4)/(t^2 - 7)$.**
4. **Find $f(0)$ and $f(-\frac{1}{2})$ for $f(x) = \sqrt{2x+1}$.**

SOLUTIONS

1. $f(4) = 3(4) - 2 = 10 \qquad f(-6) = 3(-6) - 2 = -20$
   $f(0) = 3(0) - 2 = -2$
2. $f(-10) = \sqrt{17} \qquad f(6) = \sqrt{17} \qquad f(\pi^{23}) = \sqrt{17}$
3. $h(0) = \dfrac{2(0) + 4}{0^2 - 7} = -\dfrac{4}{7} \qquad h(5) = \dfrac{2(5) + 4}{5^2 - 7} = \dfrac{14}{18} = \dfrac{7}{9}$
   $h(-2) = \dfrac{2(-2) + 4}{(-2)^2 - 7} = \dfrac{0}{-3} = 0$
4. $f(0) = \sqrt{2(0) + 1} = \sqrt{1} = 1$
   $f\left(-\dfrac{1}{2}\right) = \sqrt{2\left(-\frac{1}{2}\right) + 1} = \sqrt{0} = 0$

Evaluating Piecewise Functions

Piecewise-defined functions come in two parts. One part is an interval for x, the other part is the formula for computing y.

EXAMPLE 7-4

- **Evaluate $f(6)$ and $f(0)$ for $f(x) = \begin{cases} x-4, & \text{if } x < 1; \\ 2x+5 & \text{if } x \geq 1 \end{cases}$.**

This function is telling us that for x-values smaller than 1, the y-values are computed using $x - 4$. For x-values greater than or equal to 1, the y-values are computed using $2x + 5$. When asked to evaluate $f(\textit{number})$, we first decide what interval contains the number and then we compute the y-value using the formula written next to the interval.

Does $x = 6$ belong to the interval $x < 1$ or to $x \geq 1$? Since $6 \geq 1$, we use $2x + 5$ to compute y.

$$f(6) = 2(6) + 5 = 17$$

$f(0)$: Does $x = 0$ belong to the interval $x < 1$ or to $x \geq 1$? Since $0 < 1$, we use $x - 4$ to compute y.

$$f(0) = 0 - 4 = -4$$

- **Evaluate $f(-4)$, $f(0)$, $f(6)$, and $f(-2)$.**

$$f(x) = \begin{cases} x^2 & \text{if } x < -2; \\ 4x+8 & \text{if } -2 \leq x < 3; \\ 16 & \text{if } x \geq 3 \end{cases}$$

$f(-4)$: Since $x = -4$ belongs to the interval $x < -2$, we use x^2 to compute y.

$$f(-4) = (-4)^2 = 16$$

$f(0)$: Since $x = 0$ belongs to the interval $-2 \leq x < 3$, we use $4x + 8$ to compute y.

$$f(0) = 4(0) + 8 = 8$$

$f(6)$: Since $x = 6$ belongs to the interal $x \geq 3$, the y-value is 16.

$$f(6) = 16$$

$f(-2)$: Since $x = -2$ belongs to the interval $-2 \leq x < 3$, we use $4x + 8$ to compute y.

$$f(-2) = 4(-2) + 8 = 0$$

PRACTICE

1. Find $f(0)$, $f(-1)$, $f(7)$, and $f(-6)$.

$$f(x) = \begin{cases} x^2 - 2x & \text{if } x < -4; \\ x + 5 & \text{if } x \geq -4 \end{cases}$$

2. Find $f(-4)$, $f(3)$, $f(-1)$, and $f(1)$.

$$f(x) = \begin{cases} 3x^2 + 2x & \text{if } x \leq -1; \\ x + 4 & \text{if } -1 < x \leq 1; \\ 6x & \text{if } x > 1 \end{cases}$$

3. Find $f(3)$, $f(-2)$, and $f(0)$.

$$f(x) = \begin{cases} 0 & \text{if } x < 0; \\ 1 & \text{if } x \geq 0 \end{cases}$$

4. Find $f(3)$, $f(-4)$, $f(2)$, and $f(1)$.

$$f(x) \begin{cases} -2 & \text{if } x < 1; \\ 3x - 4 & \text{if } 1 \leq x < 3; \\ x^2 - 2x + 2 & \text{if } x \geq 3 \end{cases}$$

SOLUTIONS

1. $f(0) = 0 + 5 = 5$ $\quad f(-1) = -1 + 5 = 4$

   $f(7) = 7 + 5 = 12$ $\quad f(-6) = (-6)^2 - 2(-2) = 40$

2. $f(-4) = 3(-4)^2 + 2(-4) = 40$ $\quad f(3) = 6(3) = 18$

   $f(-1) = 3(-1)^2 + 2(-1) = 1$ $\quad f(1) = 1 + 4 = 5$

3. $f(3) = 1$ $\quad f(-2) = 0$ $\quad f(0) = 1$

4. $f(3) = 3^2 - 2(3) + 2 = 5$ $\quad f(-4) = -2$

   $f(2) = 3(2) - 4 = 2$ $\quad f(1) = 3(1) - 4 = -1$

Domain and Range

When y is a function of x, the domain of a function is the collection of all possible values for x. The range is the collection of all y-values. When asked to find the domain of a function, think in terms of what can and cannot be done. For now, keep in mind that we cannot divide by 0 and we cannot take an even root of a negative number. For example, in the function $y = 1/x$, we cannot let $x = 0$, so 0 is not in the domain of this function. The domain is the set of all nonzero real numbers. We might also say that the domain is $x \neq 0$.

When asked to find the domain of a function that has x in one or more denominators, we set each denominator (that has x in it) equal to zero and solve for x. These solutions are not in the domain.

EXAMPLE 7-5

Find the domain. Write the domain in interval notation.

- $y = \dfrac{2x}{x - 4}$

Because the denominator has an x in it, we set it equal to zero and solve for x. The solution to $x - 4 = 0$ is $x = 4$. The domain is all real numbers except 4. The interval notation is $(-\infty, 4) \cup (4, \infty)$.

- $y = \dfrac{2x+5}{x^2-x-6} = \dfrac{2x+5}{(x-3)(x+2)}$

$$x - 3 = 0 \qquad\qquad x + 2 = 0$$

$$x = 3 \qquad\qquad x = -2$$

The domain is all real numbers except -2 and 3. The interval notation for $x \neq -2, 3$ is $(-\infty, -2) \cup (-2, 3) \cup (3, \infty)$.

- $y = \dfrac{x^2+x-8}{x^2+1}$

Because $x^2 + 1 = 0$ has no real solution, we can let x be any real number. This means that the domain is all real numbers. The interval notation for *all real numbers* is $(-\infty, \infty)$.

- $y = \dfrac{3}{x^2-4} + 5$

$$x^2 - 4 = 0$$

$$x^2 = 4$$

$$x = \pm 2$$

The domain is all real numbers except 2 and -2: $(-\infty, -2) \cup (-2, 2) \cup (2, \infty)$.

PRACTICE

Find the domain. Write the domain in interval notation.

1. $y = \dfrac{x^2 - 3x + 5}{x + 6}$
2. $y = \dfrac{4}{x^2 + 2x - 8} + 12x$
3. $y = \dfrac{6x}{4x^2 + 1}$
4. $y = 3x - 6 + \dfrac{1}{x} + \dfrac{x}{x+5}$

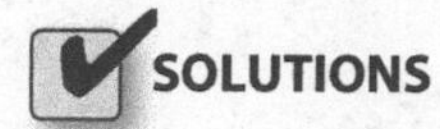

SOLUTIONS

1. **The solution to $x + 6 = 0$ is $x = -6$. The domain is all real numbers except -6: $(-\infty, -6) \cup (-6, \infty)$.**

2. Solve $x^2 + 2x - 8 = 0$

$$x^2 + 2x - 8 = (x + 4)(x - 2)$$

$$x + 4 = 0 \qquad\qquad x - 2 = 0$$

$$x = -4 \qquad\qquad x = 2$$

The domain is all real numbers except -4 and 2: $(-\infty, -4) \cup (-4, 2) \cup (2, \infty)$.

3. Solve $4x^2 + 1 = 0$

$$4x^2 + 1 = 0$$

$$x^2 = -\frac{1}{4}$$

$$x = \pm\sqrt{-\frac{1}{4}} \qquad \sqrt{-\frac{1}{4}} \text{ is not a real number.}$$

There are no real solutions to $4x^2 + 1 = 0$, so the domain is all real numbers: $(-\infty, \infty)$.

4. The domain is all real numbers except 0 and -5 (from $x + 5 = 0$): $(-\infty, -5) \cup (-5, 0) \cup (0, \infty)$.

Functions that have a variable under an even root also might have limited domains. We can find the domain of these functions by setting the expression under the root sign greater than or equal to zero and solving the inequality.

EXAMPLE 7-6

Find the domain. Write the domain in interval notation.

- $y = \sqrt{x - 6}$

$$x - 6 \geq 0$$

$$x \geq 6 \qquad \text{The domain is } x \geq 6\text{: } [6, \infty).$$

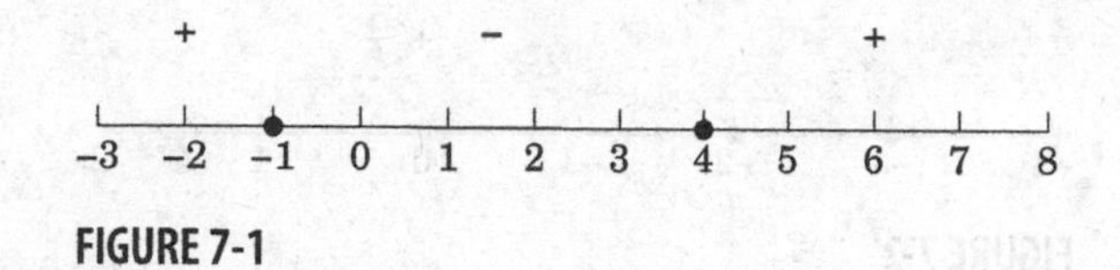

FIGURE 7-1

- $y = \sqrt{16 - 4x}$

$$16 - 4x \geq 0$$

$$-4x \geq -16 \quad \text{Reverse the sign when dividing by } -4.$$

$$x \leq 4 \quad \text{The domain is } x \leq 4\text{: } (-\infty, 4].$$

- $y = \sqrt[4]{x^2 - 3x - 4}$

$$x^2 - 3x - 4 \geq 0 \text{ becomes } (x - 4)(x + 1) \geq 0$$

$$x - 4 = 0 \qquad x + 1 = 0$$

$$x = 4 \qquad x = -1$$

See Fig. 7-1 for the sign graph for this inequality.

Because the inequality is "$\geq$," we want the intervals marked with a plus sign. The domain is $x \leq -1$ or $x \geq 4$: $(-\infty, -1] \cup [4, \infty)$.

- $y = \sqrt[4]{x^2 + 3}$

$$x^2 + 3 \geq 0$$

The inequality $x^2 + 3 \geq 0$ is true for all real numbers, making the domain all real numbers: $(-\infty, \infty)$.

- $y = \sqrt[3]{5x - 4}$

Because we can take odd roots of negative numbers, the domain for this function is all real numbers: $(-\infty, \infty)$.

PRACTICE

Find the domain. Write the domain in interval notation.

1. $y = \sqrt{6x - 8}$
2. $y = \sqrt[5]{4x + 9}$
3. $y = \sqrt{x^2 + 5x + 6}$
4. $y = \sqrt[6]{x^2 + 1}$

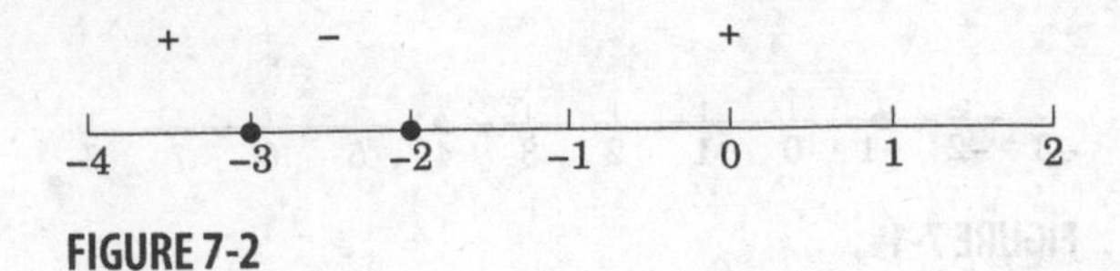

FIGURE 7-2

SOLUTIONS

1. $$6x - 8 \geq 0$$

$$x \geq \frac{8}{6} = \frac{4}{3} \qquad \text{The domain is } x \geq \frac{4}{3}\text{: } \left[\frac{4}{3}, \infty\right).$$

2. Because the fifth root is odd, the domain is all real numbers: $(-\infty, \infty)$.

3. $x^2 + 5x + 6 \geq 0$ becomes $(x + 2)(x + 3) \geq 0$.

$$x + 2 = 0 \qquad\qquad x + 3 = 0$$

$$x = -2 \qquad\qquad x = -3$$

See the sign graph in Fig. 7-2. The domain is $(-\infty, -3] \cup [-2, \infty)$.

4. Because $x^2 + 1 \geq 0$ is true for all real numbers, the domain is all real numbers: $(-\infty, \infty)$.

Some functions are combinations of different kinds of functions. The domain of a combination of two or more kinds of functions is the set of all x-values that are possible for each part. The function

$$y = \frac{\sqrt{x + 4}}{x + 3}$$

is made up of the parts $\sqrt{x + 4}$ and $\frac{1}{x+3}$. For $\sqrt{x + 4}$, we need $x \geq -4$. For $\frac{1}{x+3}$, we need $x \neq -3$.

As we can see from the shaded region in Fig. 7-3, the domain for the function is $[-4, -3) \cup (-3, \infty)$.

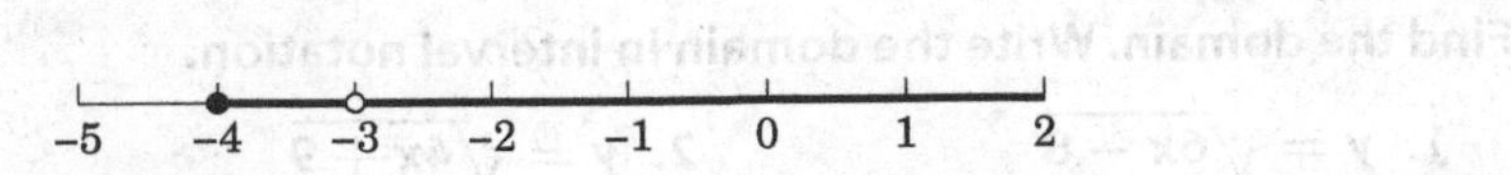

FIGURE 7-3

EXAMPLE 7-7

Find the domain. Write the domain in interval notation.

- $y = \dfrac{x}{\sqrt[4]{x-7}}$

Because $\sqrt[4]{x-7}$ is in the denominator, $\sqrt[4]{x-7}$ cannot be zero. Because $x-7$ is under an even root, it cannot be negative. Putting these two together means that $x-7>0$ (instead of $x-7 \geq 0$). The domain of this function is $x>7$: $(7, \infty)$.

PRACTICE

Find the domain. Write the domain in interval notation.

1. $y = \dfrac{\sqrt{2x+5}}{x-6}$
2. $y = \dfrac{1}{\sqrt{3-2x}}$
3. $y = \dfrac{\sqrt[4]{x-2}}{x^2-x-12}$
4. $y = \sqrt{x^2+3x-18} + \dfrac{1}{x-5}$

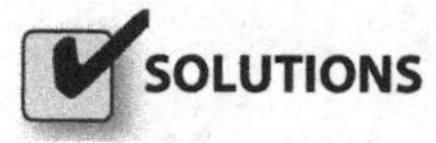

SOLUTIONS

1. **These regions are sketched in Fig. 7-4.**

$$2x+5 \geq 0 \quad \text{and} \quad x-6 \neq 0$$

$$x \geq -\frac{5}{2} \qquad\qquad x \neq 6$$

The domain is $[-\frac{5}{2}, 6) \cup (6, \infty)$.

2. $$3-2x>0$$

$$x<\frac{3}{2}$$

The domain is $(-\infty, \frac{3}{2})$.

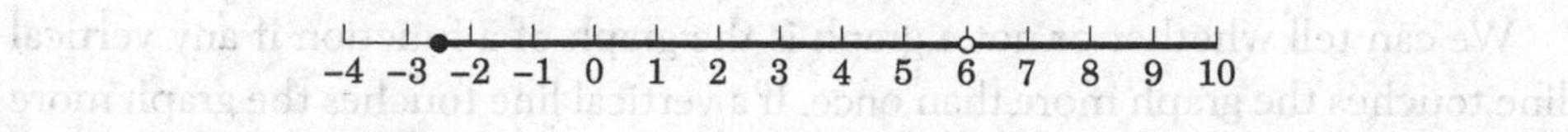

FIGURE 7-4

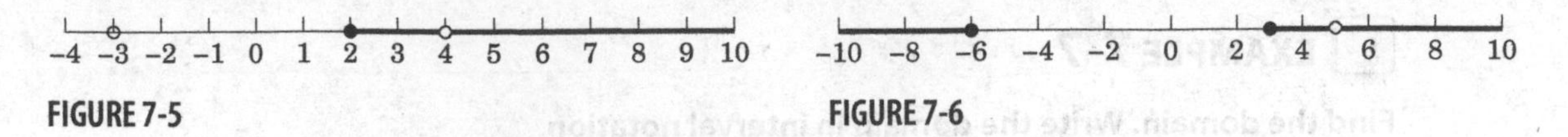

FIGURE 7-5

FIGURE 7-6

3. These regions are sketched in Fig. 7-5.

$$x - 2 \geq 0 \quad \text{and} \quad x^2 - x - 12 \neq 0$$

$$x \geq 2 \qquad (x-4)(x+3) \neq 0$$

$$x - 4 \neq 0 \quad x + 3 \neq 0$$

$$x \neq 4 \text{ and } x \neq -3$$

The domain is $[2, 4) \cup (4, \infty)$. (Because $x \geq 2$, $x = -3$ is not in the domain, anyway.)

4. These regions are sketched in Fig. 7-6.

$$x^2 + 3x - 18 \geq 0 \text{ and } x - 5 \neq 0$$

$$(x+6)(x-3) \geq 0 \qquad x \neq 5$$

The domain is $(-\infty, -6] \cup [3, 5) \cup (5, \infty)$.

Functions and Their Graphs

Reading graphs, sketching graphs by hand, and sketching graphs using graphing calculators are all important in today's algebra courses. We will concentrate on reading graphs in this section. Later, we learn strategies for sketching the graphs for different families of functions.

A graph can give us a great deal of information about a function. First, it can tell us if the graph is the graph of a function. Remember, if y is a function of x, then each x-value has exactly one y-value. What if we have a graph where an x-value has two or more y-values? A vertical line through that particular x-value touches the graph at more than one point. For example, a vertical line would touch the graph at both $(4, 2)$ and $(4, -2)$ for the graph in Fig. 7-7.

We can tell whether or not a graph is the graph of a function if any vertical line touches the graph more than once. If a vertical line touches the graph more than once (as is the case with the graph in Fig. 7-7), then the graph is not the

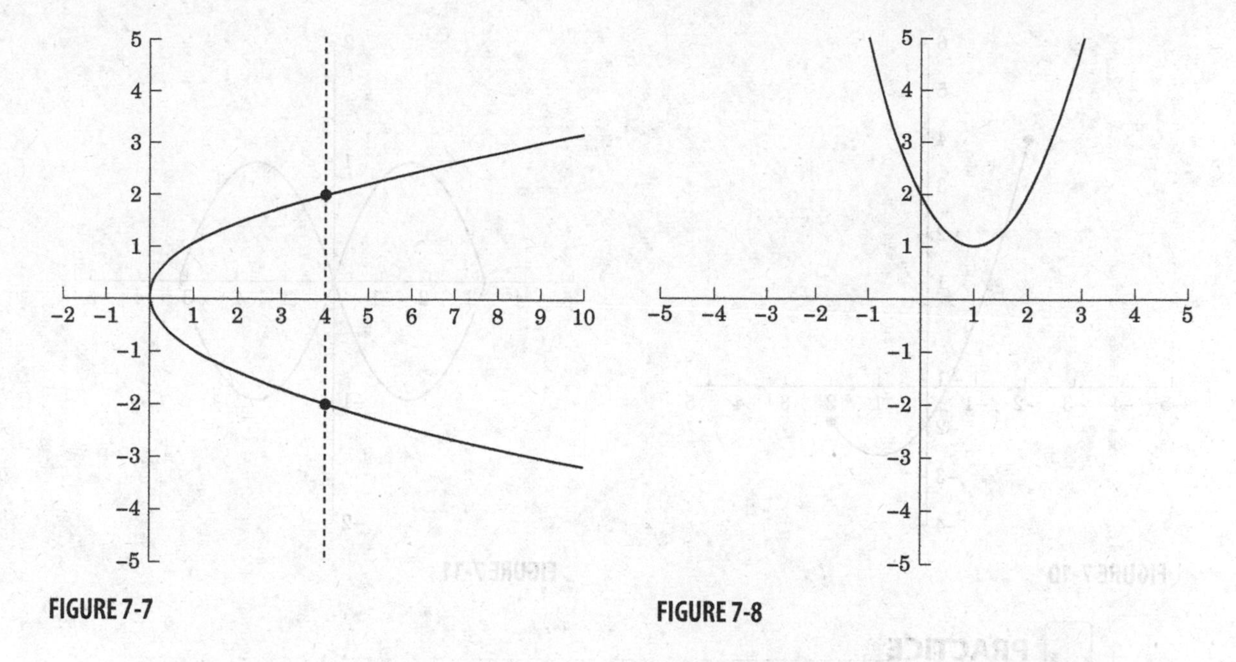

FIGURE 7-7

FIGURE 7-8

graph of a function. If every vertical line touches the graph at most once, then the graph is the graph of a function. This is called the *Vertical Line Test*. The graphs in Figs. 7-8 and 7-9 pass the Vertical Line Test, so the graphs are graphs of functions.

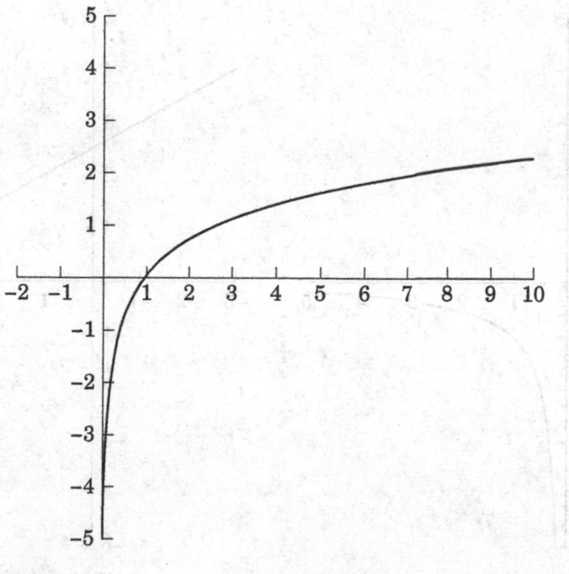

FIGURE 7-9

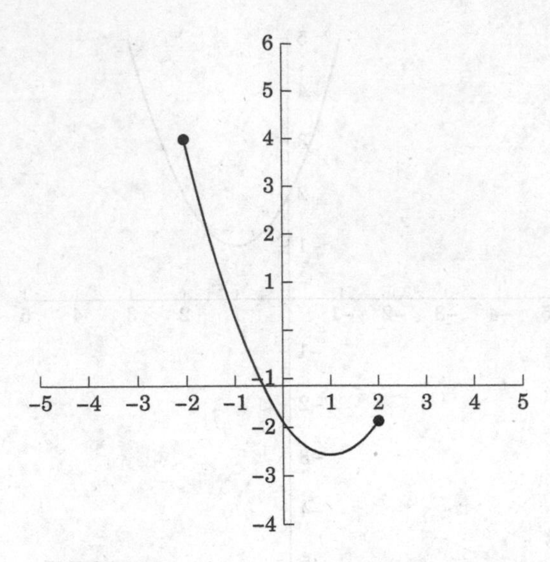

FIGURE 7-10

FIGURE 7-11

PRACTICE

Use the Vertical Line Test to determine which of the graphs in Figs. 7-10–7-14 are graphs of functions.

1. **See Fig. 7-10.**
2. **See Fig. 7-11.**
3. **See Fig. 7-12.**
4. **See Fig. 7-13.**
5. **See Fig. 7-14.**

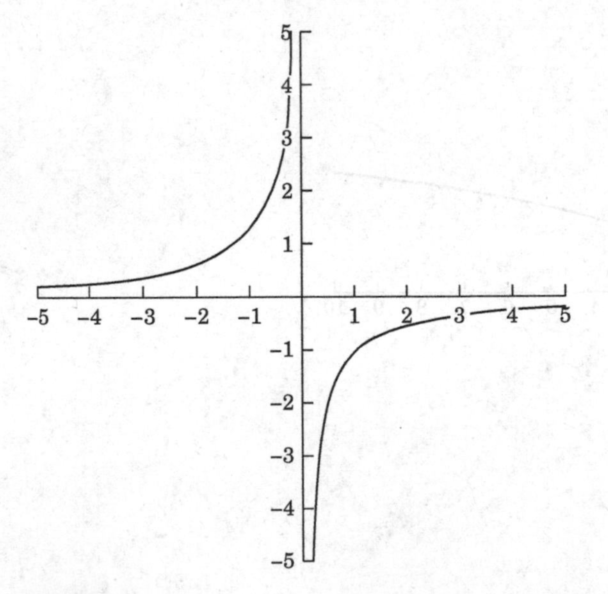

FIGURE 7-12

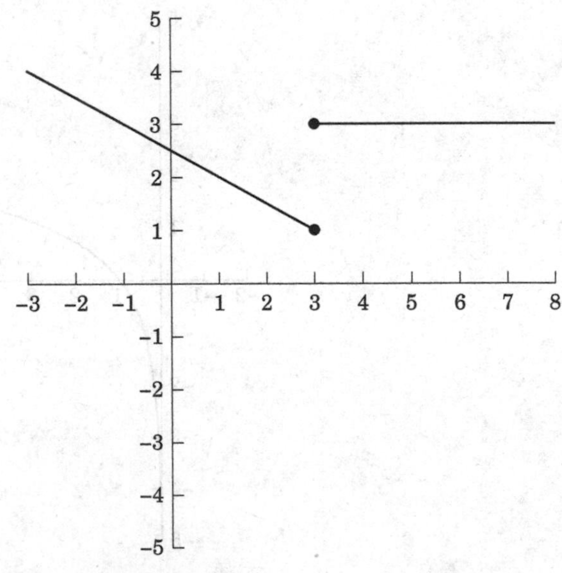

FIGURE 7-13

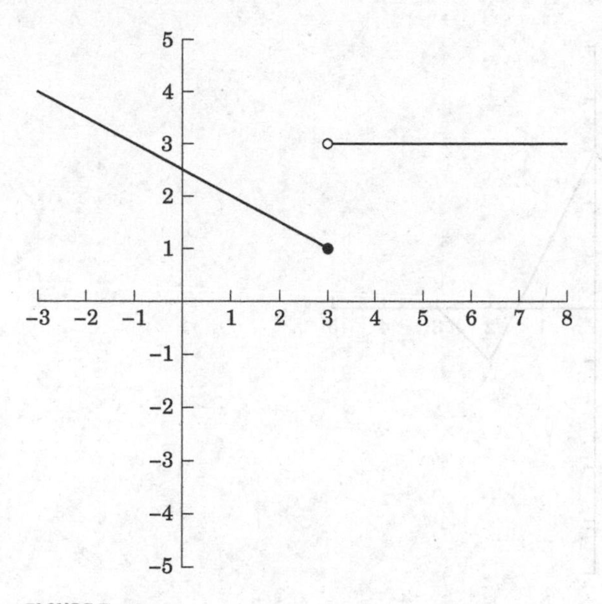

FIGURE 7-14

FIGURE 7-15

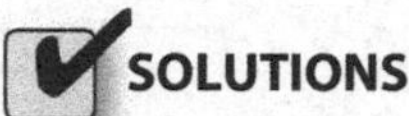

SOLUTIONS

1. Yes 2. No 3. Yes 4. No (two points for $x = 3$)

5. Yes (only one point for $x = 3$)

Graphs are also useful for evaluating functions. Remember that points on the graph are pairs of numbers, x (the horizontal distance from the origin) and y (the vertical distance from the origin). Normally, y and $f(x)$ are the same. The point $(-1, 1)$ on the graph in Fig. 7-15 means that $f(-1) = 1$. The point $(2, 4)$ on the graph means that $f(2) = 4$. What is $f(0)$? In other words, when $x = 0$, what is y? Because the point $(0, 0)$ is on the graph, $f(0) = 0$.

EXAMPLE 7-8

Refer to Fig. 7-16.

- **Find $f(-3)$.**

Another way of saying, "Find $f(-3)$" is saying, "What is the y-value for the point on the graph for $x = -3$?" The point $(-3, 2)$ is on the graph, so $f(-3) = 2$.

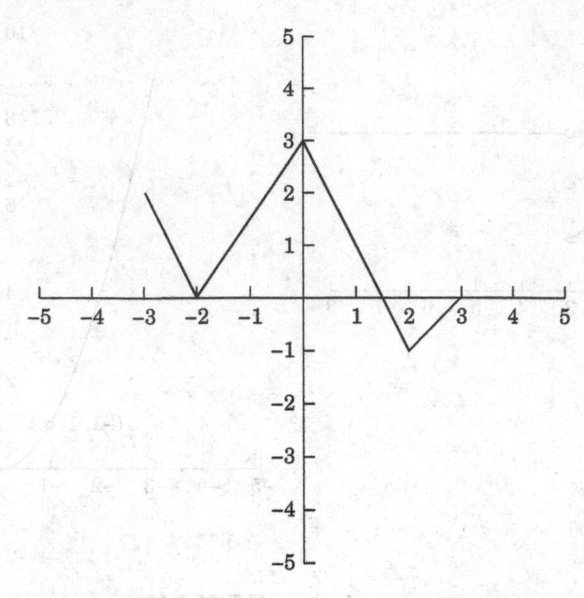

FIGURE 7-16

- Find $f(-2)$.

We look for the point on the graph for $x = -2$. The point $(-2, 0)$ is on the graph, so $f(-2) = 0$.

- Find $f(0)$.

We look for the point on the graph for $x = 0$. The point $(0, 3)$ is on the graph, so $f(0) = 3$.

PRACTICE

Refer to Fig. 7-17.

1. Find $f(0)$. 2. Find $f(3)$. 3. Find $f(-2)$. 4. Find $f(-3)$.

SOLUTIONS

1. The point $(0, 2)$ is on the graph, so $f(0) = 2$.
2. The point $(3, 4)$ is on the graph, so $f(3) = 4$.

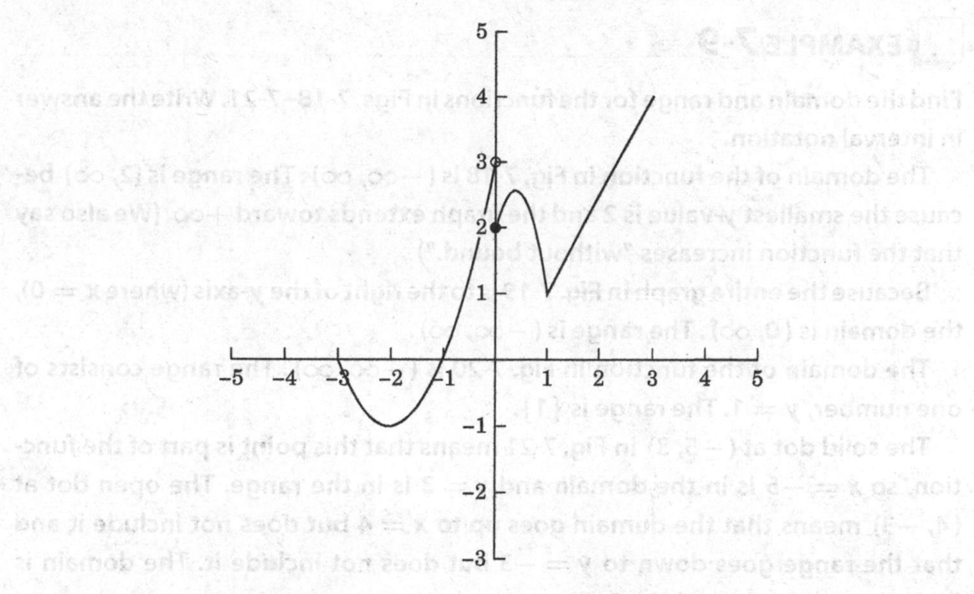

FIGURE 7-17

3. The point $(-2, -1)$ is on the graph, so $f(-2) = -1$.

4. The point $(-3, 0)$ is on the graph, so $f(-3) = 0$.

Finding the Domain and Range Graphically

The graph of a function can tell us what its domain and range are. Remember that the domain of a function is the set of x-values that can be used in the function. We can find the domain from the graph of the function by observing how far the graph extends horizontally. The range of a function is the set of y-values. We can find the range from the graph by observing how far the graph extends vertically.

A note about the graphs in this book (and in many books): Assume that the ends of the graph have arrows unless an endpoint is marked with a solid dot or an open dot.

EXAMPLE 7-9

Find the domain and range for the functions in Figs. 7-18–7-21. Write the answer in interval notation.

The domain of the function in Fig. 7-18 is $(-\infty, \infty)$. The range is $[2, \infty)$ because the smallest y-value is 2 and the graph extends toward $+\infty$. (We also say that the function increases "without bound.")

Because the entire graph in Fig. 7-19 is to the right of the y-axis (where $x = 0$), the domain is $(0, \infty)$. The range is $(-\infty, \infty)$.

The domain of the function in Fig. 7-20 is $(-\infty, \infty)$. The range consists of one number, $y = 1$. The range is $\{1\}$.

The solid dot at $(-5, 3)$ in Fig. 7-21 means that this point is part of the function, so $x = -5$ is in the domain and $y = 3$ is in the range. The open dot at $(4, -3)$ means that the domain goes up to $x = 4$ but does not include it and that the range goes down to $y = -3$ but does not include it. The domain is $[-5, 4)$, and the range is $(-3, 3]$.

PRACTICE

Find the domain and range for the function. Write the answer using interval notation.

1. See Fig. 7-22.
2. See Fig. 7-23.
3. See Fig. 7-24.
4. See Fig. 7-25.
5. See Fig. 7-26.

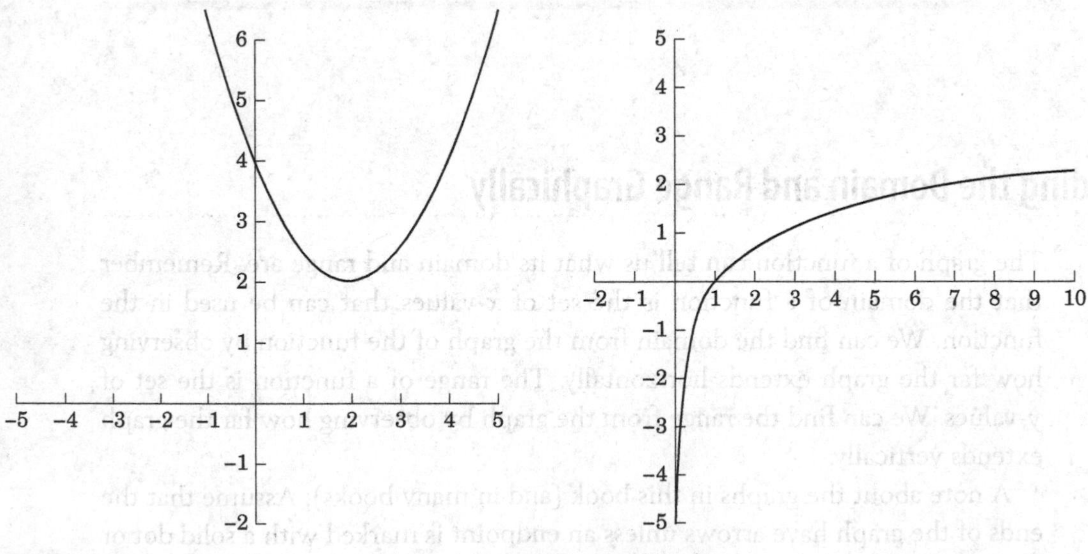

FIGURE 7-18

FIGURE 7-19

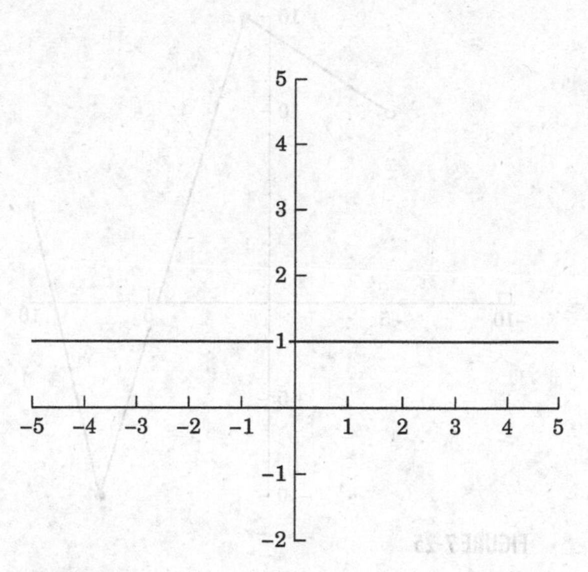

FIGURE 7-20

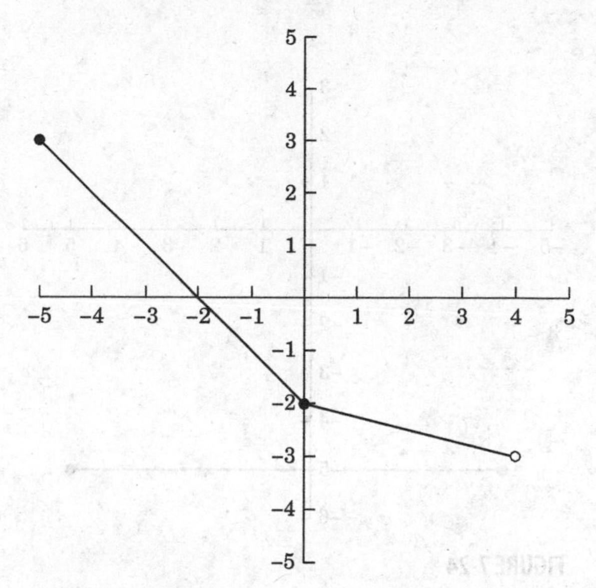

FIGURE 7-21

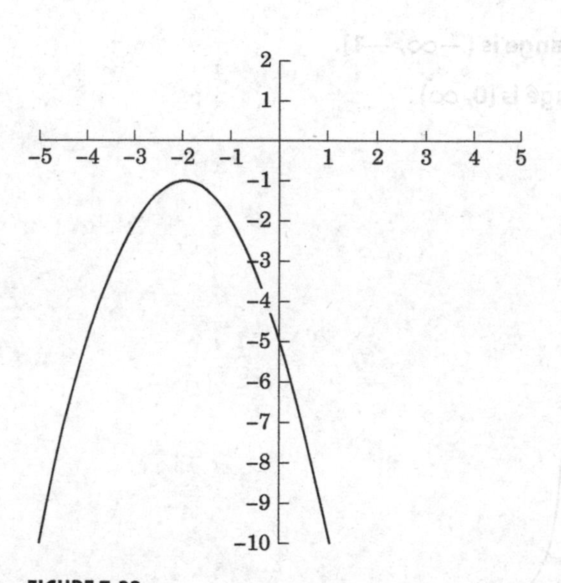

FIGURE 7-22

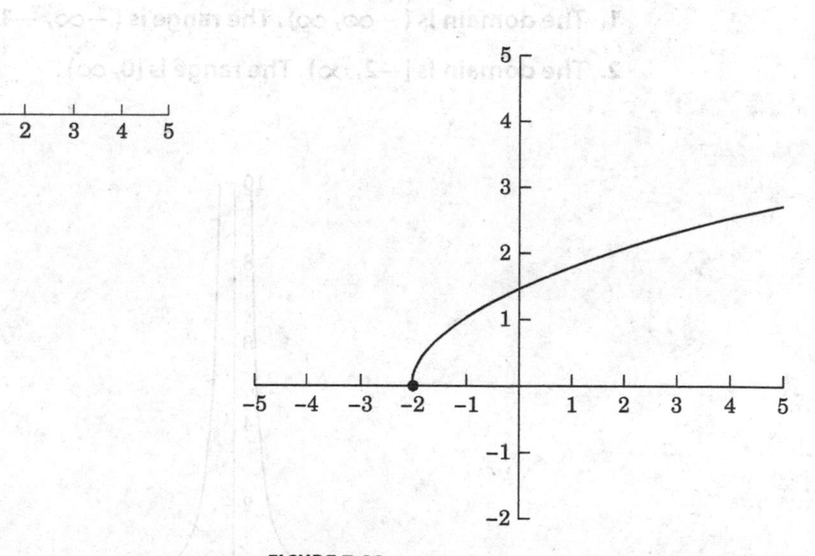

FIGURE 7-23

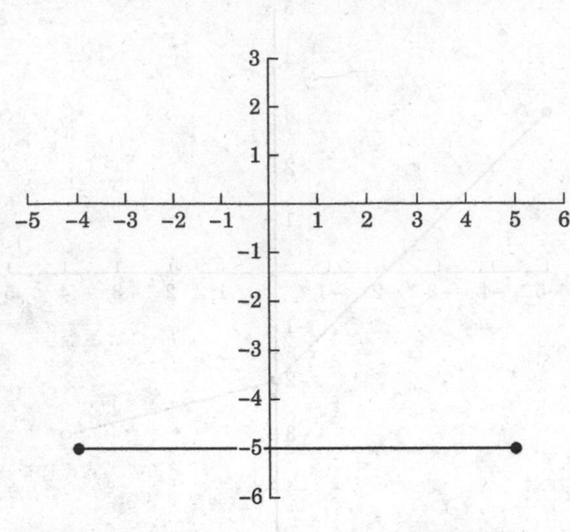

FIGURE 7-24

FIGURE 7-25

SOLUTIONS

1. The domain is $(-\infty, \infty)$. The range is $(-\infty, -1]$.
2. The domain is $[-2, \infty)$. The range is $[0, \infty)$.

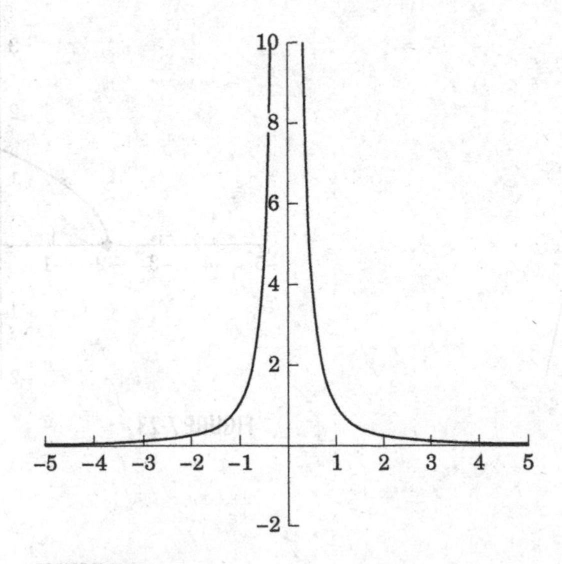

FIGURE 7-26

3. The domain is $[-4, 5]$. The range is $\{-5\}$.

4. The domain is $(-5, 10]$. The range is $[-10, 15]$.

5. The domain is $(-\infty, 0) \cup (0, \infty)$. The range is $(0, \infty)$.

Increasing Intervals and Decreasing Intervals

Graphs can tell us where functions are going up (if anywhere) and where they are going down (if anywhere). A function is said to be *increasing* on an interval if, as we move from left to right in the interval, the y-values are going up. A function is said to be *decreasing* on an interval if, as we move from left to right in the interval, the y-values are going down. A function is constant on an interval if, as we move from left to right in the interval, the y-values do not change. Refer to the graph in Fig. 7-27.

If we are anywhere to the left of $x = -1$ and move to the right, the graph is going down. We say the function is decreasing on the interval $(-\infty, -1)$. If we are anywhere between $x = -1$ and $x = 0$ and move to the right, the graph is going up. We say the function is increasing on the interval $(-1, 0)$. If we are anywhere between $x = 0$ and $x = 1$ and move to the right, the graph is going back down. We say the function is decreasing on the interval $(0, 1)$. Finally, if

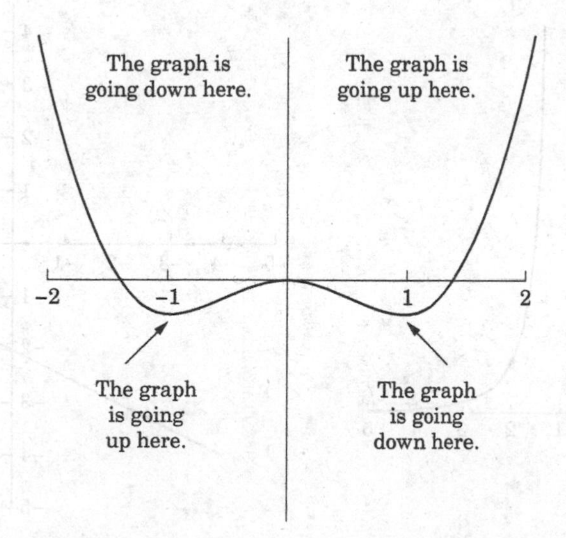

FIGURE 7-27

we are anywhere to the right of $x = 1$ and move to the right, the function is going back up. We say the function is increasing on the interval $(1, \infty)$.

EXAMPLE 7-10

Determine where the functions in Figs. 7-28–7-31 are increasing, decreasing, or constant.

This function whose graph is in Fig. 7-28 is increasing to the left of $x = 1$, $(-\infty, 1)$. It is decreasing to the right of $x = 1$, $(1, \infty)$.

No matter where we are on the graph in Fig. 7-29, as we move to the right, the graph is going up, so this function is increasing everywhere $(-\infty, \infty)$.

The function whose graph is in Fig. 7-30 is decreasing on all its domain, $(-\infty, 0)$.

The function whose graph is in Fig. 7-31 is decreasing on $(-4, -2)$ (between $x = -4$ and $x = -2$), increasing on $(-2, -1)$ (between $x = -2$ and $x = -1$), constant on $(-1, 2)$ (between $x = -1$ and $x = 2$), and decreasing on $(2, 4)$ (between $x = 2$ and $x = 4$).

You might wonder why we use parentheses for these intervals. Actually, there is no agreement at this level whether to use parentheses or brackets. Technically, the function is neither increasing nor decreasing at the "turning around points," a concept that is introduced in calculus.

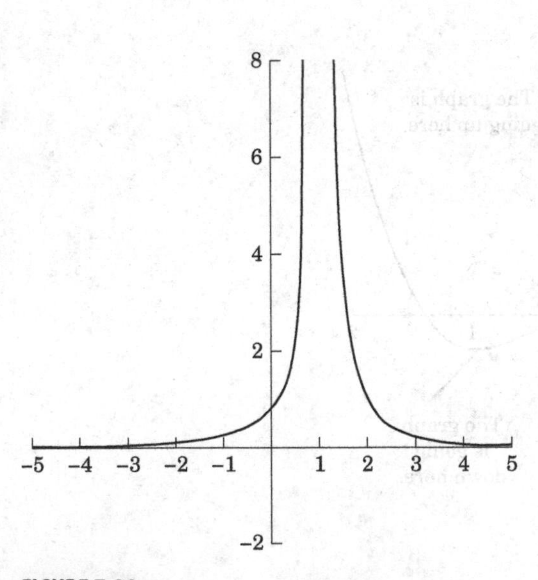

FIGURE 7-28

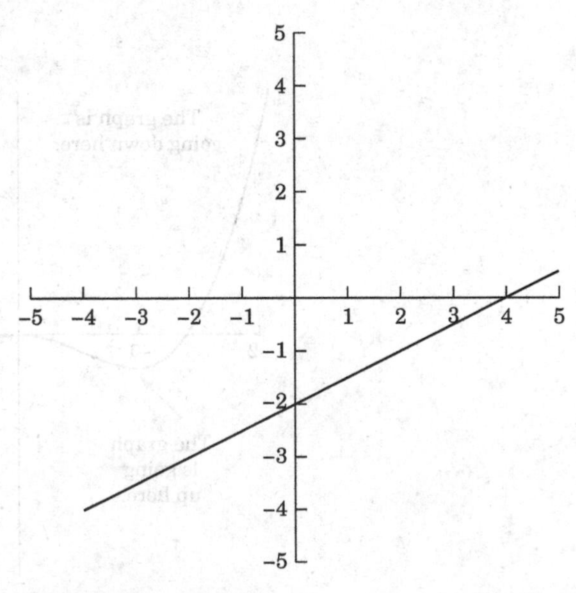

FIGURE 7-29

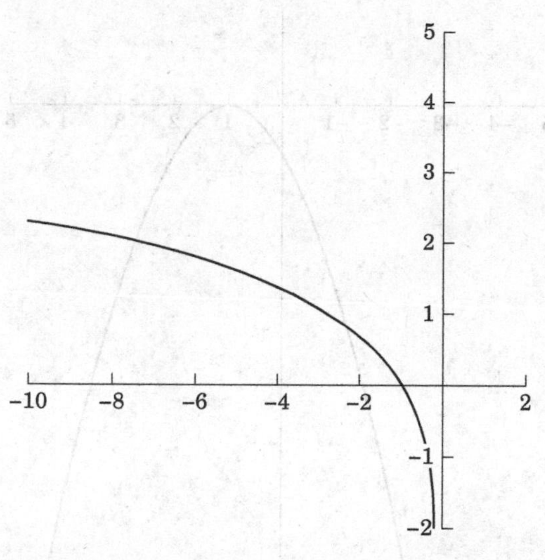

FIGURE 7-30

FIGURE 7-31

PRACTICE

Determine where the following functions are increasing, decreasing, or constant.

1. See Fig. 7-32. 2. See Fig. 7-33. 3. See Fig. 7-34. 4. See Fig. 7-35.

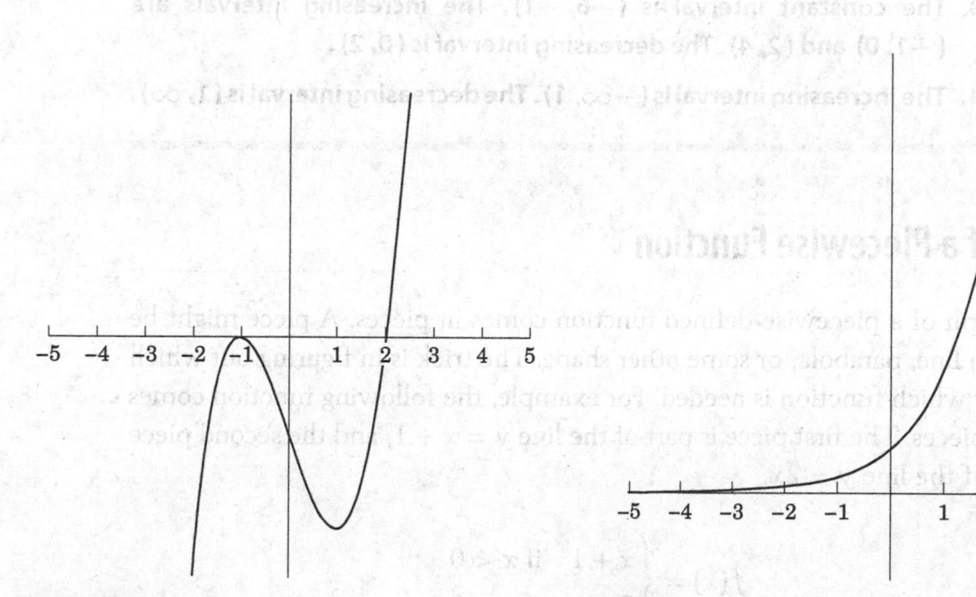

FIGURE 7-32

FIGURE 7-33

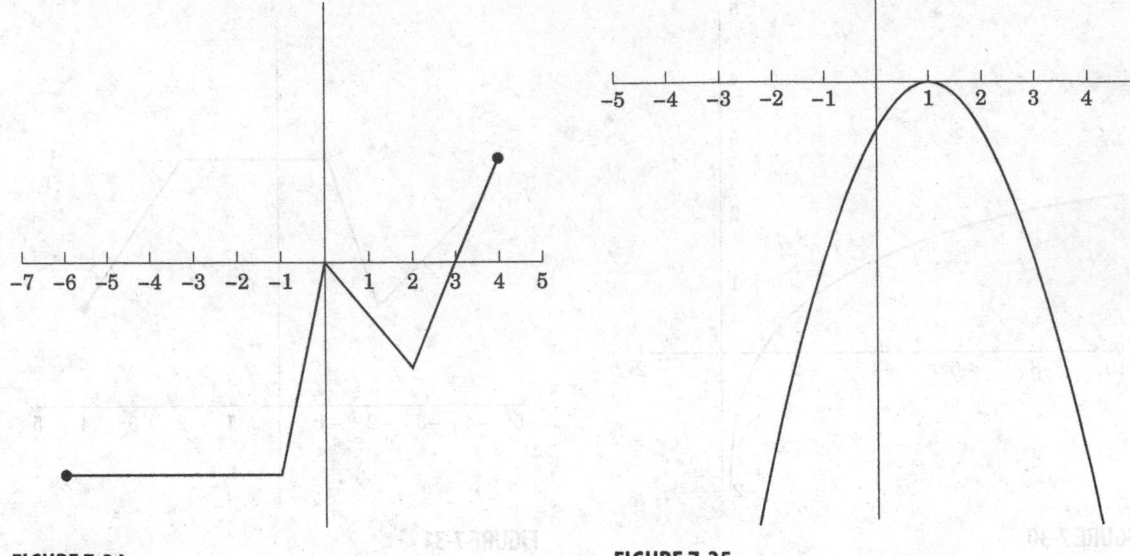

FIGURE 7-34

FIGURE 7-35

SOLUTIONS

1. **The increasing intervals are $(-\infty, -1)$ and $(1, \infty)$. The decreasing interval is $(-1, 1)$.**
2. **The function is increasing everywhere, $(-\infty, \infty)$.**
3. **The constant interval is $(-6, -1)$. The increasing intervals are $(-1, 0)$ and $(2, 4)$. The decreasing interval is $(0, 2)$.**
4. **The increasing interval is $(-\infty, 1)$. The decreasing interval is $(1, \infty)$.**

The Graph of a Piecewise Function

The graph of a piecewise-defined function comes in pieces. A piece might be part of a line, parabola, or some other shape. The trick is in figuring out which piece of which function is needed. For example, the following function comes in two pieces. The first piece is part of the line $y = x + 1$, and the second piece is part of the line $y = 2x$.

$$f(x) = \begin{cases} x + 1 & \text{if } x < 0 \\ 2x & \text{if } x \geq 0 \end{cases}$$

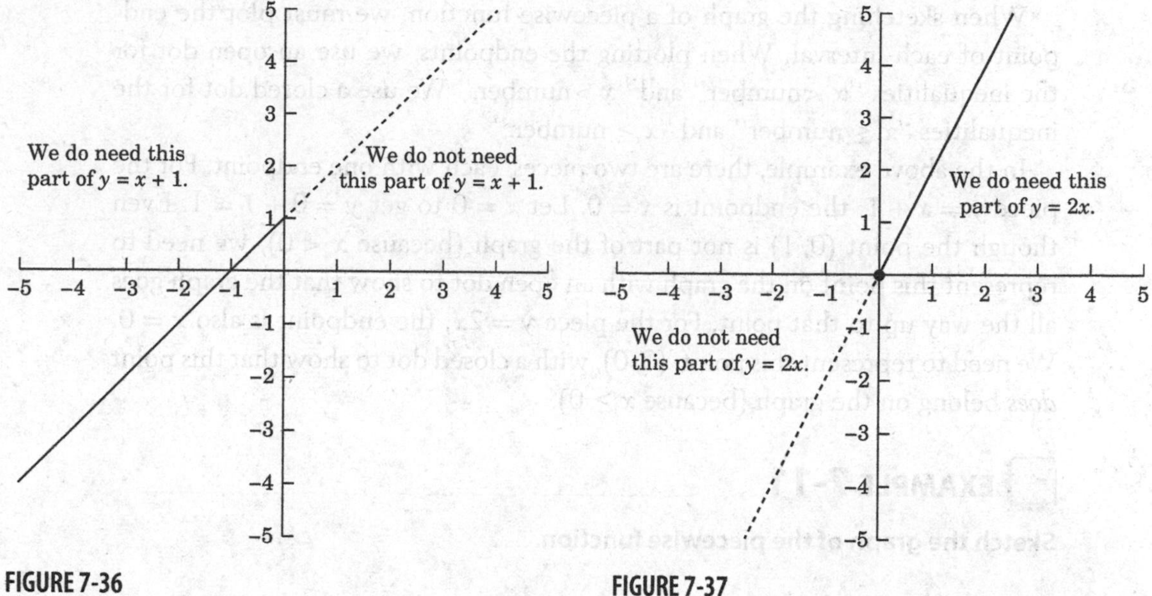
We do need this
part of y = x + 1.
We do not need
this part of y = x + 1.
We do need this
part of y = 2x.
We do not need
this part of y = 2x.

FIGURE 7-36

FIGURE 7-37

Because "$x < 0$" is written to the right of "$x + 1$," the part of the line $y = x + 1$ we need is to the left of $x = 0$. (See Fig. 7-36.)

Because "$x \geq 0$" is written to the right of "$2x$," the part of the line $y = 2x$ we need is to the right of $x = 0$. (See Fig. 7-37.)

The graph of the function is shown in Fig. 7-38.

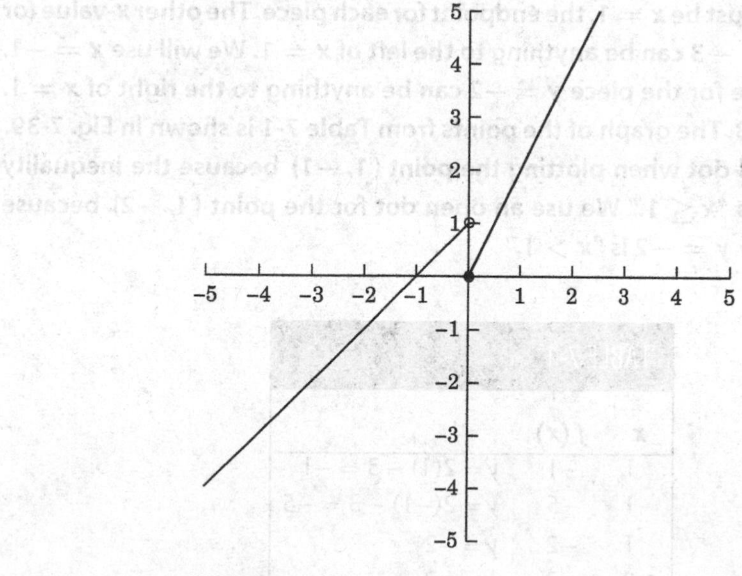

FIGURE 7-38

When sketching the graph of a piecewise function, we must plot the endpoint of each interval. When plotting the endpoints, we use an open dot for the inequalities "$x <$number" and "$x >$number." We use a closed dot for the inequalities "$x \leq$ number" and "$x \geq$ number."

In the above example, there are two pieces, each with one endpoint. For the piece $y = x + 1$, the endpoint is $x = 0$. Let $x = 0$ to get $y = 0 + 1 = 1$. Even though the point (0, 1) is not part of the graph (because $x < 0$), we need to represent this point on the graph with an open dot to show that the graph goes all the way up to that point. For the piece $y = 2x$, the endpoint is also $x = 0$. We need to represent this point, (0, 0), with a closed dot to show that this point *does* belong on the graph (because $x \geq 0$).

EXAMPLE 7-11

Sketch the graph of the piecewise function.

- $f(x) = \begin{cases} 2x - 3 & \text{if } x \leq 1; \\ -2 & \text{if } x > 1 \end{cases}$

This graph comes in two pieces. One piece is part of the line $y = 2x - 3$, and the other piece is part of the horizontal line $y = -2$. We will start by making a table of values, part of the table for $y = 2x - 3$ and the other part for $y = -2$. Because each piece is a line, we only need to plot two points for each piece. One of these points must be $x = 1$, the endpoint for each piece. The other x-value for the piece $y = 2x - 3$ can be anything to the left of $x = 1$. We will use $x = -1$. The other x-value for the piece $y = -2$ can be anything to the right of $x = 1$. We will use $x = 3$. The graph of the points from Table 7-1 is shown in Fig. 7-39.

We use a solid dot when plotting the point $(1, -1)$ because the inequality for $y = 2x - 3$ is "$x \leq 1$." We use an open dot for the point $(1, -2)$ because the inequality for $y = -2$ is "$x > 1$."

TABLE 7-1

x	$f(x)$	
1	−1	$y = 2(1) - 3 = -1$
−1	−5	$y = 2(-1) - 3 = -5$
1	−2	$y = -2$
3	−2	$y = -2$

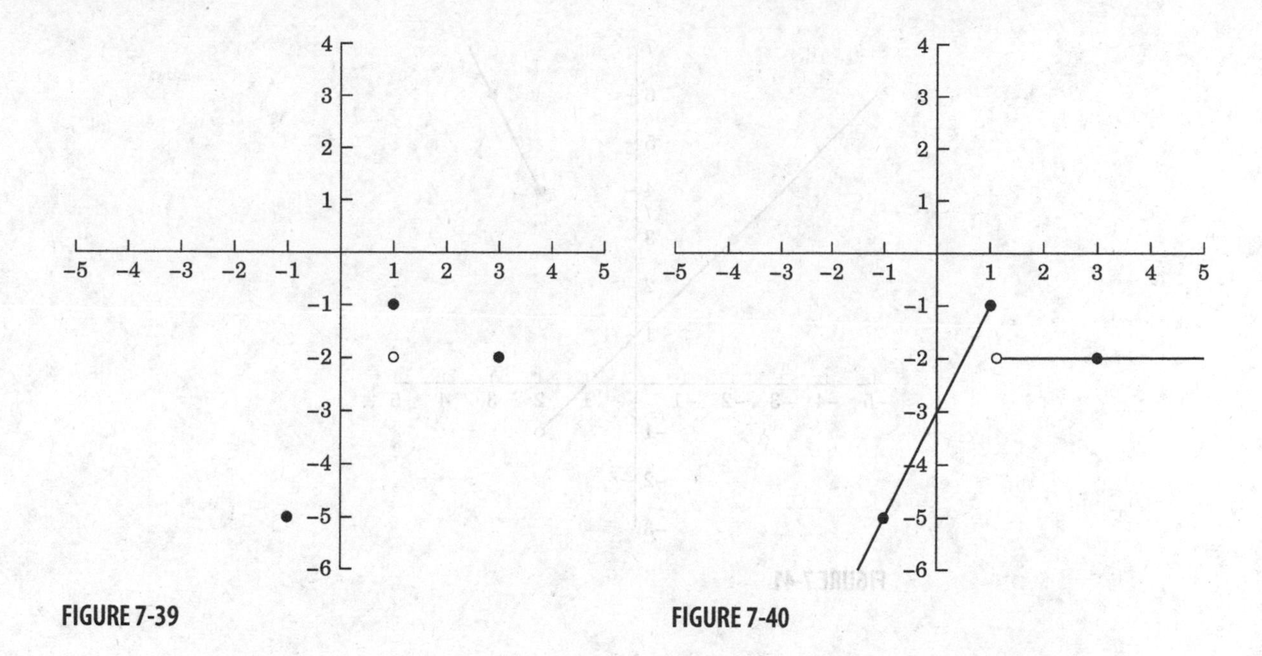

FIGURE 7-39

FIGURE 7-40

Now we draw a line starting at $(1, -1)$ through $(-1, -5)$ and another line starting at $(1, -2)$ through $(3, -2)$. See the graph of the function in Fig. 7-40.

- $f(x) = \begin{cases} 1 - x & \text{if } x < 2; \\ 2x & \text{if } x \geq 2 \end{cases}$

Each piece of this function is part of a line, so we plot two points for each piece. For the piece $y = 1 - x$, we plot the point for $x = 2$ (because it is an endpoint) and any point to the left of $x = 2$. We plot a point for $x = -2$. For the piece $y = 2x$, we plot the point for $x = 2$ (because it is an endpoint) and any point to the right of $x = 2$. We plot a point for $x = 3$. The table of values is in Table 7-2, and the graph is shown in Fig. 7-41.

TABLE 7-2

x	$f(x)$	
2	−1	$y = 1 - 2 = -1$
−2	3	$y = 1 - (-2) = 3$
2	4	$y = 2(2) = 4$
3	6	$y = 2(3) = 6$

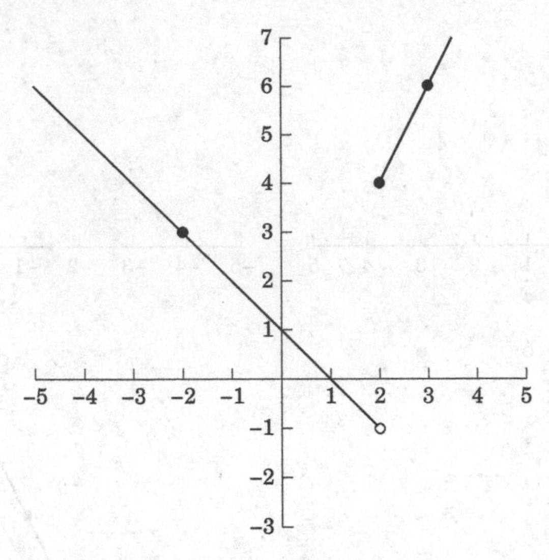

FIGURE 7-41

PRACTICE

Sketch the graph of the piecewise function.

1. $f(x) = \begin{cases} \frac{1}{2}x + 1 & \text{if } x \leq 0; \\ x - 2 & \text{if } x > 0 \end{cases}$

2. $f(x) = \begin{cases} -3 & \text{if } x \leq 2; \\ 2x - 5 & \text{if } x > 2 \end{cases}$

3. $f(x) = \begin{cases} -x & \text{if } x < 0; \\ x & \text{if } x \geq 0 \end{cases}$ (This is another way of writing the function $f(x) = |x|$.)

SOLUTIONS

1. See Fig. 7-42. 2. See Fig. 7-43. 3. See Fig. 7-44.

One or both pieces of the functions in this section are parts of quadratic functions. Our knowledge of parabolas will help to graph these piecewise functions. At first, it might be easier to sketch the graph of the entire parabola and then erase the part that we do not need.

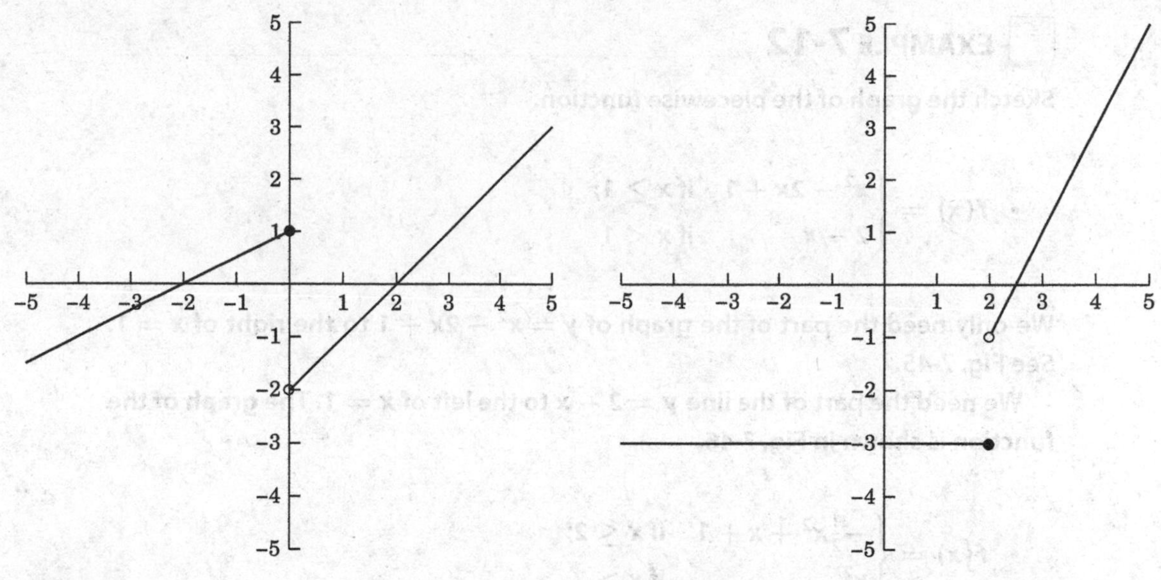

FIGURE 7-42

FIGURE 7-43

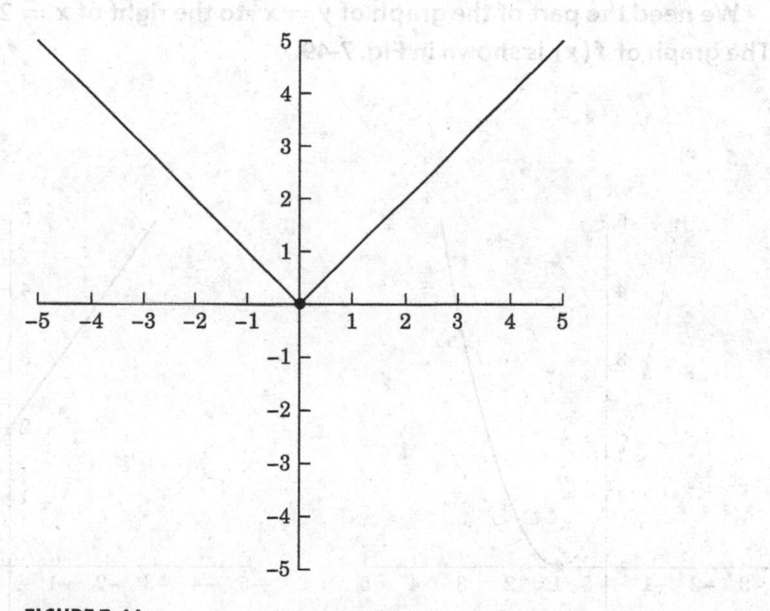

FIGURE 7-44

EXAMPLE 7-12

Sketch the graph of the piecewise function.

- $f(x) = \begin{cases} x^2 - 2x + 1 & \text{if } x \geq 1; \\ 2 - x & \text{if } x < 1 \end{cases}$

We only need the part of the graph of $y = x^2 - 2x + 1$ to the right of $x = 1$. See Fig. 7-45.

We need the part of the line $y = 2 - x$ to the left of $x = 1$. The graph of the function is shown in Fig. 7-46.

- $f(x) = \begin{cases} -\frac{1}{2}x^2 + x + 1 & \text{if } x \leq 2; \\ x^2 & \text{if } x > 2 \end{cases}$

We need the part of the graph of $y = -\frac{1}{2}x^2 + x + 1$ to the left of $x = 2$. See Fig. 7-47.

We need the part of the graph of $y = x^2$ to the right of $x = 2$. See Fig. 7-48. The graph of $f(x)$ is shown in Fig. 7-49.

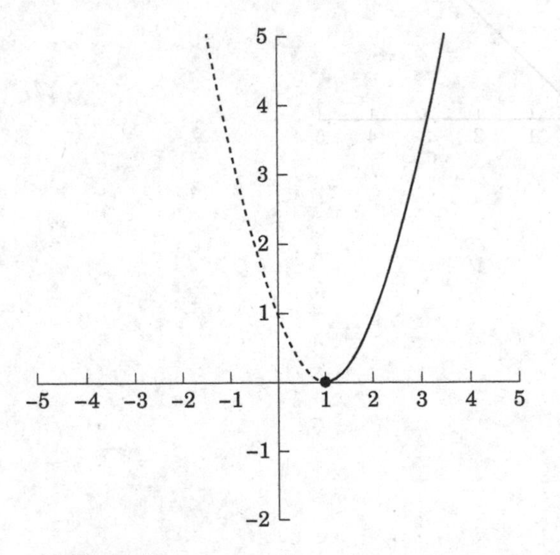

FIGURE 7-45

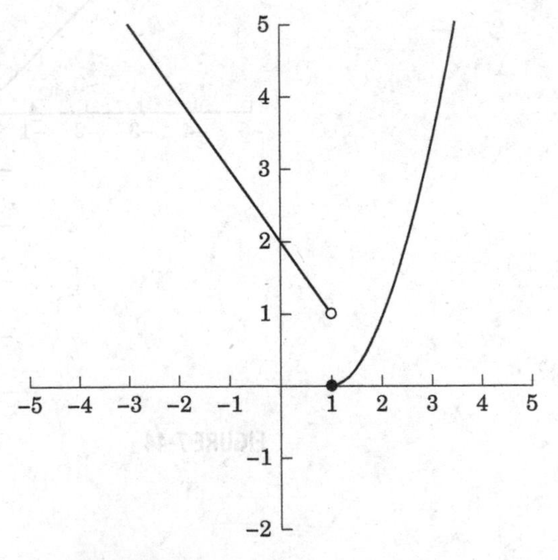

FIGURE 7-46

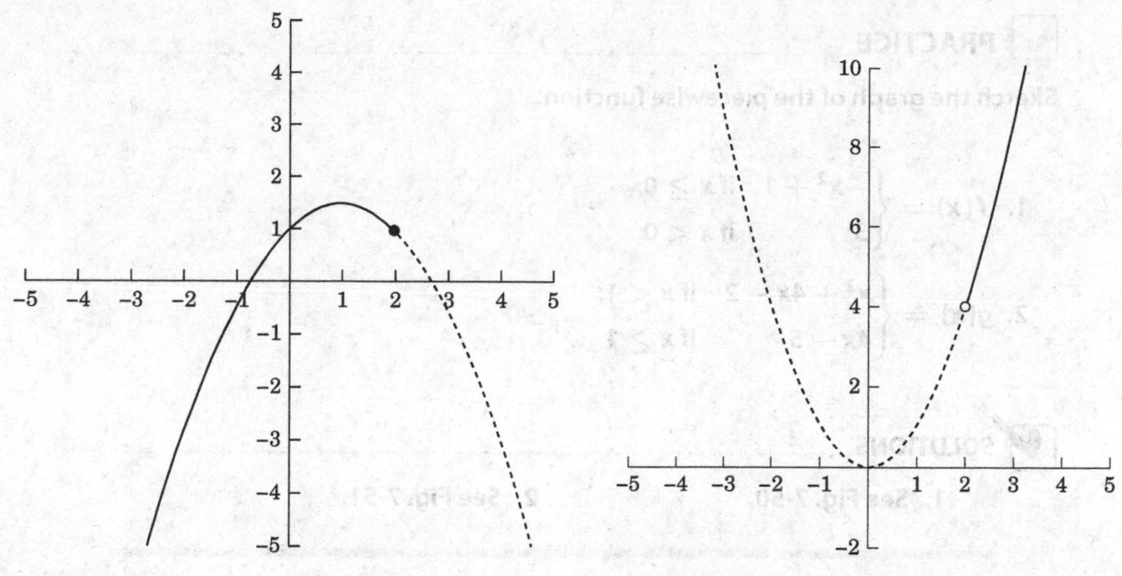

FIGURE 7-47

FIGURE 7-48

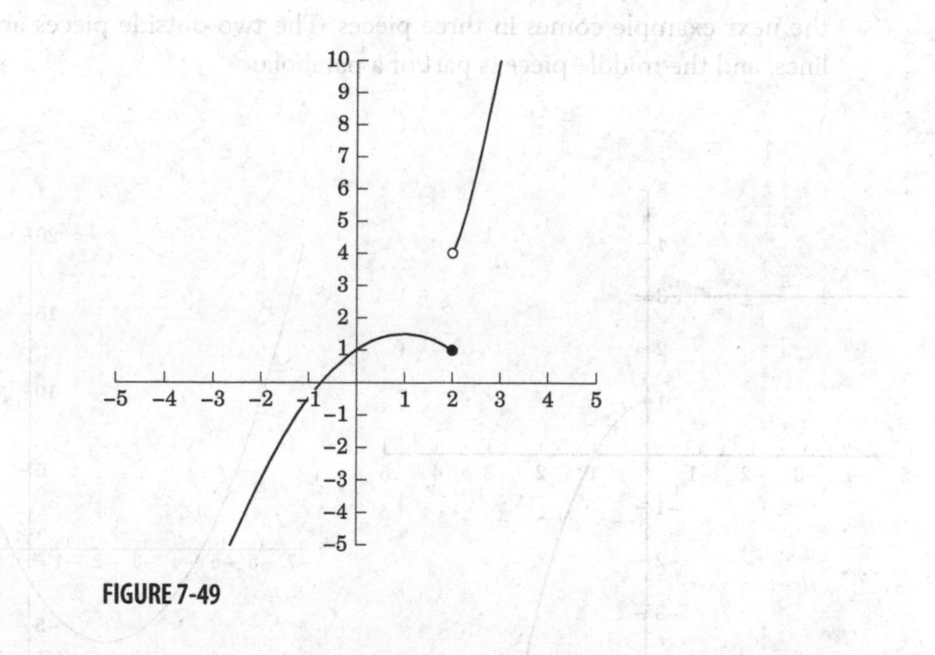

FIGURE 7-49

PRACTICE

Sketch the graph of the piecewise function.

1. $f(x) = \begin{cases} -x^2 + 1 & \text{if } x \geq 0; \\ 3 & \text{if } x < 0 \end{cases}$

2. $g(x) = \begin{cases} x^2 + 4x - 2 & \text{if } x < 1; \\ 4x - 5 & \text{if } x \geq 1 \end{cases}$

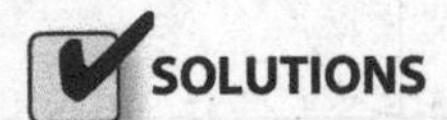

SOLUTIONS

1. See Fig. 7-50.
2. See Fig. 7-51.

Piecewise functions can come in any number of pieces. The same rules apply. The endpoints of each piece must be plotted. If the function comes in three pieces, both endpoints of the middle piece must be plotted. The function in the next example comes in three pieces. The two outside pieces are parts of lines, and the middle piece is part of a parabola.

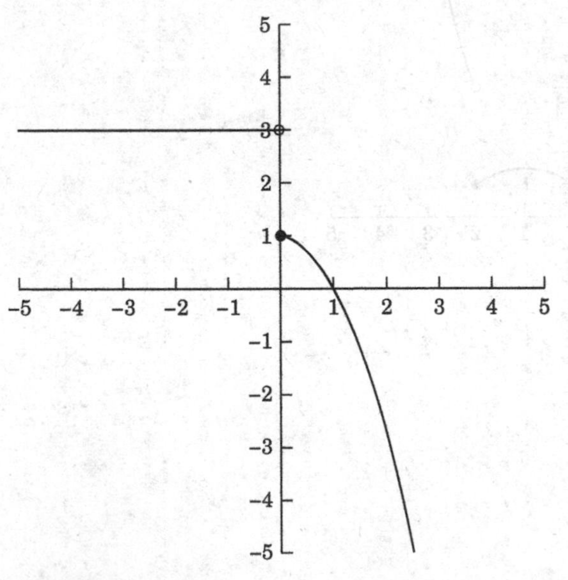

FIGURE 7-50

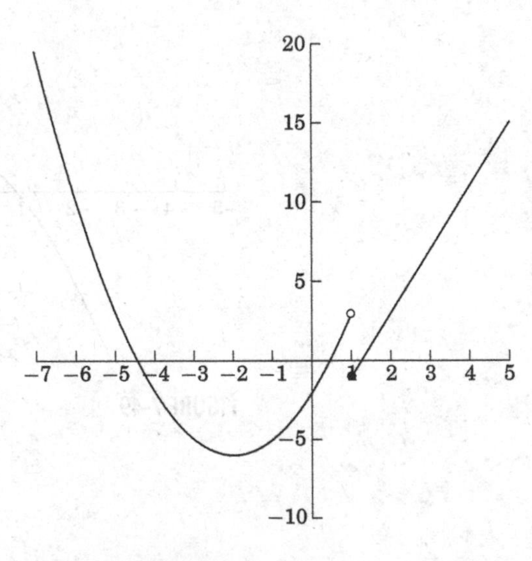

FIGURE 7-51

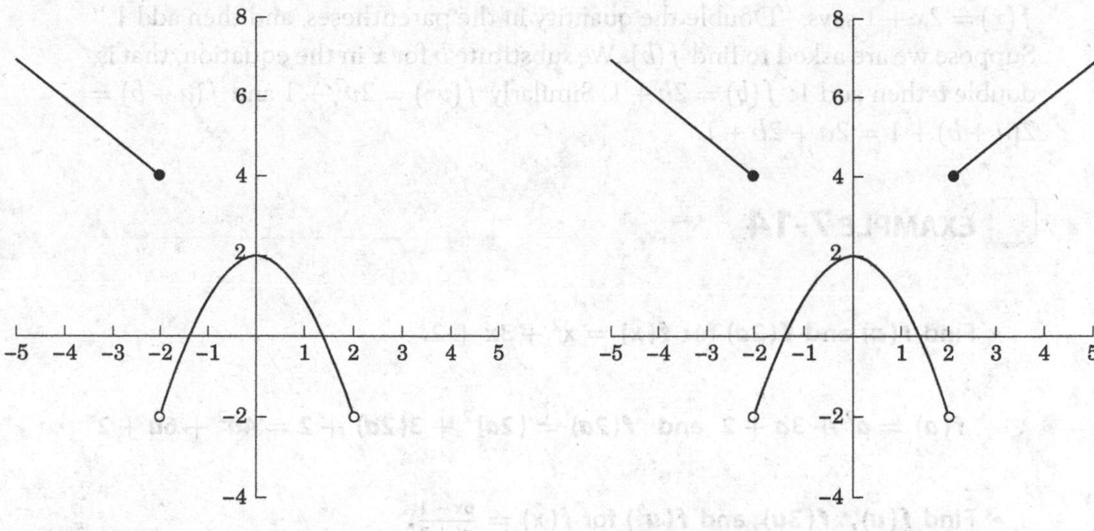

FIGURE 7-52

FIGURE 7-53

EXAMPLE 7-13

- **Sketch the graph of the piecewise function.**

$$f(x) = \begin{cases} -x+2 & \text{if } x \le -2; \\ -x^2+2 & \text{if } x - 2 < x < 2; \\ x+2 & \text{if } x \ge 2 \end{cases}$$

For the piece $y = -x + 2$, we only need to plot two points, the endpoint $x = -2$ and a point to the left of $x = -2$. The piece $y = -x^2 + 2$ is a parabola between $x = -2$ and $x = 2$. See Fig. 7-52.

The last piece, $y = x + 2$, is another line. We need to plot two points, the endpoint $x = 2$ and a point to the right of $x = 2$. The graph of the function is shown in Fig. 7-53.

More on Evaluating Functions

Functions can be evaluated at quantities other than numbers. Keep in mind that evaluating a function means to substitute whatever is in the parentheses for the variable, even if what is in the parentheses is another variable. The function

$f(x) = 2x + 1$ says, "Double the quantity in the parentheses, and then add 1." Suppose we are asked to find $f(b)$. We substitute b for x in the equation, that is, double b then add 1: $f(b) = 2b + 1$. Similarly $f(v^2) = 2v^2 + 1$ and $f(a + b) = 2(a + b) + 1 = 2a + 2b + 1$.

EXAMPLE 7-14

- Find $f(a)$ and $f(2a)$ for $f(x) = x^2 + 3x + 2$.

$$f(a) = a^2 + 3a + 2 \quad \text{and} \quad f(2a) = (2a)^2 + 3(2a) + 2 = 4a^2 + 6a + 2$$

- Find $f(u)$, $f(3u)$, and $f(u^2)$ for $f(x) = \frac{6x-1}{x^2+3}$.

$$f(u) = \frac{6u - 1}{u^2 + 3}$$

$$f(3u) = \frac{6(3u) - 1}{(3u)^2 + 3} = \frac{18u - 1}{9u^2 + 3}$$

$$f(u^2) = \frac{6u^2 - 1}{(u^2)^2 + 3} = \frac{6u^2 - 1}{u^4 + 3}$$

- Find $f(a)$, $f(a + h)$, $f(\frac{1}{a})$, and $f(-x)$ for $f(x) = \frac{1}{x-1}$.

$$f(a) = \frac{1}{a - 1}$$

$$f(a + h) = \frac{1}{(a + h) - 1} = \frac{1}{a + h - 1}$$

$$f\left(\frac{1}{a}\right) = \frac{1}{\frac{1}{a} - 1} = \frac{1}{\frac{1}{a} - \frac{a}{a}} = \frac{1}{\frac{1-a}{a}}$$

$$= 1 \div \frac{1 - a}{a} = 1 \cdot \frac{a}{1 - a} = \frac{a}{1 - a}$$

$$f(-x) = \frac{1}{-x - 1} \quad \text{or} \quad \frac{1}{-(x + 1)} = -\frac{1}{x + 1}$$

- Find $g(u)$, $g(u^2+v)$, and $g(3u-1)$ for $g(t)=12$.

Because g is a constant function, $g(t)=12$ no matter what is in the parentheses.

$$g(u)=12 \qquad g(u^2+v)=12 \qquad g(3u-1)=12$$

PRACTICE

1. Find $f(a)$, $f(2a)$, $f(a^2)$, and $f(a+h)$ for $f(x)=3x-8$.
2. Find $g(a)$, $g(a+1)$, $g(-a)$, and $g(a+h)$ for $g(t)=7$.
3. Find $f(uv)$, $f(a)$, and $f(a+h)$ for $f(x)=2x^2-x+1$.
4. Find $f(-t)$, $f(a)$, $f(a+h)$, and $f(\frac{1}{a})$ for $f(t)=\frac{3-t}{t}$.

SOLUTIONS

1. $f(a)=3a-8 \qquad f(2a)=3(2a)-8=6a-8$

$f(a^2)=3a^2-8 \qquad f(a+h)=3(a+h)-8=3a+3h-8$

2. $g(a)=7 \qquad g(a+1)=7 \qquad g(-a)=7 \qquad g(a+h)=7$

3. $f(uv)=2(uv)^2-uv+1=2u^2v^2-uv+1$

$$f(a)=2a^2-a+1$$

$$\begin{aligned}f(a+h)&=2(a+h)^2-(a+h)+1\\&=2(a+h)(a+h)-(a+h)+1\\&=2(a^2+2ah+h^2)-a-h+1\\&=2a^2+4ah+2h^2-a-h+1\end{aligned}$$

4. $$f(-t) = \frac{3-(-t)}{-t} = \frac{3+t}{-t} \text{ or } -\frac{3+t}{t}$$

$$f(a) = \frac{3-a}{a}$$

$$f(a+h) = \frac{3-(a+h)}{a+h} = \frac{3-a-h}{a+h}$$

$$f\left(\frac{1}{a}\right) = \frac{3-1/a}{1/a} = \frac{\frac{3a}{a}-\frac{1}{a}}{\frac{1}{a}}$$

$$= \frac{\frac{3a-1}{a}}{\frac{1}{a}} = \frac{3a-1}{a} \div \frac{1}{a} = \frac{3a-1}{a} \cdot \frac{a}{1} = 3a-1$$

The Difference Quotient

A very important expression in mathematics is the *difference quotient*, sometimes written as

$$\frac{f(a+h)-f(a)}{h}$$

where f is some function. In fact, the difference quotient is the basis for calculus. Algebra students work with the difference quotient so that when (and if) they take calculus, they aren't hung up on the messy algebra.

Evaluating the difference quotient is really not much more than function evaluation. First, we find $f(a)$ and $f(a+h)$ for the function given to us. Second, we perform the subtraction $f(a+h)-f(a)$ and simplify. Third, we divide this by h and simplify. The previous Practice problems gave us experience in evaluating $f(a)$ and $f(a+h)$. Now we practice finding $f(a+h)-f(a)$.

EXAMPLE 7-15

Find $f(a+h) - f(a)$.

- $f(x) = 3x + 5$

$$f(a) = 3a+5 \text{ and } f(a+h) = 3(a+h)+5 = 3a+3h+5$$

$$f(a+h) - f(a) = \overbrace{3a+3h+5}^{f(a+h)} - (\overbrace{3a+5}^{f(a)})$$
$$= 3a+3h+5-3a-5$$
$$= 3h$$

- $f(t) = t^2 + 1$

$$f(a) = a^2 + 1 \text{ and } f(a+h) = (a+h)^2 + 1 = a^2 + 2ah + h^2 + 1$$

$$f(a+h) - f(a) = \overbrace{a^2 + 2ah + h^2 + 1}^{f(a+h)} - (\overbrace{a^2+1}^{f(a)})$$
$$= a^2 + 2ah + h^2 + 1 - a^2 - 1$$
$$= 2ah + h^2$$

- $f(x) = 6$

$$f(a) = 6 \text{ and } f(a+h) = 6$$

$$f(a+h) - f(a) = 6 - 6 = 0$$

- $f(t) = \dfrac{1}{t+3}$

$$f(a) = \frac{1}{a+3} \text{ and } f(a+h) = \frac{1}{a+h+3}$$

$$f(a+h) - f(a) = \frac{1}{a+h+3} - \frac{1}{a+3} \quad \text{Get a common denominator.}$$
$$= \frac{a+3}{a+3} \cdot \frac{1}{a+h+3} - \frac{a+h+3}{a+h+3} \cdot \frac{1}{a+3}$$
$$= \frac{a+3-(a+h+3)}{(a+h+3)(a+3)} \quad \text{Leave the denominator factored.}$$
$$= \frac{a+3-a-h-3}{(a+h+3)(a+3)} = \frac{-h}{(a+h+3)(a+3)}$$

PRACTICE

Find $f(a+h) - f(a)$.

1. $f(x) = 3x - 4$
2. $f(x) = x^2 + 5$
3. $f(x) = x^2 - 3x - 6$
4. $f(t) = -19$
5. $f(t) = \dfrac{1}{t}$

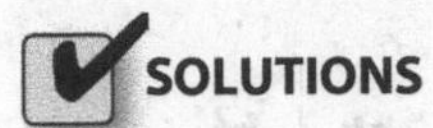

SOLUTIONS

1. $f(a+h) = 3(a+h) - 4$ and $f(a) = 3a - 4$

$$\begin{aligned} f(a+h) - f(a) &= 3(a+h) - 4 - (3a - 4) \\ &= 3a + 3h - 4 - 3a + 4 \\ &= 3h \end{aligned}$$

2. $f(a+h) = (a+h)^2 + 5$ and $f(a) = a^2 + 5$

$$\begin{aligned} f(a+h) - f(a) &= (a+h)^2 + 5 - (a^2 + 5) \\ &= a^2 + 2ah + h^2 + 5 - a^2 - 5 \\ &= 2ah + h^2 \end{aligned}$$

3. $f(a+h) = (a+h)^2 - 3(a+h) - 6$ and $f(a) = a^2 - 3a - 6$

$$\begin{aligned} f(a+h) - f(a) &= (a+h)^2 - 3(a+h) - 6 - (a^2 - 3a - 6) \\ &= a^2 + 2ah + h^2 - 3a - 3h - 6 - a^2 + 3a + 6 \\ &= 2ah + h^2 - 3h \end{aligned}$$

4. $f(a+h) = -19$ and $f(a) = -19$

$$f(a+h) - f(a) = -19 - (-19) = -19 + 19 = 0$$

5. $f(a+h) = \dfrac{1}{a+h}$ and $f(a) = \dfrac{1}{a}$

$$\begin{aligned} f(a+h) - f(a) &= \frac{1}{a+h} - \frac{1}{a} \\ &= \frac{a}{a} \cdot \frac{1}{a+h} - \frac{a+h}{a+h} \cdot \frac{1}{a} \\ &= \frac{a}{a(a+h)} - \frac{a+h}{a(a+h)} = \frac{a-(a+h)}{a(a+h)} \\ &= \frac{a-a-h}{a(a+h)} = \frac{-h}{a(a+h)} \end{aligned}$$

The only steps remaining in evaluating the difference quotient is to divide the difference $f(a+h) - f(a)$ by h. The following examples and Practice problems are from the previous section.

EXAMPLE 7-16

Evaluate $\frac{f(a+h)-f(a)}{h}$.

- $f(x) = 3x + 5$

We found that $f(a+h) - f(a) = 3h$.

$$\frac{f(a+h) - f(a)}{h} = \frac{3h}{h} = 3$$

- $f(t) = t^2 + 1$

We found that $f(a+h) - f(a) = 2ah$.

$$\frac{f(a+h) - f(a)}{h} = \frac{2ah}{h} = 2a$$

- $f(x) = 6$

We found that $f(a+h) - f(a) = 0$.

$$\frac{f(a+h) - f(a)}{h} = \frac{0}{h} = 0$$

- $f(t) = \dfrac{1}{t+3}$

We found that $f(a+h) - f(a) = \frac{-h}{(a+h+3)(a+3)}$.

$$\frac{f(a+h) - f(a)}{h} = \frac{\frac{-h}{(a+h+3)(a+3)}}{h}$$

$$= \frac{-h}{(a+h+3)(a+3)} \div h = \frac{-h}{(a+h+3)(a+3)} \cdot \frac{1}{h}$$

$$= \frac{-1}{(a+h+3)(a+3)}$$

PRACTICE

Evaluate $\frac{f(a+h) - f(a)}{h}$. The first five functions are the same as in the previous Practice problems.

1. $f(x) = 3x - 4$
2. $f(x) = x^2 + 5$
3. $f(x) = x^2 - 3x - 6$
4. $f(t) = -19$
5. $f(t) = \dfrac{1}{t}$
6. $f(x) = 3x^2 - 5x + 2$

SOLUTIONS

1. We found that $f(a+h) - f(a) = 3h$.

$$\frac{f(a+h) - f(a)}{h} = \frac{3h}{h} = 3$$

2. We found that $f(a+h) - f(a) = 2ah + h^2$.

$$\frac{f(a+h) - f(a)}{h} = \frac{2ah + h^2}{h} = \frac{h(2a+h)}{h} = 2a + h$$

3. We found that $f(a+h) - f(a) = 2ah + h^2 - 3h$.

$$\frac{f(a+h) - f(a)}{h} = \frac{2ah + h^2 - 3h}{h} = \frac{h(2a+h-3)}{h} = 2a + h - 3$$

4. We found that $f(a+h) - f(a) = 0$.

$$\frac{f(a+h) - f(a)}{h} = \frac{0}{h} = 0$$

5. We found that $f(a+h) - f(a) = \frac{-h}{a(a+h)}$

$$\frac{f(a+h) - f(a)}{h} = \frac{\frac{-h}{a(a+h)}}{h} = \frac{-h}{a(a+h)} \div h$$

$$= \frac{-h}{a(a+h)} \cdot \frac{1}{h} = \frac{-1}{a(a+h)}$$

6.

$$\frac{f(a+h) - f(a)}{h} = \frac{3(a+h)^2 - 5(a+h) + 2 - (3a^2 - 5a + 2)}{h}$$

$$= \frac{3a^2 + 6ah + 3h^2 - 5a - 5h + 2 - 3a^2 + 5a - 2}{h}$$

$$= \frac{6ah + 3h^2 - 5h}{h} = \frac{h(6a + 3h - 5)}{h}$$

$$= 6a + 3h - 5$$

The difference quotient is really nothing more than the slope of the line containing the points $(a, f(a))$ and $(a+h, f(a+h))$. Remember that the slope formula for the line containing the points (x_1, y_1) and (x_2, y_2) is

$$m = \frac{y_2 - y_1}{x_2 - x_1}$$

In the difference quotient, $x_1 = a$, $y_1 = f(a)$, $x_2 = a + h$, and $y_2 = f(a+h)$.

$$m = \frac{y_2 - y_1}{x_2 - x_1} = \frac{f(a+h) - f(a)}{a + h - a} = \frac{f(a+h) - f(a)}{h}$$

Summary

In this chapter, we learned how to

- *Determine whether or not an equation gives y as a function of x.* An equation gives y as a function of x if *every* value for x has *exactly* one y-value. A y-value can appear any number of times though. If we can find even one x-value that has more than one y-value, then the equation is not a function. Usually, when asked to determine if an equation does give y as a function of x, we solve the equation for y. If this equation gives a formula for y that is unique, then we conclude the equation is a function. For example, the equation $y = \pm\sqrt{x+1}$ is not a function, but the equation $y = \sqrt{x+1}$ is a function. We can tell from the graph of an equation whether or not it is a function with the Vertical Line Test. If any vertical line intersects (touches/crosses) the graph at most one time, then the equation is a function. If a vertical line intersects the graph more than once, then the equation is not a function.

- *Evaluate a function.* We evaluate a function by substituting a value for x in the equation and calculating the y-value. On the graph of a function, we locate the point having a given x-coordinate. The y-coordinate is the value of the function there. The statement "$f(a) = b$" means that when $x = a$, $y = b$, that is, the point (a, b) is on the graph of the function.
- *Determine the domain and range of a function.* The domain of a function is the set of all possible values for x. Values of x that cause a 0 in the denominator or a negative number under an even root are not in the domain of a function. The range of a function is the set of all y-values. We can determine the domain (and sometimes the range) algebraically. We can determine the domain and range from the graph of its equation. We find the domain by observing how far the graph extends horizontally. We find the range by observing how far the graph extends vertically.
- *Determine where a function is increasing and where it is decreasing.* We can tell from the graph of a function where it is increasing and where it is decreasing by observing where its graph goes up (as we move from left to right) and where it goes down. If a function neither goes up nor goes down on an interval of x, the function is constant on the interval.
- *Work with piecewise functions.* A piecewise function has more than one formula for calculating y-values. These formulas depend on where x is. We evaluate a piecewise function at an x-value by determining which interval contains the x-value and then using the formula written next to the interval. We can sketch the graph of a piecewise function by plotting each formula for y and then erasing the parts of the graph that are not valid for the function. For example, for the graph of $f(x) = \begin{cases} x-2 & \text{if } x < -3; \\ 5x & \text{if } x \geq -3 \end{cases}$, we can plot both lines $y = x - 2$ and $y = 5x$ and erase the part of the line $y = x - 2$ that lies to the right of $x = -3$ and erasing part of the line

$y = 5x$ to the left of $x = -3$. We must plot the endpoint (in this case, $x = -3$) using an open dot for the inequalities $<$ and $>$ and a solid dot for the inequalities $\leq$ and $\geq$.

- *Calculate the difference quotient.* The difference quotient is $\frac{f(a+h) - f(a)}{h}$. When we are asked to calculate the difference quotient for a given $f(x)$, we first evaluate $f(a)$ and $f(a+h)$ (by letting $x = a$ and $x = a + h$). We then simplify $f(a+h) - f(a)$ and divide by h.

QUIZ

1. **Evaluate $f(5)$ for $f(x) = \sqrt{x+4}$.**

   A. ± 3 B. 3 C. $\sqrt{x+4}$ D. 9

2. **Evaluate $f(3)$ for $f(x) = \begin{cases} x^2 & \text{if } x < 0; \\ 8 & \text{if } x \geq 0 \end{cases}$.**

   A. 9 B. 8, 9 C. 8 D. 24

3. **What is the domain for the function $f(x) = \frac{x-1}{x^2-4}$?**

   A. $(-\infty, -2) \cup (-2, 1) \cup (1, 2) \cup (2, \infty)$
   B. $(-\infty, -2) \cup (-2, 1) \cup (1, \infty)$
   C. $(-\infty, -2) \cup (1, \infty)$
   D. $(-\infty, -2) \cup (-2, 2) \cup (2, \infty)$

4. **Does the equation $y = |x|$ give y as a function of x?**

   A. Yes B. No C. It cannot be determined.

5. **What is the domain for the function $f(x) = \sqrt{x^2 - 25}$?**

   A. $(-5, \infty)$
   B. $(5, \infty)$
   C. $(-\infty, -5) \cup (5, \infty)$
   D. $(-\infty, -5] \cup [5, \infty)$

6. **Is the graph shown in Fig. 7-54 the graph of a function?**

   A. Yes B. No C. It cannot be determined.

7. **Evaluate $f(2u^2)$ for $f(x) = \frac{x^2}{x-4}$.**

   A. $\frac{2u^2}{2u-4}$ B. $\frac{4u^2}{4u^2-4}$ C. $\frac{2u^2}{2u^2-4}$ D. $\frac{4u^4}{2u^2-4}$

8. **Evaluate $f(a+h) - f(a)$ for $f(x) = 3x^2 - 5$.**

   A. $6ah + 3h^2$
   B. $2ah + 3h^2 - 10$
   C. $6ah + 3h^2 - 10$
   D. $6ah + 3h^2 + 10$

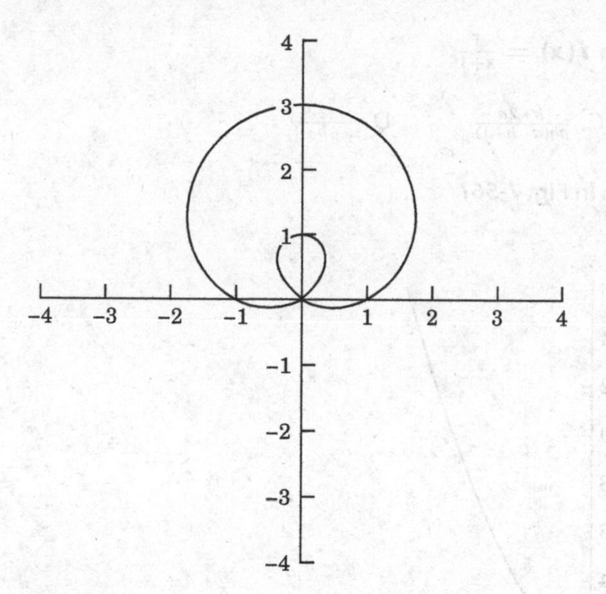

FIGURE 7-54

FIGURE 7-55

9. **For what interval(s) of x is the function shown in Fig. 7-55 increasing?**

A. $(-2, -\frac{1}{2})$ B. $(-\frac{1}{2}, 2)$ C. $(-2, 2)$ D. $(-\frac{1}{2}, \frac{1}{2}) \cup (\frac{3}{2}, 2)$

10. **Refer to Fig. 7-55. What is $f(\frac{1}{2})$?**

A. 0 B. 1 C. $\frac{1}{2}$ D. 2

11. **Refer to Fig. 7-55. What is the range for this function?**

A. $[-2, 2]$ B. $[0, 2]$ C. $[-1, 2]$ D. $[0, -2]$

12. **In the equation $y^2 - 4x^2 + 2x = 8$, is y a function of x?**

A. Yes B. No C. It cannot be determined.

13. **What is the domain for the function $f(x) = \frac{\sqrt{x-9}}{x-12}$?**

A. $(9, 12) \cup (12, \infty)$ B. $(9, 12)$ C. $[9, 12) \cup (12, \infty)$ D. $[9, \infty)$

14. Evaluate $\frac{f(a+h)-f(a)}{h}$ for the function $f(x) = \frac{1}{x+1}$.

A. $\frac{h-2a}{ah(a+h+1)}$ B. $\frac{-1}{a(a+h+1)}$ C. $\frac{h+2a}{ah(a+h+1)}$ D. $\frac{2}{a+h+1}$

15. The graph of which function is given in Fig. 7-56?

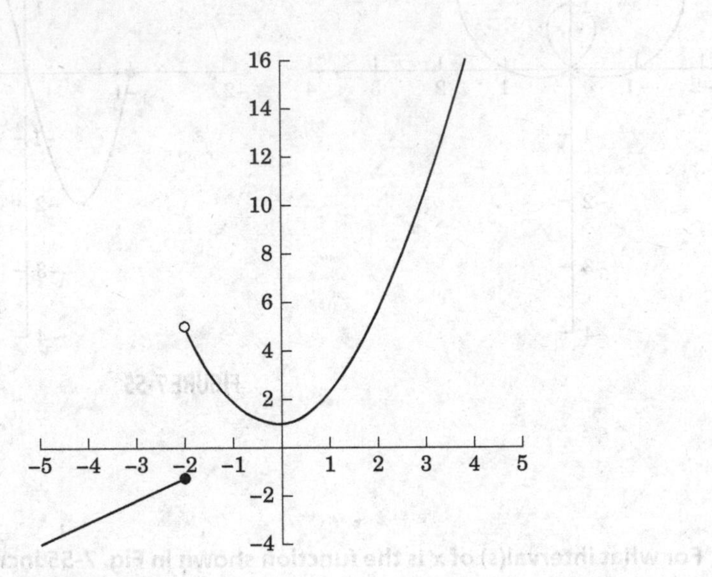

FIGURE 7-56

A. $f(x) = \begin{cases} x-1 & \text{if } x \geq -2; \\ x^2-1 & \text{if } x < 2 \end{cases}$ B. $f(x) = \begin{cases} x+1 & \text{if } x \geq -2; \\ x^2+1 & \text{if } x < 2 \end{cases}$

C. $f(x) = \begin{cases} x-1 & \text{if } x \leq -2; \\ x^2-1 & \text{if } x > 2 \end{cases}$ D. $f(x) = \begin{cases} x+1 & \text{if } x \leq -2; \\ x^2+1 & \text{if } x > 2 \end{cases}$

chapter 8

Quadratic Functions

Our work with parabolas in Chap. 5 was actually work with quadratic functions. In this chapter, we work with the range of a quadratic function, the maximum or minimum value of a quadratic function, as well as applications of quadratic functions. Some important facts about the graph of a quadratic function are outlined in Table 8-1.

CHAPTER OBJECTIVES

In this chapter, you will

- Review the vertex formula for a parabola
- Find the range of a quadratic function
- Maximize or minimize a quadratic function
- Solve applied problems represented by a quadratic function

TABLE 8-1

For the quadratic function $f(x) = ax^2 + bx + c$ and $f(x) = a(x - h)^2 + k$, the domain is $(-\infty, \infty)$.

If $a > 0$, the parabola opens up and the function decreases on the interval $(-\infty, h)$ and increases on the interval (h, ∞).

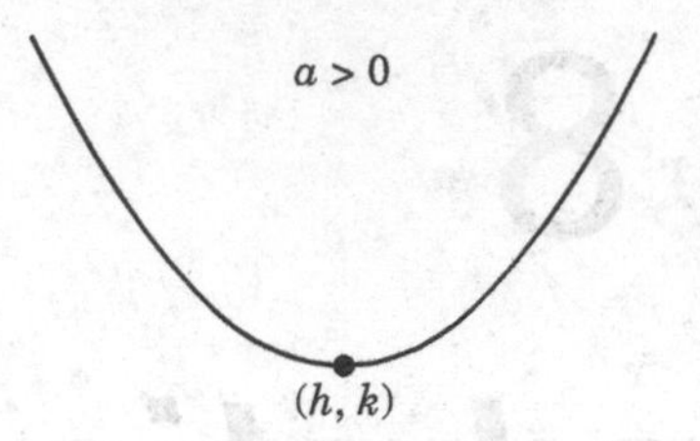

If $a < 0$, the parabola opens down and the function increases on the interval $(-\infty, h)$ and decreases on the interval (h, ∞).

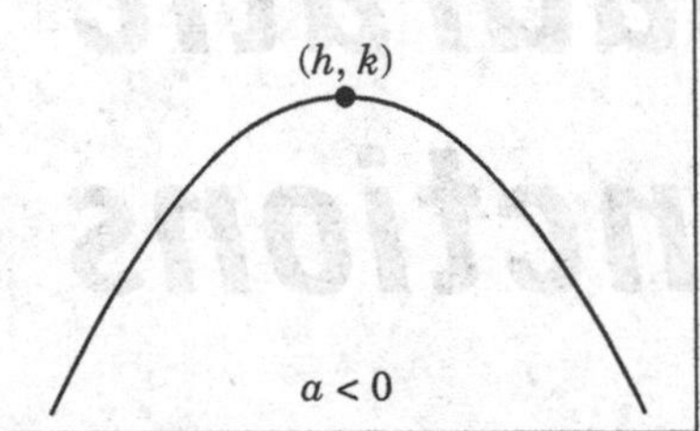

A Review of a Parabola's Vertex

Our work with quadratic functions in this chapter revolves around the vertex for its graph, so we begin with a review for finding the vertex. If the equation is in standard form, $f(x) = a(x - h)^2 + k$, the vertex is easy to find: (h, k). If the equation is in general form, $f(x) = ax^2 + bx + c$, we can find the vertex in one of two ways: by using the formula $h = -\frac{b}{2a}$ or by completing the square to rewrite the function in standard form.

EXAMPLE 8-1

Find the vertex for the graph of the quadratic function using the fact that $h = \frac{-b}{2a}$ and $k = f(h)$.

- $f(x) = 3x^2 - 6x + 1$

$a = 3$, $b = -6$, and $h = \frac{-b}{2a} = \frac{-(-6)}{2(3)} = 1$. We find k by evaluating $f(1) = 3(1)^2 - 6(1) + 1 = -2$. The vertex is $(1, -2)$.

- $f(x) = -x^2 + 4$

$a = -1$, $b = 0$, $h = \frac{-b}{2a} = \frac{-0}{2(-1)} = 0$. We find k by evaluating $f(0) = -0^2 + 4 = 4$. The vertex is $(0, 4)$.

- $f(x) = \frac{2}{3}x^2 - 2x - 6$

$a = \frac{2}{3}$, $b = -2$

$$h = \frac{-b}{2a} = \frac{-(-2)}{2\left(\frac{2}{3}\right)} = \frac{1}{\frac{2}{3}} = \frac{3}{2}$$

We find k by evaluating $f(\frac{3}{2})$.

$$f\left(\frac{3}{2}\right) = \frac{2}{3}\left(\frac{3}{2}\right)^2 - 2\left(\frac{3}{2}\right) - 6 = -\frac{15}{2}$$

The vertex is $(\frac{3}{2}, -\frac{15}{2})$.

PRACTICE

Find the vertex for the graph of the quadratic function using the fact that $h = \frac{-b}{2a}$ and $k = f(h)$.

1. $f(x) = x^2 + 8x + 3$
2. $f(x) = -5x^2 - 4x - 2$
3. $h(t) = -\frac{1}{2}t^2 + 3t + 5$
4. $r(x) = -0.001x^2 + 2x - 100$

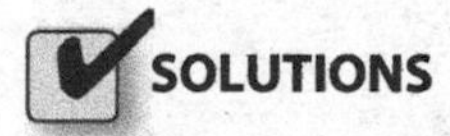

SOLUTIONS

1. The vertex is $(-4, -13)$.

$$h = \frac{-b}{2a} = \frac{-8}{2(1)} = -4$$

$$k = f(-4) = (-4)^2 + 8(-4) + 3 = -13$$

2. The vertex is $\left(-\frac{2}{5}, -\frac{6}{5}\right)$.

$$h = \frac{-b}{2a} = \frac{-(-4)}{2(-5)} = \frac{4}{-10} = -\frac{2}{5}$$

$$k = f\left(-\frac{2}{5}\right) = -5\left(-\frac{2}{5}\right)^2 - 4\left(-\frac{2}{5}\right) - 2$$

$$= -5\left(\frac{4}{25}\right) + \frac{8}{5} - 2 = -\frac{6}{5}$$

3. The vertex is $\left(3, \frac{19}{2}\right)$.

$$h = \frac{-b}{2a} = \frac{-3}{2\left(-\frac{1}{2}\right)} = \frac{-3}{-1} = 3$$

$$k = h(3) = -\frac{1}{2}(3)^2 + 3(3) + 5 = -\frac{9}{2} + 9 + 5 = \frac{19}{2}$$

4. The vertex is $(1000, 900)$.

$$h = \frac{-b}{2a} = \frac{-2}{2(-0.001)} = \frac{-2}{-0.002} = 1000$$

$$k = r(1000) = -0.001(1000)^2 + 2(1000) - 100 = 900$$

The Range of a Quadratic Function

Finding the range for many functions is not easy—basically we need to look at their graph (or use calculus). But finding the range for quadratic functions is not hard. We only need to use the fact that for parabolas that open up, the vertex is the lowest point, and for parabolas that open down, the vertex is the highest point. The range of a quadratic function whose graph opens up is $[k, \infty)$. The range of a quadratic function whose graph opens down is $(-\infty, k]$.

EXAMPLE 8-2

Determine the range for the quadratic function.

- $f(x) = 3x^2 - 6x + 1$

Earlier, we found that the vertex is $(1, -2)$. Because the parabola opens up ($a = 3$ is positive), the range is $[-2, \infty)$.

- $f(x) = -x^2 + 4$

We found that the vertex is $(0, 4)$. Because the parabola opens down ($a = -1$ is negative), the range is $(-\infty, 4]$.

- $f(x) = \frac{2}{3}x^2 - 2x - 6$

We found that the vertex is $(\frac{3}{2}, -\frac{15}{2})$. Because the parabola opens up ($a = \frac{2}{3}$ is positive), the range is $[-\frac{15}{2}, \infty)$.

PRACTICE

Determine the range for the quadratic function. (These are the same functions from the previous set of Practice problems.)

1. $f(x) = x^2 + 8x + 3$
2. $f(x) = -5x^2 - 4x - 2$
3. $h(t) = -\frac{1}{2}t^2 + 3t + 5$
4. $r(x) = -0.001x^2 + 2x - 100$

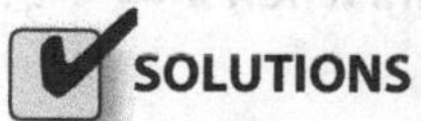

SOLUTIONS

1. **The vertex is $(-4, -13)$. Because a is positive, the parabola opens up, so the range is $[-13, \infty)$.**
2. **The vertex is $(-\frac{2}{5}, -\frac{6}{5})$. Because a is negative, the parabola opens down, so the range is $(-\infty, -\frac{6}{5}]$.**
3. **The vertex is $(3, \frac{19}{2})$. Because a is negative, the parabola opens down, so the range is $(-\infty, \frac{19}{2}]$.**
4. **The vertex is $(1000, 900)$. Because a is negative, the parabola opens down, so the range is $(-\infty, 900]$.**

The Maximum/Minimum of a Quadratic Function

An important area of mathematics is concerned with *optimizing* situations. For example, what sales level for a product gives the most profit? What production level gives the lowest cost per unit? What shape is the strongest? While calculus is used to solve many of these problems, algebra students can optimize problems involving quadratic functions. Quadratic functions can be maximized (if the parabola opens down) or minimized (if the parabola opens up). The maximum or minimum value of a quadratic function is k, the y-coordinate of the vertex.

EXAMPLE 8-3

Find the maximum or minimum value of the quadratic function.

- $f(x) = -2x^2 - 6x + 7$

Because $a = -2$ is negative, the parabola opens down and the function has a maximum value (but no minimum value). We now find k, the maximum value of the function.

$$h = \frac{-b}{2a} = \frac{-(-6)}{2(-2)} = -\frac{3}{2}$$

$$k = f\left(-\frac{3}{2}\right) = -2\left(-\frac{3}{2}\right)^2 - 6\left(-\frac{3}{2}\right) + 7 = \frac{23}{2}$$

The maximum value of the function is $\frac{23}{2}$. This maximum occurs when $x = -\frac{3}{2}$.

- $C(q) = 0.02q^2 - 5q + 600$

Because $a = 0.02$ is positive, the parabola opens up and the function has a minimum value.

$$h = \frac{-b}{2a} = \frac{-(-5)}{2(0.02)} = 125$$

$$k = C(125) = 0.02(125)^2 - 5(125) + 600 = 287.50$$

The minimum value of the function is 287.50. This minimum occurs when $q = 125$.

PRACTICE

Find the maximum or minimum value of the quadratic function.

1. $f(x) = x^2 - 8x + 2$
2. $f(t) = -16t^2 + 48t + 25$
3. $f(x) = 100x^2 + 150x + 25$
4. $P(x) = -0.015x^2 + 0.45x + 12$

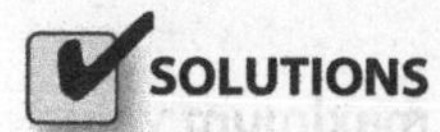

SOLUTIONS

1. Because $a = 1$ is positive, the parabola opens up, and this function has a minimum value.

$$h = \frac{-(-8)}{2(1)} = 4$$

$$k = f(4) = 4^2 - 8(4) + 2 = -14$$

The minimum value of the function is -14. The minimum occurs at $x = 4$.

2. Because $a = -16$ is negative, the parabola opens down, and this function has a maximum value.

$$h = \frac{-48}{2(-16)} = \frac{3}{2}$$

$$k = f\left(\frac{3}{2}\right) = -16\left(\frac{3}{2}\right)^2 + 48\left(\frac{3}{2}\right) + 25 = 61$$

The maximum value of this function is 61. The maximum occurs at $t = \frac{3}{2}$.

3. Because $a = 100$ is positive, the parabola opens up, and this function has a minimum value.

$$h = \frac{-150}{2(100)} = -\frac{3}{4}$$

$$k = f\left(-\frac{3}{4}\right) = 100\left(-\frac{3}{4}\right)^2 + 150\left(-\frac{3}{4}\right) + 25 = -\frac{125}{4}$$

The minimum value of the function is $-\frac{125}{4}$. The minimum value occurs at $x = -\frac{3}{4}$.

4. **Because $a = -0.015$ is negative, the parabola opens down, and this function has a maximum value.**

$$h = \frac{-0.45}{2(-0.015)} = 15$$

$$k = P(15) = -0.015(15)^2 + 0.45(15) + 12 = 15.375$$

The maximum value of the function is 15.375. The maximum value occurs at $x = 15$.

Applied Maximum/Minimum Problems

In the problems for this section, we are asked to find the maximum or minimum of a situation. The functions that model these quantities are quadratic functions, so we can answer the question(s) with one or both coordinates of the vertex. Once we have found the quadratic model, we need to think about what h and k tell us. For example, if the problem is to determine the level of production in order to maximize profit, h tells us the production level necessary to maximize profit, and k tells us what the maximum profit is. The function to be optimized will be given in the first problems. Later, we need to find the function based on information given in the problem.

EXAMPLE 8-4

- **The profit function for a product is given by $P(x) = -2x^2 + 28x + 150$, where x is the number of units sold and P is in dollars. What level of production maximizes the profit? What is the maximum profit?**

The answer to the first question is h, and the answer to the second question is k.

$$h = \frac{-28}{2(-2)} = 7$$

$$k = P(7) = -2(7)^2 + 28(7) + 150 = 248$$

The level of production that maximizes profit is 7 units, and the maximum profit is \$248.

- **The cost per unit of a product is given by the function $C(x) = \frac{1}{4}x^2 - 6x + 40$, where x is the production level (in hundreds of units) and C is the cost in thousands of dollars. What level of production yields the minimum production cost per unit?**

Which quantity answers the question—*h* or *k*? Because the production level is *x*, we must find *h*.

$$h = \frac{-(-6)}{2\left(\frac{1}{4}\right)} = \frac{6}{\frac{1}{2}} = 6 \div \frac{1}{2} = 6 \cdot 2 = 12$$

Minimize the cost per unit by producing 1200 units.

PRACTICE

1. **The daily profit for a vendor of bottled water is given by the function $P(x) = -0.001x^2 + 0.36x - 5$, where x is the number of bottles sold and P is the profit in dollars. How many bottles should be sold to maximize daily profit? What is the maximum daily profit?**
2. **The average cost per unit of a product per week is given by the function $C(x) = \frac{1}{3}x^2 - 60x + 5900$, where x is the number of units produced and C is in dollars. What is the minimum cost per unit and how many units should be produced per week to minimize the cost per unit?**
3. **The weekly revenue of a particular service offered by a company depends on the price—the higher the price, the fewer the sales, and the lower the price, the higher the sales. The function describing the revenue is $R(p) = -\frac{1}{10}p^2 + 3p + 125$, where p is the price per hour and R is sales revenue in dollars. What is the revenue-maximizing price? What is the maximum weekly revenue?**

SOLUTIONS

1. $$h = \frac{-0.36}{2(-0.001)} = 180$$

$$k = P(180) = -0.001(180)^2 + 0.36(180) - 5 = 27.4$$

The vendor should sell 180 bottles per day to maximize profit. The maximum daily profit is \$27.40.

2. $$h = \frac{-(-60)}{2\left(\frac{1}{3}\right)} = \frac{60}{\frac{2}{3}} = 60 \div \frac{2}{3} = 60 \cdot \frac{3}{2} = 90$$

$$k = C(90) = \frac{1}{3}(90)^2 - 60(90) + 5900 = 3200$$

The minimum average cost is \$3200, and 90 units should be produced to minimize the average cost.

3. $$h = \frac{-3}{2\left(\frac{-1}{10}\right)} = \frac{3}{\frac{1}{5}} = 3 \div \frac{1}{5} = 3 \cdot 5 = 15$$

$$k = R(15) = -\frac{1}{10}(15)^2 + 3(15) + 125 = 147.50$$

The revenue-maximizing price \$15 per hour. The maximum weekly revenue is \$147.50.

When an object is thrust upward and is free-falling after the initial thrust, its path is in the shape of a parabola. In the following problems, we are given the height function of these kinds of falling objects. The functions are in the form $f(x) = ax^2 + bx + c$, where x is the horizontal distance and $f(x)$ is the height. We will answer the questions, "What is the object's maximum height?" (which is k) and "How far has it traveled horizontally to reach its maximum height (which is h)?"

EXAMPLE 8-5

- **Suppose the path of a grasshopper's jump is given by the function $f(x) = \frac{-5}{216}x^2 + \frac{5}{3}x$, where both x and $f(x)$ are in inches. What is the maximum height reached by the grasshopper? How far has it traveled horizontally to reach its maximum height?**

$$h = \frac{-\frac{5}{3}}{2\left(-\frac{5}{216}\right)} = \frac{\frac{5}{3}}{\frac{5}{108}} = \frac{5}{3} \div \frac{5}{108} = \frac{5}{3} \cdot \frac{108}{5} = 36$$

$$k = f(36) = -\frac{5}{216}(36)^2 + \frac{5}{3}(36) = 30$$

The maximum height reached by the grasshopper is 30 inches and it had traveled 36 inches horizontally when it reached its maximum height.

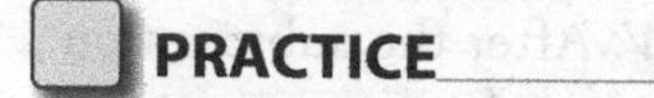

PRACTICE

1. **A child throws a ball, its path is given by the function $f(x) = -0.04x^2 + 1.5x + 3$, where x and $f(x)$ are in feet. What is the maximum height of the ball? How far has it traveled horizontally when it reaches its maximum height?**
2. **A kitten jumped to pounce on a toy mouse. The path of the kitten is given by the function $f(x) = -\frac{5}{72}x^2 + \frac{5}{3}x$, where x and $f(x)$ are in inches. How far had the kitten traveled horizontally when it reached its maximum height? What was the kitten's maximum height?**

SOLUTIONS

1. $$h = \frac{-1.5}{2(-0.04)} = 18.75$$

$$k = f(18.75) = -0.04(18.75)^2 + 1.5(18.75) + 3 = 17.0625$$

The ball's maximum height is 17.0625 feet and it had traveled 18.75 feet horizontally when it reached its maximum height.

2. $$h = \frac{-\frac{5}{3}}{2\left(-\frac{5}{72}\right)} = \frac{\frac{5}{3}}{\frac{5}{36}} = \frac{5}{3} \div \frac{5}{36} = \frac{5}{3} \cdot \frac{36}{5} = 12$$

$$k = f(12) = -\frac{5}{72}(12)^2 + \frac{5}{3}(12) = 10$$

The kitten's maximum height is 10 inches and it had traveled 12 inches horizontally when it reached its maximum height.

Quadratic functions can be used to optimize many types of geometric problems. In the following problems, a number is fixed (usually the perimeter) and we are asked to find the maximum enclosed area. The area function is quadratic, but getting to this function requires a few steps. The first few problems involve a fixed amount of fencing to be used in order to enclose a rectangular area.

When maximizing the area function, $A = LW$, we must eliminate either L or W. This is where the amount of fencing available comes in. We represent the amount of fencing available with an equation involving L and W. We then solve for either L or W and make a substitution in $A = LW$. After the substitution, we are left with a quadratic function that can be maximized.

EXAMPLE 8-6

- **A farmer has 600 feet of fencing available to enclose a rectangular pasture and then subdivide the pasture into two equal rectangular yards (see Fig. 8-1). What dimensions yield the maximum area? What is the maximum area?**

The formula for the area of a rectangle is $A = LW$. This formula has two variables (other than A) and we must reduce it to one. There is 600 feet of fencing available, so $L + W + W + W + L = 2L + 3W$ must equal 600 (see Fig. 8-2). This gives us the equation $2L + 3W = 600$. We could solve for either L or W. Here, we solve for L.

$$2L + 3W = 600$$

$$L = \frac{600 - 3W}{2} = 300 - 1.5W$$

Now we substitute $L = 300 - 1.5W$ in the formula $A = LW$.

$$A = LW = (300 - 1.5W)W = 300W - 1.5W^2 = -1.5W^2 + 300W$$

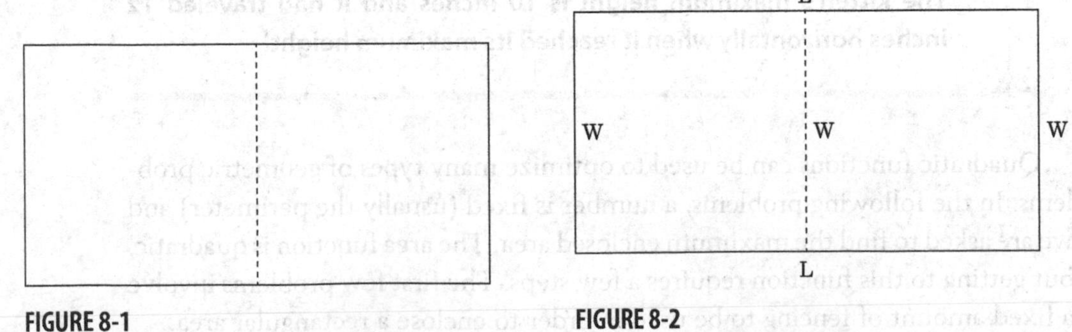
FIGURE 8-1

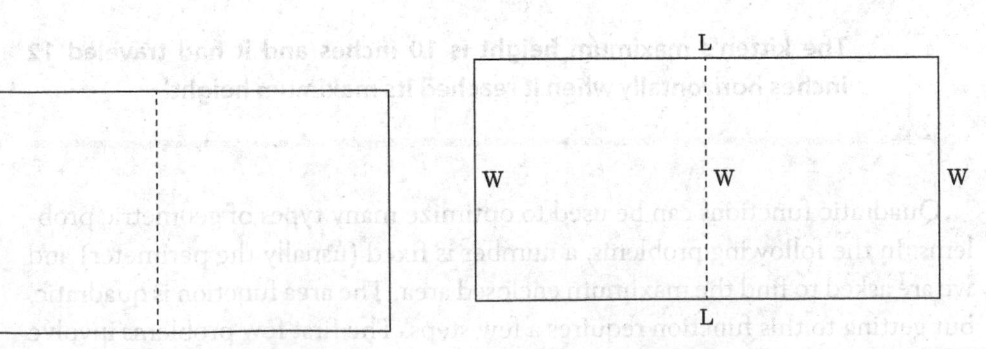

FIGURE 8-2

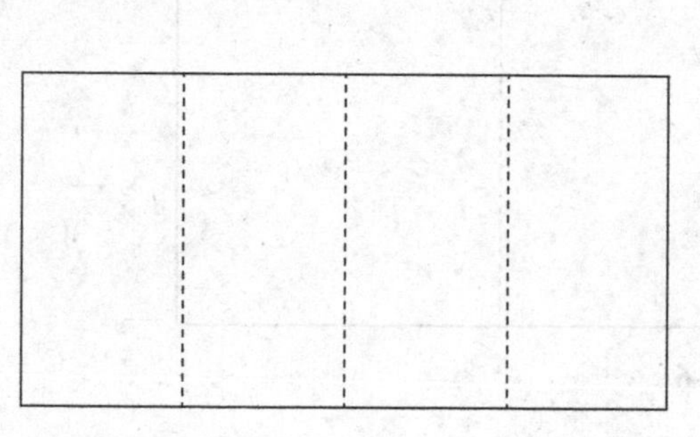

FIGURE 8-3

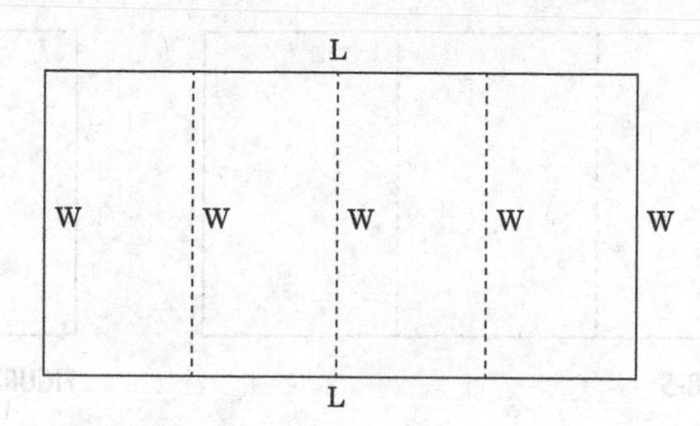

FIGURE 8-4

This quadratic function has a maximum value, the maximum area.

$$h = \frac{-300}{2(-1.5)} = 100$$

$$k = -1.5(100)^2 + 300(100) = 15{,}000$$

The maximum area is 15,000 square feet. This occurs when the width is 100 feet and the length is $300 - 1.5(100) = 150$ feet.

- **A zoo has 1100 meters of fencing available to create four rectangular pens. (See Fig. 8-3.) What dimensions enclose the maximum area? What is the maximum area?**

We label the figure with L and W (see Fig. 8-4).

We want to maximize the area, $A = LW$. The 1100 meters of fencing must be divided among 2 Ls and 5 Ws, so we have the equation $2L + 5W = 1100$. We solve for either L or W and use the new equation to reduce the number of variables in $A = LW$.

$$2L + 5W = 1100$$

$$L = \frac{1100 - 5W}{2} = 550 - 2.5W$$

$A = LW$ becomes $A = (550 - 2.5W)W = 550W - 2.5W^2 = -2.5W^2 + 550W$

$$h = \frac{-550}{2(-2.5)} = 110$$

$$k = -2.5(110)^2 + 550(110) = 30{,}250$$

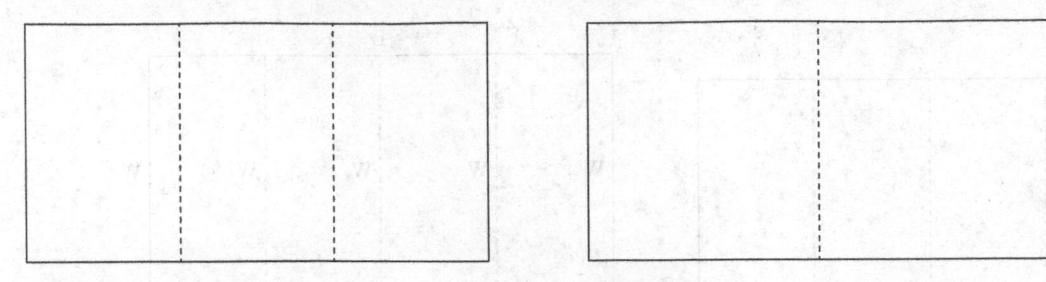

FIGURE 8-5 FIGURE 8-6

To maximize area, let $W = 110$ meters, $L = 550 - 2.5(110) = 275$ meters. The maximum area is 30,250 square meters.

PRACTICE

1. A parks department wants to enclose three adjancent rectangular playing fields. (See Fig. 8-5.) It has 1600 feet of fencing available. What dimensions yield the maximum area? What is the maximum area?
2. A farmer has 450 meters of fencing available. He decides to fence two adjacent rectangular pastures (See Fig. 8-6.) What dimensions maximize the area? What is the maximum area?

SOLUTIONS

1. We can see from Fig. 8-5 that $2L + 4W = 1600$. We solve this equation for L.

$$2L + 4W = 1600$$

$$L = \frac{1600 - 4W}{2} = 800 - 2W$$

$$A = LW = (800 - 2W)W = 800W - 2W^2$$

$$= -2W^2 + 800W$$

$$h = \frac{-800}{2(-2)} = 200$$

$$k = -2(200)^2 + 800(200) = 80{,}000$$

Maximize the area by letting $W = 200$ feet and $L = 800 - 2(200) = 400$ feet. The maximum area is 80,000 square feet.

2. **We can see from Fig. 8-6 that $2L + 3W = 450$. We solve this equation for L.**

$$2L + 3W = 450$$

$$L = \frac{450 - 3W}{2} = 225 - 1.5W$$

$$A = LW = (225 - 1.5W)W = 225W - 1.5W^2$$

$$= -1.5W^2 + 225W$$

$$h = \frac{-225}{2(-1.5)} = 75$$

$$k = -1.5(75)^2 + 225(75) = 8437.5$$

Maximize the area by letting $W = 75$ meters and $L = 225 - 1.5(75) = 112.5$ meters. The maximum area is 8437.5 square meters.

Another common problem for maximizing an enclosed area involves a rectangular region for which one side does not need to be fenced. We can see from Fig. 8-7 that "$2W + L =$ Amount of fencing." Solving this for L gives us "$L =$ Amount of fencing $- 2W$." As before, we substitute this quantity for L in $A = LW$. We then find h, which is the width that maximizes the area, and k is the maximum area.

EXAMPLE 8-7

- **A business needs to enclose an area behind its offices for storage. It has 240 feet of fencing available. If the side of the building is not fenced, what dimensions maximize the enclosed area?**

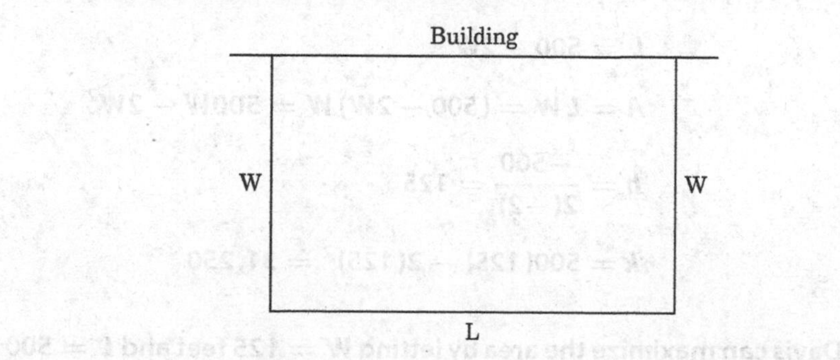

FIGURE 8-7

We see from Fig. 8-7 that $2W + L = 240$.

$$L + 2W = 240$$

$$L = 240 - 2W$$

$$A = LW = (240 - 2W)W = 240W - 2W^2$$

$$h = \frac{-240}{2(-2)} = 60$$

$$k = 240(60) - 2(60)^2 = 7200$$

Maximize the area by letting $W = 60$ feet and $L = 240 - 2(60) = 120$ feet. The maximum area is 7200 square feet.

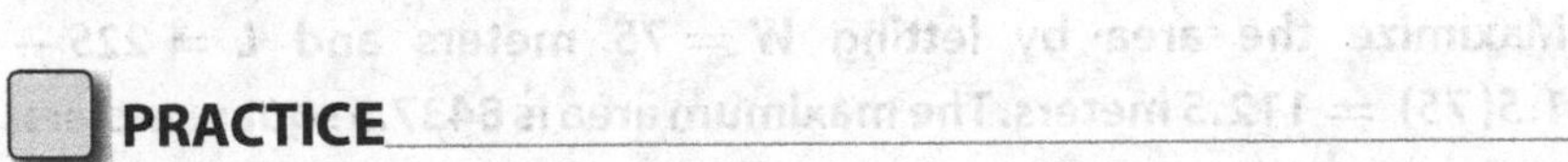

PRACTICE

1. Davis, a rancher, wants to enclose a rectangular pasture that borders a stream. He has 500 feet of fencing available and will not fence the side along the stream. What dimensions maximize the area? What is the maximum area?
2. The manager of an office complex wants to provide extra parking behind the office building. The contractor has 150 meters of fencing available. If the side along the building will not be fenced, what dimensions maximize the enclosed area? What is the maximum enclosed area?

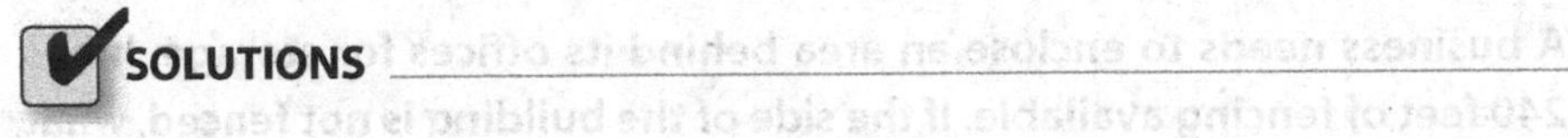

SOLUTIONS

1.
$$2W + L = 500$$

$$L = 500 - 2W$$

$$A = LW = (500 - 2W)W = 500W - 2W^2$$

$$h = \frac{-500}{2(-2)} = 125$$

$$k = 500(125) - 2(125)^2 = 31{,}250$$

Davis can maximize the area by letting $W = 125$ feet and $L = 500 - 2(125) = 250$ feet. The maximum area is 31,250 square feet.

2. $$2W + L = 150$$

$$L = 150 - 2W$$

$$A = LW = (150 - 2W)W = 150W - 2W^2$$

$$h = \frac{-150}{2(-2)} = 37.5$$

$$k = 150(37.5) - 2(37.5)^2 = 2812.5$$

Maximize the area by letting $W = 37.5$ meters and $L = 150 - 2(37.50) = 75$ meters. The maximum area is 2812.5 square meters.

Revenue-Maximizing Price

Often in business, revenue depends on the price in two ways. Obviously, if the price is raised, more money will be collected for each unit sold, but the number of units sold might drop. In general, the lower the price, the higher the demand (number of units sold), and the higher the price, the lower the demand. In the following problems, the demand for a certain price is given. We are told how many sales are lost from a price increase or how many sales are gained from a price decrease. With this information, we can find the price that maximizes revenue. Let n represent the number of increases or decreases in the price. If the price is raised in \$10 increments, then the increase in price would be $10n$ and the price would be "old price $+10n$." If five sales are lost for each \$10 increase in price, then the number sold would be "old sales level $-5n$." The total revenue would be "$R =$ (old price $+10n$)(old sales level $-5n$)." The revenue function is a quadratic function. Revenue is maximized when $n = h$ and the maximum revenue is k.

EXAMPLE 8-8

- **A store sells an average of 345 pounds of apples per day when the price is \$0.85 per pound. The manager thinks that for every increase of \$0.10 in the price, 30 fewer pounds of apples will be sold each day. What price maximizes the average revenue from the sale of apples? What is the maximum average revenue?**

Let n represent the number of \$0.10 increases in the price per pound. The price is then represented by $0.85 + 0.10n$. The number of pounds of apples sold per day would be $345 - 30n$. This makes the revenue $R = (0.85 + 0.10n)(345 - 30n)$. The vertex of this quadratic function tells us two things. First, h tells us how many times we need to raise the price by \$0.10 in order to maximize revenue, and k tells us what the maximum revenue is.

$$R = (0.85 + 0.10n)(345 - 30n) = 293.25 + 9n - 3n^2$$

$$h = \frac{-9}{2(-3)} = 1.5$$

$$k = 293.25 + 9(1.5) - 3(1.5)^2 = 300$$

The store can maximize average revenue by charging $0.85 + 0.10(1.5) = \$1$ per pound. The maximum average revenue is \$300 per day.

- **An apartment manager is leasing 60 of 75 apartments in her apartment complex with the monthly rent at \$1950. For each \$25 decrease in the monthly rent, she believes that one more apartment can be rented. What monthly rent maximizes revenue? What is the maximum monthly revenue?**

Let n represent the number of \$25 decreases in the rent. Then monthly rent is represented by $1950 - 25n$, and the number of apartments rented is $60 + n$. The revenue function is $R = (1950 - 25n)(60 + n)$. The revenue-maximizing rent is $1950 - 25h$, and k is the maximum revenue from rent.

$$R = (1950 - 25n)(60 + n) = 117{,}000 + 450n - 25n^2$$

$$h = \frac{-450}{2(-25)} = 9$$

$$k = 117{,}000 + 450(9) - 25(9)^2 = 119{,}025$$

Maximize revenue by charging $1950 - 25(9) = \$1725$ for the monthly rent. The maximum revenue is \$119,025. (Because this would have 69 apartments leased, the solution is valid.)

PRACTICE

1. **At a small college, 1200 tickets can be sold during a football game when the ticket price is \$9. The athletic director learns from a recent survey**

that for each $0.75 decrease in the ticket price, 200 more fans will attend the game. What should the ticket price be in order to maximize ticket revenue? What is the maximum ticket revenue?

2. **The owner of a concession stand sells 10,000 soft drinks for $3.70 per drink during baseball games. A survey reveals that for each $0.20 decrease in the price of the drinks, 800 more will be sold. What should the price be in order to maximize revenue? What is the maximum revenue?**
3. **The manager of an apartment complex can rent all 60 apartments in his building if the monthly rent is $2800, and for each $50 increase in the monthly rent, one tenant will be lost and will not likely be replaced. What should the monthly rent be to maximize revenue? What is the maximum revenue?**

SOLUTIONS

1. Let n represent the number of $0.75 decreases in the ticket price. This makes the new ticket price $9 - 0.75n$ and the number of tickets sold $1200 + 200n$. Ticket revenue is $R = (9 - 0.75n)(1200 + 200n)$.

$$R = (9 - 0.75n)(1200 + 200n) = 10{,}800 + 900n - 150n^2$$

$$h = \frac{-900}{2(-150)} = 3 \qquad k = 10{,}800 + 900(3) - 150(3)^2 = 12{,}150$$

Maximize ticket revenue by charging $9.00 − 0.75(3) = $6.75 per ticket. The maximum ticket revenue is $12,150.

2. Let n represent the number of $0.20 decreases in the drink price. This makes the new drink price $3.70 - 0.20n$ and the number of drinks sold $10{,}000 + 800n$. Revenue is $R = (3.70 - 0.20n)(10{,}000 + 800n)$.

$$R = (3.70 - 0.20n)(10{,}000 + 800n) = 37{,}000 + 960n - 160n^2$$

$$h = \frac{-960}{2(-160)} = 3 \qquad k = 37{,}000 + 960(3) - 160(3)^2 = 38{,}440$$

Maximize revenue by charging $3.70 − 0.20(3) = $3.10 per drink. The maximum revenue is $38,440.

3. Let n represent the number \$50 increases in the monthly rent. This makes the monthly rent $2800 + 50n$ and the number of tenants $60 - 1n = 60 - n$. Monthly revenue is $R = (2800 + 50n)(60 - n)$.

$$R = (2800 + 50n)(60 - n) = 168{,}000 + 200n - 50n^2$$

$$h = \frac{-200}{2(-50)} = 2 \quad k = 168{,}000 + 200(2) - 50(2)^2 = 168{,}200$$

Maximize revenue by charging $\$2800 + 50(2) = \$2900$ monthly rent. The maximum revenue is \$168,200.

Maximizing/Minimizing Other Functions

Algebra students can use graphing calculators to approximate the maximum and/or minimum values of other kinds of functions. For example, the volume of a certain box is given by the function $V = 4x^3 - 40x^2 + 100x$, where x is the height (in inches) and V is the volume (in cubic inches) of the box, and conditions make it necessary for $0 < x < 5$. The graph of this function is shown in Fig. 8-8.

Because the domain of this applied function is (0, 5), we only care about the part of the graph between $x = 0$ and $x = 5$ (see Fig. 8-9).

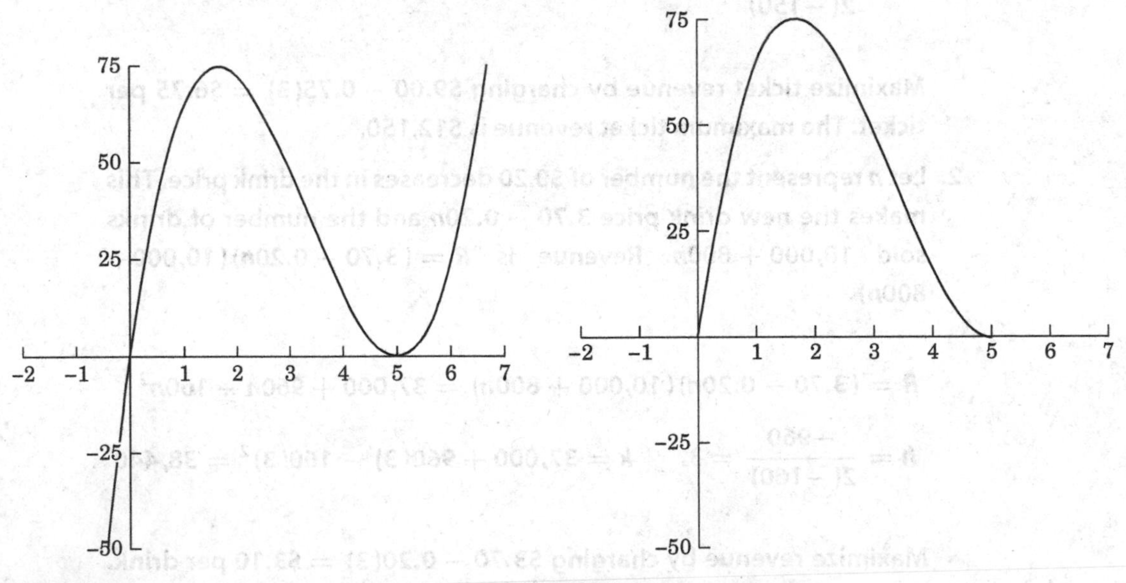

FIGURE 8-8

FIGURE 8-9

We can use a graphing calculator to approximate the highest point, (1.67, 74.07). The maximum volume is approximately 74.07 cubic inches and the height at which the box's volume is maximum is about 1.67 inches. Calculus is necessary to find the exact values.

Summary

In this chapter, we learned how to

- *Determine the range for a quadratic function.* While we can determine the range for a function by examining its graph, we can determine the range for a quadratic function, $f(x) = a(x - h)^2 + k$, algebraically. If the graph of a quadratic function (a parabola) opens up, the range of the function is $[k, \infty)$, where (h, k) is the vertex. The range of a quadratic function whose graph opens down is $(-\infty, k]$. The graph opens up if a is positive and down if a is negative. If the function is written in the form $f(x) = ax^2 + bx + c$, we can find (h, k) by completing the square or by using the formula for h: $h = -\frac{b}{2a}$ and then $k = f(h)$.
- *Maximize or minimize a quadratic function.* A quadratic function whose graph opens up has a minimum value (k), and a quadratic function whose graph opens down has a maximum value (again, k).
- *Maximize or minimize applied problems.* If an applied problem is represented by a quadratic function, we can maximize or minimize the situation. If we are maximizing an enclosed rectangular area, we begin with the area formula, $A = LW$. We eliminate either L or W from information provided in the problem (e.g., a certain amount of fencing is available), leaving us with a quadratic function that has a maximum value. We worked with revenue problems involving a change in price. We are told how a change in price affects the number of units sold. We represent the price change with an, where n is the number of price increases or decreases of a dollars. The number sold either increases or decreases by a certain amount. Once we represent the price and the number sold in terms of n, we replace p and q (price and quantity) in the formula $R = pq$ and are left with a quadratic function that has a maximum value.

QUIZ

1. What is the range of the function $f(x) = -4x^2 + 12x - 5$?

A. $[-32, \infty)$ B. $(-\infty, -32]$ C. $[4, \infty)$ D. $(-\infty, 4]$

2. What is the minimum value of the function $f(x) = 0.02x^2 + 6x - 10$?

A. 1340 B. -460 C. -280 D. 1220

3. The profit function for a product is $P(x) = -0.5x^2 + 16x - 86$, where x is the number of units sold (in thousands) and P is the profit (in thousands of dollars). What is the maximum profit?

A. \$39,000 B. \$40,000 C. \$41,000 D. \$42,000

4. A school board is preparing plans for a new middle school. Officials plan to enclose two adjacent rectangular playing fields with fencing that is already in storage. If they have 480 meters of fencing available, what is the maximum area that can be enclosed?

A. 9200 m^2 B. 9400 m^2 C. 9600 m^2 D. 9800 m^2

5. The revenue of a certain product depends on the amount spent on advertising. The function that gives revenue (in thousands of dollars) in terms of amount spent on advertising (in thousands of dollars) is $R(x) = -1.875x^2 + 45x - 100$. How much should be spent on advertising in order to maximize revenue?

A. \$12,000 B. \$13,000 C. \$14,000 D. \$15,000

6. What is the range for the function $f(x) = 0.04x^2 - 20x + 100$?

A. $(-\infty, -2400]$ B. $[-2400, \infty)$ C. $(-\infty, 250]$ D. $[250, \infty)$

7. For the quadratic function $f(x) = -3(x + 4)^2 + 15$, where does the maximum or minimum value of the function occur?

A. The maximum value of the function occurs at $x = 4$.

B. The maximum value of the function occurs at $x = -4$.

C. The minimum value of the function occurs at $x = 4$.

D. The minimum value of the function occurs at $x = -4$.

8. Cheyenne sells cotton candy at a school carnival. She averages 240 cones per day when the price is $1.50. According to research conducted by an economics class, for each $0.25 increase in the price, 20 fewer cones would be sold. Under these conditions, what is Cheyenne's maximum revenue?

A. $405 B. $415 C. $425 D. $430

chapter 9

Transformations and Combinations

Sketching the graph of a function is an important skill in algebra. In this chapter, we learn how changes to a function's equation affect its graph. We then learn about the graphs of several basic functions, such as the cubic function, square root function, and absolute value function. For example, we have already studied three types of equations and how to find information in the equation about its graph: circles, lines, and parabolas. From the equation of a circle, $(x-h)^2+(y-k)^2=r^2$, we know that its center is at (h, k) and its radius is r. From the equation of a line, $y=mx+b$, we know its slope is m and its y-intercept is b. From the equation of a quadratic function, $y=a(x-h)^2+k$, we know that its vertex is (h, k) and whether it opens up (if a is positive) or down (if a is negative).

CHAPTER OBJECTIVES

In this chapter, you will

- Learn how changes to a function affect its graph
- Determine whether or not a function is even or odd
- Combine functions arithmetically

- Find the composition of one function with another
- Determine the domain for the composition of two functions

Later in the chapter, we learn how to combine functions arithmetically (the sum, difference, product, and quotient of two functions). We also learn another combination of functions that is less familiar: function composition.

Transformations

Let us begin with a closer look at quadratic functions. The graph of every quadratic function is more or less the graph of $y = x^2$. For example, the vertex for the function $y = (x-2)^2$ is (2, 0). Another way of looking at this is to say that the vertex moved from (0, 0) to (2, 0). That is, the vertex moved to the right 2 units. The solid graph in Fig. 9-1 is the graph of $y = x^2$ and the dashed graph is the graph of $y = (x-2)^2$.

The function $y = (x-2)^2$ is really the function $f(x) = x^2$ evaluated at $x - 2$: $f(x-2) = (x-2)^2$. Evaluating *any* function at $x - 2$ shifts the entire graph 2 units to the right. For any positive number h, the graph $y = f(x-h)$ shifted to the right h units, no matter what function $f(x)$ is.

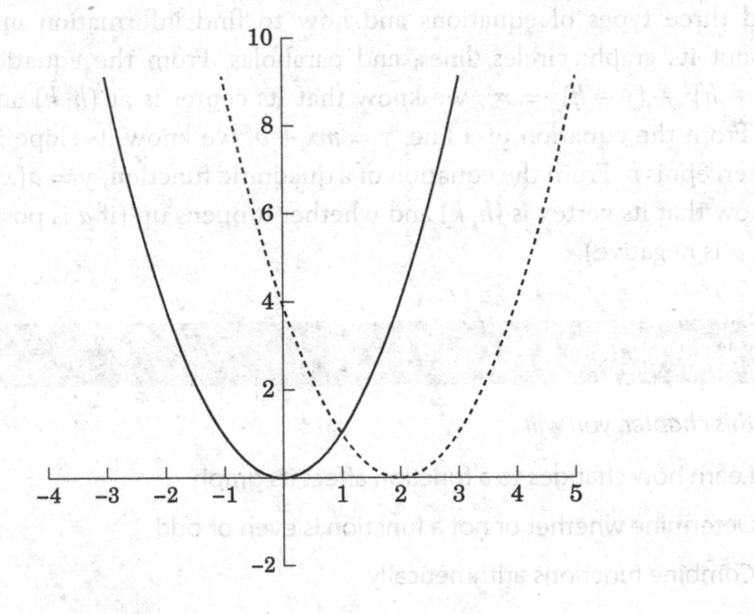

FIGURE 9-1

EXAMPLE 9-1

- **The graph of $y = f(x - 5)$ is the graph of $y = f(x)$ shifted to the right 5 units.**
- **The graph of $y = f(x - 20)$ is the graph of $y = f(x)$ shifted to the right 20 units.**
- **The graph of $y = f(x - \frac{1}{2})$ is the graph of $y = f(x)$ shifted to the right $\frac{1}{2}$ units.**

What would be the effect on $f(x)$ by adding positive h to x? The vertex for $y = (x + 3)^2$ is $(-3, 0)$. This shifts the graph of $f(x) = x^2$ to the *left* 3 units. The graph of any function $y = f(x + h)$ is the graph of $y = f(x)$ shifted to the left h units.

EXAMPLE 9-2

- **The graph of $y = f(x + 12)$ is the graph of $y = f(x)$ shifted to the left 12 units.**
- **The graph of $y = g(x + 3)$ is the graph of $y = g(x)$ shifted to the left 3 units.**

The vertex for the quadratic function, $y = x^2 + 2$, $(0, 2)$, has been shifted up 2 units from the vertex of $f(x) = x^2$. In general, adding a positive number to a function has the effect of shifting its graph upward. Subracting a positive number from a function has the effect of shifting its graph downward. If k is a positive number, the graph of $y = f(x) + k$ is the graph of $y = f(x)$ shifted up k units, and the graph of $y = f(x) - k$ is the graph of $y = f(x)$ shifted down k units.

EXAMPLE 9-3

- **The graph of $y = f(x) - 4$ is the graph of $y = f(x)$ shifted down 4 units.**
- **The graph of $y = h(x) + 9$ is the graph of $y = h(x)$ shifted up 9 units.**

PRACTICE

Compare the graph of the function with the graph of $y = f(x)$.

1. $y = f(x + 5)$ **2.** $y = f(x - 0.10)$ **3.** $y = f(x - 35)$
4. $y = f(x) + 1$ **5.** $y = f(x) + 15$ **6.** $y = f(x) - 8$
7. $y = f(x) - 1$

SOLUTIONS

1. **The graph of $y = f(x + 5)$ is the graph of $y = f(x)$ shifted to the left 5 units.**
2. **The graph of $y = f(x - 0.10)$ is the graph of $y = f(x)$ shifted to the right 0.10 units.**
3. **The graph of $y = f(x - 35)$ is the graph of $y = f(x)$ shifted to the right 35 units.**
4. **The graph of $y = f(x) + 1$ is the graph of $y = f(x)$ shifted up 1 unit.**
5. **The graph of $y = f(x) + 15$ is the graph of $y = f(x)$ shifted up 15 units.**
6. **The graph of $y = f(x) - 8$ is the graph of $y = f(x)$ shifted down 8 units.**
7. **The graph of $y = f(x) - 1$ is the graph of $y = f(x)$ shifted down 1 unit.**

Functions can have a combination of vertical and horizontal shifts. If h and k are positive numbers, the graph of $y = f(x - h) + k$ is the graph of $y = f(x)$ shifted to the right h units and up k units. The graph of $y = f(x + k) - h$ is the graph of $y = f(x)$ shifted to the left k units and down h units.

EXAMPLE 9-4

- **The graph of $y = (x - 2)^2 + 1$ is the graph of $f(x) = x^2$ shifted to the right 2 units and up 1 unit.**

- **The graph of $y = f(x+2)+3$ is the graph of $y = f(x)$ shifted to the left 2 units and up 3 units.**

PRACTICE

Compare the graph of the function with the graph of $y = f(x)$.

1. $y = f\left(x+\frac{1}{2}\right)+3$ **2.** $y = f(x-4)-5$ **3.** $y = f(x+6)-8$
4. $y = f(x-1)+9$

SOLUTIONS

1. **The graph of $y = f(x+\frac{1}{2})+3$ is the graph of $y = f(x)$ shifted to the left $\frac{1}{2}$ units and up 3 units.**
2. **The graph of $y = f(x-4)-5$ is the graph of $y = f(x)$ shifted to the right 4 units and down 5 units.**
3. **The graph of $y = f(x+6)-8$ is the graph of $y = f(x)$ shifted to the left 6 units and down 8 units.**
4. **The graph of $y = f(x-1)+9$ is the graph of $y = f(x)$ shifted to the right 1 unit and up 9 units.**

In the next two sections, we compare the graph of a function with its transformation. The solid graph is the graph of $f(x)$, and the dashed graphs are the transformations of $f(x)$.

EXAMPLE 9-5

Compare the graph of the function with the graph of $y = f(x)$ and then write the transformed function.

- **The dashed graph in Fig. 9-2 is the graph of $y = f(x)$ shifted to the right 1 unit, so this is the graph of $y = f(x-1)$.**
- **The dashed graph in Fig. 9-3 is the graph of $y = f(x)$ shifted to the right 2 units and down 1 unit, so this is the graph of $y = f(x-2)-1$.**

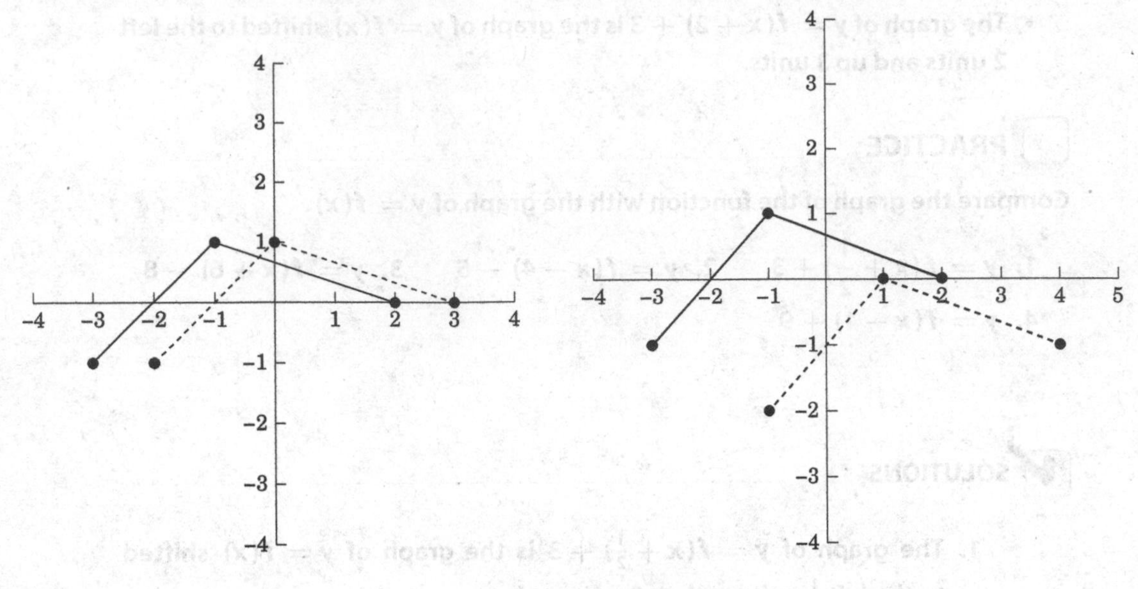

FIGURE 9-2

FIGURE 9-3

PRACTICE

Compare the graph of the function with the graph of $y = f(x)$, and then write the transformed function.

1. **See Fig. 9-4.**
2. **See Fig. 9-5.**
3. **See Fig. 9-6.**
4. **See Fig. 9-7.**

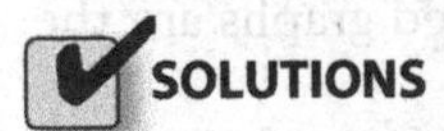

SOLUTIONS

1. **The dashed graph is the graph of $y = f(x)$ shifted down 1 unit. It is the graph of $y = f(x) - 1$.**
2. **The dashed graph is the graph of $y = f(x)$ shifted to the right 2 units and up 1 unit. It is the graph of $y = f(x - 2) + 1$.**
3. **The dashed graph is the graph of $f(x)$ shifted to the left 2 units. It is the graph of $y = f(x + 2)$.**
4. **The dashed graph is the graph of $y = f(x)$ shifted to the left 1 unit and down 2 units. It is the graph of $y = f(x + 1) - 2$.**

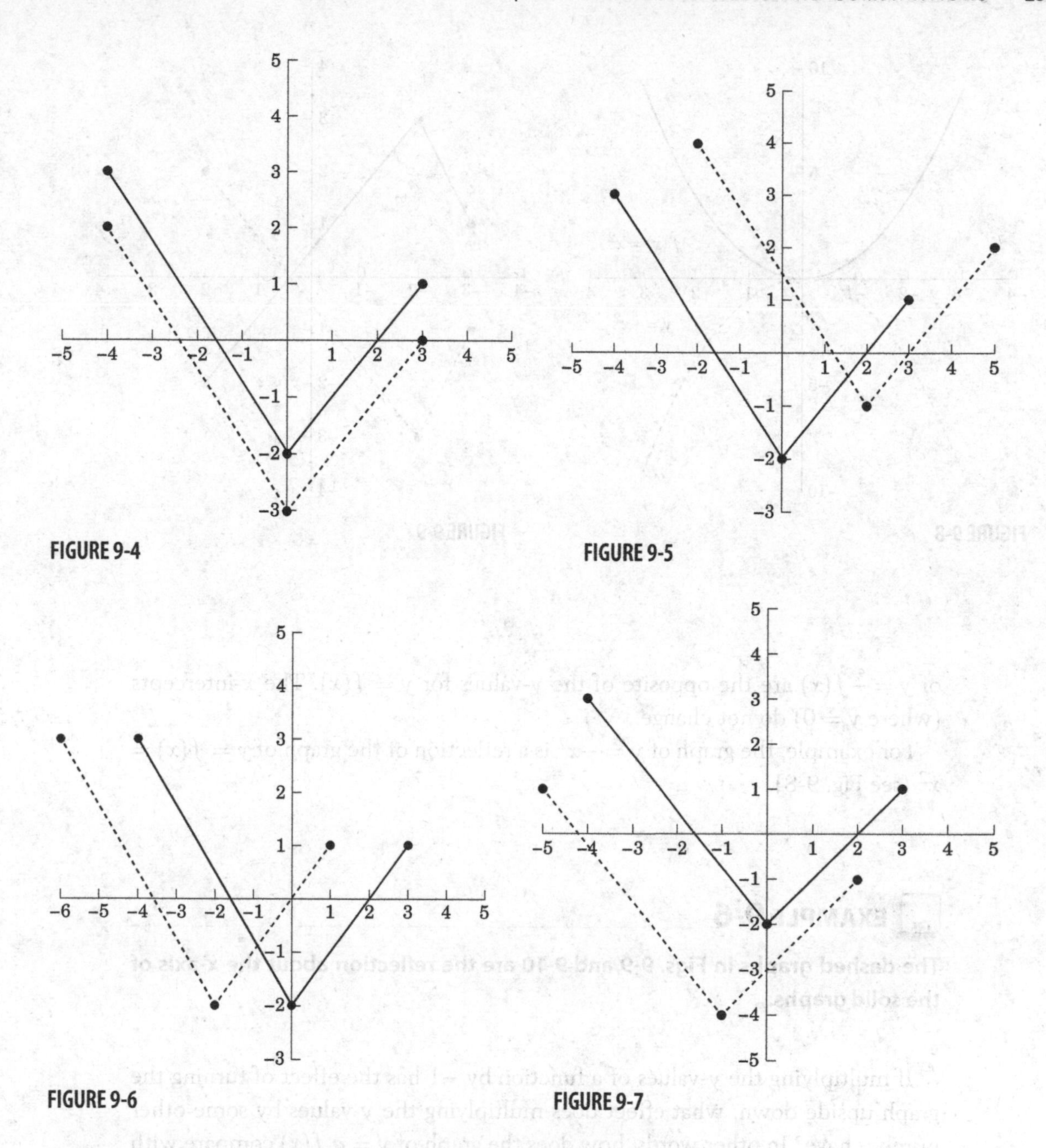

FIGURE 9-4

FIGURE 9-5

FIGURE 9-6

FIGURE 9-7

Reflections, and Vertical Stretching and Compressing

If we multiply every y-value in a function by -1, the effect on the graph is to "flip" the graph upside down or, in more technical terms, "reflect about the x-axis." With these reflections, the x-values do not change, but the y-values

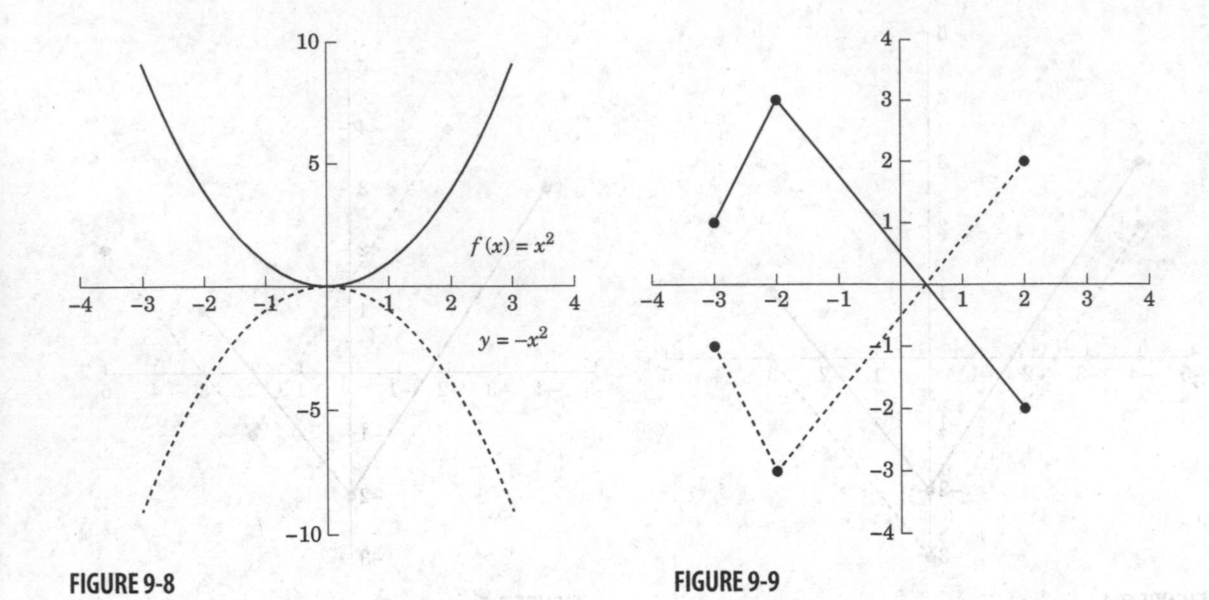

FIGURE 9-8

FIGURE 9-9

of $y = -f(x)$ are the opposite of the y-values for $y = f(x)$. The x-intercepts (where $y = 0$) do not change.

For example, the graph of $y = -x^2$ is a reflection of the graph of $y = f(x) = x^2$ (see Fig. 9-8).

EXAMPLE 9-6

The dashed graphs in Figs. 9-9 and 9-10 are the reflection about the *x*-axis of the solid graphs.

If multiplying the y-values of a function by -1 has the effect of turning the graph upside down, what effect does multiplying the y-values by some other number have? In other words, how does the graph of $y = a\,f(x)$ compare with the graph of $y = f(x)$? It depends on a. If a is larger than 1, the graph of $y = a\,f(x)$ is vertically stretched. For example, the graph of $y = 50\,f(x)$ is stretched more than the graph of $y = 3\,f(x)$. If a is between 0 and 1, then the graph of $y = a\,f(x)$ is vertically compressed or flattened. The graph of $y = \frac{1}{10}\,f(x)$ is flattened more than the graph of $y = \frac{2}{3}\,f(x)$.

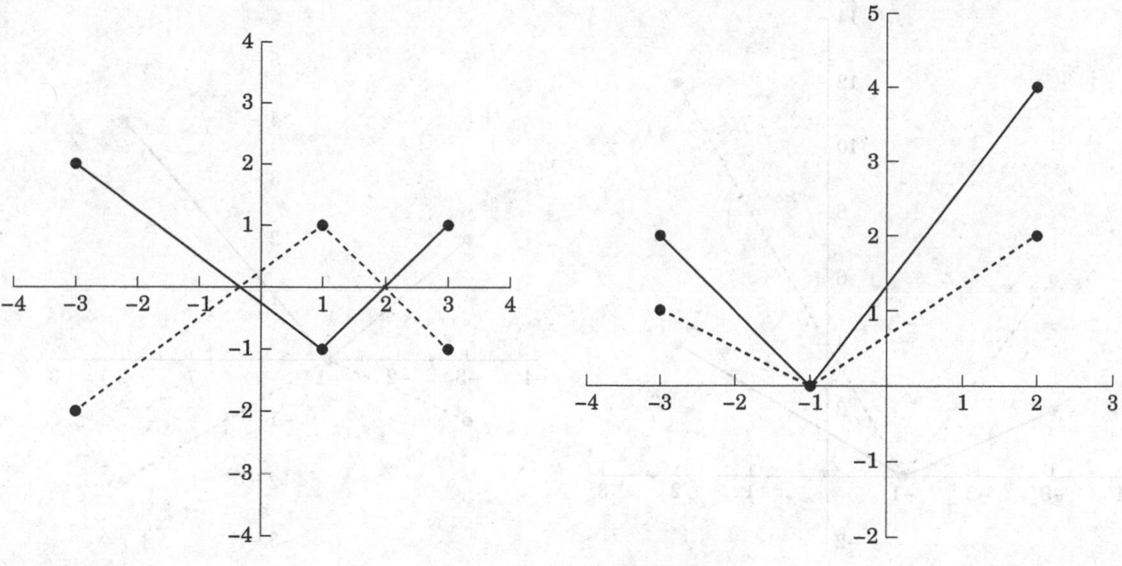

FIGURE 9-10

FIGURE 9-11

EXAMPLE 9-7

- **The dashed graph in Fig. 9-11 is the graph of $y = \frac{1}{2} f(x)$. The y-value for each point on the dashed graph is half of the corresponding y-value in the solid graph. For example, the point $(-3, 2)$ on the graph of $f(x)$ is moved to $(-3, 1)$ on the graph of $y = \frac{1}{2} f(x)$. The point $(2, 4)$ on the solid graph is moved to $(2, 2)$ on the dashed graph.**
- **The dashed graph in Fig. 9-12 is the graph of $y = 3 f(x)$. The y-values for each point on the dashed graph are three times the y-values on the solid graph. For example, the point $(-3, 2)$ on the solid graph is moved to $(-3, 6)$ on the dashed graph. The point $(-1, 0)$ on the solid graph did not move because $3 \cdot 0 = 0$.**

When a is a negative number, other than -1, the effect of $y = a\,f(x)$ is a combination of the changes above. First, the graph is turned upside down (reflected about the x-axis). Then it is either vertically compressed or stretched. In Figs. 9-13–9-20 the solid graph is the graph of $y = f(x)$ and the dashed graphs are the graphs of $y = a\,f(x)$.

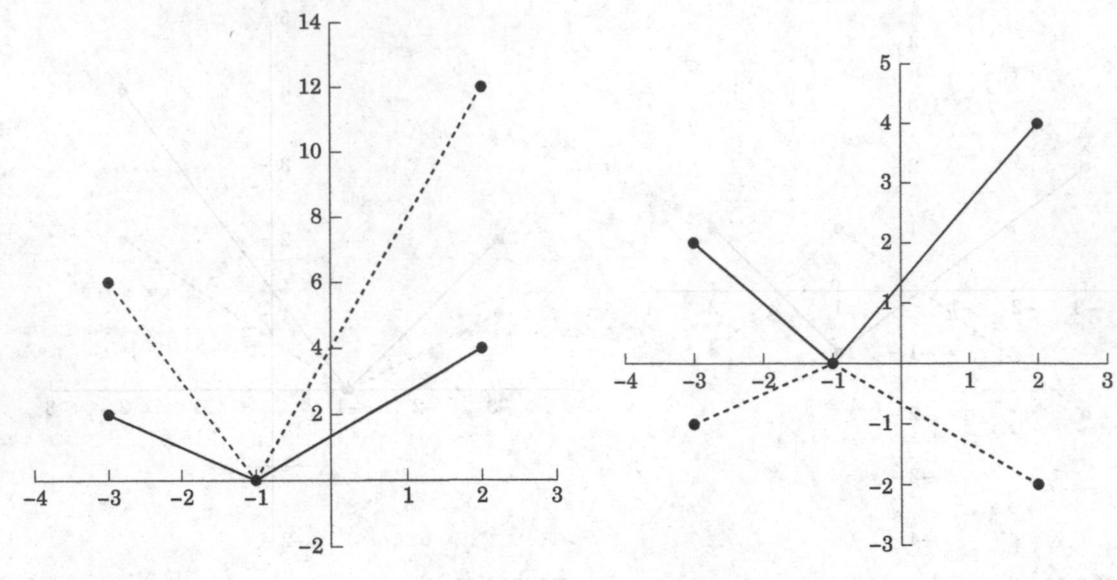

FIGURE 9-12

FIGURE 9-13

EXAMPLE 9-8

- **The dashed graph in Fig. 9-13 is the graph of $y = -\frac{1}{2}\ f(x)$. The point $(-3, 2)$ on the solid graph is moved to $(-3, -1)$ because $2 \cdot (-\frac{1}{2}) = -1$. The point $(2, 4)$ on the solid graph is moved to $(2, -2)$ because $4 \cdot (-\frac{1}{2}) = -2$. The point $(-1, 0)$ on the solid graph did not move because $0 \cdot (-\frac{1}{2}) = 0$.**
- **The dashed graph in Fig. 9-14 is the graph of $y = -3\ f(x)$. Notice that the points $(2, 4)$ and $(-3, 2)$ have moved to $(2, -3(4))$ and $(-3, -3(2))$. That is, the point (x, y) on the graph of $y = f(x)$ moves to $(x, -3y)$ for the graph of $y = -3\ f(x)$.**

To summarize, if $a > 1$, the graph of $y = a\ f(x)$ is vertically stretched. If $0 < a < 1$, the graph of $y = a\ f(x)$ is vertically flattened. If $a < -1$, the graph is reflected about the x-axis and vertically stretched. If $-1 < a < 0$, the graph is reflected about the x-axis and vertically flattened (or compressed).

PRACTICE

For Problems 1-4, determine whether the dashed graph is vertically stretched or a flattened version of the solid graph and whether or not it is reflected about the *x*-axis.

1. **See Fig. 9-15.** 2. **See Fig. 9-16.** 3. **See Fig. 9-17.** 4. **See Fig. 9-18.**

5. **The solid graphs in Figs. 9-19 and 9-20 are the graphs of $y = f(x)$. One of the dashed graphs is the graph of $y = \frac{3}{2} f(x)$ and the other is the graph of $y = 4\ f(x)$. Which graph is the graph of $y = \frac{3}{2} f(x)$? Which is the graph of $y = 4\ f(x)$?**

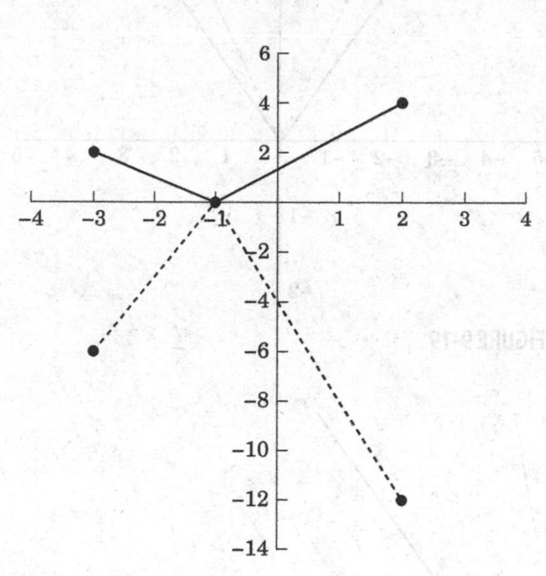

FIGURE 9-14

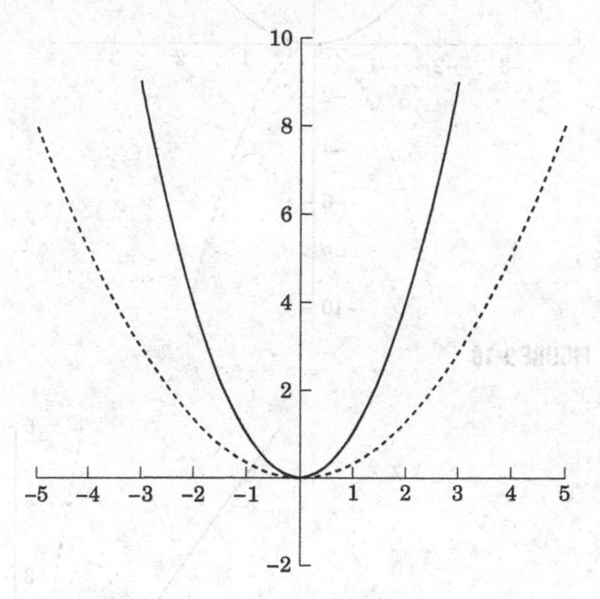

FIGURE 9-15

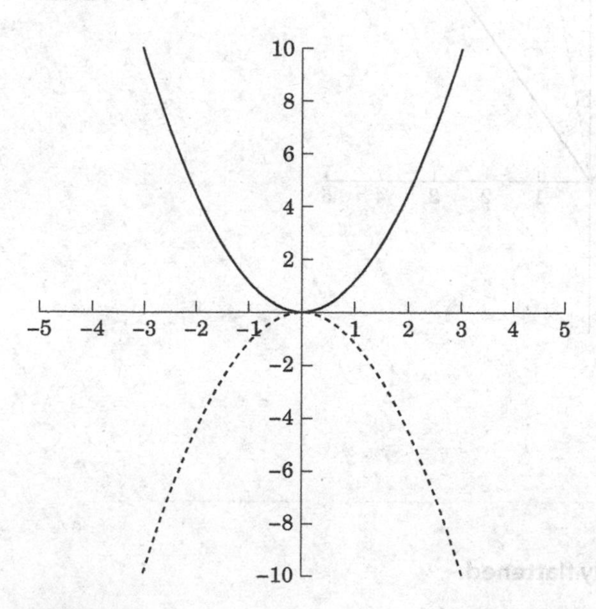

FIGURE 9-16

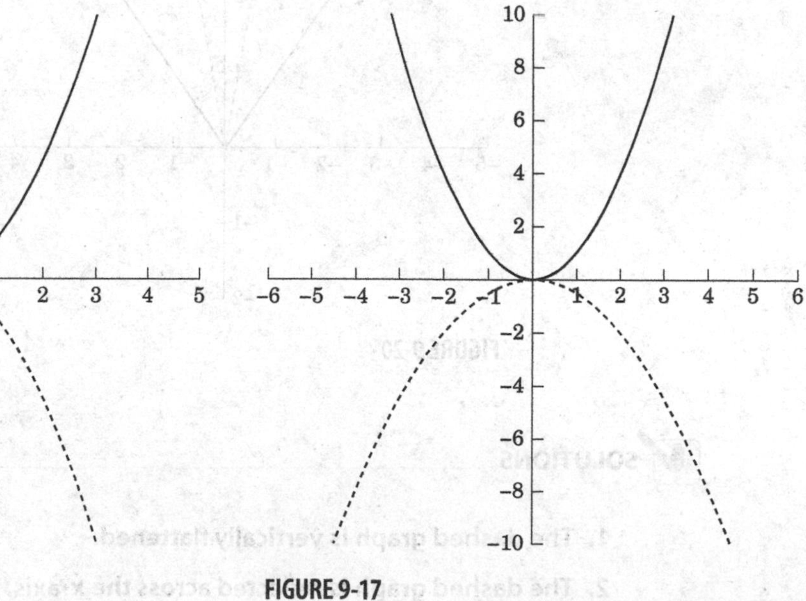

FIGURE 9-17

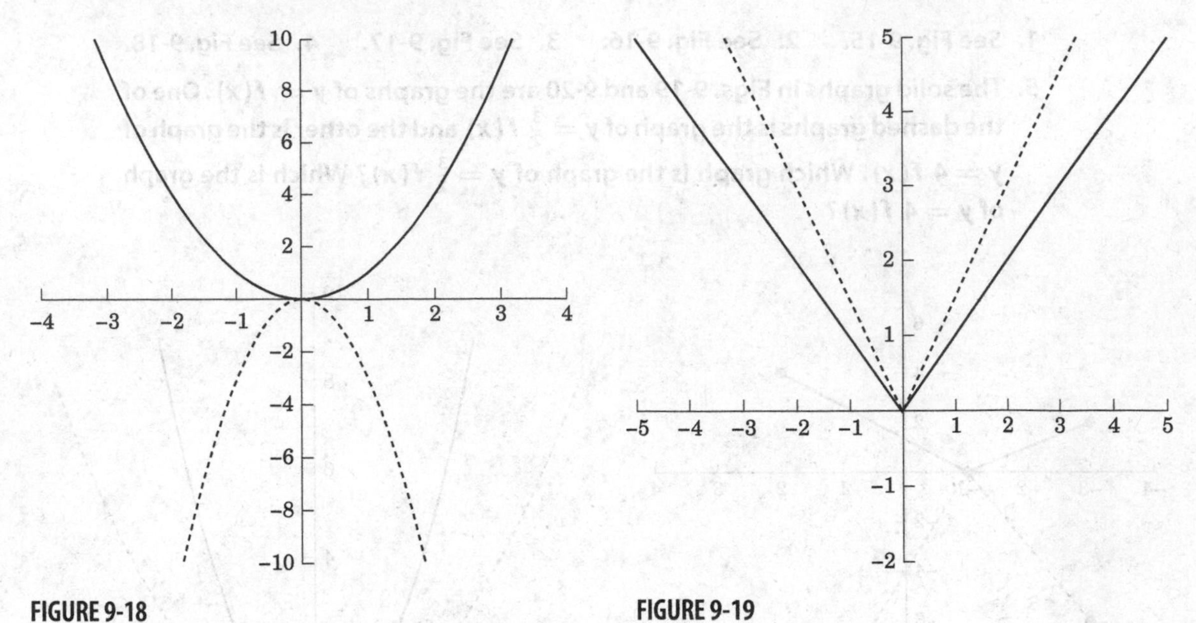

FIGURE 9-18

FIGURE 9-19

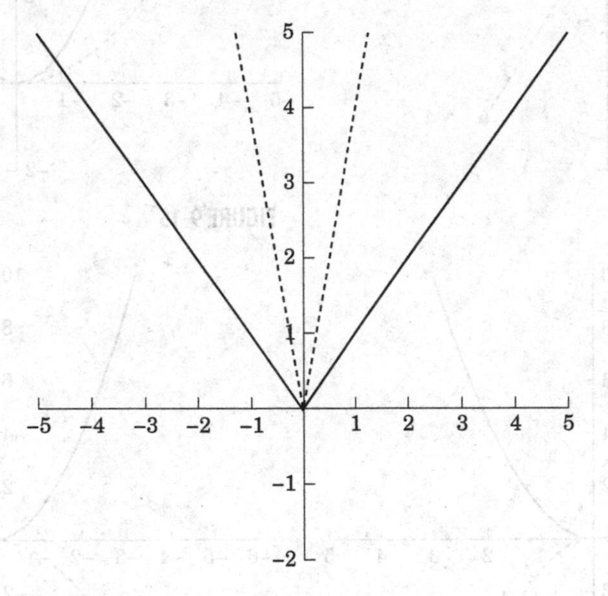

FIGURE 9-20

SOLUTIONS

1. **The dashed graph is vertically flattened.**
2. **The dashed graph is reflected across the *x*-axis.**

3. **The dashed graph is vertically flattened and is reflected about the *x*-axis.**

4. **The dashed graph is vertically stretched and is reflected about the *x*-axis.**

5. **The dashed graph in Fig. 9-19 is the graph of $y = \frac{3}{2}f(x)$. The dashed graph in Fig. 9-20 is the graph of $y = 4f(x)$.**

Sketching the Graph of a Transformation

We now look at one more transformation, $f(-x)$. The transformation $-f(x)$ turns the graph upside down. The transformation $f(-x)$ turns the graphs sideways or "is reflected about the y-axis." The solid graph in Fig. 9-21 is the graph of $y = f(x)$, and the dashed graph is the graph of $y = f(-x)$. We can get the graph of $y = f(-x)$ by replacing the x-values with their opposites. For example, the point (2, 4) on $f(x)$ is replaced by (−2, 4) for the graph of $y = f(-x)$.

We are ready to sketch transformations of a given graph. The graph of $y = f(x)$ is given in Fig. 9-22, and we are asked to sketch some transformation. Some of the transformations can be done with no extra work, but we need to

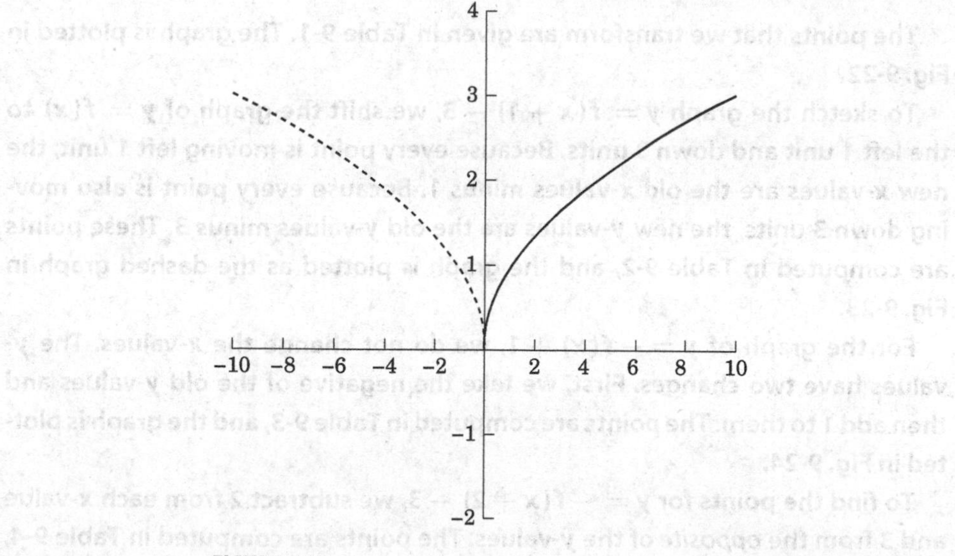

FIGURE 9-21

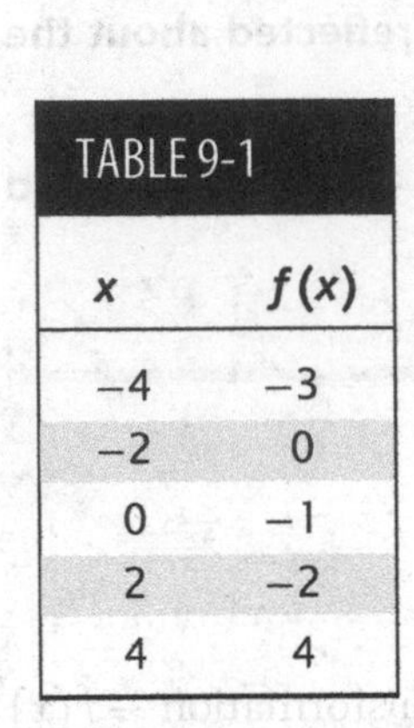

TABLE 9-1

x	$f(x)$
−4	−3
−2	0
0	−1
2	−2
4	4

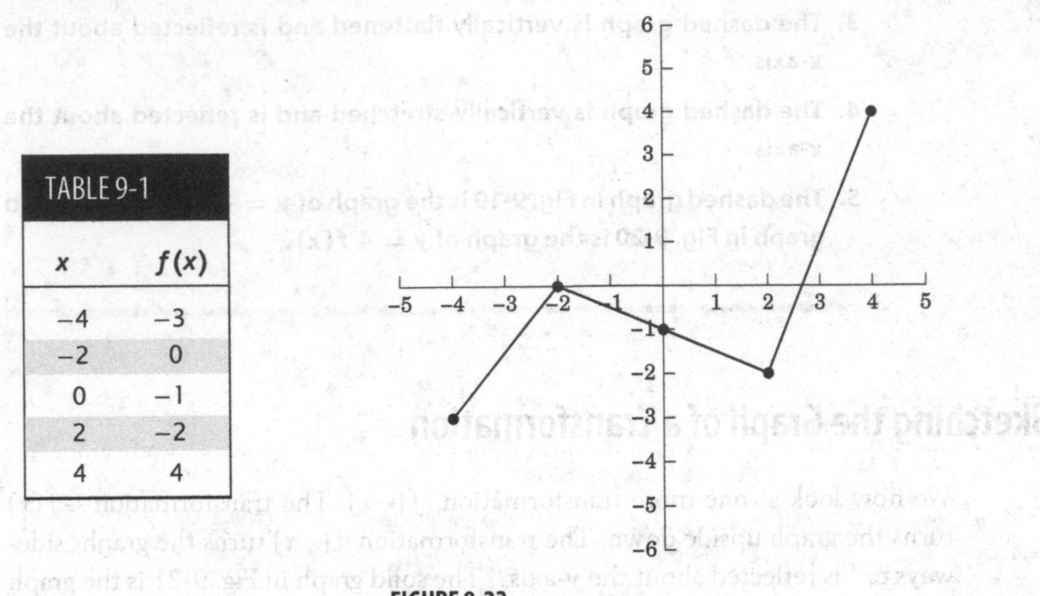

FIGURE 9-22

be careful with others. To help with the more complicated transformations, we use tables of values.

EXAMPLE 9-9

- **Sketch the transformations $y = f(x+1) - 3$, $y = -f(x) + 1$, $y = -f(x+2) - 3$, $y = 2f(x)$, $y = f(-x)$, and $y = f(-x) - 1$.**

The points that we transform are given in Table 9-1. The graph is plotted in Fig. 9-22.

To sketch the graph $y = f(x+1) - 3$, we shift the graph of $y = f(x)$ to the left 1 unit and down 3 units. Because every point is moving left 1 unit, the new x-values are the old x-values minus 1. Because every point is also moving down 3 units, the new y-values are the old y-values minus 3. These points are computed in Table 9-2, and the graph is plotted as the dashed graph in Fig. 9-23.

For the graph of $y = -f(x) + 1$, we do not change the x-values. The y-values have two changes. First, we take the negative of the old y-values and then add 1 to them. The points are computed in Table 9-3, and the graph is plotted in Fig. 9-24.

To find the points for $y = -f(x+2) - 3$, we subtract 2 from each x-value and 3 from the *opposite* of the y-values. The points are computed in Table 9-4, and the graph is sketched in Fig. 9-25.

TABLE 9-2		
$x-1$	$y-3$	Plot This Point
$-4-1=-5$	$-3-3=-6$	$(-5,-6)$
$-2-1=-3$	$0-3=-3$	$(-3,-3)$
$0-1=-1$	$-1-3=-4$	$(-1,-4)$
$2-1=1$	$-2-3=-5$	$(1,-5)$
$4-1=3$	$4-3=1$	$(3,1)$

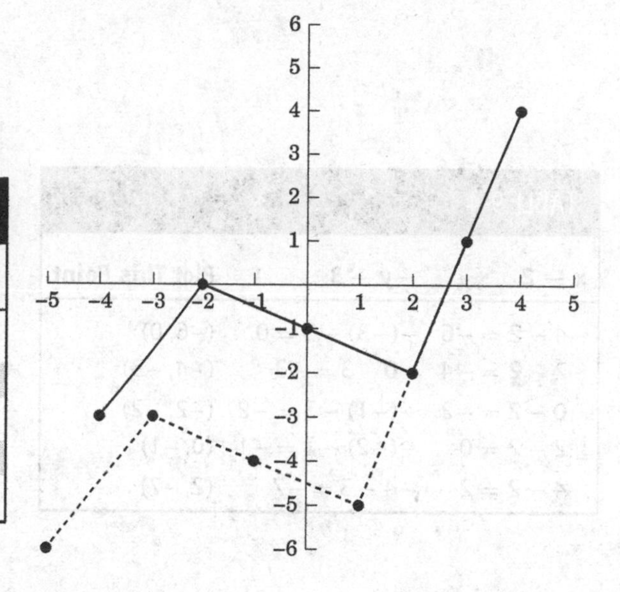

FIGURE 9-23

TABLE 9-3		
x	$-y+1$	Plot This Point
-4	$-(-3)+1=4$	$(-4,4)$
-2	$-0+1=1$	$(-2,1)$
0	$-(-1)+1=2$	$(0,2)$
2	$-(-2)+1=3$	$(2,3)$
4	$-4+1=-3$	$(4,-3)$

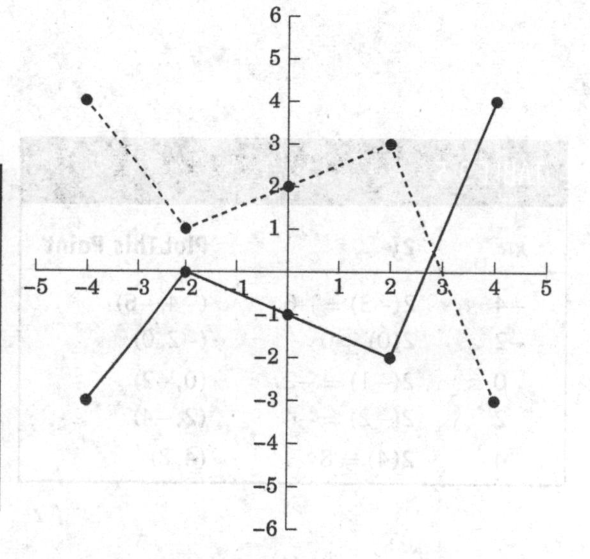

FIGURE 9-24

We sketch the graph of $y = 2\,f(x)$ by multiplying each y-value by 2. The x-values do not change. The points are computed in Table 9-5, and the graph is plotted in Fig. 9-26.

TABLE 9-4

$x-2$	$-y-3$	Plot This Point
$-4-2=-6$	$-(-3)-3=0$	$(-6,0)$
$-2-2=-4$	$-0-3=-3$	$(-4,-3)$
$0-2=-2$	$-(-1)-3=-2$	$(-2,-2)$
$2-2=0$	$-(-2)-3=-1$	$(0,-1)$
$4-2=2$	$-4-3=-7$	$(2,-7)$

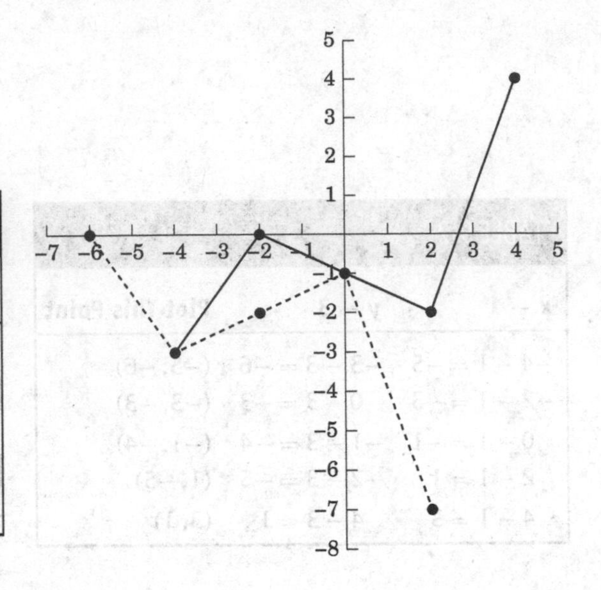

FIGURE 9-25

TABLE 9-5

x	$2y$	Plot This Point
-4	$2(-3)=-6$	$(-4,-6)$
-2	$2(0)=0$	$(-2,0)$
0	$2(-1)=-2$	$(0,-2)$
2	$2(-2)=-4$	$(2,-4)$
4	$2(4)=8$	$(4,8)$

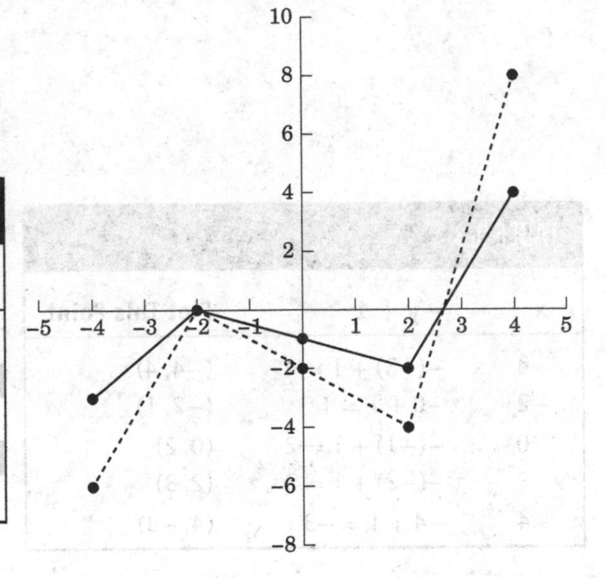

FIGURE 9-26

We sketch the graph of $y = f(-x)$ by replacing each x-value with its opposite. The y-values do not change. The points are computed in Table 9-6, and the graph is sketched in Fig. 9-27.

TABLE 9-6

$-x$	y	Plot This Point
$-(-4) = 4$	-3	$(4, -3)$
$-(-2) = 2$	0	$(2, 0)$
$-0 = 0$	-1	$(0, -1)$
-2	-2	$(-2, -2)$
-4	4	$(-4, 4)$

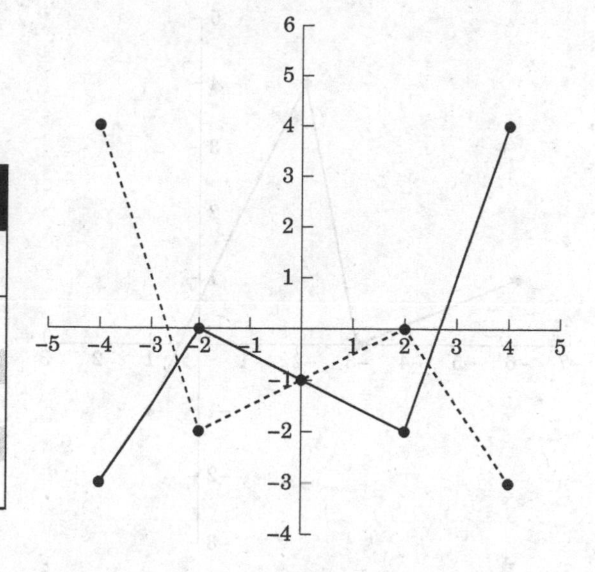

FIGURE 9-27

TABLE 9-7

$-x$	$y - 1$	Plot This Point
$-(-4) = 4$	$-3 - 1 = -4$	$(4, -4)$
$-(-2) = 2$	$0 - 1 = -1$	$(2, -1)$
$-0 = 0$	$-1 - 1 = -2$	$(0, -2)$
-2	$-2 - 1 = -3$	$(-2, -3)$
-4	$4 - 1 = 3$	$(-4, 3)$

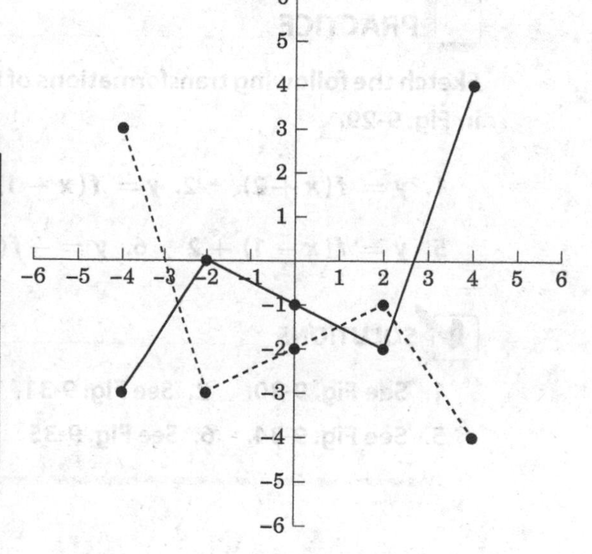

FIGURE 9-28

We sketch the graph of $y = f(-x) - 1$ by taking the the opposite of each x-value and by subtracting 1 from each y-value. The points are computed in Table 9-7, and the graph is sketched in Fig. 9-28.

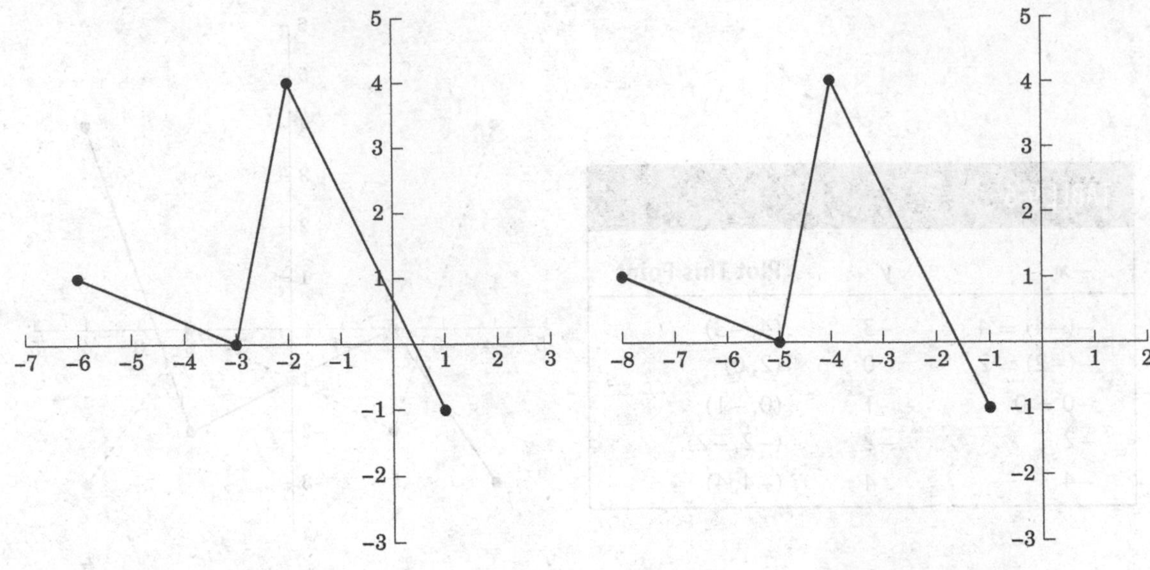

FIGURE 9-29

FIGURE 9-30

PRACTICE

Sketch the following transformations of the function $f(x)$ whose graph is given in Fig. 9-29.

1. $y = f(x+2)$ 2. $y = f(x-1)$ 3. $y = \frac{2}{3} f(x)$ 4. $y = f(-x)$

5. $y = f(x-1)+2$ 6. $y = -f(x)+1$ 7. $y = 2\,f(x-2)$

SOLUTIONS

1. See Fig. 9-30. 2. See Fig. 9-31. 3. See Fig. 9-32. 4. See Fig. 9-33.

5. See Fig. 9-34. 6. See Fig. 9-35. 7. See Fig. 9-36.

Special Functions

There are several families of functions whose graphs college algebra students should know. They are summarized in Table 9-8.

Once we know the basic shape of the graphs of these functions, we can use what we learned earlier in this chapter to sketch the graphs of many functions

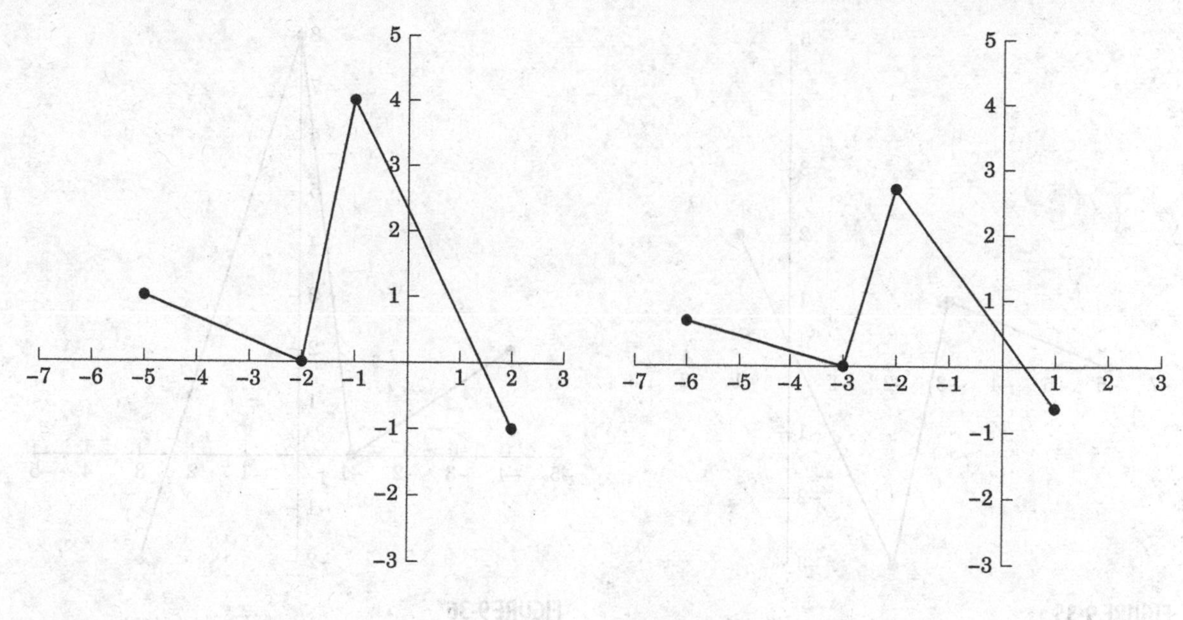

FIGURE 9-31

FIGURE 9-32

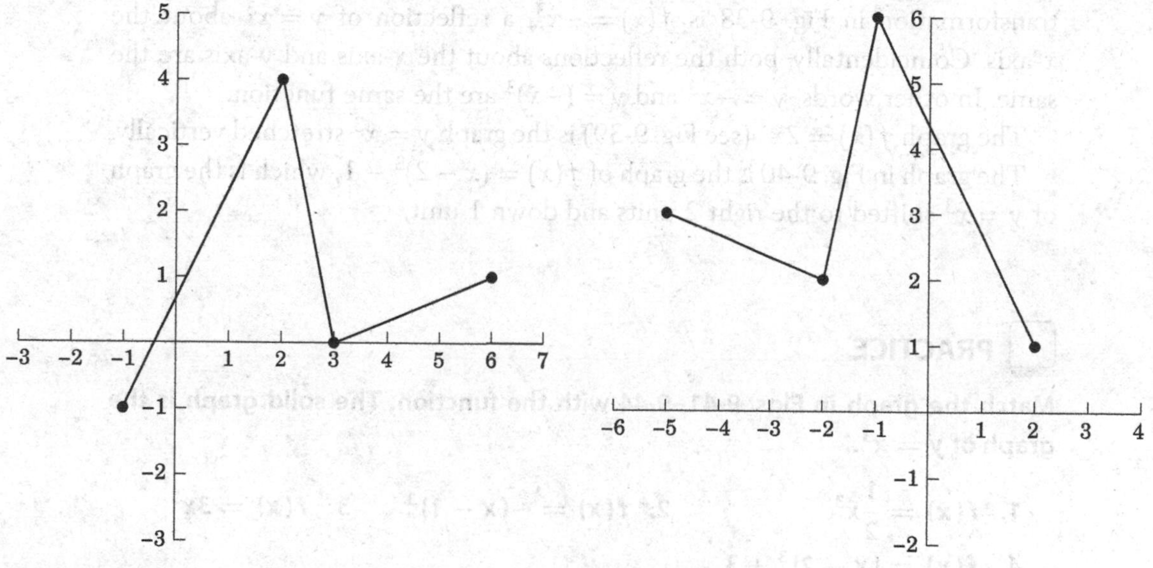

FIGURE 9-33

FIGURE 9-34

with only a little work. This information can help to use a graphing calculator more effectively, too.

Fig. 9-37 is the graph of $y = x^3$. The solid graphs in Figs. 9-38–9-44 are the graphs of $y = x^3$ and the dashed graphs are transformations of $y = x^3$. The

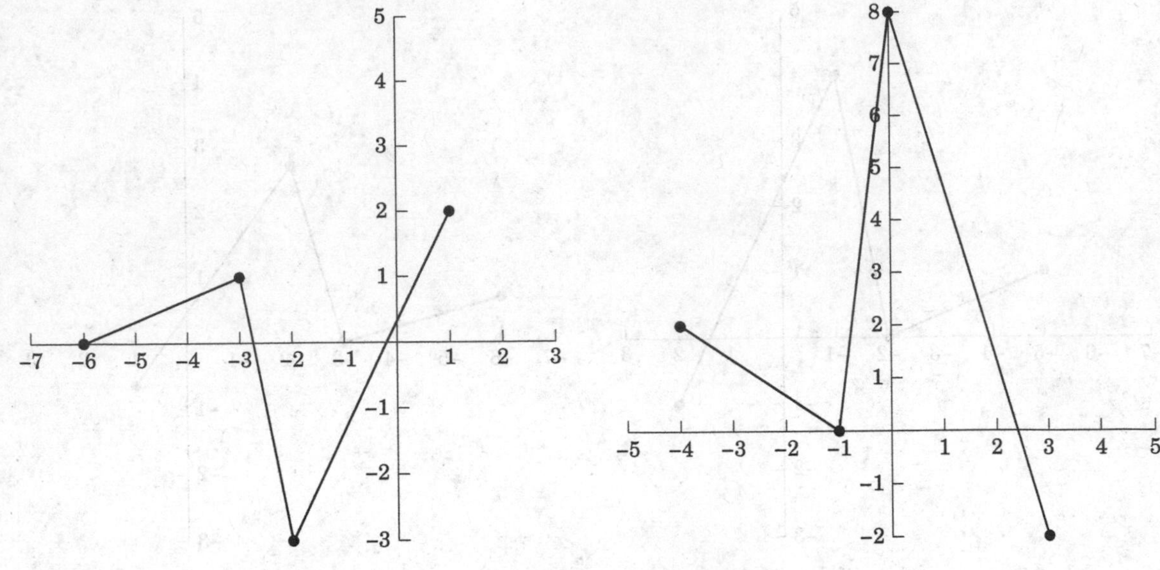

FIGURE 9-35

FIGURE 9-36

transformation in Fig. 9-38 is $f(x) = -x^3$, a reflection of $y = x^3$ about the x-axis. Coincidentally, both the reflections about the x-axis and y-axis are the same. In other words, $y = -x^3$ and $y = (-x)^3$ are the same function.

The graph $f(x) = 2x^3$ (see Fig. 9-39) is the graph $y = x^3$ stretched vertically.

The graph in Fig. 9-40 is the graph of $f(x) = (x - 2)^3 - 1$, which is the graph of $y = x^3$ shifted to the right 2 units and down 1 unit.

PRACTICE

Match the graph in Figs. 9-41–9-44 with the function. The solid graph is the graph of $y = x^3$.

1. $f(x) = \frac{1}{2}x^3$
2. $f(x) = -(x - 1)^3$
3. $f(x) = 3x^3$
4. $f(x) = (x - 2)^3 + 3$

SOLUTIONS

1. See Fig. 9-43 2. See Fig. 9-44 3. See Fig. 9-41 4. See Fig. 9-42

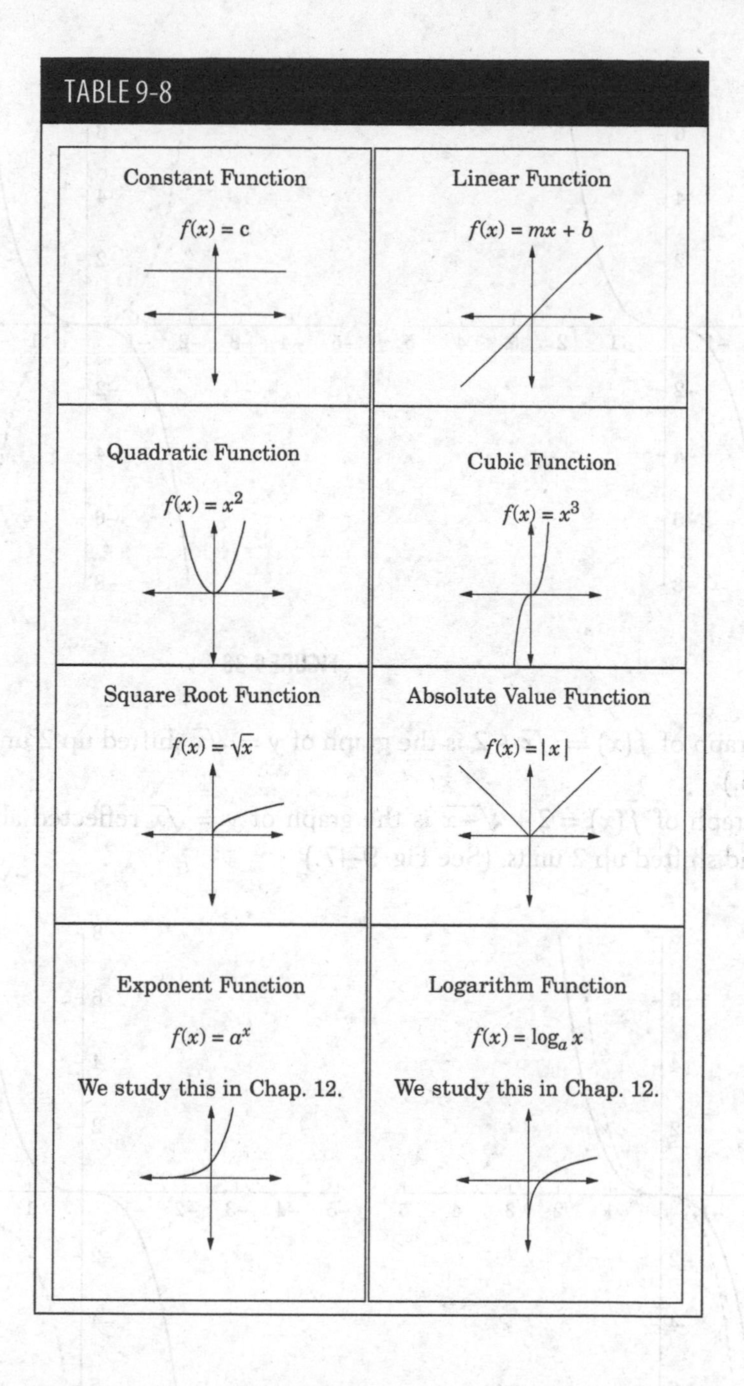

TABLE 9-8

Constant Function $f(x) = c$	Linear Function $f(x) = mx + b$
Quadratic Function $f(x) = x^2$	Cubic Function $f(x) = x^3$
Square Root Function $f(x) = \sqrt{x}$	Absolute Value Function $f(x) = \lvert x \rvert$
Exponent Function $f(x) = a^x$ We study this in Chap. 12.	Logarithm Function $f(x) = \log_a x$ We study this in Chap. 12.

The graph of $y = \sqrt{x}$ is part of a parabola. Imagine turning a parabola on its side and cutting off the bottom half. What would be left is the graph of $y = \sqrt{x}$. The graph of $y = \sqrt{x}$ is in Fig. 9-45. The solid graphs in Figs. 9-46–9-54 are the graphs of $y = \sqrt{x}$. The dashed graphs are transformations.

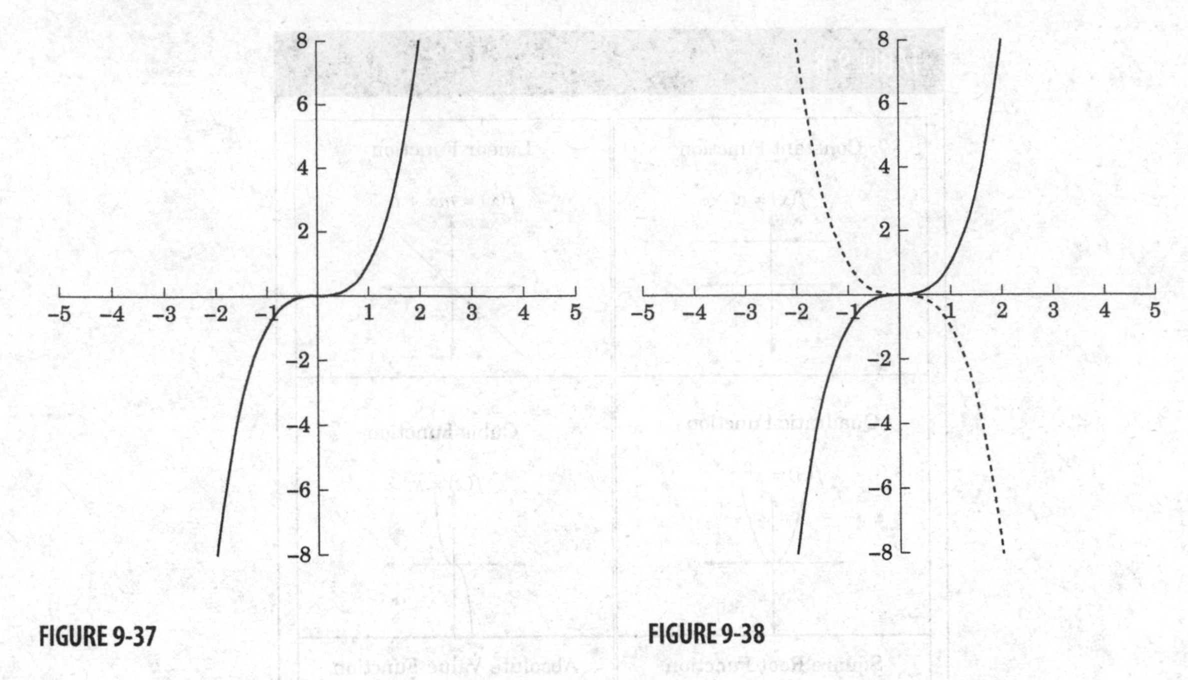

FIGURE 9-37

FIGURE 9-38

The graph of $f(x) = \sqrt{x} + 2$ is the graph of $y = \sqrt{x}$ shifted up 2 units. (See Fig. 9-46.)

The graph of $f(x) = 2 + \sqrt{-x}$ is the graph of $y = \sqrt{x}$ reflected about the y-axis and shifted up 2 units. (See Fig. 9-47.)

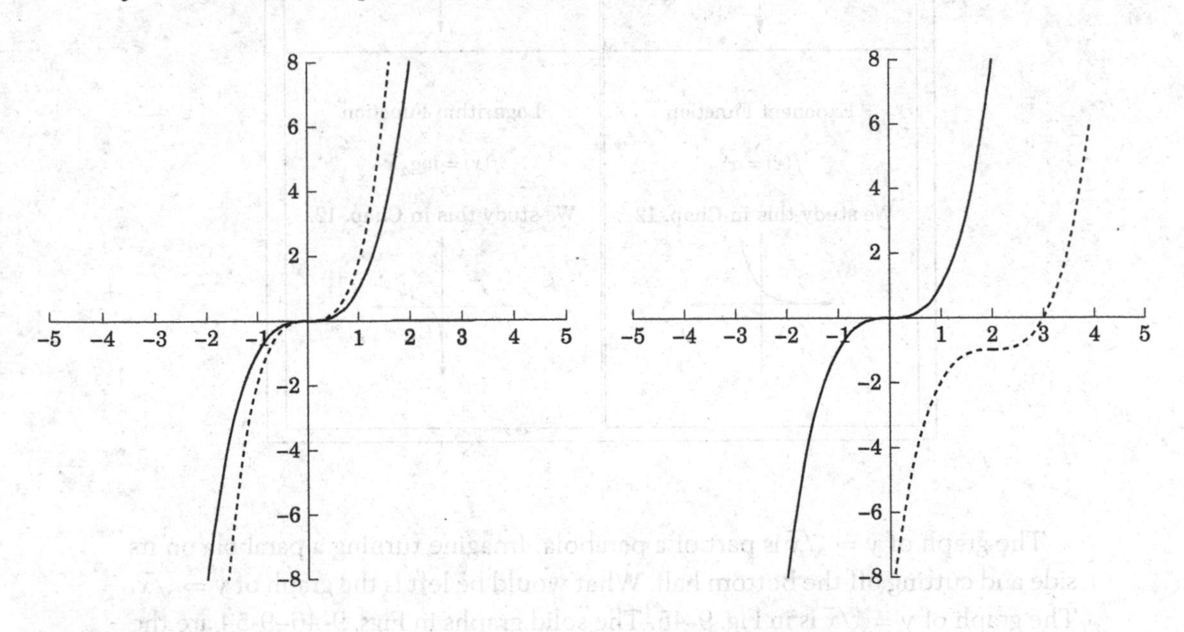

FIGURE 9-39

FIGURE 9-40

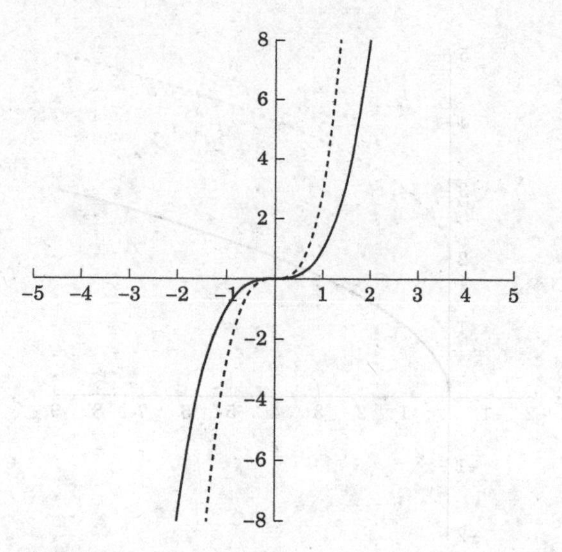

FIGURE 9-41

FIGURE 9-42

The graph of $f(x) = -\sqrt{x}$ is the graph of $y = \sqrt{x}$ reflected about the x-axis. (See Fig. 9-48.)

The graph of $f(x) = \sqrt{-(x-2)}$ is the graph of $y = \sqrt{x}$ first reflected across the y-axis and then shifted to the right 2 units. (See Fig. 9-49.)

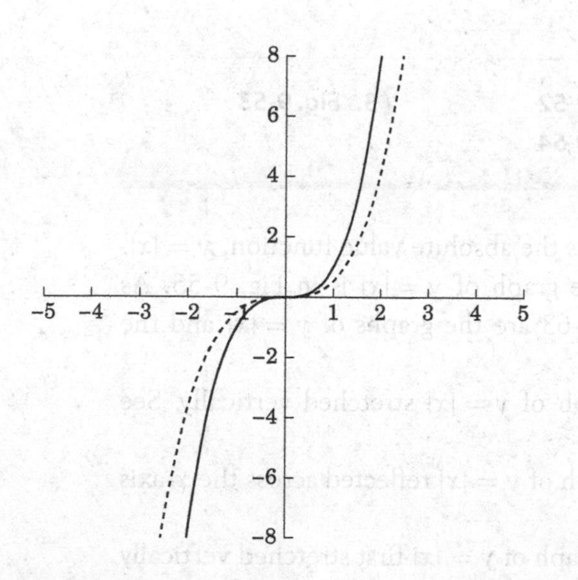

FIGURE 9-43

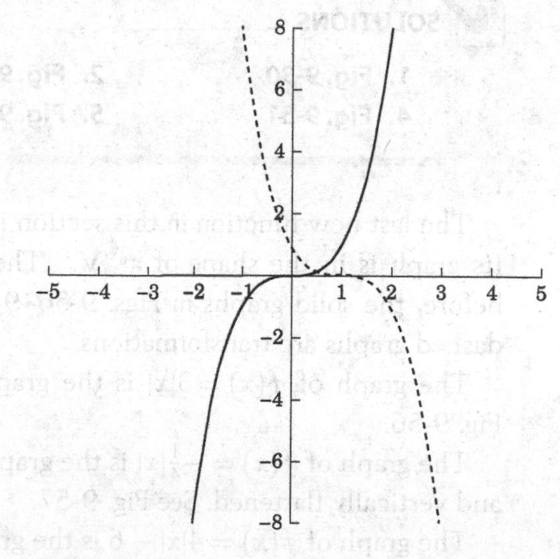

FIGURE 9-44

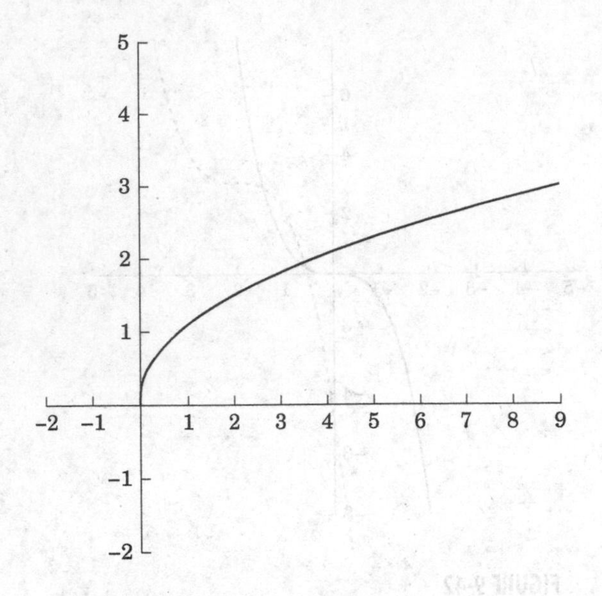

FIGURE 9-45

FIGURE 9-46

PRACTICE

Match the graph in Figs. 9-50–9-54 with the function.

1. $f(x) = \sqrt{1-x}$
2. $f(x) = -3 + \sqrt{x}$
3. $f(x) = 4\sqrt{x}$
4. $f(x) = 1 - \sqrt{x}$
5. $f(x) = \frac{1}{2}\sqrt{x}$

SOLUTIONS

1. Fig. 9-50
2. Fig. 9-52
3. Fig. 9-53
4. Fig. 9-51
5. Fig. 9-54

The last new function in this section is the absolute value function, $y = |x|$. Its graph is in the shape of a "V." The graph of $y = |x|$ is in Fig. 9-55. As before, the solid graphs in Figs. 9-56–9-63 are the graphs of $y = |x|$ and the dashed graphs are transformations.

The graph of $f(x) = 3|x|$ is the graph of $y = |x|$ stretched vertically. See Fig. 9-56.

The graph of $f(x) = -\frac{1}{4}|x|$ is the graph of $y = |x|$ reflected across the x-axis and vertically flattened. See Fig. 9-57.

The graph of $f(x) = 4|x| - 6$ is the graph of $y = |x|$ first stretched vertically and then shifted down 6 units. See Fig. 9-58.

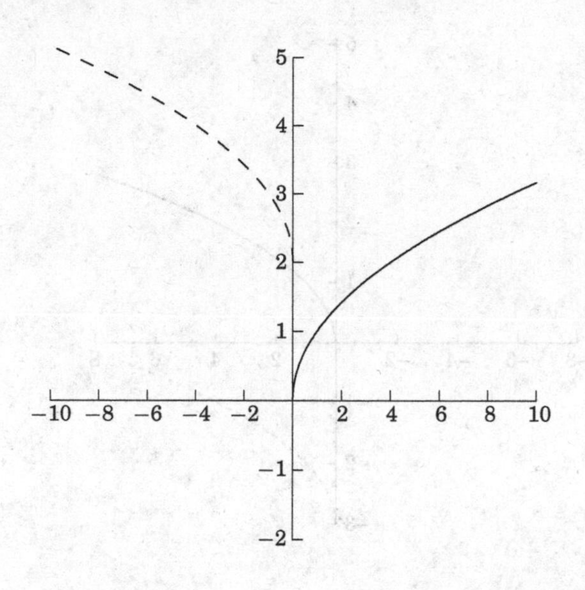

FIGURE 9-47

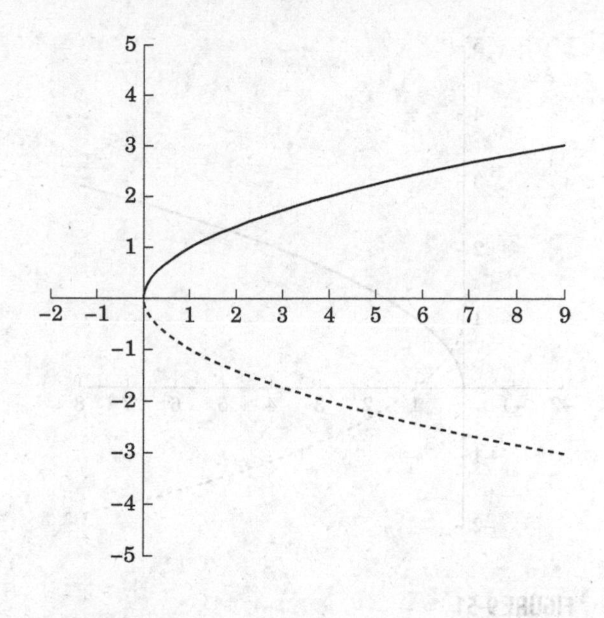

FIGURE 9-48

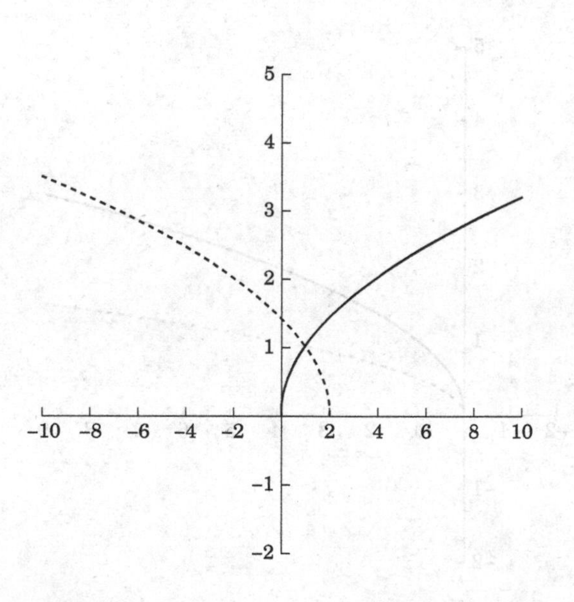

FIGURE 9-49

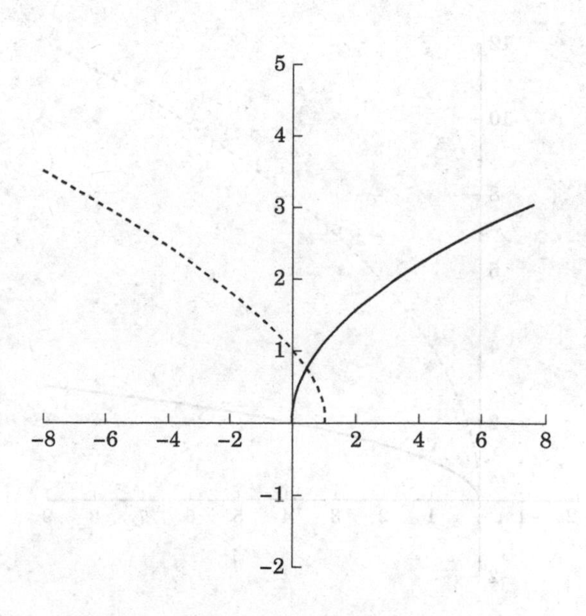

FIGURE 9-50

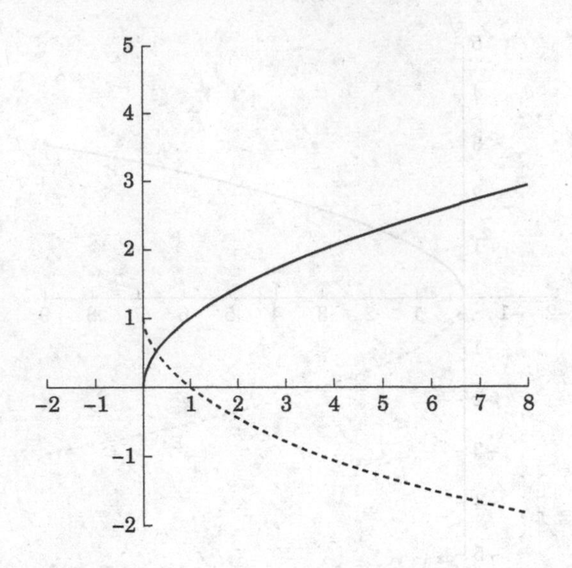

FIGURE 9-51

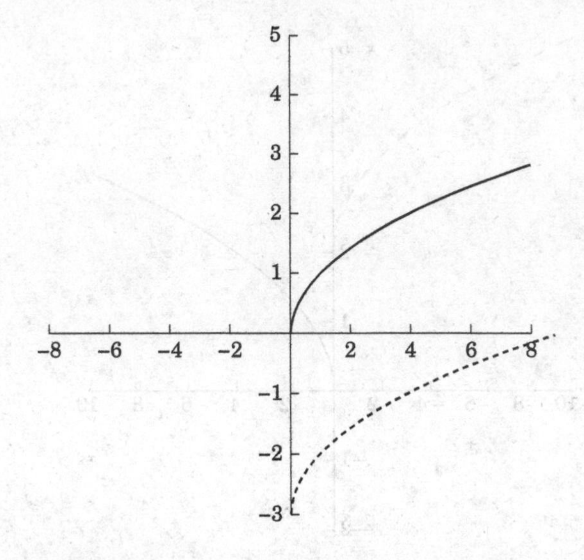

FIGURE 9-52

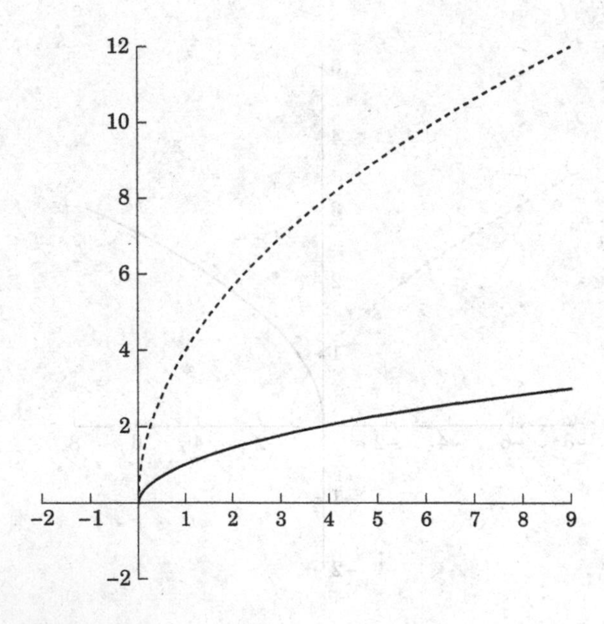

FIGURE 9-53

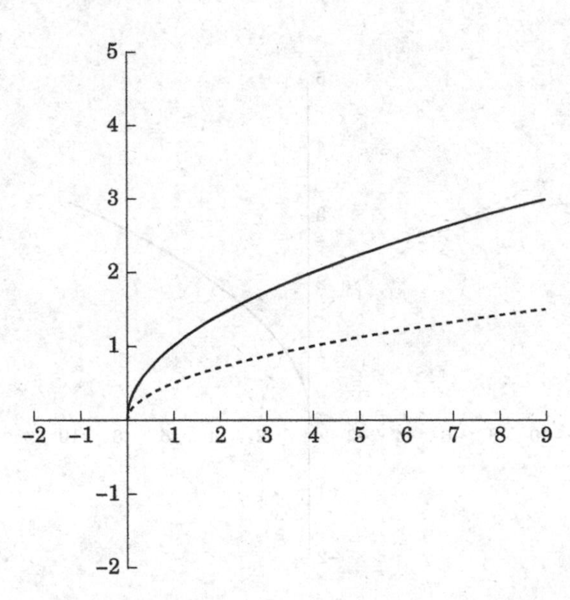

FIGURE 9-54

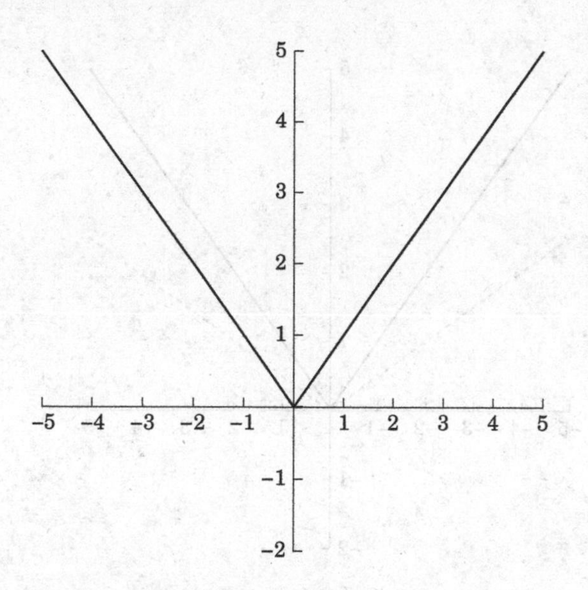

FIGURE 9-55

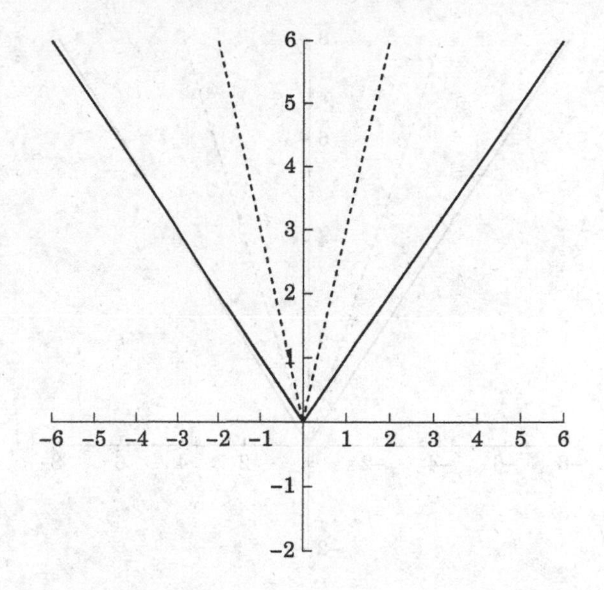

FIGURE 9-56

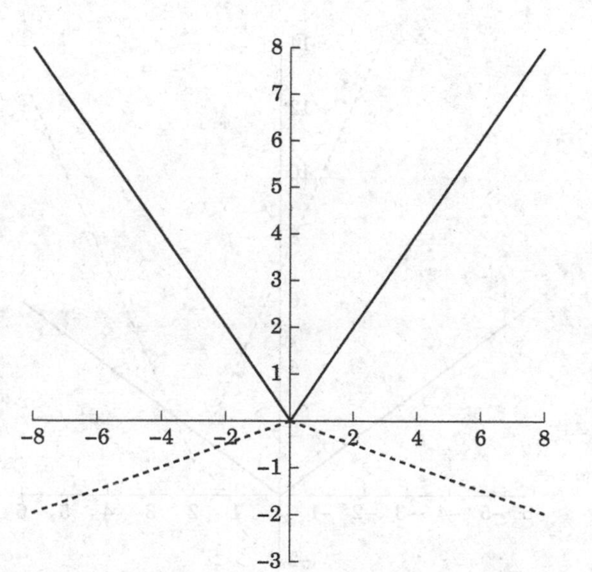

FIGURE 9-57

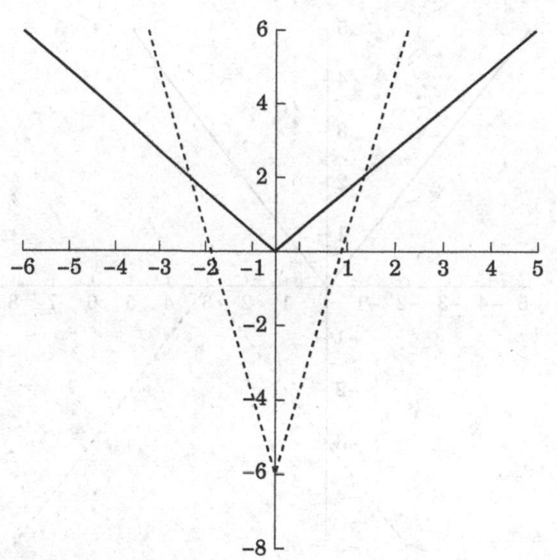

FIGURE 9-58

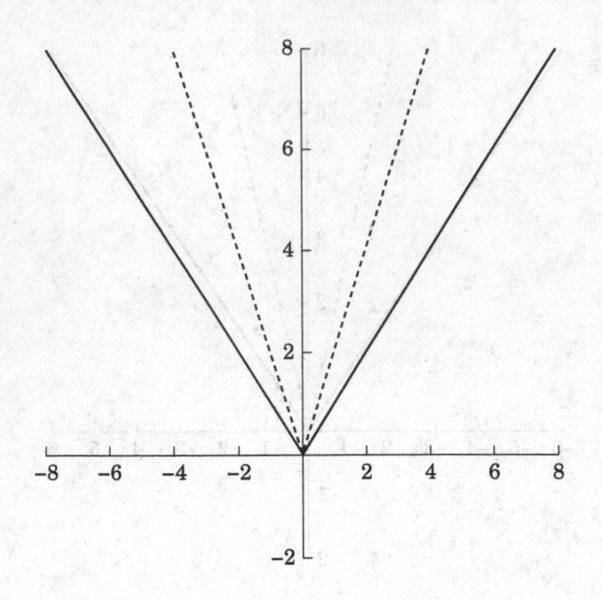

FIGURE 9-59

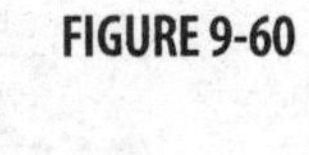

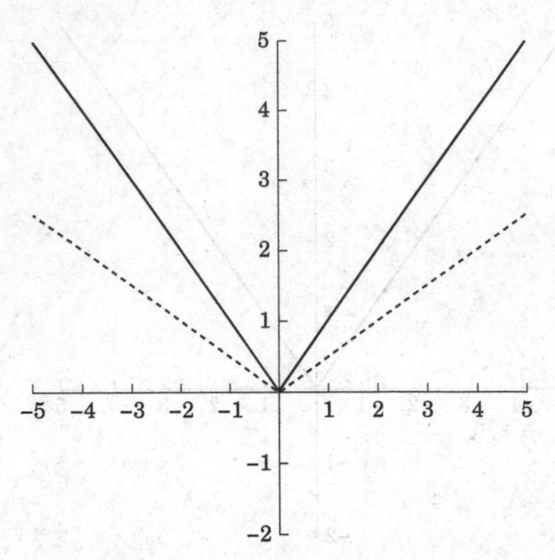

FIGURE 9-60

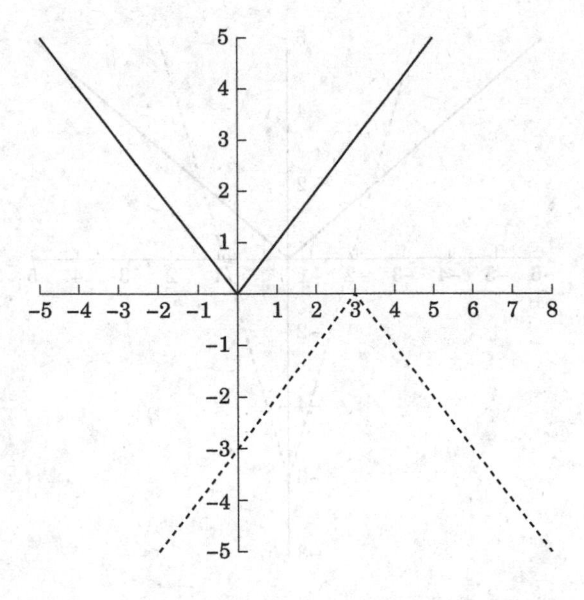

FIGURE 9-61

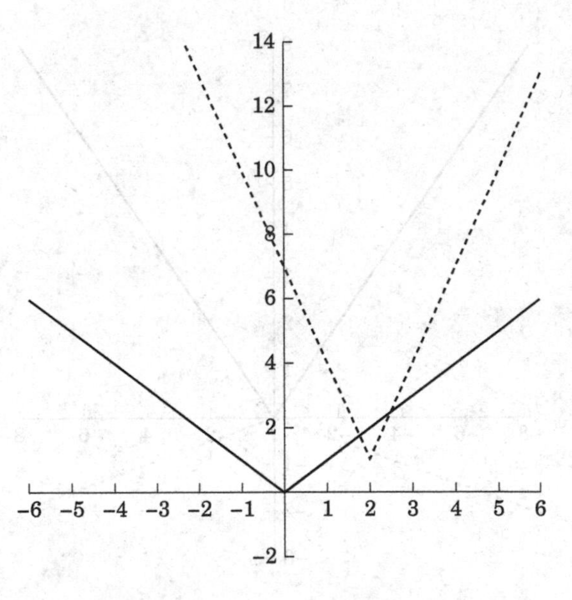

FIGURE 9-62

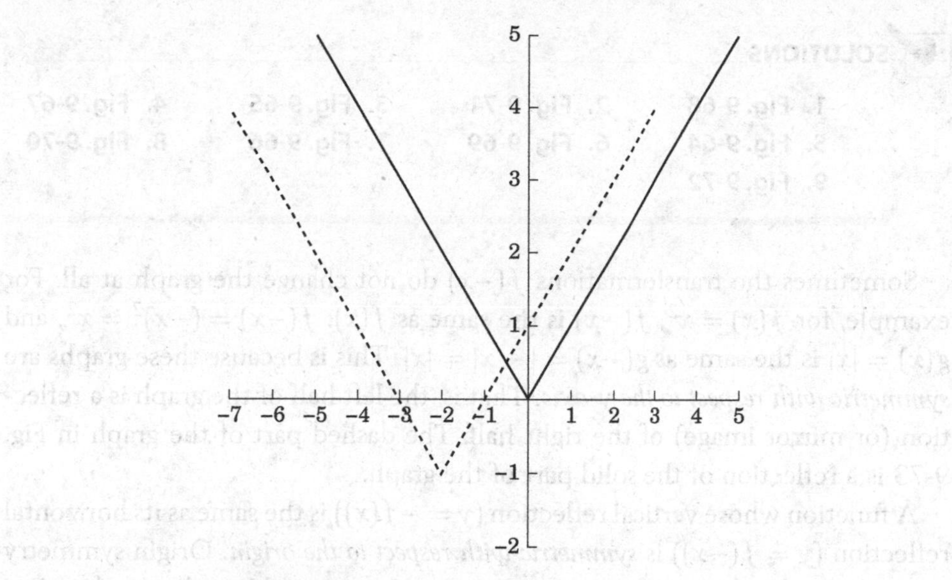

FIGURE 9-63

PRACTICE

Match the graph in Figs. 9-59–9-63 with the function.

1. $f(x) = -|x - 3|$ 2. $f(x) = |x + 2| - 1$ 3. $f(x) = \frac{1}{2}|x|$
4. $f(x) = 2|x|$ 5. $f(x) = 3|x - 2| + 1$

SOLUTIONS

1. Fig. 9-61 2. Fig. 9-63 3. Fig. 9-60 4. Fig. 9-59
5. Fig. 9-62

The next set of Practice problems is another set of matching problems, but the reference graphs are not given. These are transformations of $y = x^2$, $y = x^3$, $y = \sqrt{x}$, $y = |x|$.

PRACTICE

Match the graph in Figs. 9-64–9-72 with its function.

1. $f(x) = (x - 1)^3$ 2. $f(x) = 4\sqrt{x}$ 3. $f(x) = -x^2$
4. $f(x) = |x + 2| + 2$ 5. $f(x) = (x + 1)^3 - 4$ 6. $f(x) = 2 + \sqrt{x}$
7. $f(x) = \sqrt{x - 3}$ 8. $f(x) = \frac{1}{2}x^2 - 4$ 9. $f(x) = -|x| - 4$

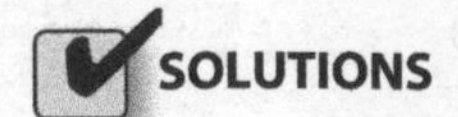

1. Fig. 9-68
2. Fig. 9-71
3. Fig. 9-65
4. Fig. 9-67
5. Fig. 9-64
6. Fig. 9-69
7. Fig. 9-66
8. Fig. 9-70
9. Fig. 9-72

Sometimes the transformations $f(-x)$ do not change the graph at all. For example, for $f(x) = x^2$, $f(-x)$ is the same as $f(x)$: $f(-x) = (-x)^2 = x^2$, and $g(x) = |x|$ is the same as $g(-x) = |-x| = |x|$. This is because these graphs are *symmetric with respect to the y-axis*. That is, the left half of the graph is a reflection (or mirror image) of the right half. The dashed part of the graph in Fig. 9-73 is a reflection of the solid part of the graph.

A function whose vertical reflection ($y = -f(x)$) is the same as its horizontal reflection ($y = f(-x)$) is *symmetric with respect to the origin*. Origin symmetry is a little harder to see than y-axis symmetry. Imagine folding the graph in Fig. 9-74 along the x-axis then again along the y-axis, the upper right-hand part of the graph would coincide with the lower left-hand part of the graph.

The graph of $y^2 = x$ in Fig. 9-75 has x-axis symmetry. This symmetry is not as important as y-axis symmetry and origin symmetry because only one function has this kind of symmetry ($y = 0$).

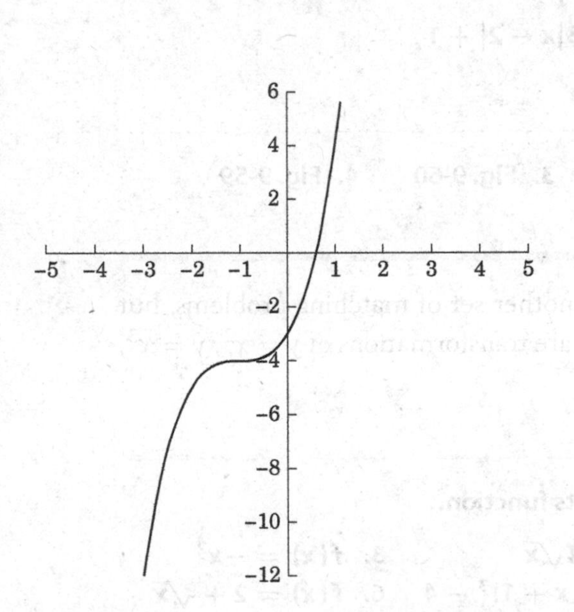

FIGURE 9-64

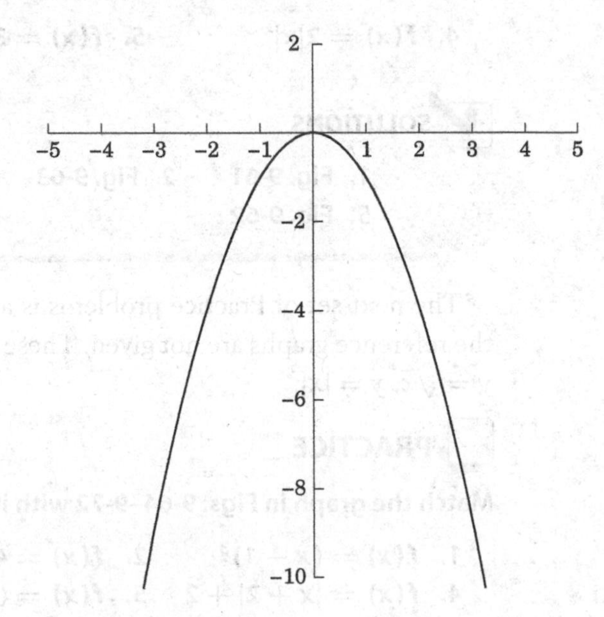

FIGURE 9-65

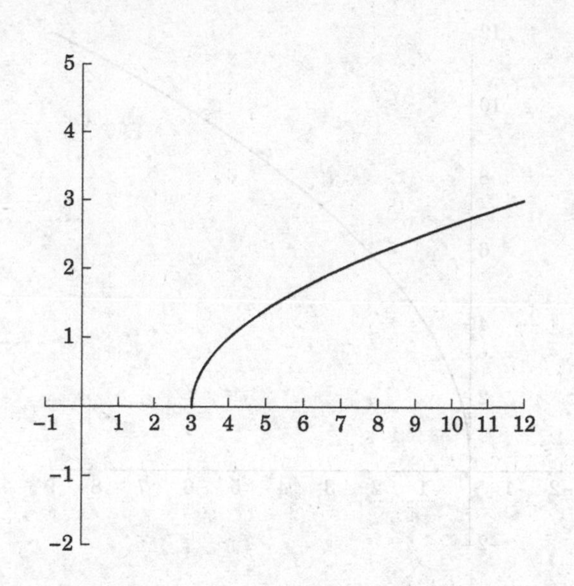

FIGURE 9-66

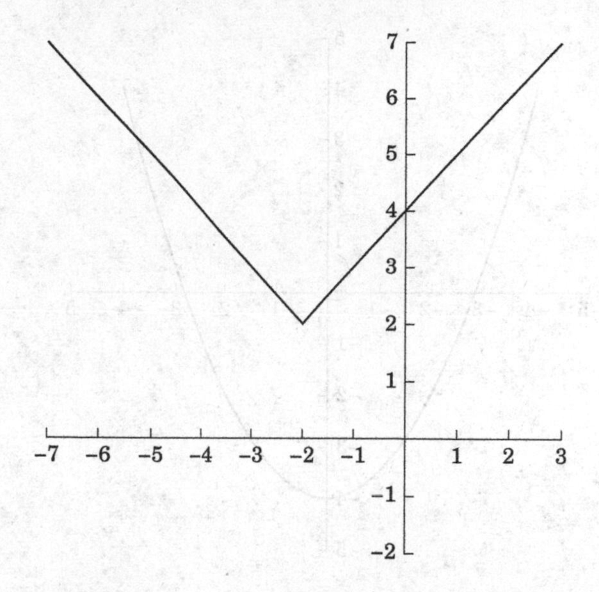

FIGURE 9-67

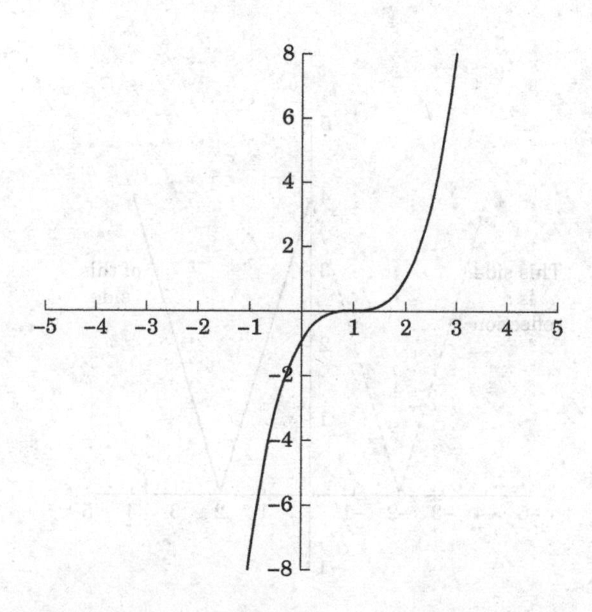

FIGURE 9-68

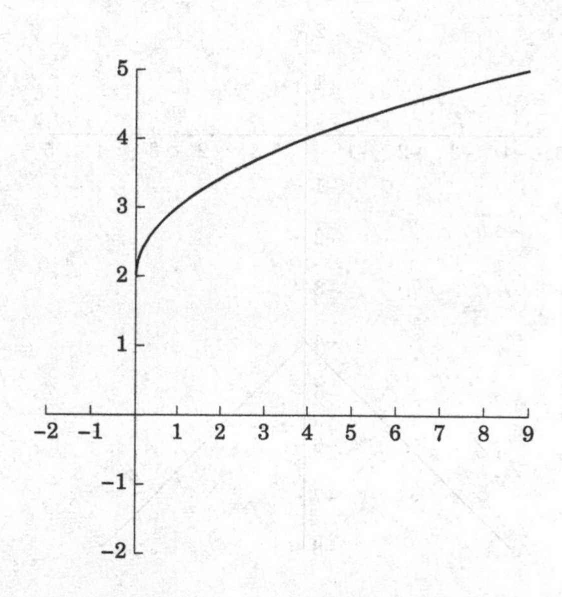

FIGURE 9-69

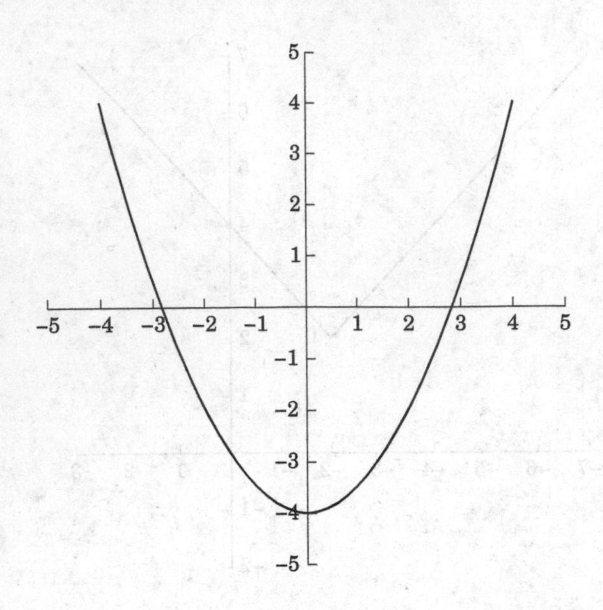

FIGURE 9-70

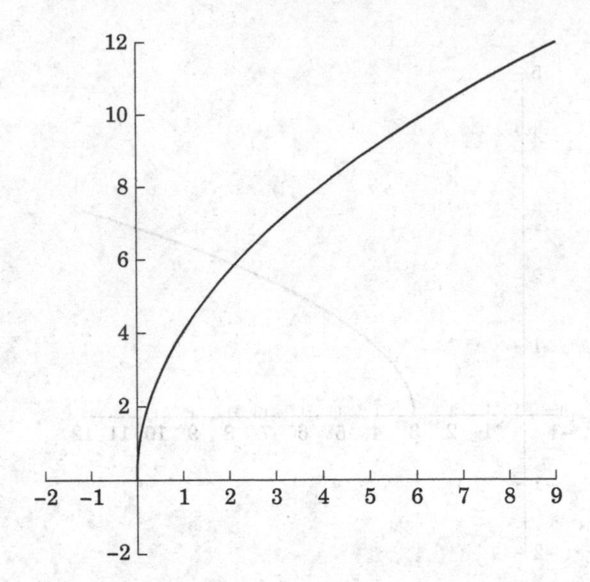

FIGURE 9-71

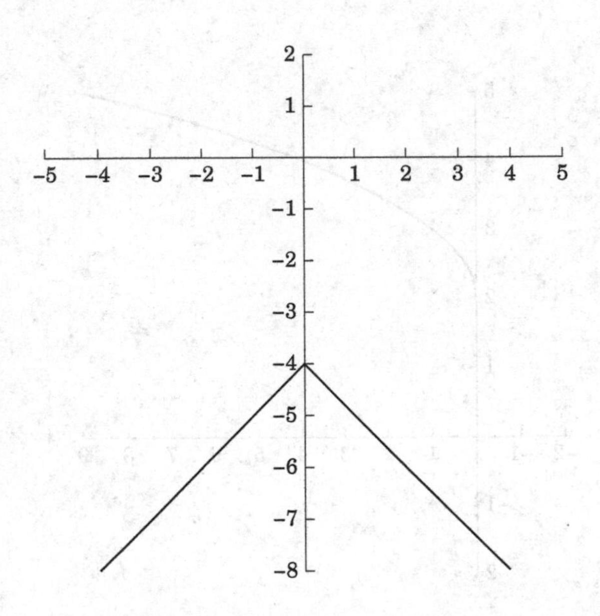

FIGURE 9-72

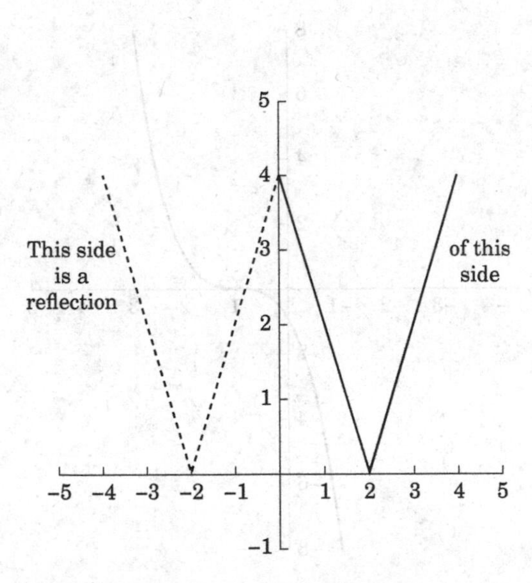

FIGURE 9-73

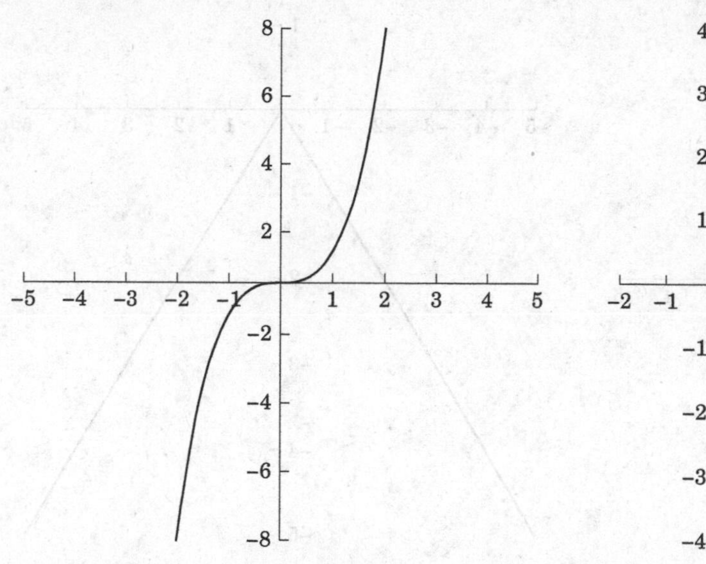

FIGURE 9-74

FIGURE 9-75

PRACTICE

Determine whether the graphs in Figs. 9-76–9-81 are symmetric with respect to the *y*-axis, *x*-axis, or origin or none of these.

1. See Fig. 9-76.

2. See Fig. 9-77.

3. See Fig. 9-78.

4. See Fig. 9-79.

5. See Fig. 9-80.

6. See Fig. 9-81.

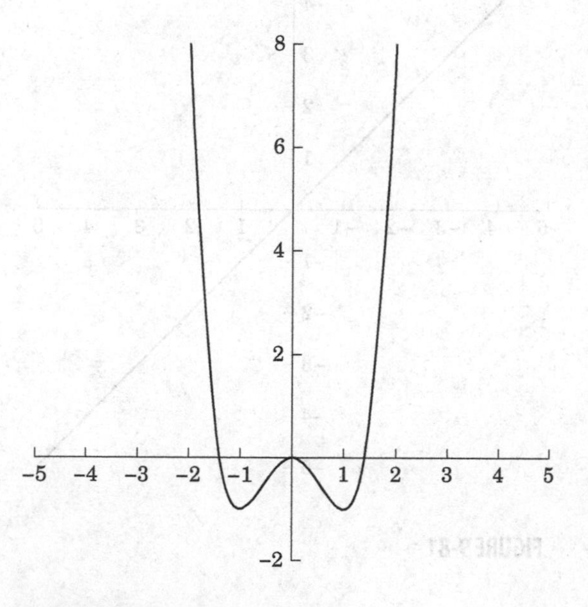

FIGURE 9-76

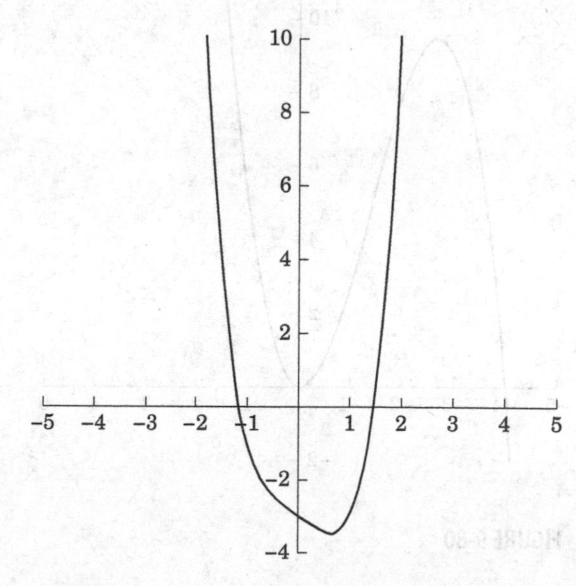

FIGURE 9-77

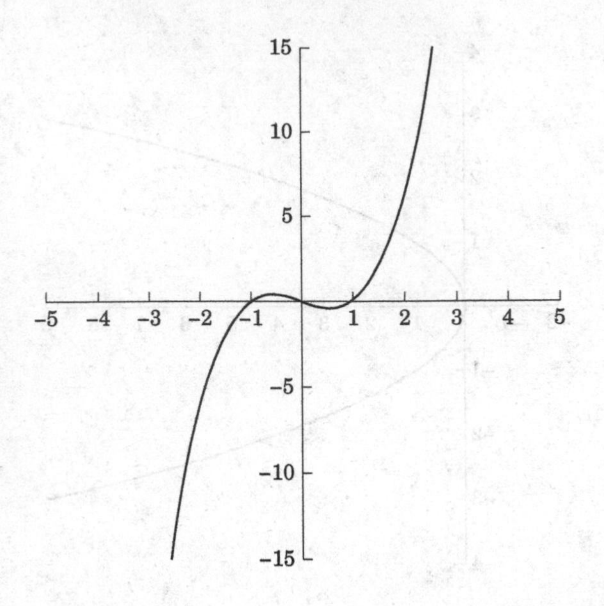

FIGURE 9-78

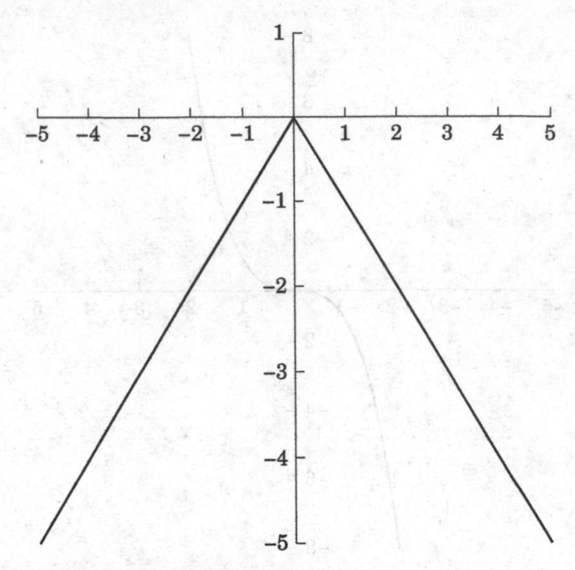

FIGURE 9-79

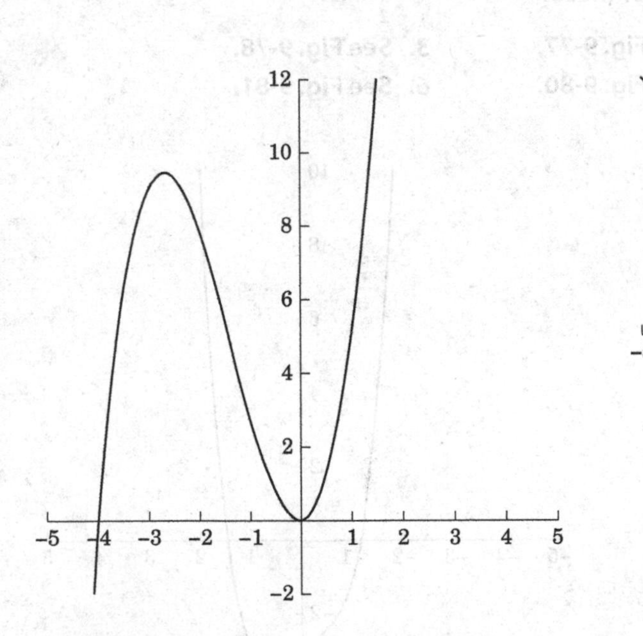

FIGURE 9-80

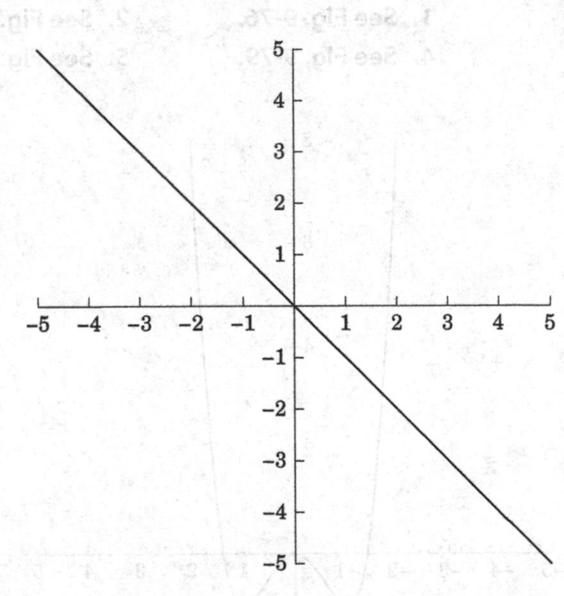

FIGURE 9-81

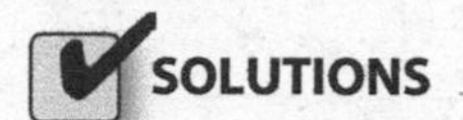

1. **y-axis symmetry**
2. **No symmetry**
3. **Origin symmetry**
4. **y-axis symmetry**
5. **No symmetry**
6. **Origin symmetry**

Even/Odd Functions

We can tell if the graph of a function has y-axis symmetry or origin symmetry by looking at its equation. If we evaluate the function at $-x$ (replace x with $-x$) and the y-values do not change, then the function has y-axis symmetry. Knowing that the graph of a function has y-axis symmetry is very useful when sketching the graph by hand. This is because if (x, y) is on the graph, then $(-x, y)$ is also on the graph. For example, if a function is symmetric with respect to the y-axis and the point $(2, 3)$ is on the graph, then we automatically know that $(-2, 3)$ is also on the graph. The graph of a function has origin symmetry when evaluating the function at $-x$ changes the sign on the y-values too. If a function is symmetric with respect to the origin and the point $(2, 3)$ is on the graph, then the point $(-2, -3)$ is also on the graph.

Functions whose graphs have y-axis symmetry are called *even* functions. Functions whose graphs have origin symmetry are called *odd* functions. A function is even if $f(-x) = f(x)$. This is the mathematical notation for the fact that replacing an x-value with its opposite does not change the y-value. A function is odd if $f(-x) = -f(x)$. This is the mathematical notation for the fact that replacing an x-value with its opposite replaces the y-value with its opposite.

Because so many important functions involve x to powers, the following facts are useful.

$$a(-x)^{\text{even power}} = ax^{\text{even power}} \quad \text{and} \quad a(-x)^{\text{odd power}} = -ax^{\text{odd power}}$$

The product of an even number of negative numbers is positive, and the product of an odd number of negative numbers is negative.

EXAMPLE 9-10

Simplify.

- $2(-x)^3 = -2x^3$
- $5(-x)^6 = 5x^6$
- $-3(-x)^7 = 3x^7$
- $-4(-x)^8 = -4x^8$
- $1-(-x) = 1+x$
- $1-(-x)^2 = 1-x^2$
- $1-(-x)^3 = 1+x^3$
- $(-x)^2-(-x)^3 = x^2+x^3$

PRACTICE

Simplify.

1. $(-x)^3$
2. $(-x)^2$
3. $4(-x)^{100}$
4. $7(-x)^{15}$
5. $-2(-x)^5$
6. $-8(-x)^4$
7. $(-x)^3+2(-x)^2+4(-x)+1$

SOLUTIONS

1. $(-x)^3 = -x^3$
2. $(-x)^2 = x^2$
3. $4(-x)^{100} = 4x^{100}$
4. $7(-x)^{15} = -7x^{15}$
5. $-2(-x)^5 = 2x^5$
6. $-8(-x)^4 = -8x^4$
7. $(-x)^3+2(-x)^2+4(-x)+1 = -x^3+2x^2-4x+1$

Evaluating a function at $-x$ is often the most difficult part of determining whether a function is even, odd, or neither. Once we have evaluated a function at $-x$ and simplified it, we compare the expression both to $f(x)$ and to $-f(x)$. If the simplified equation is the same as $f(x)$, then the function is even. If it is the same as $-f(x)$, then the function is odd.

EXAMPLE 9-11

Determine whether the function is even, odd, or neither.

- $f(x) = x^3 + x$

First we multiply both sides of $f(x)$ by -1 so that we can compare $f(-x)$ to $f(x)$ and to $-f(x)$.

$$-f(x) = -(x^3+x) = -x^3 - x$$

Now we find and simplify $f(-x)$ and compare it to $f(x) = x^3 + x$ and $-f(x) = -x^3 - x$.

$$f(-x) = (-x)^3 + (-x)$$
$$= -x^3 - x$$

Because $-f(x)$ and $f(-x)$ are the same, $f(x)$ is an odd function.

- $f(x) = 5x^3 - 4$

We multiply both sides of the equation by -1 to find $-f(x)$.

$$-f(x) = -(5x^3 - 4) = -5x^3 + 4$$

Now we find and simplify $f(-x)$.

$$f(-x) = 5(-x)^3 - 4 = -5x^3 - 4$$

Because $f(-x)$ is not the same as $f(x)$ and $-f(x)$, $f(x)$ is neither even nor odd.

- $f(x) = -6x^2 + 1$

$$-f(x) = -(-6x^2 + 1) = 6x^2 - 1$$
$$f(-x) = -6(-x)^2 + 1 = -6x^2 + 1$$

Since $f(x)$ and $f(-x)$ are the same, $f(x)$ is an even function.

- $f(x) = -9$

Because the y-value is -9 no matter what x is, $f(-x) = -9$, making $f(x)$ an even function.

- $f(x) = |2x| + 3$

$$-f(x) = -(|2x| + 3) = -|2x| - 3$$
$$f(-x) = |2(-x)| + 3 = |-2x| + 3 = |2x| + 3 \qquad (|-x| = |x|)$$

As $f(x)$ and $f(-x)$ are the same, $f(x)$ is an even function.

- $g(x) = \dfrac{3}{x^4 - x^2 + 2}$

$$-g(x) = -\frac{3}{x^4 - x^2 + 2}$$

$$g(-x) = \frac{3}{(-x)^4 - (-x)^2 + 2} = \frac{3}{x^4 - x^2 + 2}$$

As $g(x)$ and $g(-x)$ are the same, $g(x)$ is an even function.

PRACTICE

Determine whether the function is even, odd, or neither.

1. $f(x) = 2x^5 - 6x^3 + x$
2. $f(x) = 8 - x^2$
3. $g(x) = x^3 - x^2 + x - 1$
4. $f(x) = 2x + 1$
5. $h(x) = \dfrac{-8}{x+1}$
6. $f(x) = \dfrac{-2x}{5x^3 + x}$
7. $g(x) = \sqrt{x^2 + 4}$
8. $h(x) = |3x| - 8$
9. $f(x) = 25$

SOLUTIONS

1. $f(x) = 2x^5 - 6x^3 + x$

$$-f(x) = -(2x^5 - 6x^3 + x) = -2x^5 + 6x^3 - x$$

$$f(-x) = 2(-x)^5 - 6(-x)^3 + (-x) = -2x^5 + 6x^3 - x$$

$-f(x)$ and $f(-x)$ are the same, so $f(x)$ is an odd function.

2. $f(x) = 8 - x^2$

$$-f(x) = -(8 - x^2) = -8 + x^2$$

$$f(-x) = 8 - (-x)^2 = 8 - x^2$$

$f(x)$ and $f(-x)$ are the same, so $f(x)$ is an even function.

3. $g(x) = x^3 - x^2 + x - 1$

$$-g(x) = -(x^3 - x^2 + x - 1) = -x^3 + x^2 - x + 1$$

$$g(-x) = (-x)^3 - (-x)^2 + (-x) - 1 = -x^3 - x^2 - x - 1$$

$g(-x)$ is neither the same as $g(x)$ nor $-g(x)$, so $g(x)$ is neither even nor odd.

4. $f(x) = 2x + 1$

$$-f(x) = -(2x + 1) = -2x - 1$$

$$f(-x) = 2(-x) + 1 = -2x + 1$$

$f(-x)$ is neither the same as $f(x)$ nor $-f(x)$, so $f(x)$ is neither even nor odd.

5. $h(x) = \dfrac{-8}{x+1}$

$$-h(x) = -\left(\frac{-8}{x+1}\right) = \frac{8}{x+1}$$

$$h(-x) = \frac{-8}{-x+1}$$

$h(-x)$ is neither the same as $h(x)$ nor $-h(x)$, so $h(x)$ is neither even nor odd.

6. $f(x) = \dfrac{-2x}{5x^3 + x}$

$$-f(x) = -\frac{-2x}{5x^3 + x} = \frac{2x}{5x^3 + x}$$

Factor negative in denominator

$$f(-x) = \frac{-2(-x)}{5(-x)^3 + (-x)} = \frac{2x}{-5x^3 - x} = \frac{2x}{-(5x^3 + x)} = \frac{-2x}{5x^3 + x}$$

$f(-x)$ is the same as $f(x)$, so $f(x)$ is an even function.

7. $g(x) = \sqrt{x^2 + 4}$

$$-g(x) = -\sqrt{x^2 + 4}$$

$$g(-x) = \sqrt{(-x)^2 + 4} = \sqrt{x^2 + 4}$$

$g(-x)$ and $g(x)$ are the same, so $g(x)$ is an even function.

8. $h(x) = |3x| - 8$

$-h(x) = -(|3x| - 8) = -|3x| + 8$

$h(-x) = |3(-x)| - 8 = |-3x| - 8 = |3x| - 8$

$h(-x)$ and $h(x)$ are the same, so $h(x)$ is an even function.

9. $f(x) = 25$

Because $f(x) = 25$ for *every* x, $f(-x) = 25$ also, so $f(x)$ is an even function.

Combining Functions

The vast majority of functions studied in algebra and calculus are some combination of only a handful of basic functions, most of them introduced in this book. The most obvious combination of two or more functions are arithmetic combinations: sums, differences, products, and quotients. Suppose two functions $f(x)$ and $g(x)$ are given.

- $(f+g)(x)$ means $f(x)+g(x)$.
- $(f-g)(x)$ means $f(x)-g(x)$.
- $(fg)(x)$ means $f(x)g(x)$.
- $\frac{f}{g}(x)$ means $\frac{f(x)}{g(x)}$.

EXAMPLE 9-12

Find and simplify (if necessary) $(f+g)(x)$, $(f-g)(x)$, $(fg)(x)$, and $\frac{f}{g}(x)$.

- **$f(x) = 3x - 4$ and $g(x) = x^2 + x$**

$$(f+g)(x) = (3x-4) + (x^2+x)$$
$$= x^2 + 4x - 4$$
$$(f-g)(x) = (3x-4) - (x^2+x) = 3x - 4 - x^2 - x$$
$$= -x^2 + 2x - 4$$
$$(fg)(x) = (3x-4)(x^2+x) = 3x^3 + 3x^2 - 4x^2 - 4x$$
$$= 3x^3 - x^2 - 4x$$
$$\frac{f}{g}(x) = \frac{3x-4}{x^2+x}$$

PRACTICE

Find and simplify (if necessary) $(f+g)(x)$, $(f-g)(x)$, $(fg)(x)$, and $\frac{f}{g}(x)$.

1. $f(x) = x + 6$ and $g(x) = -2x + 4$
2. $f(x) = \sqrt{x+6}$ and $g(x) = x - 2$

SOLUTIONS

1. $(f+g)(x) = (x+6) + (-2x+4) = -x + 10$

$$(f-g)(x) = (x+6) - (-2x+4) = x + 6 + 2x - 4 = 3x + 2$$

$$(fg)(x) = (x+6)(-2x+4) = -2x^2 + 4x - 12x + 24$$

$$= -2x^2 - 8x + 24$$

$$\frac{f}{g}(x) = \frac{x+6}{-2x+4}$$

2. $(f+g)(x) = \sqrt{x+6} + x - 2$

$$(f-g)(x) = \sqrt{x+6} - (x-2) = \sqrt{x+6} - x + 2$$

$$(fg)(x) = \sqrt{x+6}(x-2) = (x-2)\sqrt{x+6}$$

$$\frac{f}{g}(x) = \frac{\sqrt{x+6}}{x-2}$$

Function Composition

Two (or more) functions can be combined by composing one function with another. The basic idea behind function composition is that one function is evaluated at another. For example, if $f(x) = 4x + 7$ and $g(x) = 2x - 3$, then "to evaluate f at g" means to compute $f(2x-3)$. Remember that to "evaluate a function" means to substitute the quantity in the parentheses for x.

$$f(2x-3) = 4(2x-3) + 7 = 8x - 12 + 7 = 8x - 5$$

The notation for this operation is $(f \circ g)(x)$. By definition, $(f \circ g)(x)$ means $f(g(x))$. This operation is not commutative. That is, $(f \circ g)(x)$ is usually not the same as $(g \circ f)(x)$.

EXAMPLE 9-13

Find $(f \circ g)(x)$ and $(g \circ f)(x)$.

- $f(x) = x^2 + 4x - 3$ and $g(x) = 2x - 5$

$(f \circ g)(x) = f(g(x))$ — This is the definition of $(f \circ g)(x)$.

$= f(2x - 5)$ — Replace $g(x)$ with $2x - 5$.

$= (2x - 5)^2 + 4(2x - 5) - 3$ — Replace x with $2x - 5$ in $f(x)$.

$$= (2x - 5)(2x - 5) + 4(2x - 5) - 3$$

$$= 4x^2 - 20x + 25 + 8x - 20 - 3$$

$$= 4x^2 - 12x + 2$$

$(g \circ f)(x) = g(f(x))$ — This is the definition of $(g \circ f)(x)$.

$= g(x^2 + 4x - 3)$ — Replace $f(x)$ with $x^2 + 4x - 3$.

$= 2(x^2 + 4x - 3) - 5$ — Replace x with $x^2 + 4x - 3$ in $g(x)$.

$$= 2x^2 + 8x - 6 - 5$$

$$= 2x^2 + 8x - 11$$

- $f(x) = 8 - 5x$ and $g(x) = x + 4$

$$(f \circ g)(x) = f(g(x)) = f(x + 4)$$

$$= 8 - 5(x + 4) = 8 - 5x - 20 = -5x - 12$$

$$(g \circ f)(x) = g(f(x)) = g(8 - 5x)$$

$$= 8 - 5x + 4 = -5x + 12$$

- $f(x) = \sqrt{x}$ and $g(x) = x^2 + 2x + 2$

$$(f \circ g)(x) = f(g(x)) = f(x^2 + 2x + 2) = \sqrt{x^2 + 2x + 2}$$

$$(g \circ f)(x) = g(f(x)) = f(\sqrt{x}) = (\sqrt{x})^2 + 2\sqrt{x} + 2 = x + 2\sqrt{x} + 2$$

- $f(x) = 16x - 1$ and $g(x) = \dfrac{1}{x + 2}$

$$(f \circ g)(x) = f(g(x)) = f\left(\frac{1}{x+2}\right)$$

$$= 16\left(\frac{1}{x+2}\right) - 1 = \frac{16}{x+2} - 1$$

$$(g \circ f)(x) = g(f(x)) = g(16x - 1)$$

$$= \frac{1}{(16x - 1) + 2} = \frac{1}{16x + 1}$$

PRACTICE

Find $(f \circ g)(x)$ and $(g \circ f)(x)$.

1. $f(x) = x + 2$ and $g(x) = x^2 - 4$
2. $f(x) = x^2$ and $g(x) = \sqrt{2x - 4}$
3. $f(x) = 3x^2 + x$ and $g(x) = \dfrac{1}{x}$
4. $f(x) = \dfrac{1}{x}$ and $g(x) = \dfrac{2}{x - 1}$

SOLUTIONS

1. $(f \circ g)(x) = f(g(x)) = f(x^2 - 4) = (x^2 - 4) + 2 = x^2 - 2$

$$(g \circ f)(x) = g(f(x)) = g(x + 2) = (x + 2)^2 - 4$$

$$= (x + 2)(x + 2) - 4 = x^2 + 4x + 4 - 4 = x^2 + 4x$$

2. $(f \circ g)(x) = f(g(x)) = f(\sqrt{2x - 4}) = (\sqrt{2x - 4})^2 = 2x - 4$

$$(g \circ f)(x) = g(f(x)) = g(x^2) = \sqrt{2x^2 - 4}$$

3. $$(f \circ g)(x) = f(g(x)) = f\left(\frac{1}{x}\right) = 3\left(\frac{1}{x}\right)^2 + \frac{1}{x} = \frac{3}{x^2} + \frac{1}{x}$$

$$(g \circ f)(x) = g(f(x)) = g(3x^2 + x) = \frac{1}{3x^2 + x}$$

4. $$(f \circ g)(x) = f(g(x)) = f\left(\frac{2}{x-1}\right) = \frac{1}{\frac{2}{x-1}} = 1 \div \frac{2}{x-1}$$

$$= 1 \cdot \frac{x-1}{2} = \frac{x-1}{2}$$

$$(g \circ f)(x) = g(f(x)) = g\left(\frac{1}{x}\right) = \frac{2}{\frac{1}{x} - 1} = \frac{2}{\frac{1}{x} - \frac{x}{x}}$$

$$= \frac{2}{\frac{1-x}{x}} = 2 \div \frac{1-x}{x} = 2 \cdot \frac{x}{1-x} = \frac{2x}{1-x}$$

There is no reason a function cannot be evaluated at itself. In other words, sometimes we are asked to compute $(f \circ f)(x)$ for some function $f(x)$. For example, suppose $f(x) = 3x^2 + 5$.

$$(f \circ f)(x) = f(f(x)) = f(3x^2 + 5) = 3(3x^2 + 5)^2 + 5$$

$$= 3(3x^2 + 5)(3x^2 + 5) + 5 = 3(9x^4 + 30x^2 + 25) + 5 = 27x^4 + 90x^2 + 80$$

Function Composition for a Single Value

At times we only need to compose functions at a single x-value. For example, if $f(x) = 8 - 5x$ and $g(x) = x + 4$, we might only need to find $(f \circ g)(x)$ for $x = 2$. To do this, let $x = 2$ in $g(x)$: $g(2) = 2 + 4 = 6$. Now let $x = 6$ in $f(x)$: $f(6) = 8 - 5(6) = -22$. We have just found that $(f \circ g)(2) = -22$. Of course, if we know that $(f \circ g)(x) = -5x - 10$ (as we computed earlier), we could evaluate $(f \circ g)(2)$ by letting $x = 2$ in $-5x - 12$. For the following examples and Practice problems, both $(f \circ g)(x)$ and $(g \circ f)(x)$ were computed in the previous section. You could use this fact to check our answers.

EXAMPLE 9-14

- **Find $(f \circ g)(1)$ and $(g \circ f)(-2)$ for $f(x) = x^2 + 4x - 3$ and $g(x) = 2x - 5$.**

First find $g(1)$: $g(1) = 2(1) - 5 = -3$. Now let $x = -3$ in $f(x)$: $f(-3) = (-3)^2 + 4(-3) - 3 = -6$, so $(f \circ g)(1) = f(g(1)) = f(-3) = -6$.

$$(g \circ f)(-2) = g(f(-2))$$

$$= g(-7) \qquad f(-2) = (-2)^2 + 4(-2) - 3 = -7$$

$$= -19 \qquad g(-7) = 2(-7) - 5 = -19$$

- **Find $(f \circ g)(0)$ and $(g \circ f)(9)$ for $f(x) = \sqrt{x}$ and $g(x) = x^2 + 2x + 2$.**

$$(f \circ g)(0) = f(g(0))$$

$$= f(2) \qquad g(0) = 0^2 + 2(0) + 2 = 2$$

$$= \sqrt{2}$$

$$(g \circ f)(9) = g(f(9)) = g(3) \qquad 9 = \sqrt{9} = 3$$

$$= 3^2 + 2(3) + 2 = 17$$

PRACTICE

1. **Find $(f \circ g)(-3)$ and $(g \circ f)(5)$ for $f(x) = x + 2$ and $g(x) = x^2 - 4$.**
2. **Find $(f \circ g)(5)$ and $(g \circ f)(2)$ for $f(x) = x^2$ and $g(x) = \sqrt{2x - 4}$.**
3. **Find $(f \circ g)(-2)$ and $(g \circ f)(2)$ for $f(x) = \dfrac{1}{x}$ and $g(x) = \dfrac{2}{x - 1}$.**

SOLUTIONS

1. $(f \circ g)(-3) = f(g(-3))$

$= f(5) \qquad g(-3) = (-3)^2 - 4 = 5$

$= 5 + 2 = 7$

$(g \circ f)(5) = g(f(5))$

$= g(7) \qquad f(5) = 5 + 2 = 7$

$= 7^2 - 4 = 45$

2. $(f \circ g)(5) = f(g(5))$

$= f(\sqrt{6}) \qquad g(5) = \sqrt{2(5) - 4} = \sqrt{6}$

$= (\sqrt{6})^2 = 6$

$(g \circ f)(2) = g(4) \qquad f(2) = 2^2 = 4$

$= \sqrt{2(4) - 4} = \sqrt{4} = 2$

3. $(f \circ g)(-2) = f(g(-2))$

$= f\left(-\frac{2}{3}\right) \qquad g(-2) = \frac{2}{-2-1} = -\frac{2}{3}$

$= \frac{1}{-\frac{2}{3}} = 1 \div -\frac{2}{3} = 1 \cdot -\frac{3}{2} = -\frac{3}{2}$

$(g \circ f)(2) = g(f(2))$

$= g\left(\frac{1}{2}\right) \qquad f(2) = \frac{1}{2}$

$= \frac{2}{\frac{1}{2} - 1} = \frac{2}{-\frac{1}{2}}$

$= 2 \div \left(-\frac{1}{2}\right) = 2 \cdot (-2) = -4$

Graphs can be used to evaluate the composition of functions at a particular x-value. For $(f \circ g)(x)$, the y-value of $g(x)$ becomes the x-value for $f(x)$. In other words, $(f \circ g)(x)$ is the y-value for $f(x)$ whose x-value is $g(x)$. For example, if (a, b) is on the graph of $g(x)$ and (b, c) is on the graph of $f(x)$, then $f(g(a)) = f(b) = c$.

EXAMPLE 9-15

The solid graph in Fig. 9-82 is the graph of $f(x)$, and the dashed graph is the graph of $g(x)$.

- **Find $(f \circ g)(1)$, $(f \circ g)(-1)$, $(g \circ f)(0)$.**

For $(f \circ g)(1) = f(g(1))$, we look on the graph of $g(x)$ for the point whose x-coordinate is 1. This point is $(1, -2)$. Now we use $y = -2$ as $x = -2$ in $f(x)$. We look on the graph of $f(x)$ for the point whose x-coordinate is -2. That point is $(-2, 5)$. This means that $(f \circ g)(1) = 5$.

For $(f \circ g)(-1) = f(g(-1))$, we look on the graph of $g(x)$ for the point whose x-coordinate is -1. This point is $(-1, 2)$. Now we use $y = 2$ as $x = 2$ in

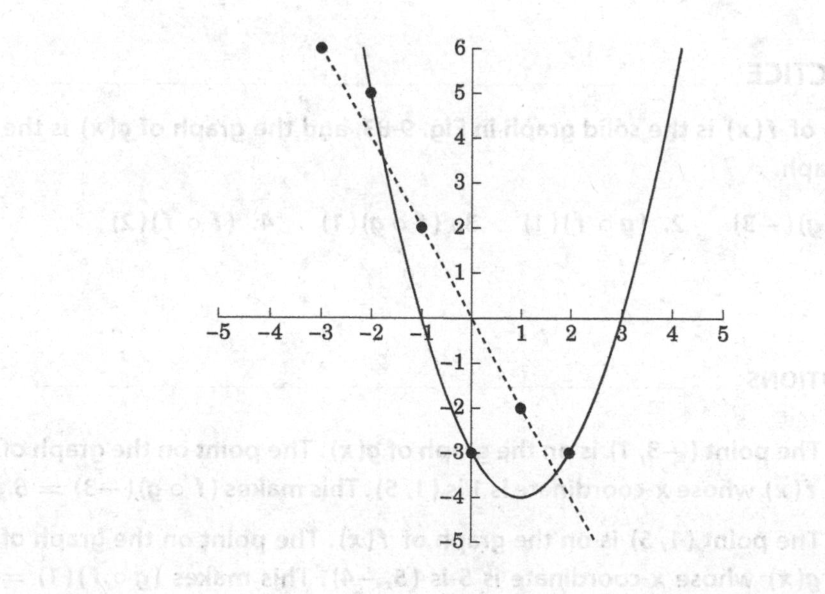

FIGURE 9-82

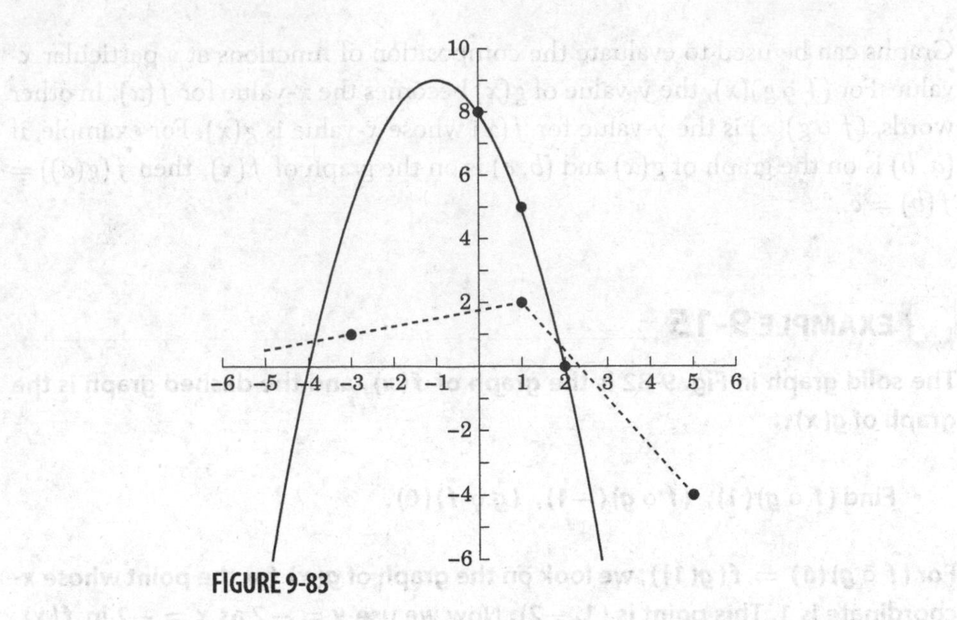

FIGURE 9-83

$f(x)$. We look on the graph of $f(x)$ for the point whose x-coordinate is 2. That point is $(2, -3)$. This means that $(f \circ g)(-1)$ is -3.

For $(g \circ f)(0) = g(f(0))$, we look on the graph of $f(x)$ for the point whose x-coordinate is 0. This point is $(0, -3)$. Now we look on the graph of $g(x)$ for the point whose x-coordinate is -3. That point is $(-3, 6)$. This means that $(g \circ f)(0) = 6$.

PRACTICE

The graph of $f(x)$ is the solid graph in Fig. 9-83, and the graph of $g(x)$ is the dashed graph.

1. $(f \circ g)(-3)$ 2. $(g \circ f)(1)$ 3. $(f \circ g)(1)$ 4. $(f \circ f)(2)$

SOLUTIONS

1. **The point $(-3, 1)$ is on the graph of $g(x)$. The point on the graph of $f(x)$ whose x-coordinate is 1 is $(1, 5)$. This makes $(f \circ g)(-3) = 5$.**
2. **The point $(1, 5)$ is on the graph of $f(x)$. The point on the graph of $g(x)$ whose x-coordinate is 5 is $(5, -4)$. This makes $(g \circ f)(1) = -4$.**

3. **The point $(1, 2)$ is on the graph of $g(x)$. The point on the graph of $f(x)$ whose x-coordinate is 2 is $(2, 0)$. This makes $(f \circ g)(1) = 0$.**

4. **The point $(2, 0)$ is on the graph of $f(x)$. The point on the graph of $f(x)$ whose x-coordinate is 0 is $(0, 8)$. This makes $(f \circ f)(2) = 8$.**

The Domain for the Composition of Functions

Finding the domain for the composition of functions is a little more complicated than finding the domain for the other combinations. For example, $x = -1$ is in the domain of both $f(x) = \frac{1}{x}$ and $g(x) = x + 1$ but not in the domain of $(f \circ g)(x)$. Why not? We cannot let $f(x)$ be evaluated at $x = 0$, and $g(-1) = -1 + 1 = 0$, so $(f \circ g)(-1) = f(g(-1)) = f(0) = \frac{1}{0}$ is not defined.

For any functions $f(x)$ and $g(x)$, the domain of $(f \circ g)(x)$ is the domain of $g(x)$ after deleting any x-values whose y-values are not allowed in $f(x)$. In the above example, $x = -1$ is in the domain of $g(x)$, but $x = g(-1)$ is not in the domain of $f(x)$. To find the domain of $(f \circ g)(x)$, we first find the domain of $g(x)$. Next, we evaluate $f(x)$ at $g(x)$. Before simplifying, see which, if any, x-values need to be removed from the domain of $g(x)$.

EXAMPLE 9-16

Find the domain of $(f \circ g)(x)$.

- $f(x) = \dfrac{1}{x-1}$ **and** $g(x) = x^2$

The domain for $g(x)$ is all x. Are there any x-values that need to be removed? This means, are there any y-values for $g(x)$ that cause 0 in the denominator for $f(x)$? We answer this question by evaluating $f(g(x))$.

$$f(g(x)) = f(x^2) = \frac{1}{x^2 - 1}$$

Because $x = 1$ and $x = -1$ cause the denominator to be 0, we must remove them from the domain of $g(x)$. The domain of $(f \circ g)(x)$ is all x except 1 and -1. In interval notation, this set is $(-\infty, -1) \cup (-1, 1) \cup (1, \infty)$.

- $f(x) = x^2$ and $g(x) = \sqrt{x+1}$

The domain for $g(x)$ is all $x \geq -1$. Are there any x-values that need to be removed from $[-1, \infty)$? We need for $(f \circ g)(x) = f(g(x)) = f(\sqrt{x+1}) = (\sqrt{x+1})^2$ to be defined. Since $\sqrt{x+1}$ is defined for all $x \geq -1$, we do not need to remove any numbers from $[-1, \infty)$, so the domain for $(f \circ g)(x)$ is $[-1, \infty)$.

- $f(x) = \dfrac{1}{x+1}$ and $g(x) = \dfrac{1}{x-1}$

The domain for $g(x)$ is all x except 1.

$$(f \circ g)(x) = f(g(x)) = \frac{1}{\frac{1}{x-1} + 1}$$

We cannot allow the denominator, $\dfrac{1}{x-1} + 1$, to be 0, so we need to remove any x-value from the domain of $g(x)$ that makes the equation $\frac{1}{x-1} + 1 = 0$ true.

$$\frac{1}{x-1} + 1 = 0$$

$$\frac{1}{x-1} + 1 \cdot \frac{x-1}{x-1} = 0$$

$$\frac{1 + (x-1)}{x} = \frac{x}{x-1} = 0$$

The fraction $\frac{x}{x-1}$ is 0 when the numerator, x, is 0. This makes the domain for $(f \circ g)(x)$ all x except 1 and 0. In interval notation, this is $(-\infty, 0) \cup (0, 1) \cup (1, \infty)$.

This is a good example of why we need to look at $(f \circ g)(x)$ *before* it is simplified because $(f \circ g)(x)$ simplifies to $\frac{x-1}{x}$. This simplification hides the fact that we cannot allow x to equal 1.

PRACTICE

Find the domain for $(f \circ g)(x)$. Give solutions in interval notation.

1. $f(x) = 4x - 5$ and $g(x) = \sqrt{x+3}$
2. $f(x) = x^2$ and $g(x) = \sqrt{3x+5}$
3. $f(x) = \dfrac{1}{x+1}$ and $g(x) = x - 1$
4. $f(x) = \dfrac{1}{x}$ and $g(x) = \sqrt{x}$

SOLUTIONS

1. **The domain for $g(x)$ is $x \geq -3$. Evaluate $(f \circ g)(x)$: $f(\sqrt{x+3}) = 4\sqrt{x+3} - 5$. We do not need to remove any other x-values, so the domain is $[-3, \infty)$.**
2. **The domain for $g(x)$ is $x \geq -\frac{5}{3}$. Evaluate $(f \circ g)(x)$: $f(\sqrt{3x+5}) = (\sqrt{3x+5})^2 = 3x+5$. We do not need to remove any other x-values. The domain is $[-\frac{5}{3}, \infty)$.**
3. **The domain for $g(x)$ is all x. We evaluate $(f \circ g)(x)$.**

$$(f \circ g)(x) = f(g(x)) = f(x-1) = \frac{1}{(x-1)+1} = \frac{1}{x}$$

We cannot allow x to be 0, so we need to remove $x = 0$ from the domain of $g(x)$. The domain is $(-\infty, 0) \cup (0, \infty)$.

4. The domain for $g(x)$ is $x \geq 0$. Evaluate $(f \circ g)(x)$.

$$(f \circ g)(x) = f(\sqrt{x}) = \frac{1}{\sqrt{x}}$$

We cannot allow $\sqrt{x} = 0$, so we need to remove $x = 0$ from the domain of $g(x)$. The domain is $(0, \infty)$.

Summary

In this chapter, we learned how to

- *Change the graph of a function by changing its equation.* If we add a constant, k, to every y-value of a function, the graph is shifted vertically k units. Adding a constant h to every value of x shifts the graph horizontally h units. Multiplying every value of y by -1 reflects the graph across the x-axis, and multiplying every value of x by -1 reflects the graph across the y-axis. Multiplying every y-value by a constant a (except -1) vertically stretches or compresses the graph.
- *Determine whether a function is even, odd, or neither.* A function is even if its graph is symmetric with respect to the y-axis and is odd if its graph is symmetric with respect to the origin. We can algebraically determine whether or not a function is even or odd by evaluating the function at $-x$ and comparing the result to both $f(x)$ and $-f(x)$. If $f(-x) = f(x)$ (the y-values for x and $-x$ is the same), then the function is even. If $f(-x) = -f(x)$ (the y-values for x and $-x$ are opposite), then the function is odd.
- *Combine functions arithmetically.* Two functions can be combined arithmetically by simply performing arithmetic on their equations. That is,

$$(f \pm g)(x) = f(x) \pm g(x) \qquad (fg)(x) = f(x) \cdot g(x) \qquad \frac{f}{g}(x) = \frac{f(x)}{g(x)}$$

- *Find the composition of one function with another.* The composition $(f \circ g)(x)$ means to evaluate the f function with the g function. That is, we replace x in $f(x)$ with the formula for $g(x)$. If we want to find $(f \circ g)(x)$ for a single x-value (either algebraically or graphically), we find the y-value for x in the g function and use this number as the x-value in the f function.

- *Determine the domain for the composition of two functions.* Using the fact that the y-value for $g(x)$ becomes the x-value for $f(x)$ in $(f \circ g)(x)$, we determine the domain of $(f \circ g)(x)$ by removing from the domain of $g(x)$ any x-value whose y-value is not allowed into $f(x)$.

QUIZ

1. **The graph of $y = -f(x-2) - 4$ is the graph of $y = f(x)$**

   A. Shifted to the right 2 units, reflected across the y-axis, and shifted down 4 units

   B. Shifted to the left 2 units, reflected across the y-axis, and shifted down 4 units

   C. Shifted to the right 2 units, reflected across the x-axis, and shifted down 4 units

   D. Shifted to the left 2 units, reflected across the x-axis, and shifted down 4 units

2. **The solid graph in Fig. 9-84 is the graph of $y = |x|$. The dashed graph is the graph of which function?**

   A. $y = |x+2| - 4$ B. $y = |x-2| - 4$ C. $y = |x+2| + 4$

   D. $y = |x+2| + 4$

3. **Is the function $f(x) = \frac{1}{x^2-4}$ even, odd, or neither?**

   A. Even B. Odd C. Neither

   D. It is impossible to tell without the graph.

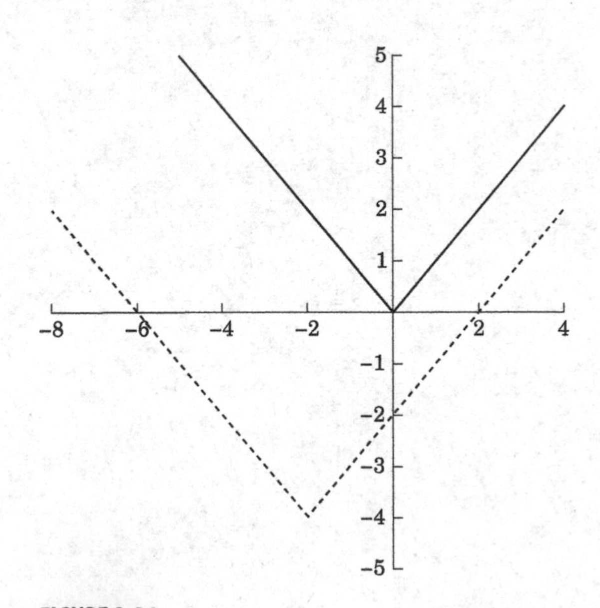

FIGURE 9-84

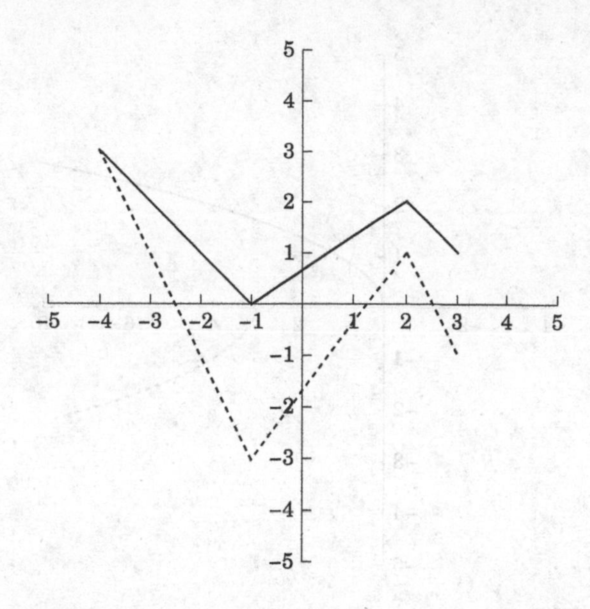

FIGURE 9-85

4. **Let $f(x) = \sqrt{x}$ and $g(x) = x^2 + 4$. Find $(f \circ g)(x)$.**

A. $\sqrt{x}(x^2 + 4)$ B. $x + 4$ C. $\sqrt{x^2 + 4}$ D. $x + 2$

5. **The solid graph in Fig. 9-85 is the graph of $y = f(x)$. The dashed graph is the graph of which function?**

A. $y = 3f(x) - 2$ B. $y = 2f(x) - 3$ C. $y = f(x - 1) - 2$
D. $y = f(-x) - 3$

6. **The graph of $y = \frac{1}{2}f(-x)$ is the graph of $y = f(x)$.**

A. Reflected across the y-axis and vertically stretched
B. Reflected across the y-axis and vertically compressed
C. Reflected across the x-axis and vertically stretched
D. Reflected across the x-axis and vertically compressed

7. **Is the function $f(x) = 4x^3 + 2x$ even, odd, or neither?**

A. Even B. Odd C. Neither
D. It is impossible to tell without the graph.

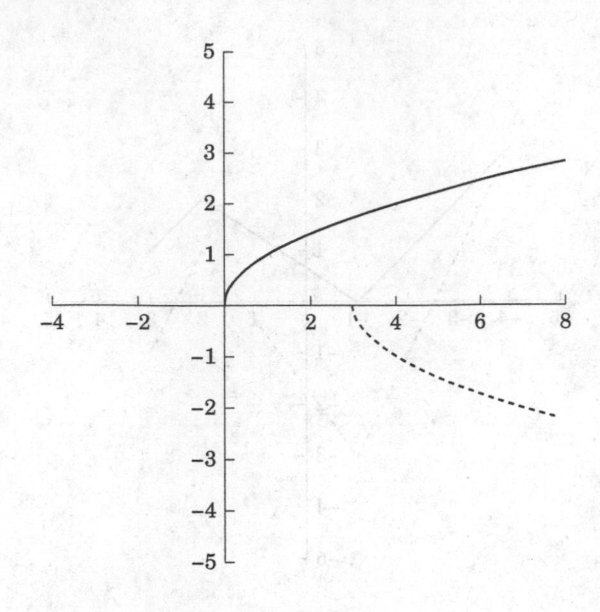

FIGURE 9-86

8. **Let $f(x) = \frac{1}{x+4}$ and $g(x) = 3x - 2$. Find $(f \circ g)(x)$.**

   A. $\dfrac{3x-2}{x+4}$ B. $\dfrac{3}{x+4} - 2$ C. $\dfrac{x+4}{3x-2}$ D. $\dfrac{1}{3x+2}$

9. **What is the domain of $(f \circ g)$ if $f(x) = \dfrac{1}{x-5}$ and $g(x) = \dfrac{1}{x}$?**

   A. $x \neq 5,\ x \neq 0$ B. $x \neq 0$ only C. $x \neq 5$ only D. $x \neq 0,\ x \neq \frac{1}{5}$

10. **The solid graph in Fig. 9-86 is the graph of $y = \sqrt{x}$, the dashed graph is the graph of which function?**

    A. $y = -\sqrt{x-3}$ B. $y = \sqrt{3-x}$ C. $y = -\sqrt{x+3}$ D. $y = \sqrt{x+3}$

11. **Let $f(x) = x^2 - 4$ and $g(x) = 2x - 4$. Find $(f \circ g)(5)$.**

    A. 38 B. 32 C. 672 D. 96

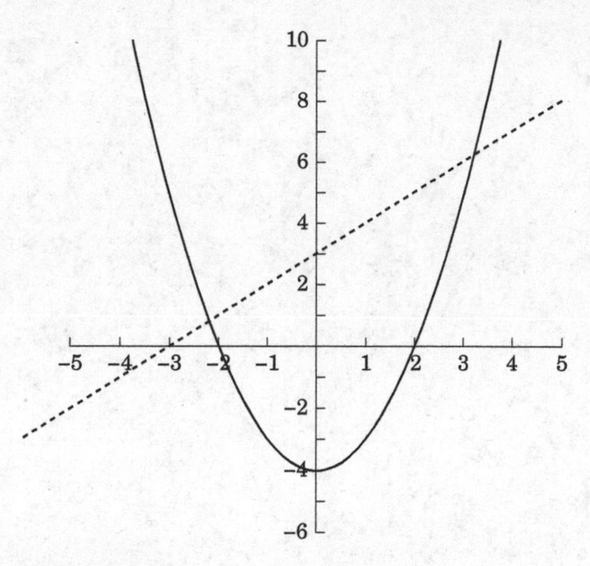

FIGURE 9-87

12. The solid graph in Fig. 9-87 is the graph of $y = f(x)$. The dashed graph is the graph of $y = g(x)$. Use these graphs to evaluate $(f \circ g)(-2)$.

A. 1 B. 0 C. 3 D. –3

chapter 10

Polynomial Functions

We work with polynomial functions in this chapter. This is a deep subject, which is why the chapter is so long. As we have with other functions, we learn how to use clues in a function's equation to help us sketch the graph of a polynomial function. Because the factors of a polynomial help us with the graph of a polynomial function, we spend a lot of time learning strategies for factoring a polynomial. Because these strategies involve polynomial division, we learn the Division Algorithm for polynomials. Later in the chapter, we will work with complex numbers and the role they play in the important *Fundamental Theorem of Algebra*.

CHAPTER OBJECTIVES

In this chapter, you will

- Determine the end behavior of a polynomial function
- Understand the relationship between x-intercepts for the graph of a polynomial function and its factors
- Sketch the graph of a polynomial function
- Perform polynomial division
- Use polynomial division to factor a polynomial

- Perform arithmetic with complex numbers
- Find complex zeros for a polynomial

Introduction to Polynomial Functions

A polynomial function is a function in the form $f(x) = a_n x^n + a_{n-1}x^{n-1} + \ldots + a_1 x + a_0$, where each a_i is a real number and the powers on x are whole numbers. The number a_i is called a *coefficient*. For example, in the polynomial function $f(x) = -2x^3 + 5x^2 - 4x + 8$, the coefficients are $-2,\ 5,\ -4$, and 8. The *constant* term (the term with no variable) is 8. The powers on x are 3, 2, and 1. The *degree* of the polynomial (and polynomial function) is the highest power on x. In this example, the degree is 3. Quadratic functions are degree 2. Linear functions of the form $f(x) = mx + b$ are degree 1. The constant function, $f(x) = c$, is a polynomial of degree 0 because $x^0 = 1$.

The *leading term* of a polynomial (and polynomial function) is the term having x to the highest power. Usually, but not always, the leading term is written first. The *leading coefficient* is the coefficient on the leading term. In our example, the leading term is $-2x^3$, and the leading coefficient is -2. The graph of any polynomial either goes up on both ends, goes down on both ends, or goes up on one end and down on the other. This is called the *end behavior* of the graph. We can determine the end behavior of the graph of a polynomial simply by examining its leading term. The figures in Table 10-1 illustrate the end behavior for any polynomial function. The shape of the dashed part of the graph depends on the individual function.

If the degree of the polynomial is an even number, both ends of its graph go up or they go down (as in Figs. 10-1 and 10-2). If the degree of the polynomial is odd, one end of its graph goes up and the other down (as in Figs. 10-3 and 10-4). These facts are summarized in Table 10-2.

EXAMPLE 10-1

Match the end behavior of the graph of the function with one of the graphs in Figs. 10-1–10-4.

- $f(x) = 4x^5 + 6x^3 - 2x^2 + 8x + 11$

We only need to look at the leading term, $4x^5$. The degree, 5, is odd, and the leading coefficient, 4, is positive, the end behavior of the graph of this function looks like the one in Fig. 10-3.

TABLE 10-1

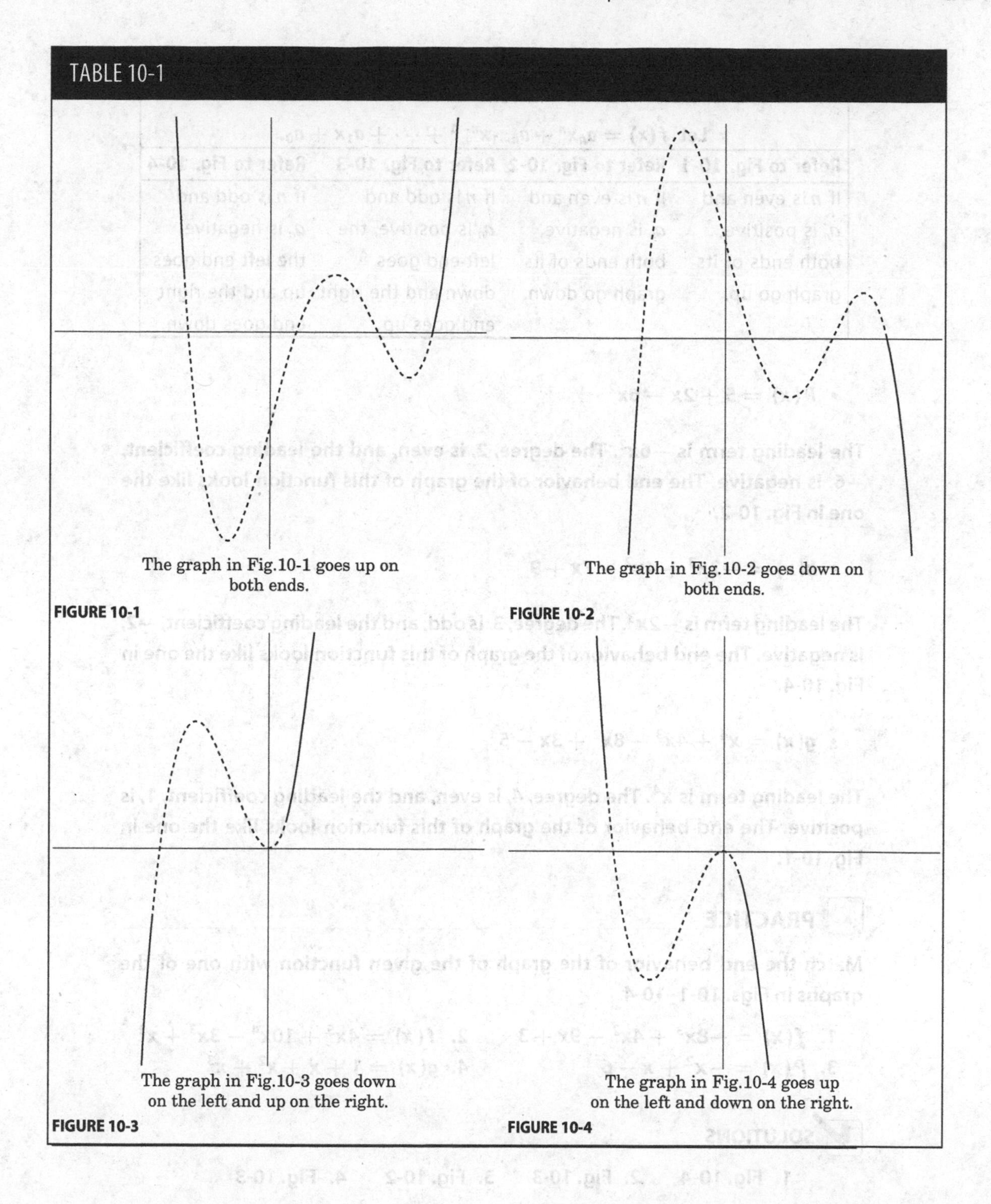

The graph in Fig.10-1 goes up on both ends.

FIGURE 10-1

The graph in Fig.10-2 goes down on both ends.

FIGURE 10-2

The graph in Fig.10-3 goes down on the left and up on the right.

FIGURE 10-3

The graph in Fig.10-4 goes up on the left and down on the right.

FIGURE 10-4

TABLE 10-2

Let $f(x) = a_nx^n + a_{n-1}x^{n-1} + \cdots + a_1x + a_0$.			
Refer to Fig. 10-1	**Refer to Fig. 10-2**	**Refer to Fig. 10-3**	**Refer to Fig. 10-4**
If n is even and a_n is positive, both ends of its graph go up.	If n is even and a_n is negative, both ends of its graph go down.	If n is odd and a_n is positive, the left end goes down and the right end goes up.	If n is odd and a_n is negative, the left end goes up and the right end goes down.

- $P(x) = 5 + 2x - 6x^2$

The leading term is $-6x^2$. The degree, 2, is even, and the leading coefficient, -6, is negative. The end behavior of the graph of this function looks like the one in Fig. 10-2.

- $h(x) = -2x^3 + 4x^2 - 7x + 9$

The leading term is $-2x^3$. The degree, 3, is odd, and the leading coefficient, -2, is negative. The end behavior of the graph of this function looks like the one in Fig. 10-4.

- $g(x) = x^4 + 4x^3 - 8x^2 + 3x - 5$

The leading term is x^4. The degree, 4, is even, and the leading coefficient, 1, is positive. The end behavior of the graph of this function looks like the one in Fig. 10-1.

PRACTICE

Match the end behavior of the graph of the given function with one of the graphs in Figs. 10-1–10-4.

1. $f(x) = -8x^3 + 4x^2 - 9x + 3$
2. $f(x) = 4x^5 + 10x^4 - 3x^3 + x^2$
3. $P(x) = -x^2 + x - 6$
4. $g(x) = 1 + x + x^2 + x^3$

SOLUTIONS

1. Fig. 10-4 2. Fig. 10-3 3. Fig. 10-2 4. Fig. 10-3

The x-intercepts for the graph of a polynomial function (if it has any) are important points on its graph. In fact, for many polynomials, the x-intercepts almost completely determine what the equation of the function looks like. Remember that the x-intercept of any graph is where the graph crosses or touches the x-axis. This happens when the y-coordinate of the point is 0. Earlier, we found the x-intercepts for many quadratic functions—by factoring and setting each factor equal to 0. This is how we find the x-intercepts for any polynomial function. This is not always easy to do. In fact, some polynomials are so hard to factor that the best we can do is approximate the x-intercepts (using graphing calculators or calculus manipulations). This is not the case for the polynomials in this book, however. Most polynomials in this chapter can be factored using the techniques that we cover here.

Because an x-intercept for $f(x) = a_nx^n + a_{n-1}x^{n-1} + \ldots + a_1x + a_0$ is a solution to the equation $0 = a_nx^n + a_{n-1}x^{n-1} + \ldots + a_1x + a_0$, x-intercepts are also called *zeros* of the polynomial. All of the following statements have the same meaning for a polynomial. Let c be a real number, and let $P(x)$ be a polynomial function.

1. c is an x-intercept of the graph of $P(x)$.
2. c is a zero for $P(x)$.
3. $x - c$ is a factor of $P(x)$.

EXAMPLE 10-2

- **$x - 1$ is a factor means that 1 is an x-intercept and a zero.**
- **$x + 5$ is a factor means that -5 is an x-intercept and a zero.**
- **x is a factor means that 0 is an x-intercept and a zero.**
- **3 is a zero means that $x - 3$ is a factor and 3 is an x-intercept.**

We can find the zeros of a function (or at least the approximate zeros) by looking at its graph.

EXAMPLE 10-3

Because the x-intercepts of the graph in Fig. 10-5 are 2 and -2, we know that $x - 2$ and $x + 2$, which is $x - (-2)$, are factors of the polynomial.

The graph of the polynomial function in Fig. 10-6 has x-intercepts of -1, 1, and 2, so the factors are $x - 1$, $x - 2$, and $x + 1$, as $x - (-1)$.

The x-intercepts for the graph in Fig. 10-7 are -3, 0, and 2, making $x + 3$, $x - 2$, and x (as $x - 0$) factors of the polynomial.

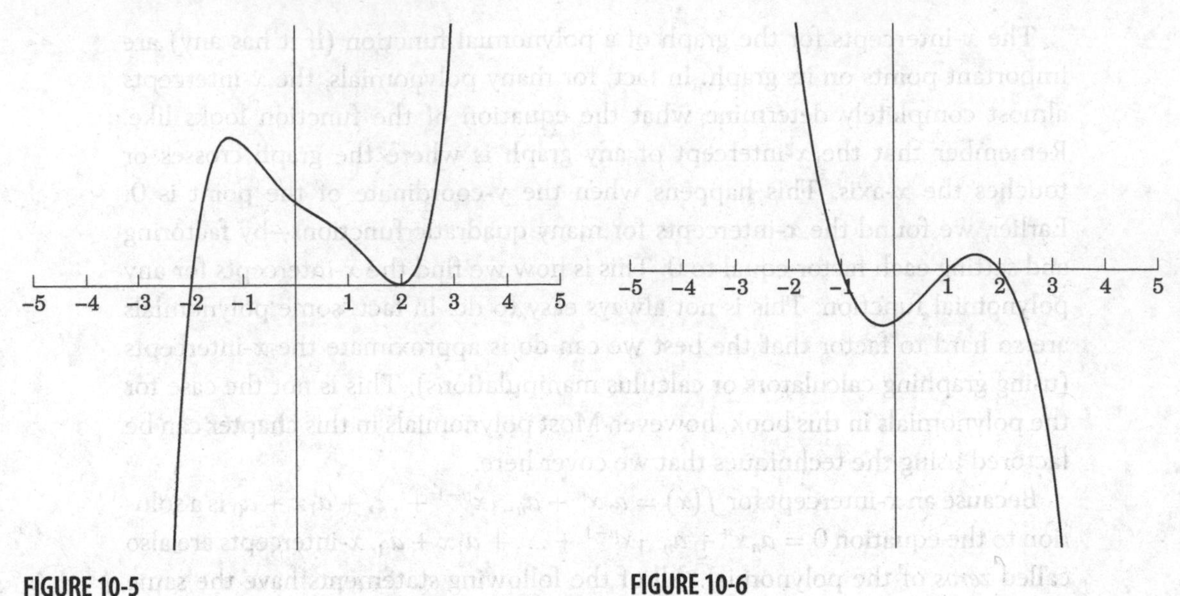

FIGURE 10-5

FIGURE 10-6

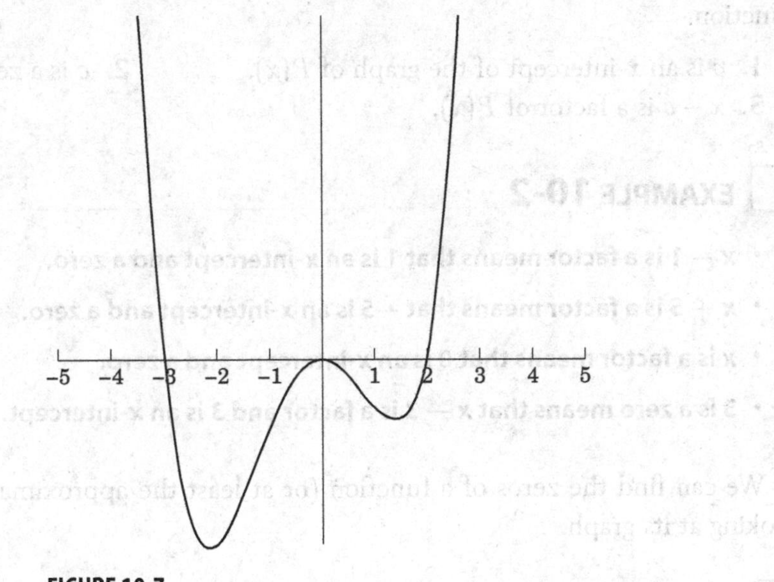

FIGURE 10-7

PRACTICE

Identify the *x*-intercepts and factors for the polynomial function whose graphs are given.

1. Fig. 10-8 2. Fig. 10-9 3. Fig. 10-10 4. Fig. 10-11 5. Fig. 10-12

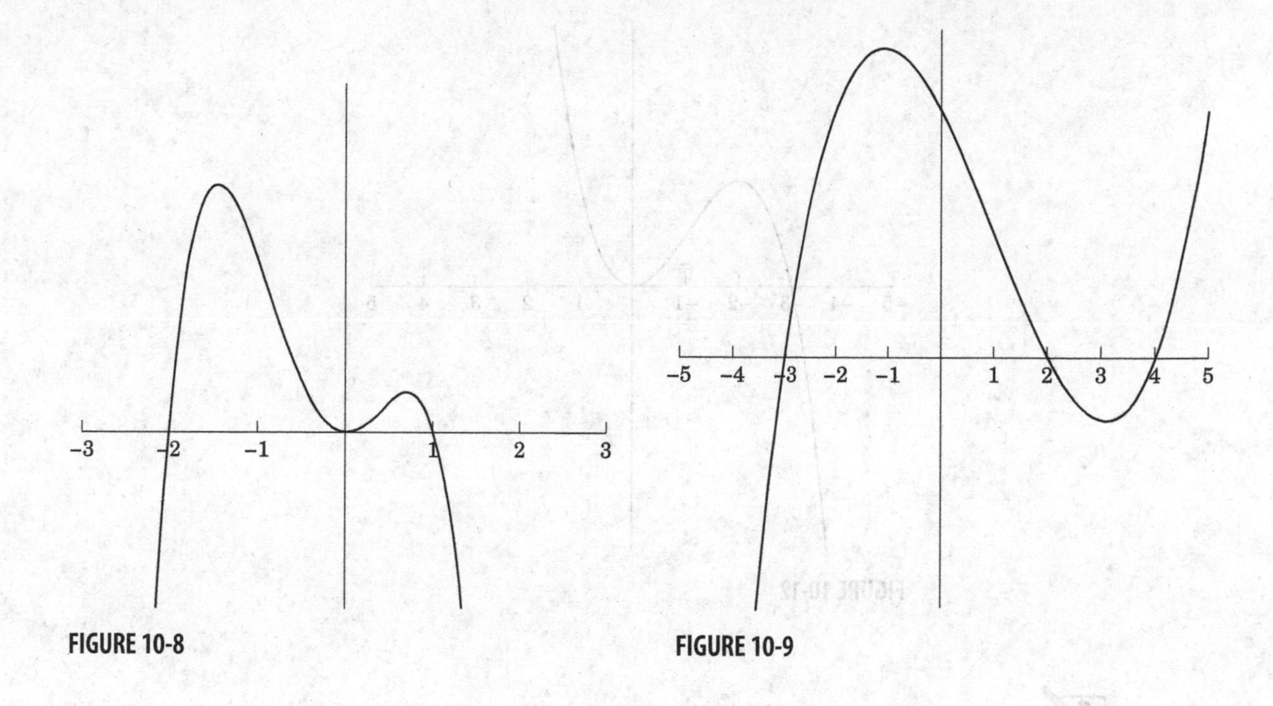

FIGURE 10-8

FIGURE 10-9

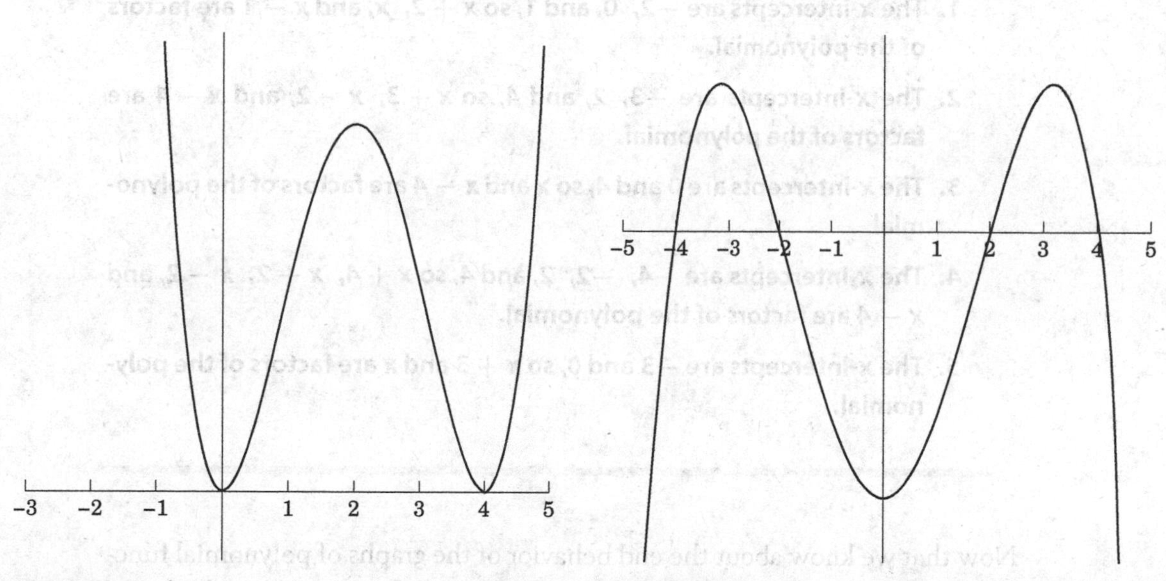

FIGURE 10-10

FIGURE 10-11

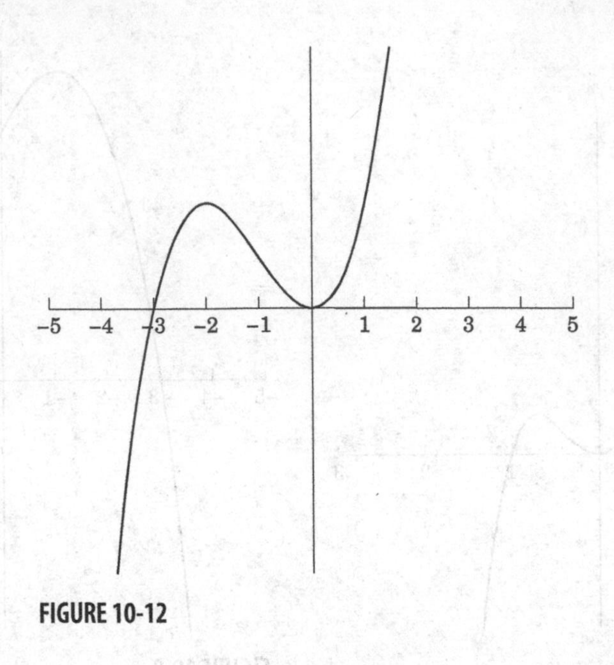

FIGURE 10-12

1. The x-intercepts are -2, 0, and 1, so $x+2$, x, and $x-1$ are factors of the polynomial.

2. The x-intercepts are -3, 2, and 4, so $x+3$, $x-2$, and $x-4$ are factors of the polynomial.

3. The x-intercepts are 0 and 4, so x and $x-4$ are factors of the polynomial.

4. The x-intercepts are -4, -2, 2, and 4, so $x+4$, $x+2$, $x-2$, and $x-4$ are factors of the polynomial.

5. The x-intercepts are -3 and 0, so $x+3$ and x are factors of the polynomial.

Now that we know about the end behavior of the graphs of polynomial functions and the relationship between x-intercepts and factors, we can look at a polynomial and have a pretty good idea of what its graph looks like. In the next set of examples and Practice problems, we match the graphs from the previous section with their polynomial functions.

EXAMPLE 10-4

Match the functions with the graphs in Figs. 10-5–10-7.

- $f(x) = \frac{1}{10}x^2(x+3)(x-2) = \frac{1}{10}x^4 + \frac{1}{10}x^3 - \frac{3}{5}x^2$

Because $f(x)$ is a polynomial whose degree is even and whose leading coefficient is positive, we look for a graph that goes up on the left and on the right. Because the factors are x^2, $x+3$, and $x-2$, we also look for a graph with x-intercepts 0, -3, and 2. The graph in Fig. 10-7 satisfies these conditions.

- $g(x) = -\frac{1}{2}(x-1)(x-2)(x+1) = -\frac{1}{2}x^3 + x^2 + \frac{1}{2}x - 1$

Because $g(x)$ is a polynomial whose degree is odd and whose leading coefficient is negative, we look for a graph that goes up on the left and down on the right. The factors are $x-1$, $x-2$, and $x+1$, we also look for a graph with 1, 2, and -1 as x-intercepts. The graph in Fig. 10-6 satisifies these conditions.

- $P(x) = (x^2+2)(x-2)^2(x+2) = x^5 - 2x^4 - 2x^3 + 4x^2 - 8x + 16$

Because $P(x)$ is a polynomial whose degree is odd and whose leading term is positive, we look for a graph that goes down on the left and up on the right. Although x^2+2 is a factor, there is no x-intercept from this factor (this is because $x^2+2=0$ has no real number solution). The x-intercepts are 2 and -2. The graph in Fig. 10-5 satisfies these conditions.

PRACTICE

Match the polynomial function with one of the graphs in Figs. 10-8–10-12.

1. $f(x) = -\frac{1}{8}(x+4)(x+2)(x-2)(x-4) = -\frac{1}{8}x^4 + \frac{5}{2}x^2 - 8$
2. $g(x) = x^2(x-4)^2 = x^4 - 8x^3 + 16x^2$
3. $P(x) = -\frac{1}{2}x^2(x+2)(x-1) = -\frac{1}{2}x^4 - \frac{1}{2}x^3 + x^2$
4. $Q(x) = x^2(x+3) = x^3 + 3x^2$
5. $R(x) = \frac{1}{2}(x+3)(x-2)(x-4) = \frac{1}{2}x^3 - \frac{3}{2}x^2 - 5x + 12$

SOLUTIONS

1. Fig. 10-11
2. Fig. 10-10
3. Fig. 10-8
4. Fig. 10-12
5. Fig. 10-9

Sketching Graphs of Polynomials

To sketch the graph of most polynomial functions accurately, we need to use calculus (don't let that scare you—the calculus part is easier than the algebra part!). We can still get a pretty good graph using algebra alone. In general, we plot the intercepts, a point between consecutive x-intercepts, and two points (one on each side) that show the end behavior of the graph. For example, suppose the graph of a polynomial function $y = f(x)$ has three x-intercepts, A, B, and C (in order, from left to right). Once we have filled in Table 10-3, we would plot the points and draw a smooth curve through them.

EXAMPLE 10-5

- $f(x) = \frac{1}{2}x(x+1)(x-3)$

The x-intercepts are 0, −1, and 3. We use $x = -2$ for the point to the left of the smallest x-intercept, $x = -0.5$ for the point between the x-intercepts −1 and 0, $x = 1.5$ for the point between the x-intercepts 0 and 3, and $x = 4$ for the

TABLE 10-3

x	$f(x)$
Any point to the left of A	
A	0
Any point between A and B	
B	0
Any point between B and C	
C	0
Any point to the right of C	

TABLE 10-4

x	$f(x)$
−2	−5
−1	0
−0.5	0.4375
0	0
1.5	−2.8125
3	0
4	10

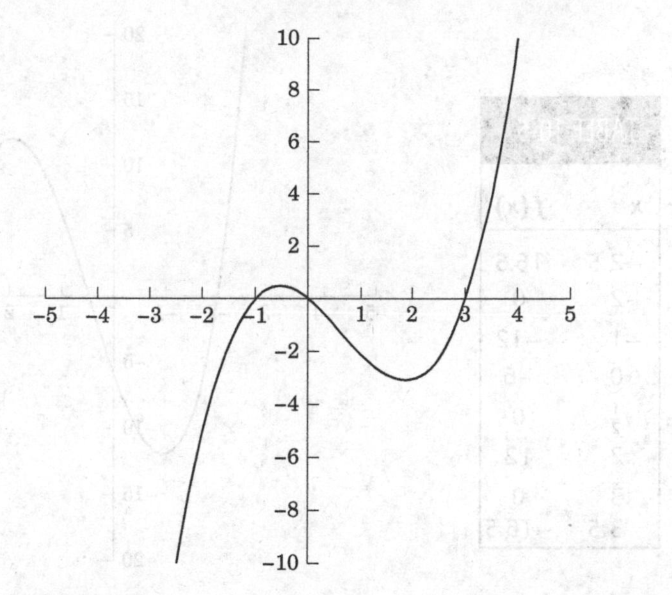

FIGURE 10-13

point to the right of the x-intercept 3. The points that we plot are in Table 10-4, and the graph is in Fig. 10-13.

- $f(x) = -(2x - 1)(x + 2)(x - 3)$

The x-intercepts are $\frac{1}{2}$ (from $2x - 1 = 0$), -2, and 3. In addition to the x-intercepts, we plot the points for $x = -2.5$ (to the left of $x = -2$), $x = -1$ (between $x = -2$ and $x = \frac{1}{2}$), $x = 2$ (between $x = \frac{1}{2}$ and $x = 3$), and $x = 3.5$ (to the right of $x = 3$). The points are computed in Table 10-5, and the graph is sketched in Fig. 10-14. The reason we used $x = -2.5$ instead of $x = -3$ and $x = 3.5$ instead of $x = 4$ is that their y-values were too large for our graph.

PRACTICE

Sketch the graph.

1. $f(x) = \dfrac{1}{2}x(x - 2)(x + 2)$

2. $g(x) = -\dfrac{1}{2}(x + 3)(x - 1)(x - 3)$

3. $h(x) = -\dfrac{1}{10}(x + 4)(x + 1)(x - 2)(x - 3)$

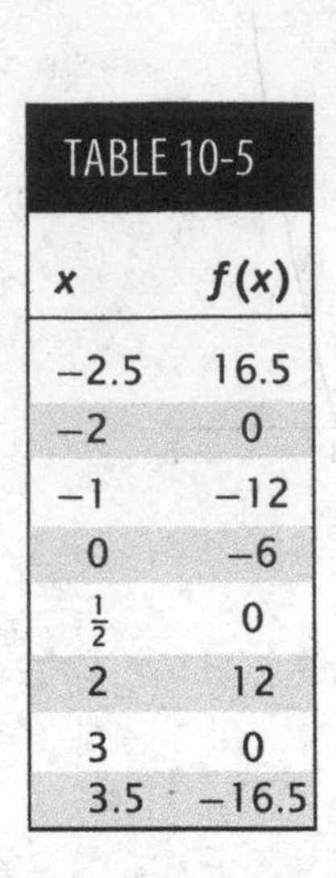

TABLE 10-5

x	$f(x)$
−2.5	16.5
−2	0
−1	−12
0	−6
$\frac{1}{2}$	0
2	12
3	0
3.5	−16.5

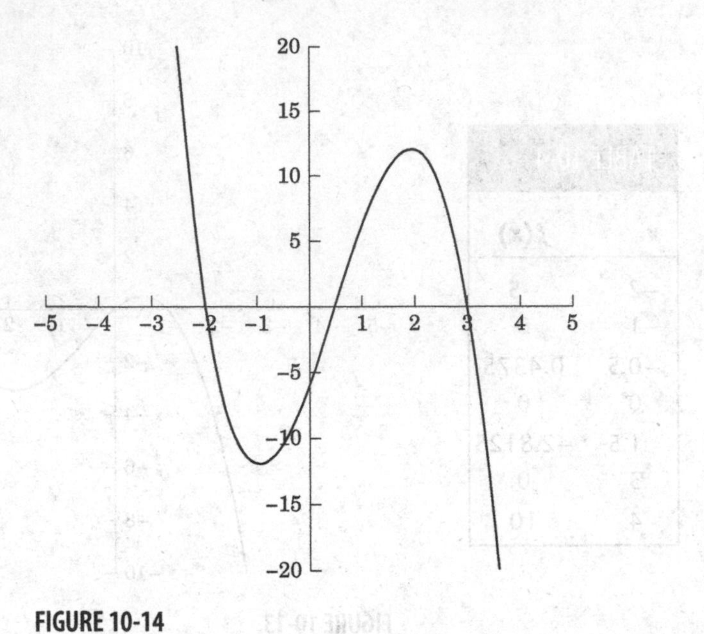

FIGURE 10-14

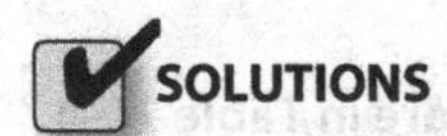

SOLUTIONS

1. Fig. 10-15 2. Fig. 10-16 3. Fig. 10-17

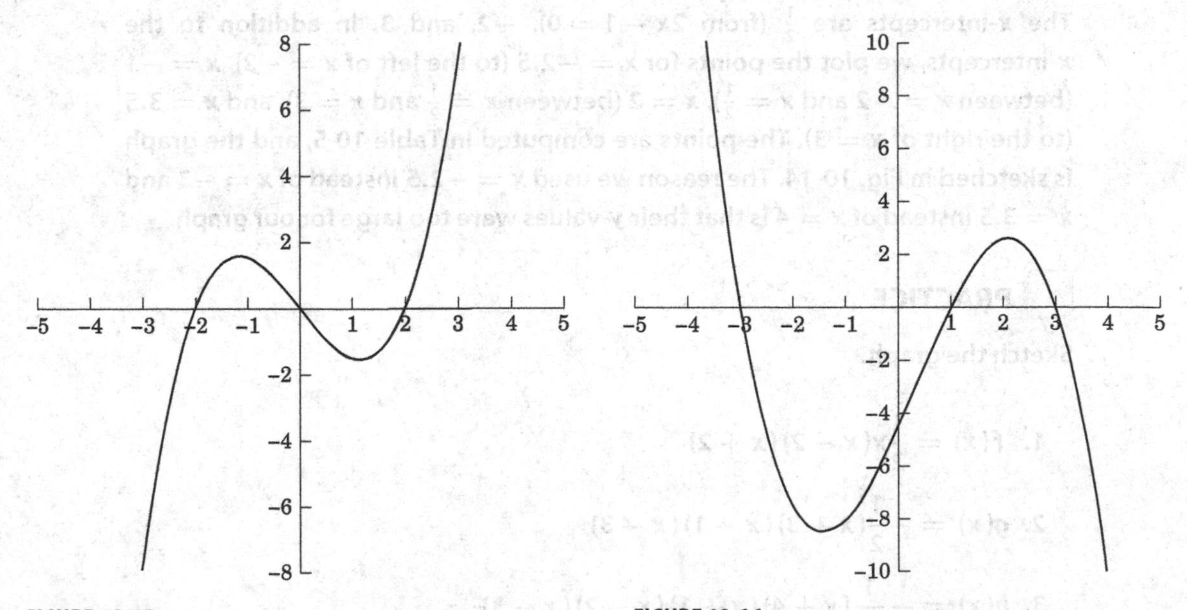

FIGURE 10-15

FIGURE 10-16

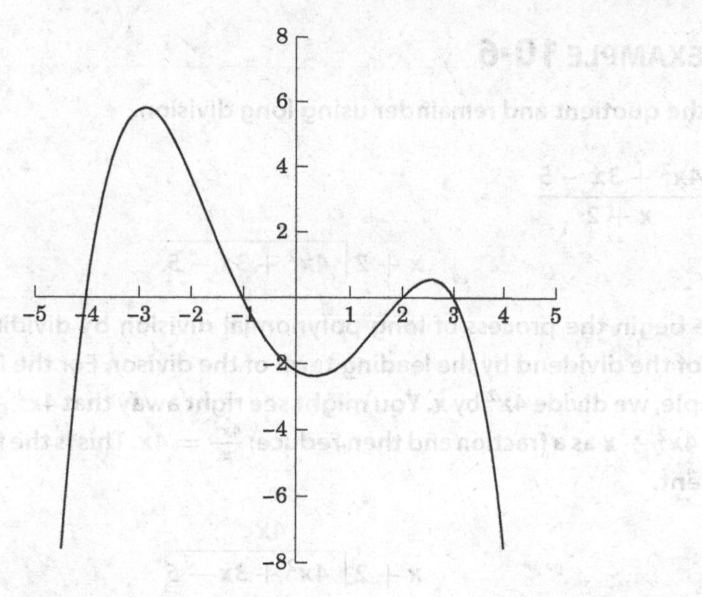

FIGURE 10-17

Polynomial Division

We now work with polynomial division, a skill that we will need later to help us locate the zeros of a polynomial function. One polynomial can be divided by another in much the same way one whole number can be divided by another. When we take the quotient of two whole numbers (where the divisor is not 0), we have a quotient and a remainder. The same happens when we take the quotient of two polynomials.

According to the Division Algorithm for polynomials, for any polynomials $f(x)$ and $g(x)$ (with $g(x)$ not the zero function)

$$\frac{f(x)}{g(x)} = q(x) + \frac{r(x)}{g(x)},$$

where $q(x)$ is the quotient (which might be 0) and $r(x)$ is the remainder, which is either 0 or has degree *strictly* less than the degree of $g(x)$. Multiplying both sides of the equation by $g(x)$ to clear the fraction gives us $f(x) = g(x)q(x) + r(x)$. Let us begin with long division of polynomials.

$$\begin{array}{r|l} & \quad q(x) \\ \hline g(x) & f(x) \\ & \\ \hline & \quad\quad r(x) \end{array}$$

EXAMPLE 10-6

Find the quotient and remainder using long division.

- $\dfrac{4x^2+3x-5}{x+2}$

$$x+2\overline{\big)\,4x^2+3x-5}$$

We begin the process of long polynomial division by dividing the leading term of the dividend by the leading term of the divisor. For the first step in this example, we divide $4x^2$ by x. You might see right away that $4x^2 \div x$ is $4x$. If not, write $4x^2 \div x$ as a fraction and then reduce: $\frac{4x^2}{x} = 4x$. This is the first term of the quotient.

$$\begin{array}{r|l} & 4x \\ \hline x+2 & 4x^2+3x-5 \end{array}$$

Multiply $4x$ by the divisor: $4x(x+2) = 4x^2+8x$. Subtract this from the first two terms of the dividend. Be careful to subtract all of $4x^2+8x$, not just $4x^2$.

$$\begin{array}{r|l} & 4x \\ \hline x+2 & 4x^2+3x-5 \\ & -(4x^2+8x) \\ \hline & \quad\;\; -5x \end{array}$$

Bring down the next term.

$$\begin{array}{r|l} & 4x \\ \hline x+2 & 4x^2+3x-5 \\ & -(4x^2+8x) \\ \hline & \quad\;\; -5x-5 \end{array}$$

Start the process again with $-5x \div x = -5$. Multiply $x+2$ by -5: $-5(x+2) = -5x-10$.
Subtract this from $-5x-5$.

$$\begin{array}{r|l} & 4x-5 \\ \hline x+2 & 4x^2+3x-5 \\ & -(4x^2+8x) \\ \hline & \quad\;\; -5x-5 \\ & \;\; -(-5x-10) \\ \hline & \qquad\qquad 5 \end{array}$$

We are finished because $5 \div x = \frac{5}{x}$ cannot be a term in the quotient polynomial. The remainder is 5 and the quotient is $4x - 5$.

- **$x^2 + 2x - 3 \overline{)3x^4 + 5x^3 - 4x^2 + 7x - 1}$**

Divide $3x^4$ by x^2 to get the first term of the quotient: $\frac{3x^4}{x^2} = 3x^2$. Multiply $x^2 + 2x - 3$ by $3x^2$: $3x^2(x^2 + 2x - 3) = 3x^4 + 6x^3 - 9x^2$. Subtract this from the first three terms in the dividend.

$$\begin{array}{r|l} & 3x^2 \\ \hline x^2 + 2x - 3 & 3x^4 + 5x^3 - 4x^2 + 7x - 1 \\ & -(3x^4 + 6x^3 - 9x^2) \\ \hline & \quad - x^3 + 5x^2 \end{array}$$

Bring down $7x$ and divide $-x^3$ by x^2 to get the second term in the quotient: $\frac{-x^3}{x^2} = -x$. Multiply $x^2 + 2x - 3$ by $-x$: $-x(x^2 + 2x - 3) = -x^3 - 2x^2 + 3x$. Subtract this from $-x^3 + 5x^2 + 7x$.

$$\begin{array}{r|l} & 3x^2 - x \\ \hline x^2 + 2x - 3 & 3x^4 + 5x^3 - 4x^2 + 7x - 1 \\ & -(3x^4 + 6x^3 - 9x^2) \\ & \quad - x^3 + 5x^2 + 7x \\ & \quad -(-x^3 - 2x^2 + 3x) \\ \hline & \qquad 7x^2 + 4x \end{array}$$

Bring down -1 and divide $7x^2$ by x^2 to get the third term in the quotient: $\frac{7x^2}{x^2} = 7$. Multiply $x^2 + 2x - 3$ by 7: $7(x^2 + 2x - 3) = 7x^2 + 14x - 21$. Subtract this from $7x^2 + 4x - 1$.

$$\begin{array}{r|l} & 3x^2 - x + 7 \\ \hline x^2 + 2x - 3 & 3x^4 + 5x^3 - 4x^2 + 7x - 1 \\ & -(3x^4 + 6x^3 - 9x^2) \\ \hline & \quad - x^3 + 5x^2 + 7x \\ & \quad -(-x^3 - 2x^2 + 3x) \\ \hline & \qquad 7x^2 + 4x - 1 \\ & \qquad -(7x^2 + 14x - 21) \\ \hline & \qquad\quad -10x + 20 \end{array}$$

Because $\frac{-10x}{x^2}$ cannot be a term in the quotient polynomial, we are finished. The quotient is $3x^2 - x + 7$, and the remainder is $-10x + 20$.

Polynomial division is a little trickier when the leading coefficient of the divisor is not 1. One reason is that the terms of the quotient are harder to find and are likely to be fractions.

EXAMPLE 10-7

Find the quotient and remainder using long division.

- $\dfrac{x^2 - x + 2}{2x - 1}$

Find the first term in the quotient by dividing the first term of the dividend by the first term in the divisor: $\frac{x^2}{2x} = \frac{x}{2} = \frac{1}{2}x$.

$$\begin{array}{r|l} & \frac{1}{2}x \\ 2x - 1 & x^2 - x + 2 \\ & -(x^2 - \frac{1}{2}x) \\ \hline & -\frac{1}{2}x + 2 \end{array}$$

The second term in the quotient is

$$\frac{-\frac{1}{2}x}{2x} = \frac{-\frac{1}{2}}{2} = -\frac{1}{2} \div 2 = -\frac{1}{2} \cdot \frac{1}{2} = -\frac{1}{4}$$

Multiply $2x - 1$ by $-\frac{1}{4}$: $-\frac{1}{4}(2x - 1) = -\frac{1}{2}x + \frac{1}{4}$.

$$\begin{array}{r|l} & \frac{1}{2}x - \frac{1}{4} \\ 2x - 1 & x^2 - x + 2 \\ & -(x^2 - \frac{1}{2}x) \\ \hline & -\frac{1}{2}x + 2 \\ & -(-\frac{1}{2}x + \frac{1}{4}) \\ \hline & \frac{7}{4} \end{array}$$

The quotient is $\frac{1}{2}x - \frac{1}{4}$, and the remainder is $\frac{7}{4}$.

PRACTICE

Use long division to find the quotient and remainder.

1. $\dfrac{x^3 - 6x^2 + 12x - 4}{x + 2}$
2. $(6x^3 - 2x^2 + 5x - 1) \div (x^2 + 3x + 2)$
3. $\dfrac{x^5 + 3x^4 - 3x^3 - 3x^2 + 19x - 13}{x^2 + 2x - 3}$
4. $(x^3 - x^2 + 2x + 5) \div (3x - 4)$

SOLUTIONS

1.

$$
\begin{array}{r|l}
 & x^2 - 8x + 28 \\
\hline
x + 2 & x^3 - 6x^2 + 12x - 4 \\
 & -(x^3 + 2x^2) \\
\hline
 & -8x^2 + 12x \\
 & -(-8x^2 - 16x) \\
\hline
 & 28x - 4 \\
 & -(28x + 56) \\
\hline
 & -60
\end{array}
$$

The quotient is $x^2 - 8x + 28$, and the remainder is -60.

2.

$$
\begin{array}{r|l}
 & 6x - 20 \\
\hline
x^2 + 3x + 2 & 6x^3 - 2x^2 + 5x - 1 \\
 & -(6x^3 + 18x^2 + 12x) \\
\hline
 & -20x^2 - 7x - 1 \\
 & -(-20x^2 - 60x - 40) \\
\hline
 & 53x + 39
\end{array}
$$

The quotient is $6x - 20$, and the remainder is $53x + 39$.

3.

$$
\begin{array}{r|l}
 & x^3 + x^2 - 2x + 4 \\
\hline
x^2 + 2x - 3 & x^5 + 3x^4 - 3x^3 - 3x^2 + 19x - 13 \\
 & -(x^5 + 2x^4 - 3x^3) \\
\cline{2-2}
 & x^4 + 0x^3 - 3x^2 \\
 & -(x^4 + 2x^3 - 3x^2) \\
\cline{2-2}
 & -2x^3 + 0x^2 + 19x \\
 & -(-2x^3 - 4x^2 + 6x) \\
\cline{2-2}
 & 4x^2 + 13x - 13 \\
 & -(4x^2 + 8x - 12) \\
\cline{2-2}
 & 5x - 1
\end{array}
$$

The quotient is $x^3 + x^2 - 2x + 4$, and the remainder is $5x - 1$.

4.

$$
\begin{array}{r|l}
 & \frac{1}{3}x^2 + \frac{1}{9}x \\
\hline
3x - 4 & x^3 - x^2 + 2x + 5 \\
 & -(x^3 - \frac{4}{3}x^2) \\
\cline{2-2}
 & \frac{1}{3}x^2 + 2x \\
 & -(\frac{1}{3}x^2 - \frac{4}{9}x) \\
\cline{2-2}
 & \frac{22}{9}x + 5
\end{array}
$$

$$\frac{\frac{22}{9}x}{3x} = \frac{\frac{22}{9}}{3} = \frac{22}{9} \cdot \frac{1}{3} = \frac{22}{27}$$

$$\frac{22}{27}(3x - 4) = \frac{22}{9}x - \frac{88}{27}$$

$$
\begin{array}{r|l}
 & \frac{1}{3}x^2 + \frac{1}{9}x + \frac{22}{27} \\
\hline
3x - 4 & x^3 - x^2 + 2x + 5 \\
 & -(x^3 - \frac{4}{3}x) \\
\cline{2-2}
 & \frac{1}{3}x^2 + 2x \\
 & -(\frac{1}{3}x^2 - \frac{4}{9}x) \\
\cline{2-2}
 & \frac{22}{9}x + 5 \\
 & -(\frac{22}{9}x - \frac{88}{27}) \\
\cline{2-2}
 & \frac{223}{27}
\end{array}
$$

The quotient is $\frac{1}{3}x^2 + \frac{1}{9}x + \frac{22}{27}$, and the remainder is $\frac{223}{27}$.

It is important that every power of x, from the highest power to the constant term, be represented in the polynomial. Although it is possible to perform long division without all powers represented, it is very easy to make an error. Also, it is not possible to perform synthetic division (later in this chapter) without a coefficient for *every* term. If a power of x is not written, we need to rewrite the polynomial (the dividend, divisor, or both) using a coefficient of 0 on the missing powers. For example, we would write $x^3 - 1$ as $x^3 + 0x^2 + 0x - 1$.

EXAMPLE 10-8

- $(x^3 - 8) \div (x + 1)$

Rewrite as $(x^3 + 0x^2 + 0x - 8) \div (x + 1)$.

$$
\begin{array}{r|l}
 & x^2 - x + 1 \\
\hline
x + 1 & x^3 + 0x^2 + 0x - 8 \\
 & -(x^3 + x^2) \\
 & \quad - x^2 + 0x \\
 & \quad -(- x^2 - x) \\
 & \qquad x - 8 \\
 & \qquad -(x + 1) \\
 & \qquad\quad -9
\end{array}
$$

The quotient is $x^2 - x + 1$, and the remainder is -9.

PRACTICE

1. $\dfrac{x^3 - 1}{x - 1}$

2. $\dfrac{3x^4 - x^2 + 1}{x^2 - 2}$

SOLUTIONS

1.

$$
\begin{array}{r|l}
 & x^2 + x + 1 \\
\hline
x - 1 & x^3 + 0x^2 + 0x - 1 \\
 & -(x^3 - x^2) \\
 & \quad x^2 + 0x \\
 & \quad -(x^2 - x) \\
 & \qquad x - 1 \\
 & \qquad -(x - 1) \\
 & \qquad\quad 0
\end{array}
$$

The quotient is $x^2 + x + 1$, and the remainder is 0.

2.
$$\begin{array}{r|l} & 3x^2+5 \\ x^2+0x-2 & 3x^4+0x^3-x^2+0x+1 \\ & -(3x^4+0x^3-6x^2) \\ \hline & 5x^2+0x+1 \\ & -(5x^2+0x-10) \\ \hline & 11 \end{array}$$

The quotient is $3x^2 + 5$, and the remainder is 11.

Synthetic Division

Synthetic division of polynomials is much easier to perform than long division. It only works when the divisor is of a certain form though. Here, we will use synthetic division when the divisor is of the form "$x -$ number" or "$x +$ number." We begin by learning how to set up the problems.

For a problem of the form

$$\frac{a_n x^n + a_{n-1}x^{n-1} + \cdots + a_1 x + a_0}{x - c} \quad \text{or}$$

$$(a_n x^n + a_{n-1}x^{n-1} + \cdots + a_1 x + a_0) \div (x - c),$$

we write

$$c \,\big|\, \overline{a_n \;\; a_{n-1} \;\; \ldots \;\; a_1 \;\; a_0}\,.$$

Every power of x must be represented by a coefficient, even if it is 0.

EXAMPLE 10-9

Set up the division problems for synthetic division.

- $\dfrac{4x^3 - 5x^2 + x - 8}{x - 2}$

The coefficients of the dividend are 4, −5, 1, and −8. Because the divisor is $x - 2$, $c = 2$.

$$2 \,\big|\, \overline{4 \;\; -5 \;\; 1 \;\; -8}$$

- $\dfrac{x^3 - 2x + 1}{x - 4}$

We need to think of $x^3 - 2x + 1$ as $x^3 + 0x^2 - 2x + 1$. The coefficients are 1, 0, −2, and 1. The divisor is $x - 4$, so $c = 4$.

$$4 \,\big|\; 1 \quad 0 \quad -2 \quad 1$$

- $\dfrac{3x^4 - x^2 + 2x + 9}{x + 5}$

We can think of $3x^4 - x^2 + 2x + 9$ as $3x^4 + 0x^3 - x^2 + 2x + 9$ and $x + 5$ as $x - (-5)$. The coefficients are 3, 0, −1, 2, and 9, and $c = -5$.

$$-5 \,\big|\; 3 \quad 0 \quad -1 \quad 2 \quad 9$$

PRACTICE

Set up the problems for synthetic division.

1. $\dfrac{5x^3 + x^2 - 3x + 4}{x - 2}$
2. $\dfrac{x^4 - x^3 + 3x - 10}{x - 6}$
3. $\dfrac{x^3 + 2x^2 + x - 8}{x + 3}$
4. $(x^3 + 8) \div (x + 2)$

SOLUTIONS

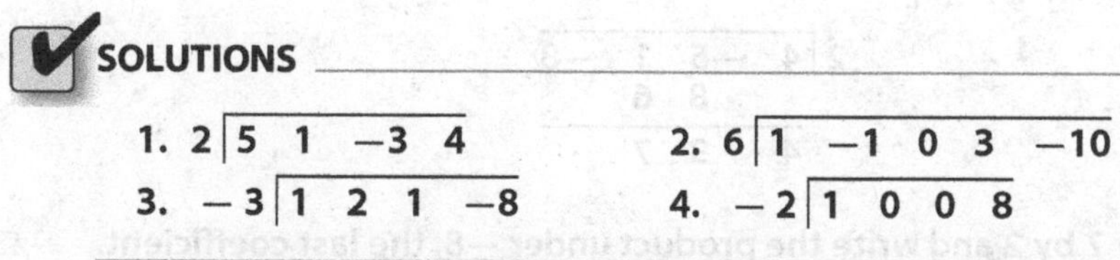

1. $2 \,\big|\; 5 \quad 1 \quad -3 \quad 4$
2. $6 \,\big|\; 1 \quad -1 \quad 0 \quad 3 \quad -10$
3. $-3 \,\big|\; 1 \quad 2 \quad 1 \quad -8$
4. $-2 \,\big|\; 1 \quad 0 \quad 0 \quad 8$

We are ready to learn the steps in synthetic division. The tedious work of long division is reduced to a few steps and some simple arithmetic. The strategy is outlined in the next example.

EXAMPLE 10-10

Find the quotient and remainder using synthetic division.

- $\dfrac{4x^3 - 5x^2 + x - 8}{x - 2}$

$$2 \,\big|\; 4 \quad -5 \quad 1 \quad -8$$

We begin by bringing down the first coefficient.

$$\begin{array}{r|rrrr} 2 & 4 & -5 & 1 & -8 \\ & & & & \\ \hline & 4 & & & \end{array}$$

We multiply this coefficient by 2 (the c) and write the product under -5, the next coefficient.

$$\begin{array}{r|rrrr} 2 & 4 & -5 & 1 & -8 \\ & & 8 & & \\ \hline & 4 & & & \end{array}$$

We add -5 and 8 and write the sum under 8.

$$\begin{array}{r|rrrr} 2 & 4 & -5 & 1 & -8 \\ & & 8 & & \\ \hline & 4 & 3 & & \end{array}$$

We multiply 3 by 2 and write the product under 1, the next coefficient.

$$\begin{array}{r|rrrr} 2 & 4 & -5 & 1 & -8 \\ & & 8 & 6 & \\ \hline & 4 & 3 & & \end{array}$$

We add 1 and 6 and write the sum under 6.

$$\begin{array}{r|rrrr} 2 & 4 & -5 & 1 & -8 \\ & & 8 & 6 & \\ \hline & 4 & 3 & 7 & \end{array}$$

We multiply 7 by 2 and write the product under -8, the last coefficient.

$$\begin{array}{r|rrrr} 2 & 4 & -5 & 1 & -8 \\ & & 8 & 6 & 14 \\ \hline & 4 & 3 & 7 & \end{array}$$

We add -8 and 14 and write the sum under 14. This is the last step.

$$\begin{array}{r|rrrr} 2 & 4 & -5 & 1 & -8 \\ & & 8 & 6 & 14 \\ \hline & 4 & 3 & 7 & 6 \end{array}$$

The numbers in the bottom row are the coefficients of the quotient and the remainder. Recall that the degree of the remainder is smaller than the degree

of the divisor. Because the divisor is $x - 2$, its degree is 1. This means that the remainder has to be a constant (which is a term of degree 0). It also means that the degree of the quotient is exactly one less degree than the degree of the dividend. In this example, the degree of the dividend is 3, so the degree of the quotient is 2. The last number on the bottom row is the remainder. The numbers before it are the coefficients of the quotient in order from the highest degree to the lowest. The remainder in this example is 6. The coefficients of the quotient are 4, 3, and 7. The quotient is $4x^2 + 3x + 7$.

EXAMPLE 10-11

- $(3x^4 - x^2 + 2x + 9) \div (x + 5)$

$$\begin{array}{r|rrrrr} -5 & 3 & 0 & -1 & 2 & 9 \end{array}$$

We bring down 3, the first coefficient and then multiply it by -5. Next, we write $3(-5) = -15$ under 0.

$$\begin{array}{r|rrrrr} -5 & 3 & 0 & -1 & 2 & 9 \\ & & -15 & & & \\ \hline & 3 & & & & \end{array}$$

We add $0 + (-15) = -15$. Next, we multiply -15 by -5 and write $(-15)(-5) = 75$ under -1.

$$\begin{array}{r|rrrrr} -5 & 3 & 0 & -1 & 2 & 9 \\ & & -15 & 75 & & \\ \hline & 3 & -15 & & & \end{array}$$

We add -1 and 75. After we multiply $-1 + 75 = 74$ by -5, we write $(74)(-5) = -370$ under 2.

$$\begin{array}{r|rrrrr} -5 & 3 & 0 & -1 & 2 & 9 \\ & & -15 & 75 & -370 & \\ \hline & 3 & -15 & 74 & & \end{array}$$

We add 2 to -370 and then multiply $2 + (-370) = -368$ by -5 and write $(-368)(-5) = 1840$ under 9.

$$\begin{array}{r|rrrrr} -5 & 3 & 0 & -1 & 2 & 9 \\ & & -15 & 75 & -370 & 1840 \\ \hline & 3 & -15 & 74 & -368 & \end{array}$$

We add 9 to 1840 and write $9 + 1840 = 1849$ under 1840. This is the last step.

$$\begin{array}{r|rrrrr} -5 & 3 & 0 & -1 & 2 & 9 \\ & & -15 & 75 & -370 & 1840 \\ \hline & 3 & -15 & 74 & -368 & 1849 \end{array}$$

The dividend has degree 4, so the quotient has degree 3. The quotient is $3x^3 - 15x^2 + 74x - 368$, and the remainder is 1849.

Still Struggling

For this method to work, the divisor must be in the form $x - c$. For example, if the divisor is $x - 4$, we write 4 outside the division symbol. If the divisor is $x + 6$, we write -6 outside the division symbol.

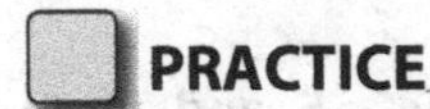

PRACTICE

Find the quotient and remainder using synthetic division.

1. $\dfrac{5x^3 + x^2 - 3x + 4}{x - 2}$

2. $\dfrac{x^3 + 2x^2 + x - 8}{x + 3}$

3. $(x^3 + 8) \div (x + 2)$

4. $\dfrac{-3x^4 + 6x^3 + 4x^2 + 9x - 11}{x + 1}$

5. $\dfrac{2x^3 + x^2 - 4x - 12}{x + \frac{1}{2}}$

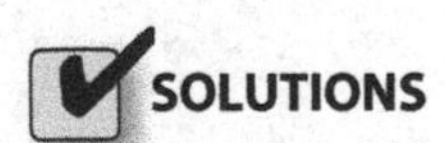

SOLUTIONS

1.

$$\begin{array}{r|rrrr} 2 & 5 & 1 & -3 & 4 \\ & & 10 & 22 & 38 \\ \hline & 5 & 11 & 19 & 42 \end{array}$$

The quotient is $5x^2 + 11x + 19$, and the remainder is 42.

2.

$$\begin{array}{r|rrrr} -3 & 1 & 2 & 1 & -8 \\ & & -3 & 3 & -12 \\ \hline & 1 & -1 & 4 & -20 \end{array}$$

The quotient is $x^2 - x + 4$, and the remainder is -20.

3.

$$\begin{array}{r|rrrr} -2 & 1 & 0 & 0 & 8 \\ & & -2 & 4 & -8 \\ \hline & 1 & -2 & 4 & 0 \end{array}$$

The quotient is $x^2 - 2x + 4$, and the remainder is 0.

4.

$$\begin{array}{r|rrrrr} -1 & -3 & 6 & 4 & 9 & -11 \\ & & 3 & -9 & 5 & -14 \\ \hline & -3 & 9 & -5 & 14 & -25 \end{array}$$

The quotient is $-3x^3 + 9x^2 - 5x + 14$, and the remainder is -25.

5.

$$\begin{array}{r|rrrr} -\frac{1}{2} & 2 & 1 & -4 & -12 \\ & & -1 & 0 & 2 \\ \hline & 2 & 0 & -4 & -10 \end{array}$$

The quotient is $2x^2 - 4$, and the remainder is -10.

When dividing a polynomial $f(x)$ by $x - c$, the remainder tells us two things. If we get a remainder of 0, then both the divisor, $x - c$, and quotient are factors of $f(x)$. In Practice problem 3 above, we had $(x^3 + 8) \div (x + 2) = x^2 - 2x + 4$, with a remainder of 0. This means that $x^3 + 8 = (x + 2)(x^2 - 2x + 4)$. Another fact we get from the remainder is that $f(c) = \text{remainder}$.

$f(x) = (x - c)q(x) + \text{remainder}$ This is the Division Algorithm.

$f(c) = (c - c)q(c) + \text{remainder}$ Evaluate f at $x = c$.

$f(c) = 0q(c) + \text{remainder}$ $c - c = 0$

$f(c) = \text{remainder}$ $0 \cdot q(c) = 0$

The fact that $f(c)$ is the remainder is called the *Remainder Theorem*. It is useful when trying to evaluate complicated polynomials. We can also use this fact to check our work, when doing synthetic division and long division (providing the divisor is $x - c$).

EXAMPLE 10-12

- $(x^3 - 6x^2 + 4x - 5) \div (x - 3)$

By the Remainder Theorem, the remainder should be $3^3 - 6(3^2) + 4(3) - 5 = -20$.

3	1	−6	4	−5
		3	−9	−15
	1	−3	−5	−20

Use synthetic division and the Remainder Theorem to evaluate $f(c)$.

- $f(x) = 14x^3 - 16x^2 + 10x + 8; c = 1$

We first perform synthetic division with $x - 1$.

1	14	−16	10	8
		14	−2	8
	14	−2	8	16

The remainder is 16, so $f(1) = 16$.

PRACTICE

Use synthetic division and the Remainder Theorem to evaluate $f(c)$.

1. $f(x) = 6x^4 - 8x^3 + x^2 + 2x - 5; c = -2$
2. $f(x) = 3x^3 + 7x^2 - 3x + 4; c = \frac{2}{3}$
3. $f(x) = -4x^3 + 5x^2 - 3x + 4; c = -\frac{1}{2}$

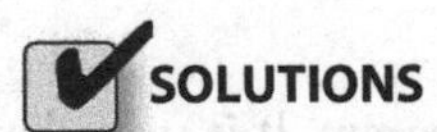

SOLUTIONS

1.

−2	6	−8	1	2	−5
		−12	40	−82	160
	6	−20	41	−80	155

The remainder is 155, so $f(-2) = 155$.

2.

$$\begin{array}{r|rrrr} \frac{2}{3} & 3 & 7 & -3 & 4 \\ & & 2 & 6 & 2 \\ \hline & 3 & 9 & 3 & 6 \end{array}$$

The remainder is 6, so $f(\frac{2}{3}) = 6$.

3.

$$\begin{array}{r|rrrr} -\frac{1}{2} & -4 & 5 & -3 & 4 \\ & & 2 & -\frac{7}{2} & \frac{13}{4} \\ \hline & -4 & 7 & -\frac{13}{2} & \frac{29}{4} \end{array}$$

The remainder is $\frac{29}{4}$, so $f(-\frac{1}{2}) = \frac{29}{4}$.

Synthetic Division and Factoring

We now can use synthetic division and the Remainder Theorem to help us factor polynomials. Suppose $x = c$ is a zero for a polynomial $f(x)$. Let us see what happens when we divide $f(x)$ by $x - c$.

$$f(x) = (x - c)q(x) + r(x)$$

Because $x = c$ is a zero, the remainder is 0, so $f(x) = (x - c)q(x) + 0$, which means $f(x) = (x - c)q(x)$. The next step in completely factoring $f(x)$ is factoring $q(x)$, if necessary.

EXAMPLE 10-13

Completely factor the polynomial.

- **$f(x) = x^3 - 4x^2 - 7x + 10$; $c = 1$ is a zero.**

We use the fact that $c = 1$ is a zero to begin. We use synthetic division to divide $f(x)$ by $x - 1$.

$$\begin{array}{r|rrrr} 1 & 1 & -4 & -7 & 10 \\ & & 1 & -3 & -10 \\ \hline & 1 & -3 & -10 & 0 \end{array}$$

The quotient is $x^2 - 3x - 10$. We now have $f(x)$ partially factored.

$$f(x) = x^3 - 4x^2 - 7x + 10 = (x-1)(x^2 - 3x - 10)$$

Because the quotient is quadratic, we can factor it directly or use the quadratic formula.

$$x^2 - 3x - 10 = (x-5)(x+2)$$

Now we have the complete factorization of $f(x)$.

$$f(x) = x^3 - 4x^2 - 7x + 10 = (x-1)(x-5)(x+2)$$

- $R(x) = x^3 - 2x + 1$; $c = 1$ is a zero.

$$\begin{array}{r|rrrr} 1 & 1 & 0 & -2 & 1 \\ & & 1 & 1 & -1 \\ \hline & 1 & 1 & -1 & 0 \end{array}$$

$$R(x) = x^3 - 2x + 1 = (x-1)(x^2 + x - 1)$$

We now use the quadratic formula to find the zeros of $x^2 + x - 1$.

$$x = \frac{-1 \pm \sqrt{1^2 - 4(1)(-1)}}{2(1)} \frac{-1 \pm \sqrt{5}}{2} = \frac{-1+\sqrt{5}}{2}, \frac{-1-\sqrt{5}}{2}$$

The factors for these zeros are $x - \frac{-1+\sqrt{5}}{2}$ and $x - \frac{-1-\sqrt{5}}{2}$.

$$R(x) = (x-1)\left(x - \frac{-1+\sqrt{5}}{2}\right)\left(x - \frac{-1-\sqrt{5}}{2}\right)$$

PRACTICE

Completely factor the polynomials.

1. $f(x) = x^3 + 2x^2 - x - 2$; $c = 1$ is a zero.
2. $h(x) = x^3 + x^2 - 30x - 72$; $c = -4$ is a zero.
3. $P(x) = x^3 - 5x^2 + 5x + 3$; $c = 3$ is a zero.

SOLUTIONS

1.

$$\begin{array}{r|rrrr} 1 & 1 & 2 & -1 & -2 \\ & & 1 & 3 & 2 \\ \hline & 1 & 3 & 2 & 0 \end{array}$$

$$f(x) = (x-1)(x^2+3x+2) = (x-1)(x+1)(x+2)$$

2.

$$\begin{array}{r|rrrr} -4 & 1 & 1 & -30 & -72 \\ & & -4 & 12 & 72 \\ \hline & 1 & -3 & -18 & 0 \end{array}$$

$$h(x) = (x+4)(x^2-3x-18) = (x+4)(x-6)(x+3)$$

3.

$$\begin{array}{r|rrrr} 3 & 1 & -5 & 5 & 3 \\ & & 3 & -6 & -3 \\ \hline & 1 & -2 & -1 & 0 \end{array}$$

$$P(x) = (x-3)(x^2-2x-1)$$

In order to factor x^2-2x-1, we must first find its zeros. We use the quadratic formula.

$$x = \frac{-(-2) \pm \sqrt{(-2)^2 - 4(1)(-1)}}{2(1)} = \frac{2 \pm \sqrt{8}}{2} = \frac{2 \pm 2\sqrt{2}}{2}$$

$$= \frac{2(1 \pm \sqrt{2})}{2} = 1 \pm \sqrt{2} = 1+\sqrt{2}, 1-\sqrt{2}$$

Because $x = 1+\sqrt{2}$ is a zero, $x-(1+\sqrt{2}) = x-1-\sqrt{2}$ is a factor.
Because $x = 1-\sqrt{2}$ is a zero, $x-(1-\sqrt{2}) = x-1+\sqrt{2}$ is a factor.

$$P(x) = (x-3)(x-1-\sqrt{2})(x-1+\sqrt{2})$$

In the above examples and Practice problems, a zero was given to help us get started. Usually, we have to find a starting point ourselves. The *Rational Zero Theorem* gives us a place to start. The Rational Zero Theorem says that if a polynomial function $f(x)$, with integer coefficients, has a rational number

p/q as a zero, then p is a divisor of the constant term and q is a divisor of the leading coefficient. Not all polynomials have rational zeros, but most of those in algebra courses do.

We use the Rational Zero Theorem to find a list of candidates for zeros. These candidates are rational numbers whose numerators divide the polynomial's constant term and whose denominators divide its leading coefficient. Once we have this list, we try each number in the list to see which, if any, are zeros. Once we have found a zero, we can begin the factorization process.

EXAMPLE 10-14

List the possible rational zeros according to the Rational Zero Theorem.

- $f(x) = 4x^3 + 6x^2 - 2x + 9$

The numerators in our list are the divisors of 9: 1, 3, and 9 as well as their negatives, −1, −3, and −9. The denominators are the divisors of 4: 1, 2, and 4. The list of possible rational zeros is

$$\frac{1}{1}, \frac{3}{1}, \frac{9}{1}, -\frac{1}{1}, -\frac{3}{1}, -\frac{9}{1}, \frac{1}{2}, \frac{3}{2}, \frac{9}{2}, -\frac{1}{2}, -\frac{3}{2}, -\frac{9}{2}, \frac{1}{4}, \frac{3}{4}, \frac{9}{4}, -\frac{1}{4}, -\frac{3}{4}, -\frac{9}{4}$$

Normally, we would not write a fraction with 1 as a denominator. This list could be written with a little less effort as ± 1, ± 3, ± 9, $\pm\frac{1}{2}$, $\pm\frac{3}{2}$, $\pm\frac{9}{2}$, $\pm\frac{1}{4}$, $\pm\frac{3}{4}$, and $\pm\frac{9}{4}$. We only need to list the numerators with negative numbers and not the denominators. The reason is that no new numbers are added to the list, only duplicates of numbers are already there. For example, $\frac{-1}{2}$ and $\frac{1}{-2}$ are the same number.

- $g(x) = 6x^4 - 5x^3 + 2x - 8$

The possible numerators are the divisors of 8: ± 1, ± 2, ± 4, and ± 8. The possible denominators are the divisors of 6: 1, 2, 3, and 6. The list of possible rational zeros is

$$\pm 1, \pm 2, \pm 4, \pm 8, \pm\frac{1}{2}, \pm\frac{2}{2}, \pm\frac{4}{2}, \pm\frac{8}{2}, \pm\frac{1}{3}, \pm\frac{2}{3}, \pm\frac{4}{3}, \pm\frac{8}{3}, \pm\frac{1}{6}, \pm\frac{2}{6}, \pm\frac{4}{6}, \pm\frac{8}{6}$$

There are several duplicates on this list, which happens when the constant term and leading coefficient have common factors. The duplicates do not really hurt anything, but they could waste our time when checking the list for zeros.

PRACTICE

List the candidates for rational zeros. Do not try to find the zeros.

1. $f(x) = 3x^4 + 8x^3 - 11x^2 + 3x + 4$
2. $f(x) = x^3 - 1$
3. $g(x) = x^5 - x^3 + x - 10$
4. $P(x) = 6x^4 - 24$

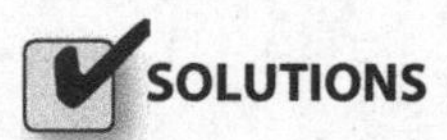

SOLUTIONS

1. **Possible numerators:** $\pm, 1, \pm 2, \pm 4$

   Possible denominators: $1, 3$

   Possible rational zeros: $\pm 1,\ \pm 2,\ \pm 4,\ \pm\frac{1}{3},\ \pm\frac{2}{3}, \pm\frac{4}{3}$

2. **Possible numerators:** ± 1

   Possible denominator: 1

   Possible rational zeros: ± 1

3. **Possible numerators:** $\pm 1,\ \pm 2,\ \pm 5,\ \pm 10$

   Possible denominator: 1

   Possible rational zeros: $\pm 1,\ \pm 2,\ \pm 5,\ \pm 10$

4. **Possible numerators:** $\pm 1, \pm 2, \pm 3,\ \pm 4,\ \pm 6,\ \pm 8, \pm 12, \pm 24$

   Possible denominators: $1,\ 2,\ 3,\ 6$

   Possible rational zeros (with duplicates omitted): $\pm 1,\ \pm 2,\ \pm 3,\ \pm 4,\ \pm 6,\ \pm 8, \pm 12,\ \pm 24,\ \pm\frac{1}{2}, \pm\frac{3}{2}, \pm\frac{1}{3}, \pm\frac{2}{3}, \pm\frac{4}{3}, \pm\frac{8}{3}, \pm\frac{1}{6}$

Now that we have a starting place, we can factor many polynomials. Here is the strategy. First we see if the polynomial can be factored directly. If not, we list the possible rational zeros. We then try the numbers in this list, one at a time, until we find a zero. Once we have found a zero, we use polynomial division (long division or synthetic division) to find the quotient. Next, we factor the quotient. If the quotient is a quadratic factor, we either factor it directly or use the quadratic formula to find its zeros. If the quotient is a polynomial of degree 3 or higher, we need to start over to factor the quotient. Eventually, the quotient is a quadratic factor.

EXAMPLE 10-15

Completely factor the polynomial.

- $P(x) = x^3 + 5x^2 - x - 5$

While this polynomial factors using factoring by grouping, we use the above strategy. The possible rational zeros are ± 1 and ± 5.

$$P(1) = 1^3 + 5(1)^2 - 1 - 5 = 0$$

Now that we know that $x = 1$ is a zero, we use synthetic division to find the quotient for $(x^3 + 5x^2 - x - 5) \div (x - 1)$.

$$\begin{array}{r|rrrr} 1 & 1 & 5 & -1 & -5 \\ & & 1 & 6 & 5 \\ \hline & 1 & 6 & 5 & 0 \end{array}$$

$$P(x) = x^3 + 5x^2 - x - 5 = (x - 1)(x^2 + 6x + 5) = (x - 1)(x + 1)(x + 5)$$

- $f(x) = 3x^4 - 2x^3 - 7x^2 - 2x$

First we factor x from each term: $f(x) = x(3x^3 - 2x^2 - 7x - 2)$. The possible rational zeros for $3x^3 - 2x^2 - 7x - 2$ are ± 1, ± 2, $\pm \frac{1}{3}$, and $\pm \frac{2}{3}$.

$$3(1)^3 - 2(1)^2 - 7(1) - 2 \neq 0 \quad \text{and} \quad 3(-1)^3 - 2(-1)^2 - 7(-1) - 2 = 0$$

We now use synthetic division to find the quotient for $(3x^3 - 2x^2 - 7x - 2) \div (x + 1)$.

$$\begin{array}{r|rrrr} -1 & 3 & -2 & -7 & -2 \\ & & -3 & 5 & 2 \\ \hline & 3 & -5 & -2 & 0 \end{array}$$

The quotient is $3x^2 - 5x - 2$, which factors into $(3x + 1)(x - 2)$.

$$f(x) = 3x^4 - 2x^3 - 7x^2 - 2x = x(3x^3 - 2x^2 - 7x - 2)$$
$$= x(x + 1)(3x^2 - 5x - 2) = x(x + 1)(3x + 1)(x - 2)$$

- $h(x) = 3x^3 + 4x^2 - 18x + 5$

The possible rational zeros are $\pm 1, \ \pm 5, \ \pm \frac{1}{3}$, and $\pm \frac{5}{3}$.

$$h(1) = 3(1^3) + 4(1^2) - 18(1) + 5 \neq 0;$$
$$h(-1) = 3(-1)^3 + 4(-1)^2 - 18(-1) + 5 \neq 0$$
$$h(5) = 3(5^3) + 4(5^2) - 18(5) + 5 \neq 0$$

Continuing in this way, we see that $h(-5) \neq 0, \ h(\frac{1}{3}) \neq 0, \ h(-\frac{1}{3}) \neq 0$, and $h(\frac{5}{3}) = 0$.

$$\begin{array}{r|rrrr} \frac{5}{3} & 3 & 4 & -18 & 5 \\ & & 5 & 15 & -5 \\ \hline & 3 & 9 & -3 & 0 \end{array}$$

$$h(x) = \left(x - \frac{5}{3}\right)(3x^2 + 9x - 3) = \left(x - \frac{5}{3}\right)\overset{\text{Factor 3}}{(3)}(x^2 + 3x - 1)$$
$$= \left[3\left(x - \frac{5}{3}\right)\right](x^2 + 3x - 1) = (3x - 5)(x^2 + 3x - 1)$$

We find the zeros of $x^2 + 3x - 1$ using the quadratic formula.

$$x = \frac{-3 \pm \sqrt{3^2 - 4(1)(-1)}}{2(1)} = \frac{-3 \pm \sqrt{13}}{2} = \frac{-3 + \sqrt{13}}{2}, \ \frac{-3 - \sqrt{13}}{2}$$

$$h(x) = (3x - 5)\left(x - \frac{-3 + \sqrt{13}}{2}\right)\left(x - \frac{-3 - \sqrt{13}}{2}\right)$$

PRACTICE

Completely factor the polynomial.

1. $f(x) = x^4 - 2x^3 - 3x^2 + 8x - 4$
2. $h(x) = 2x^3 + 5x^2 - 23x + 10$
3. $P(x) = 7x^3 + 26x^2 - 15x + 2$

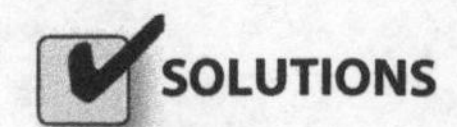

SOLUTIONS

1. The possible rational zeros are ± 1, ± 2, and ± 4. $f(1) = 0$.

$$\begin{array}{r|rrrrr} 1 & 1 & -2 & -3 & 8 & -4 \\ & & 1 & -1 & -4 & 4 \\ \hline & 1 & -1 & -4 & 4 & 0 \end{array}$$

$$f(x) = (x-1)(x^3 - x^2 - 4x + 4)$$

The possible rational zeros for $x^3 - x^2 - 4x + 4$ (which could be factored by grouping) are ± 1, ± 2, and ± 4. We try $x = 1$ again. Because $1^3 - 1^2 - 4(1) + 4 = 0$, $x = 1$ is a zero again.

$$\begin{array}{r|rrrr} 1 & 1 & -1 & -4 & 4 \\ & & 1 & 0 & -4 \\ \hline & 1 & 0 & -4 & 0 \end{array}$$

$$x^3 - x^2 - 4x + 4 = (x-1)(x^2-4) = (x-1)(x-2)(x+2)$$

$$\begin{aligned} f(x) &= (x-1)(x^3 - x^2 - 4x + 4) \\ &= (x-1)(x-1)(x-2)(x+2) \\ &= (x-1)^2(x-2)(x+2) \end{aligned}$$

2. The possible rational zeros are ± 1, ± 2, ± 5, ± 10, $\pm \frac{1}{2}$, and $\pm \frac{5}{2}$. Because $h(2) = 0$, $x = 2$ is a zero of $h(x)$.

$$\begin{array}{r|rrrr} 2 & 2 & 5 & -23 & 10 \\ & & 4 & 18 & -10 \\ \hline & 2 & 9 & -5 & 0 \end{array}$$

$$h(x) = (x-2)(2x^2 + 9x - 5) = (x-2)(2x-1)(x+5)$$

3. The possible rational zeros are ± 1, ± 2, $\pm \frac{1}{7}$, and $\pm \frac{2}{7}$. Because $P(\frac{2}{7}) = 0$, $x = \frac{2}{7}$ is a zero for $P(x)$.

$$\begin{array}{r|rrrr} \frac{2}{7} & 7 & 26 & -15 & 2 \\ & & 2 & 8 & -2 \\ \hline & 7 & 28 & -7 & 0 \end{array}$$

Factor 7

$$P(x) = \left(x - \frac{2}{7}\right)(7x^2 + 28x - 7) = \left(x - \frac{2}{7}\right)(7)(x^2 + 4x - 1)$$

$$= \left[7\left(x - \frac{2}{7}\right)\right](x^2 + 4x - 1) = (7x - 2)(x^2 + 4x - 1)$$

We use the quadratic formula to find the zeros of $x^2 + 4x - 1$.

$$x = \frac{-4 \pm \sqrt{4^2 - 4(1)(-1)}}{2(1)} = \frac{-4 \pm \sqrt{20}}{2} = \frac{-4 \pm 2\sqrt{5}}{2}$$

$$= \frac{2(-2 \pm \sqrt{5})}{2} = -2 \pm \sqrt{5} = -2 + \sqrt{5},\ -2 - \sqrt{5}$$

Now that we know the zeros, we write the function in factored form.

$$x^2 + 4x - 1 = (x - (-2 + \sqrt{5}))(x - (-2 - \sqrt{5}))$$

$$= (x + 2 - \sqrt{5})(x + 2 + \sqrt{5})$$

$$P(x) = (7x - 2)(x + 2 - \sqrt{5})(x + 2 + \sqrt{5})$$

For a polynomial such as $f(x) = 5x^3 + 20x^2 - 9x - 36$, the list of possible rational zeros is quite long—36! There are ways of getting around having to test every one of them. The fastest way is to use a graphing calculator to sketch the graph of $y = 5x^3 + 20x^2 - 9x - 36$. The graph in Fig. 10-18 appears to have an x-intercept at $x = -4$, so we begin with $c = -4$.

$$\begin{array}{r|rrrr} -4 & 5 & 20 & -9 & -36 \\ & & -20 & 0 & 36 \\ \hline & 5 & 0 & -9 & 0 \end{array}$$

$$f(x) = (x + 4)(5x^2 - 9)$$

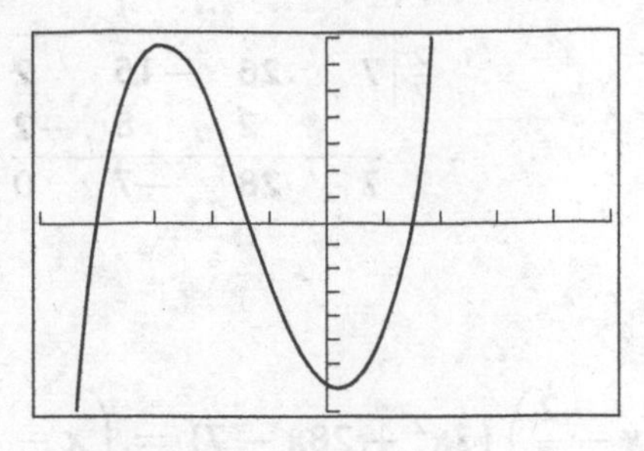

FIGURE 10-18

We now solve $5x^2 - 9 = 0$ to find the other zeros.

$$5x^2 - 9 = 0$$

$$5x^2 = 9$$

$$x^2 = \frac{9}{5}$$

$$x = \pm\sqrt{\frac{9}{5}} = \pm\frac{3}{\sqrt{5}} = \pm\frac{3}{\sqrt{5}} \cdot \frac{\sqrt{5}}{\sqrt{5}} = \pm\frac{3\sqrt{5}}{5} = \frac{3\sqrt{5}}{5}, \ -\frac{3\sqrt{5}}{5}$$

$$f(x) = 5x^3 + 20x^2 - 9x - 36 = (x+4)\left(x - \frac{3\sqrt{5}}{5}\right)\left(x + \frac{3\sqrt{5}}{5}\right)$$

Rule of Signs and Upper and Lower Bounds Theorem

There are also a couple of algebra facts that can help eliminate some of the possible rational zeros. The first we learn is *Descartes' Rule of Signs* and the second is the *Upper and Lower Bounds Theorem*. Descartes' Rule of Signs counts the number of positive zeros and negative zeros. For instance, according to the rule, $f(x) = x^3 + x^2 + 4x + 6$ has no positive zeros at all. This shrinks the list of possible rational zeros from ± 1, ± 2, ± 3, and ± 6 to $-1, -2, -3$, and -6. The Upper and Lower Bounds Theorem gives us an idea of how large (in both the positive and negative directions) the zeros can be.

Descartes' Rule of Signs counts the number of positive zeros and negative zeros by counting sign changes. The *maximum* number of positive zeros for a polynomial function is the number of sign changes in $f(x) = a_n x^n + a_{n-1}x^{n-1} + \ldots + a_1 x + a_0$. The *possible* number of positive zeros is the number of sign changes minus an even whole number. For example, if there are 5 sign changes, then there are 5 or 3 or 1 positive zeros. If there are 6 sign changes, there are 6 or 4 or 2 or 0 positive zeros. The polynomial function $f(x) = 3x^4 - 2x^3 + 7x^2 + 5x - 8$ has 3 sign changes: from 3 to −2, from −2 to 7, and from 5 to −8. There are either 3 or 1 positive zeros. Similarly, the maximim number of negative zeros is the number of sign changes in the polynomial $f(-x)$.

The possible number of negative zeros is the number of sign changes in $f(-x)$ minus an even whole number.

EXAMPLE 10-16

Use Descartes' Rule of Signs to count the possible number of positive zeros and negative zeros for the polynomial function.

- $f(x) = 5x^3 - 6x^2 - 10x + 4$

There are 2 sign changes: from 5 to −6 and from −10 to 4. This means that there are either 2 or 0 positive zeros. Before we count the possible number of negative zeros, remember from earlier that for a number a, $a(-x)^{\text{even power}} = ax^{\text{even power}}$ and $a(-x)^{\text{odd power}} = -ax^{\text{odd power}}$.

$$\begin{aligned} f(-x) &= 5(-x)^3 - 6(-x)^2 - 10(-x) + 4 \\ &= -5x^3 - 6x^2 + 10x + 4 \end{aligned}$$

There is 1 sign change, from −6 to 10, so there is exactly 1 negative zero.

- $g(x) = -x^4 + 3x^2 - 9x + 1$

There are 3 sign changes: from −1 to 3, from 3 to −9, and from −9 to 1, so there are 3 or 1 positive zeros. If we were to rewrite $g(x)$ as $g(x) = -x^4 + 0x^3 + 3x^2 - 9x + 1$, we would not consider 0 coefficients as changing signs. In other words, we ignore the 0 coefficients.

$$\begin{aligned} g(-x) &= -(-x)^4 + 3(-x)^2 - 9(-x) + 1 \\ &= -x^4 + 3x^2 + 9x + 1 \end{aligned}$$

There is 1 sign change, from −1 to 3, so there is exactly 1 negative zero.

- $P(x) = x^5 + x^3 + x + 4$

There are no sign changes, so there are no positive zeros.

$$\begin{aligned} P(-x) &= (-x)^5 + (-x)^3 + (-x) + 4 \\ &= -x^5 - x^3 - x + 4 \end{aligned}$$

There is 1 sign change, so there is exactly 1 negative zero.

One of the advantages of the sign test is that if we know that there are 2 positive zeros and we have found one of them, then we *know* that there is exactly one more.

PRACTICE

Use Descartes' Rule of Signs to count the possible number of positive zeros and the possible number of negative zeros of the polynomial function.

1. $f(x) = 2x^4 - 6x^3 - x^2 + 4x - 8$
2. $f(x) = -x^3 - x^2 + x + 1$
3. $h(x) = -x^4 - x^2 - 6$

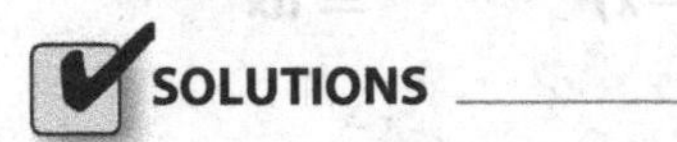

SOLUTIONS

1. **There are 3 sign changes in $f(x)$, so there are 3 or 1 positive zeros.**

$$\begin{aligned} f(-x) &= 2(-x)^4 - 6(-x)^3 - (-x)^2 + 4(-x) - 8 \\ &= 2x^4 + 6x^3 - x^2 - 4x - 8 \end{aligned}$$

There is 1 sign change in $f(-x)$, so there is exactly 1 negative zero.

2. **There is 1 sign change in $f(x)$, so there is exactly 1 positive zero.**

$$f(-x) = -(-x)^3 - (-x)^2 + (-x) + 1 = x^3 - x^2 - x + 1$$

There are 2 sign changes in $f(-x)$, so there are 2 or 0 negative zeros.

3. **There are no sign changes in $h(x)$, so there are no positive zeros.**

$$h(-x) = -(-x)^4 - (-x)^2 - 6 = -x^4 - x^2 - 6$$

There are no sign changes in $h(-x)$, so there are no negative zeros.

The Upper and Lower Bounds Theorem helps us to find an interval of x-values that contains all real zeros. It does *not* tell us what these bounds are. We make a guess as to what these bounds might be and then check them. For a negative number $x = a$, the statement "a is a lower bound for the real zeros" means that there is no number to the left of $x = a$ on the x-axis that is a zero. For a positive number $x = b$, the statement "b is an upper bound for the real zeros" means that there is no number to the right of $x = b$ on the x-axis that is a zero. That is, all the x-intercepts for the graph are between $x = a$ and $x = b$. (We only consider *real* zeros for now. Later, we will learn about *complex* zeros.)

To determine whether a negative number $x = a$ is a lower bound for a polynomial, we need to use synthetic division. If the numbers in the bottom row alternate between nonpositive and nonnegative numbers, then $x = a$ is a lower bound for the negative zeros. "Nonpositive" means 0 or a negative number, and "nonnegative" means 0 or a positive number.

To determine whether a positive number $x = b$ is an upper bound for the positive zeros, again we need to use synthetic division. If the numbers on the bottom row are all nonnegative, then $x = b$ is an upper bound for the positive zeros.

EXAMPLE 10-17

Show that the given value for a is a lower bound and for b is an upper bound for the zeros of the polynomials.

- $f(x) = x^4 + x^3 - 16x^2 - 4x + 48$; $a = -5$ and $b = 5$

$$\begin{array}{r|rrrrr} -5 & 1 & 1 & -16 & -4 & 48 \\ & & -5 & 20 & -20 & 120 \\ \hline & 1 & -4 & 4 & -24 & 168 \end{array}$$

The bottom row alternates between positive and negative numbers, so $a = -5$ is a lower bound for the negative zeros of $f(x)$.

$$\begin{array}{r|rrrrr} 5 & 1 & 1 & -16 & -4 & 48 \\ & & 5 & 30 & 70 & 330 \\ \hline & 1 & 6 & 14 & 66 & 378 \end{array}$$

The entries on the bottom row are all positive, so $b = 5$ is an upper bound for the positive zeros of $f(x)$. All the real zeros for $f(x)$ are between $x = -5$ and $x = 5$.

If 0 appears on the bottom row when testing for an upper bound, we can consider 0 to be positive. If 0 appears in the bottom row when testing for a lower bound, we can consider 0 to be negative if the previous entry is positive and to be positive if the previous entry is negative. In other words, consider 0 to be the opposite sign from the previous entry.

EXAMPLE 10-18

- $P(x) = 4x^4 + 20x^3 + 7x^2 + 3x - 6; a = -5$

-5	4	20	7	3	-6
		-20	0	-35	160
	4	0	7	-32	154

Because 0 follows a positive number, we consider 0 to be negative. This makes the bottom row alternate between nonnegative and nonpositive entries, so $a = -5$ is a lower bound for the negative zeros of $P(x)$.

- $R(x) = -x^3 + 4x^2 + 12x - 5; a = -2$

-2	-1	4	12	-5
		2	-12	0
	-1	6	0	-5

Because 0 follows a positive number, we consider 0 to be negative. The bottom row does not alternate between negative and positive entries, so $a = -2$ is not a lower bound for the negative zeros of $R(x)$.

PRACTICE

Show that the given value for a is a lower bound and for b is an upper bound for the real zeros of the polynomials.

1. $f(x) = x^3 - 6x^2 + x + 5; a = -3,\ b = 7$
2. $f(x) = x^4 - x^2 - 2; a = -2,\ b = 2$
3. $g(x) = 3x^4 + 6x^3 + 2x^2 + x - 5; a = -2,\ b = 1$

SOLUTIONS

1.

$$\begin{array}{r|rrrr} -3 & 1 & -6 & 1 & 5 \\ & & -3 & 27 & -84 \\ \hline & 1 & -9 & 28 & -79 \end{array}$$

The entries on the bottom row alternate between positive and negative (or nonnegative and nonpositive), so $a = -3$ is a lower bound for the real zeros of $f(x)$.

$$\begin{array}{r|rrrr} 7 & 1 & -6 & 1 & 5 \\ & & 7 & 7 & 56 \\ \hline & 1 & 1 & 8 & 61 \end{array}$$

The entries on the bottom are positive (nonnegative), so $b = 7$ is an upper bound for the real zeros of $f(x)$.

2.

$$\begin{array}{r|rrrrr} -2 & 1 & 0 & -1 & 0 & -2 \\ & & -2 & 4 & -6 & 12 \\ \hline & 1 & -2 & 3 & -6 & 10 \end{array}$$

The entries on the bottom row alternate between positive and negative, so $a = -2$ is a lower bound for the real zeros of $f(x)$.

$$\begin{array}{r|rrrrr} 2 & 1 & 0 & -1 & 0 & -2 \\ & & 2 & 4 & 6 & 12 \\ \hline & 1 & 2 & 3 & 6 & 10 \end{array}$$

The entries on the bottom row are all positive, so $b = 2$ is an upper bound for the real zeros of $f(x)$.

3.

$$\begin{array}{r|rrrrr} -2 & 3 & 6 & 2 & 1 & -5 \\ & & -6 & 0 & -4 & 6 \\ \hline & 3 & 0 & 2 & -3 & 1 \end{array}$$

The entries on the bottom row alternate between nonnegative and nonpositive (because 0 follows a positive number, consider it nonpositive), so $a = -2$ is a lower bound for the real zeros of $g(x)$.

$$\begin{array}{r|rrrrr} 1 & 3 & 6 & 2 & 1 & -5 \\ & & 3 & 9 & 11 & 12 \\ \hline & 3 & 9 & 11 & 12 & 7 \end{array}$$

The entries on the bottom row are all positive, so $b = 1$ is an upper bound for the real zeros of $g(x)$.

Complex Numbers

Until now, the zeros of polynomials have been real numbers. The next topic involves *complex* zeros. These zeros come from even roots of negative numbers such as $\sqrt{-1}$. Before working with complex zeros of polynomials, we first learn some complex number arithmetic.

Complex numbers are normally written in the form $a + bi$, where a and b are real numbers and $i = \sqrt{-1}$. Technically, real numbers are complex numbers where $b = 0$. A number such as $4 + \sqrt{-9}$ can be written as $4 + 3i$ because $\sqrt{-9} = \sqrt{9}\sqrt{-1} = 3i$.

EXAMPLE 10-19

Write the complex numbers in the form $a + bi$, where a and b are real numbers.

- $\sqrt{-64} = \sqrt{64}\sqrt{-1} = 8i$
- $\sqrt{-27} = \sqrt{27}\sqrt{-1} = \sqrt{27}i = \sqrt{9 \cdot 3}i = 3\sqrt{3}i$

Be careful, $\sqrt{3i} \neq \sqrt{3}i$.

- $6 + \sqrt{-8} = 6 + \sqrt{8}i = 6 + \sqrt{4 \cdot 2}i = 6 + 2\sqrt{2}i$
- $2 - \sqrt{-50} = 2 - \sqrt{50}i = 2 - \sqrt{25 \cdot 2}i = 2 - 5\sqrt{2}i$

PRACTICE

Write the complex numbers in the form $a + bi$, where a and b are real numbers.

1. $\sqrt{-25}$
2. $\sqrt{-10}$
3. $\sqrt{-24}$
4. $14 - \sqrt{-36}$

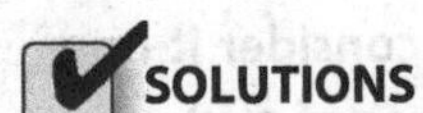

SOLUTIONS

1. $\sqrt{-25} = \sqrt{25}i = 5i$
2. $\sqrt{-10} = \sqrt{10}i$
3. $\sqrt{-24} = \sqrt{24}i = \sqrt{4 \cdot 6}i = 2\sqrt{6}i$
4. $14 - \sqrt{-36} = 14 - \sqrt{36}i = 14 - 6i$

Adding complex numbers is a matter of adding like terms. Add the real parts, a and c, and the imaginary parts, b and d.

$$(a + bi) + (c + di) = (a + c) + (b + d)i$$

EXAMPLE 10-20

Perform the addition. Write the sum in the form $a + bi$, where a and b are real numbers.

- $(3 - 5i) + (4 + 8i) = (3 + 4) + (-5 + 8)i = 7 + 3i$
- $2i - 6 + 9i = -6 + 11i$
- $4 + i - 3 - i = (4 - 3) + (1 - 1)i = 1$
- $7 - \sqrt{-18} + 3 + 5\sqrt{-2} = 7 - \sqrt{18}i + 3 + 5\sqrt{2}i$
  $= 7 - \sqrt{9 \cdot 2}i + 3 + 5\sqrt{2}i$
  $= 7 - 3\sqrt{2}i + 3 + 5\sqrt{2}i = 10 + 2\sqrt{2}i$

We subtract two complex numbers by distributing the minus sign in the parentheses and then adding the like terms.

$$a + bi - (c + di) = a + bi - c - di = (a - c) + (b - d)i$$

EXAMPLE 10-21

Perform the subtraction and write the difference in the form $a + bi$, where a and b are real numbers.

- $11 - 3i - (7 + 6i) = 11 - 3i - 7 - 6i = 4 - 9i$
- $i - (1 + i) = i - 1 - i = -1$
- $9 - (4 - i) = 9 - 4 + i = 5 + i$
- $7 + \sqrt{-8} - (1 - \sqrt{-18}) = 7 + \sqrt{8}i - 1 + \sqrt{18}i$
  $= 7 + 2\sqrt{2}i - 1 + 3\sqrt{2}i = 6 + 5\sqrt{2}i$

PRACTICE

Perform the addition. Write the sum in the form $a + bi$, where a and b are real numbers for Problems 1-5.

1. $18 - 4i + (-15) + 2i$
2. $8 - 2 + 5i$
3. $5 + i + 5 - i$
4. $7 + i + 12 + i$
5. $1 + \sqrt{-15} - 6 + 2\sqrt{-15}$

Perform the subtraction and write the difference in the form $a + bi$, where a and b are real numbers for Problems 6-8.

6. $2 + 3i - (8 + 7i)$
7. $4 + 5i - (4 - 5i)$
8. $\sqrt{-48} - (-1 - \sqrt{-75})$

SOLUTIONS

1. $18 - 4i + (-15) + 2i = 3 - 2i$
2. $8 - 2 + 5i = 6 + 5i$
3. $5 + i + 5 - i = 10 + 0i = 10$
4. $7 + i + 12 + i = 19 + 2i$
5. $1 + \sqrt{-15} + (-6) + 2\sqrt{-15} = 1 + \sqrt{15}i - 6 + 2\sqrt{15}i = -5 + 3\sqrt{15}i$
6. $2 + 3i - (8 + 7i) = 2 + 3i - 8 - 7i = -6 - 4i$
7. $4 + 5i - (4 - 5i) = 4 + 5i - 4 + 5i = 10i$
8. $$\begin{aligned}\sqrt{-48} - (-1 - \sqrt{-75}) &= \sqrt{48}i + 1 + \sqrt{75}i \\ &= \sqrt{16 \cdot 3}i + 1 + \sqrt{25 \cdot 3}i \\ &= 4\sqrt{3}i + 1 + 5\sqrt{3}i = 1 + 9\sqrt{3}i\end{aligned}$$

Multiplying complex numbers is not as straightforward as adding and subtracting complex numbers. We begin with the product of two purely imaginary numbers (numbers whose real parts are 0). Remember that $i = \sqrt{-1}$, which makes $i^2 = -1$. In most complex number multiplication problems, we have a term with i^2. We replace i^2 with -1.

EXAMPLE 10-22

- $(5i)(6i) = 30i^2 = 30(-1) = -30$
- $(2i)(-9i) = -18i^2 = -18(-1) = 18$
- $(\sqrt{-6})(\sqrt{-9}) = (\sqrt{6}i)(\sqrt{9}i) = (\sqrt{6})(3)i^2 = 3\sqrt{6}(-1) = -3\sqrt{6}$

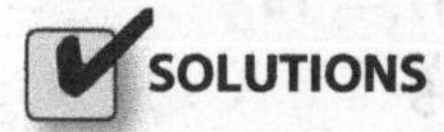

PRACTICE

Find the product.

1. $(2i)(10i)$
2. $(4\sqrt{-25})(2\sqrt{-25})$
3. $\sqrt{-3}\cdot\sqrt{-12}$
4. $\sqrt{-6}\cdot\sqrt{-15}$

SOLUTIONS

1. $(2i)(10i) = 20i^2 = 20(-1) = -20$
2. $(4\sqrt{-25})(2\sqrt{-25}) = 4(5i)[2(5i)] = 200i^2 = 200(-1) = -200$
3. $\sqrt{-3}\cdot\sqrt{-12} = \sqrt{3}i\cdot\sqrt{12}i = \sqrt{3\cdot 12}i^2 = \sqrt{36}i^2$
   $= 6(-1) = -6$
4. $\sqrt{-6}\cdot\sqrt{-15} = \sqrt{6}i\cdot\sqrt{15}i = \sqrt{6\cdot 15}i^2 = \sqrt{90}i^2$
   $= 3\sqrt{10}(-1) = -3\sqrt{10}$

We multiply two complex numbers in the form $a+bi$ using the FOIL method, substituting -1 for i^2 and then combining like terms.

EXAMPLE 10-23

Perform the multiplication. Write the product in the form $a+bi$, where a and b are real numbers.

- $(4+2i)(5+3i) = 20+12i+10i+6i^2 = 20+22i+6(-1)$
  $= 14+22i$
- $(1-i)(2+i) = 2+i-2i-i^2 = 2-i-(-1) = 3-i$
- $(8-2i)(8+2i) = 64+16i-16i-4i^2 = 64-4(-1) = 68$

PRACTICE

Perform the multiplication. Write the product in the form $a+bi$, where a and b are real numbers.

1. $(15+3i)(-2+i)$
2. $(-1+3i)(4-2i)$
3. $(3+2i)(3-2i)$
4. $(2-i)(2+i)$

SOLUTIONS

1. $(15+3i)(-2+i) = -30+15i-6i+3i^2$
$= -30+9i+3(-1) = -33+9i$
2. $(-1+3i)(4-2i) = -4+2i+12i-6i^2$
$= -4+14i-6(-1) = 2+14i$
3. $(3+2i)(3-2i) = 9-6i+6i-4i^2 = 9-4(-1) = 13$
4. $(2-i)(2+i) = 4+2i-2i-i^2 = 4-(-1) = 5$

The two complex numbers $a+bi$ and $a-bi$ are called *complex conjugates*. The only difference between a complex number and its conjugate is the sign between the real part and the imaginary part.

EXAMPLE 10-24

- **The complex conjugate of $3+2i$ is $3-2i$.**
- **The complex conjugate of $-7-i$ is $-7+i$.**
- **The complex conjugate of $10i$ (as $0+10i$) is $-10i$ (as $0-10i$).**

PRACTICE

Identify the complex conjugate.

1. $15+7i$ 2. $-3+i$ 3. $-9i$

1. **The complex conjugate of $15+7i$ is $15-7i$.**
2. **The complex conjugate of $-3+i$ is $-3-i$.**
3. **The complex conjugate of $-9i$ is $9i$.**

The product of any complex number and its conjugate is a real number.

$$(a+bi)(a-bi) = a^2 - abi + abi - b^2i^2$$
$$= a^2 - b^2(-1)$$
$$= a^2 + b^2$$

EXAMPLE 10-25

Perform the multiplication.

- $(7-2i)(7+2i)$; here, $a=7$ and $b=2$, so $a^2=49$ and $b^2=4$, making $(7-2i)(7+2i)=49+4=53$.
- $(1-i)(1+i)$; here $a=1$ and $b=1$, so $a^2=1$ and $b^2=1$, making $(1-i)(1+i)=1+1=2$.
- $(-6+3i)(-6-3i)=36+9=45$

PRACTICE

Perform the multiplication. Write the product in the form $a+bi$, where a and b are real numbers.

1. $(8-10i)(8+10i)$ 2. $(1-9i)(1+9i)$ 3. $(5-2i)(5+2i)$

SOLUTIONS

1. $(8-10i)(8+10i)=64+100=164$
2. $(1-9i)(1+9i)=1+81=82$
3. $(5-2i)(5+2i)=25+4=29$

Dividing two complex numbers can be complicated. These problems are normally written in fraction form. If the denominator is purely imaginary, we can simply multiply the fraction by $\frac{i}{i}$ and simplify.

EXAMPLE 10-26

Perform the division. Write the quotient in the form $a+bi$, where a and b are real numbers.

- $$\frac{2+3i}{i}=\frac{2+3i}{i}\cdot\frac{i}{i}=\frac{(2+3i)i}{i^2}=\frac{2i+3i^2}{i^2}=\frac{2i+3(-1)}{-1}$$
$$=\frac{-3+2i}{-1}=-(-3+2i)=3-2i$$

- $$\frac{4+5i}{2i}=\frac{4+5i}{2i}\cdot\frac{i}{i}=\frac{4i+5i^2}{2i^2}=\frac{4i+5(-1)}{2(-1)}=\frac{4i-5}{-2}$$
$$=\frac{-(-5+4i)}{2}=\frac{5-4i}{2}=\frac{5}{2}-2i$$

PRACTICE

Perform the division. Write the quotient in the form $a+bi$, where a and b are real numbers.

1. $\dfrac{12+5i}{2i}$ 2. $\dfrac{4-9i}{-3i}$ 3. $\dfrac{1+i}{i}$

SOLUTIONS

1. $$\frac{12+5i}{2i}=\frac{12+5i}{2i}\cdot\frac{i}{i}=\frac{12i+5i^2}{2i^2}=\frac{12i+5(-1)}{2(-1)}=\frac{-5+12i}{-2}$$
$$=\frac{-(-5+12i)}{2}=\frac{5-12i}{2}=\frac{5}{2}-6i$$

2. $$\frac{4-9i}{-3i}=\frac{4-9i}{-3i}\cdot\frac{i}{i}=\frac{4i-9i^2}{-3i^2}=\frac{4i-9(-1)}{-3(-1)}=\frac{9+4i}{3}$$
$$=3+\frac{4}{3}i$$

3. $$\frac{1+i}{i}=\frac{1+i}{i}\cdot\frac{i}{i}=\frac{i+i^2}{i^2}=\frac{i+(-1)}{-1}=\frac{-1+i}{-1}$$
$$=\frac{-(-1+i)}{1}=1-i$$

When the divisor (denominator) is in the form $a+bi$, multiplying the fraction by $\frac{i}{i}$ does not work.

$$\frac{2-5i}{3+6i}\cdot\frac{i}{i}=\frac{2i-5i^2}{3i+6i^2}=\frac{5+2i}{-6+3i}$$

What *does* work is to multiply the fraction by the denominator's conjugate over itself. This works because the product of any complex number and its conjugate is a real number. We use the FOIL method in the numerator (if necessary) and the fact that $(a+bi)(a-bi)=a^2+b^2$ in the denominator.

EXAMPLE 10-27

Perform the division. Write the quotient in the form $a + bi$, where a and b are real numbers.

- $\dfrac{2+7i}{6+i}$

The denominator's conjugate is $6 - i$.

$$= \frac{2+7i}{6+i} \cdot \frac{6-i}{6-i} = \frac{12-2i+42i-7i^2}{6^2+1^2} = \frac{12+40i-7(-1)}{37}$$

$$= \frac{12+40i+7}{37} = \frac{19+40i}{37} = \frac{19}{37} + \frac{40}{37}i$$

- $\dfrac{4-9i}{5-2i}$

The denominator's conjugate is $5 + 2i$.

$$= \frac{4-9i}{5-2i} \cdot \frac{5+2i}{5+2i} = \frac{20+8i-45i-18i^2}{5^2+2^2} = \frac{20-37i-18(-1)}{25+4}$$

$$= \frac{20-37i+18}{29} = \frac{38-37i}{29} = \frac{38}{29} - \frac{37}{29}i$$

PRACTICE

Perform the division. Write the quotient in the form $a + bi$, where a and b are real numbers.

1. $\dfrac{1-2i}{1-i}$
2. $\dfrac{4+2i}{1-3i}$
3. $\dfrac{8-i}{2-5i}$

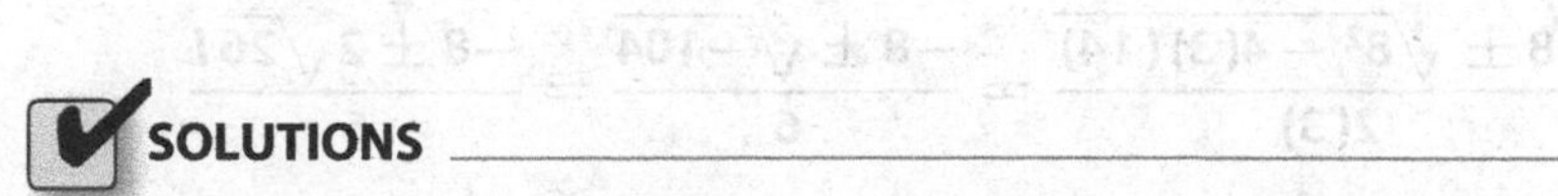

SOLUTIONS

1. $$\frac{1-2i}{1-i} = \frac{1-2i}{1-i} \cdot \frac{1+i}{1+i} = \frac{1+i-2i-2i^2}{1^2+1^2} = \frac{1-i-2(-1)}{2}$$

$$= \frac{3-i}{2} = \frac{3}{2} - \frac{1}{2}i$$

2. $$\frac{4+2i}{1-3i}=\frac{4+2i}{1-3i}\cdot\frac{1+3i}{1+3i}=\frac{4+12i+2i+6i^2}{1^2+3^2}$$
$$=\frac{4+14i+6(-1)}{10}=\frac{-2+14i}{10}=-\frac{1}{5}+\frac{7}{5}i$$

3. $$\frac{8-i}{2-5i}=\frac{8-i}{2-5i}\cdot\frac{2+5i}{2+5i}=\frac{16+40i-2i-5i^2}{2^2+5^2}$$
$$=\frac{16+38i-5(-1)}{29}=\frac{21+38i}{29}=\frac{21}{29}+\frac{38}{29}i$$

Complex Solutions to Quadratic Equations

Every quadratic equation has a solution, though the solution might not be real. The real solution, or solutions, for a quadratic equation is, or are, the x-intercept, or intercepts, for the graph of the quadratic function. The graph for $f(x)=x^2+1$ has no x-intercepts because the equation $x^2+1=0$ has no real solution; however, it *does* have two complex solutions, $\pm i$.

$$x^2+1=0$$
$$x^2=-1$$
$$x=\pm\sqrt{-1}=\pm i$$

EXAMPLE 10-28

Solve the equations and write the solutions in the form $a+bi$, where a and b are real numbers.

- $3x^2+8x+14=0$

$$x=\frac{-8\pm\sqrt{8^2-4(3)(14)}}{2(3)}=\frac{-8\pm\sqrt{-104}}{6}=\frac{-8\pm2\sqrt{26}\,i}{6}$$
$$=\frac{2(-4\pm\sqrt{26}\,i)}{6}=\frac{-4\pm\sqrt{26}\,i}{3}=-\frac{4}{3}\pm\frac{\sqrt{26}}{3}i$$
$$=-\frac{4}{3}+\frac{\sqrt{26}}{3}i,\ -\frac{4}{3}-\frac{\sqrt{26}}{3}i$$

- $9x^2 + 25 = 0$

$$9x^2 + 25 = 0$$

$$9x^2 = -25$$

$$x^2 = -\frac{25}{9}$$

$$x = \pm\sqrt{-\frac{25}{9}} = \pm\sqrt{\frac{25}{9}}\,i = \pm\frac{5}{3}i = \frac{5}{3}i,\ -\frac{5}{3}i$$

PRACTICE

Solve the equations and write the solutions in the form $a + bi$, where a and b are real numbers.

1. $x^2 + 2x + 4 = 0$
2. $x^2 + 25 = 0$
3. $9x^2 + 4 = 0$
4. $6x^2 + 8x + 9 = 0$

SOLUTIONS

1. We complete the square because it is a little easier than using the quadratic formula.

$$x^2 + 2x + 4 = 0$$

$$x^2 + 2x + \left(\frac{2}{2}\right)^2 = -4 + \left(\frac{2}{2}\right)^2$$

$$(x+1)^2 = -3$$

$$x + 1 = \pm\sqrt{-3}$$

$$x = -1 \pm \sqrt{3}\,i = -1 + \sqrt{3}\,i,\ -1 - \sqrt{3}\,i$$

2. We solve for x^2 and then take the square root of each side.

$$x^2 + 25 = 0$$

$$x^2 = -25$$

$$x = \pm\sqrt{-25} = \pm 5i = 5i,\ -5i$$

3. We solve for x^2 and then take the square root of each side.

$$9x^2 + 4 = 0$$

$$9x^2 = -4$$

$$x^2 = -\frac{4}{9}$$

$$x = \pm\sqrt{-\frac{4}{9}} = \pm\frac{2}{3}i = \frac{2}{3}i, -\frac{2}{3}i$$

4. $$x = \frac{-8 \pm \sqrt{8^2 - 4(6)(9)}}{2(6)} = \frac{-8 \pm \sqrt{-152}}{12}$$

$$= \frac{-8 \pm 2\sqrt{38}i}{12} = \frac{2(-4 \pm \sqrt{38}i)}{12} = \frac{-4 \pm \sqrt{38}i}{6}$$

$$= -\frac{4}{6} \pm \frac{\sqrt{38}}{6}i = -\frac{2}{3} \pm \frac{\sqrt{38}}{6}i = -\frac{2}{3} + \frac{\sqrt{38}}{6}i, -\frac{2}{3} - \frac{\sqrt{38}}{6}i$$

In all the previous examples and Practice problems, complex solutions to quadratic equations came in conjugate pairs. This *always* happens when the solutions to polynomial equations are complex numbers. A quadratic expression that has complex zeros is called *irreducible* (over the reals) because it cannot be factored using only real numbers. For example, the polynomial function $f(x) = x^4 - 1$ can be factored using real numbers as $(x^2 - 1)(x^2 + 1) = (x - 1)(x + 1)(x^2 + 1)$. The factor $x^2 + 1$ is irreducible because it is factored as $(x - i)(x + i)$. We can tell which quadratic factors are irreducible without having to use the quadratic formula.

We only need part of the quadratic formula, the part under the squart root sign, $b^2 - 4ac$. When this number is negative, the quadratic factor has two complex zeros, $\frac{-b \pm \sqrt{\text{negative number}}}{2a}$. When this number is positive, there are two real number solutions, $\frac{-b \pm \sqrt{\text{positive number}}}{2a}$. When this number is 0, there is one real zero, $\frac{-b \pm \sqrt{0}}{2a} = \frac{-b}{2a}$. For this reason, $b^2 - 4ac$ is called the *discriminant*.

The Fundamental Theorem of Algebra

The *Fundamental Theorem of Algebra* states that any polynomial of degree $n \geq 1$ has a zero (which might be complex). From this, we have the fact that every polynomial of degree n, with $n \geq 1$, has exactly n zeros (some might be counted more than once). Because $x = c$ is a zero implies $x - c$ is a factor, every polynomial can be completely factored in the form $a(x - c_n)(x - c_{n-1}) \ldots (x - c_1)$, where a is a real number and c_i is real or complex. Factors in the form $x - c$ are called *linear factors*. Factors such as $2x + 1$ can be written in the form $x - c$ by factoring: $2(x + \frac{1}{2})$ or $2(x - (-\frac{1}{2}))$.

To completely factor a polynomial, we must find its zeros. We can use some of the techniques that we used to find the real zeros of a polynomial: polynomial division and the quadratic formula. Note that the instructions will state "Find all real zeros, real or complex," but they would be more technically correct if they would state "Find all real zeros, real or complex but not real" as a real number is actually a complex number as well.

 **EXAMPLE 10-29**

Find all zeros, real or complex.

- $h(x) = x^4 - 16$

This is the difference of two squares.

$$x^4 - 16 = (x^2 - 4)(x^2 + 4) = (x - 2)(x + 2)(x^2 + 4)$$

The real zeros are 2 and -2. We find the complex zeros by solving $x^2 + 4 = 0$.

$$x^2 + 4 = 0$$

$$x^2 = -4$$

$$x = \pm\sqrt{-4} = \pm 2i$$

The complex zeros are $\pm 2i$.

- $P(x) = x^4 + 6x^3 + 9x^2 - 6x - 10$

The possible rational zeros are ± 1, ± 2, ± 5, and ± 10.

$$P(1) = 0$$

$$\begin{array}{r|rrrrr} 1 & 1 & 6 & 9 & -6 & -10 \\ & & 1 & 7 & 16 & 10 \\ \hline & 1 & 7 & 16 & 10 & 0 \end{array}$$

$$P(x) = (x-1)(x^3+7x^2+16x+10)$$

Because $x^3+7x^2+16x+10$ has no sign changes, there are no positive zeros; $x=-1$ is a zero for $x^3+7x^2+16x+10$.

$$\begin{array}{r|rrrr} -1 & 1 & 7 & 16 & 10 \\ & & -1 & -6 & -10 \\ \hline & 1 & 6 & 10 & 0 \end{array}$$

$$P(x) = (x-1)(x+1)(x^2+6x+10)$$

We solve $x^2+6x+10=0$ to find the complex zeros.

$$x = \frac{-6 \pm \sqrt{6^2-4(1)(10)}}{2(1)} = \frac{-6 \pm \sqrt{-4}}{2}$$

$$= \frac{-6 \pm 2i}{2} = \frac{2(-3 \pm i)}{2} = -3 \pm i$$

The zeros are ± 1 and $-3 \pm i$.

PRACTICE

Find all zeros, real or complex.

1. $f(x) = x^4 - 81$
2. $h(x) = x^3 + 13x - 34$
3. $P(x) = x^4 - 6x^3 + 29x^2 - 76x + 68$

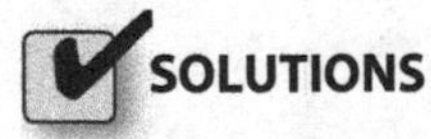

SOLUTIONS

1. $f(x) = (x^2-9)(x^2+9) = (x-3)(x+3)(x^2+9)$

$$x^2 + 9 = 0$$

$$x^2 = -9$$

$$x = \pm\sqrt{-9} = \pm 3i$$

The zeros are ± 3 and $\pm 3i$.

2. $h(2) = 0$

$$\begin{array}{r|rrrr} 2 & 1 & 0 & 13 & -34 \\ & & 2 & 4 & 34 \\ \hline & 1 & 2 & 17 & 0 \end{array}$$

$h(x) = (x-2)(x^2+2x+17)$; we now solve $x^2+2x+17=0$ using the quadratic formula.

$$x = \frac{-2 \pm \sqrt{2^2 - 4(1)(17)}}{2(1)} == \frac{-2 \pm \sqrt{-64}}{2} = \frac{-2 \pm 8i}{2}$$

$$= \frac{2(-1 \pm 4i)}{2} = -1 \pm 4i$$

The zeros are 2 and $-1 \pm 4i$.

3. $P(2) = 0$

$$\begin{array}{r|rrrrr} 2 & 1 & -6 & 29 & -76 & 68 \\ & & 2 & -8 & 42 & -68 \\ \hline & 1 & -4 & 21 & -34 & 0 \end{array}$$

$$P(x) = (x-2)(x^3 - 4x^2 + 21x - 34)$$

$x = 2$ is a zero for $x^3 - 4x^2 + 21x - 34$.

$$\begin{array}{r|rrrr} 2 & 1 & -4 & 21 & -34 \\ & & 2 & -4 & 34 \\ \hline & 1 & -2 & 17 & 0 \end{array}$$

$P(x) = (x-2)(x-2)(x^2-2x+17)$; we solve $x^2 - 2x + 17 = 0$ using the quadratic formula.

$$x = \frac{-(-2) \pm \sqrt{(-2)^2 - 4(1)(17)}}{2(1)} = \frac{2 \pm \sqrt{-64}}{2}$$

$$= \frac{2 \pm 8i}{2} = \frac{2(1 \pm 4i)}{2} = 1 \pm 4i$$

The zeros are 2 and $1 \pm 4i$.

In the next set of problems, we use polynomial division to factor a polynomial to find its zeros. We are given one complex zero, which automatically tells us the second zero: the complex conjugate is also a zero. These complex zeros give us a quadratic factor for the polynomial. Once we have this computed, we can use long division to find the quotient, which is another factor of the polynomial.

EXAMPLE 10-30

Find all zeros, real or complex.

- **$f(x) = 3x^4 + x^3 + 17x^2 + 4x + 20$; $x = 2i$ is a zero.**

Because $x = 2i$ is a zero, its conjugate, $-2i$, is another zero. This tells us that two factors are $x - 2i$ and $x + 2i$.

$$(x - 2i)(x + 2i) = x^2 + 2ix - 2ix - 4i^2 = x^2 - 4(-1) = x^2 + 4$$

We divide $f(x)$ by $x^2 + 4 = x^2 + 0x + 4$.

$$\begin{array}{r|l} & 3x^2 + x + 5 \\ \hline x^2 + 0x + 4 & 3x^4 + x^3 + 17x^2 + 4x + 20 \\ & -(3x^4 + 0x^3 + 12x^2) \\ \hline & x^3 + 5x^2 + 4x \\ & -(x^3 + 0x^2 + 4x) \\ \hline & 5x^2 + 0x + 20 \\ & -(5x^2 + 0x + 20) \\ \hline & 0 \end{array}$$

$$f(x) = (x^2 + 4)(3x^2 + x + 5)$$

Solving $3x^2 + x + 5 = 0$, we find the other solutions.

$$x = \frac{-1 \pm \sqrt{1^2 - 4(3)(5)}}{2(3)} = \frac{-1 \pm \sqrt{-59}}{6} = \frac{-1 \pm \sqrt{59}\,i}{6}$$

The zeros are $\pm 2i$ and $\frac{-1 \pm \sqrt{59}\,i}{6}$.

- **$h(x) = 2x^3 - 7x^2 + 170x - 246$; $x = 1 + 9i$ is a zero.**

Because $x = 1 + 9i$ is a zero, we know that $x = 1 - 9i$ is another a zero. We also know that $x - (1 + 9i) = x - 1 - 9i$ and $x - (1 - 9i) = x - 1 + 9i$ are factors. When we multiply these factors, we will have a quadratic factor (with only real numbers). We then divide $h(x)$ by this quadratic factor. The zero of the quotient is also a zero of $h(x)$.

$$(x - 1 - 9i)(x - 1 + 9i) = x^2 - x + 9ix - x + 1 - 9i - 9ix + 9i - 81i^2$$
$$= x^2 - 2x + 1 - 81(-1) = x^2 - 2x + 82$$

$$\begin{array}{r|l} & \quad 2x-3 \\ \hline x^2 - 2x + 82 & 2x^3 - 7x^2 + 170x - 246 \\ & -(2x^3 - 4x^2 + 164x) \\ \hline & \quad -3x^2 + 6x - 246 \\ & \quad -(-3x^2 + 6x - 246) \\ \hline & \qquad 0 \end{array}$$

$$h(x) = (2x - 3)(x^2 - 2x + 82)$$

The zeros are $1 \pm 9i$ and $\frac{3}{2}$ (from $2x - 3 = 0$).

PRACTICE

Find all zeros, real or complex.

1. **$f(x) = x^4 - x^3 + 8x^2 - 9x - 9$; $x = -3i$ is a zero.**
2. **$g(x) = x^3 - 5x^2 + 7x + 13$; $x = 3 - 2i$ is a zero.**
3. **$h(x) = x^4 - 8x^3 + 21x^2 + 32x - 100$; $x = 4 + 3i$ is a zero.**

SOLUTIONS

1. **$x = -3i$ is a zero, so $x = 3i$ is a zero also. One factor of $f(x)$ is**

$$(x - 3i)(x + 3i) = x^2 + 9 = x^2 + 0x + 9$$

$$
\begin{array}{r|l}
 & x^2 - x - 1 \\
\hline
x^2 + 0x + 9 & x^4 - x^3 + 8x^2 - 9x - 9 \\
 & -(x^4 + 0x^3 + 9x^2) \\
\hline
 & \quad -x^3 - x^2 - 9x \\
 & \quad -(-x^3 + 0x^2 - 9x) \\
\hline
 & \qquad -x^2 + 0x - 9 \\
 & \qquad -(-x^2 + 0x - 9) \\
\hline
 & \qquad\qquad 0
\end{array}
$$

$$f(x) = (x^2 + 9)(x^2 - x - 1); \text{ we now solve } x^2 - x - 1 = 0$$

$$x = \frac{-(-1) \pm \sqrt{(-1)^2 - 4(1)(-1)}}{2(1)} = \frac{1 \pm \sqrt{5}}{2}$$

The zeros are $\pm 3i$ and $\frac{1 \pm \sqrt{5}}{2}$.

2. $x = 3 - 2i$ is a zero, so $x = 3 + 2i$ is also a zero. One factor of $g(x)$ is

$$(x - (3 - 2i))(x - (3 + 2i)) = (x - 3 + 2i)(x - 3 - 2i)$$

$$= x^2 - 3x - 2ix - 3x + 9 + 6i + 2ix - 6i - 4i^2$$

$$= x^2 - 6x + 9 - 4(-1)$$

$$= x^2 - 6x + 13$$

$$
\begin{array}{r|l}
 & x + 1 \\
\hline
x^2 - 6x + 13 & x^3 - 5x^2 + 7x + 13 \\
 & -(x^3 - 6x^2 + 13x) \\
\hline
 & \qquad x^2 - 6x + 13) \\
 & \quad -(x^2 - 6x + 13) \\
\hline
 & \qquad\qquad 0
\end{array}
$$

$$g(x) = (x + 1)(x^2 - 6x + 13)$$

The zeros are -1 and $3 \pm 2i$.

3. $x = 4 + 3i$ **is a zero, so** $x = 4 - 3i$ **is also a zero. One factor of** $h(x)$ **is**

$$(x - (4 + 3i))(x - (4 - 3i)) = (x - 4 - 3i)(x - 4 + 3i)$$
$$= x^2 - 4x + 3ix - 4x + 16 - 12i - 3ix + 12i - 9i^2$$
$$= x^2 - 8x + 16 - 9(-1)$$
$$= x^2 - 8x + 25$$

$$\begin{array}{r|l} & x^2 - 4 \\ \hline x^2 - 8x + 25 & x^4 - 8x^3 + 21x^2 + 32x - 100 \\ & -(x^4 - 8x^3 + 25x^2) \\ \hline & -4x^2 + 32x - 100 \\ & -(-4x^2 + 32x - 100) \\ \hline & 0 \end{array}$$

$$h(x) = (x^2 - 4)(x^2 - 8x + 25)$$

The zeros are $4 \pm 3i$ **and** ± 2 **(from** $x^2 - 4 = 0$**).**

One consequence of the Fundamental Theorem of Algebra is that a polynomial of degree n has n zeros, though not necessarily n different zeros. For example, the polynomial $f(x) = (x - 2)^3 = (x - 2)(x - 2)(x - 2)$ has $x = 2$ as a zero three times. The number of times an x-value is a zero is called its *multiplicity*. In this example, $x = 2$ is a zero with multiplicity 3.

EXAMPLE 10-31

- $f(x) = x^4(x + 3)^2(x - 6)$

$x = 0$ **is a zero with multiplicity 4.**
$x = -3$ **is a zero with multiplicity 2.**
$x = 6$ **is a zero with multiplicity 1.**

PRACTICE

State each zero and its multiplicity.

$$f(x) = x^2(x+4)(x+9)^6(x-5)^3$$

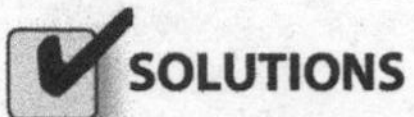

SOLUTIONS

$x = 0$ is a zero with multiplicity 2.
$x = -4$ is a zero with multiplicity 1.
$x = -9$ is a zero with multiplicity 6.
$x = 5$ is a zero with multiplicity 3.

If a real zero has *odd* multiplicity, then the graph crosses at the x-intercept. If the real zero has *even* multiplicity, then the graph does not cross at the x-intercept. See Figs. and 10-19 and 10-20.

Now, instead of finding the zeros for a given polynomial, we find a polynomial with the given zeros. Because we know the zeros, we know the factors. Once we know the factors of a polynomial, we pretty much know the polynomial.

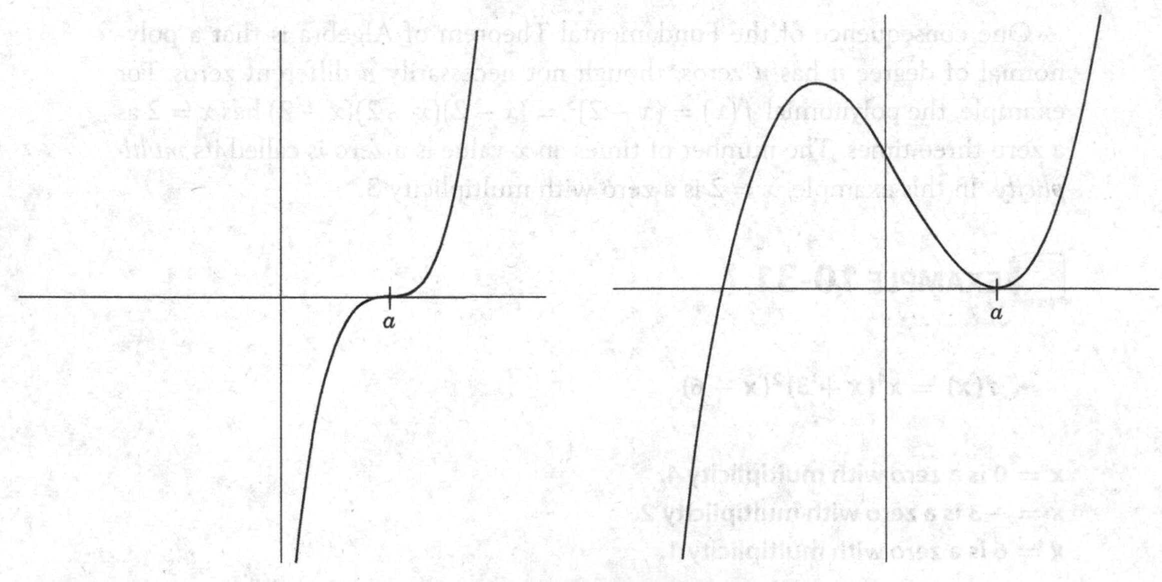

FIGURE 10-19

FIGURE 10-20

EXAMPLE 10-32

Find a polynomial with integer coefficients having the given degree and zeros.

- **Degree 3 with zeros 1, 2, and 5**

Because $x = 1$ is a zero, $x - 1$ is a factor. Because $x = 2$ is a zero, $x - 2$ is a factor. And because $x = 5$ is a zero, $x - 5$ is a factor. Such a polynomial is of the form $a(x-1)(x-2)(x-5)$, where a is some nonzero number. We want to choose a so that the coefficients are integers.

$$a(x-1)(x-2)(x-5) = a(x-1)[(x-2)(x-5)]$$

$$= a(x-1)(x^2 - 7x + 10)$$

$$= a(x^3 - 7x^2 + 10x - x^2 + 7x - 10)$$

$$= a(x^3 - 8x^2 + 17x - 10)$$

Because the coefficients are already integers, we can let $a = 1$. One polynomial of degree 3 having integer coefficients and 1, 2, and 5 as zeros is $x^3 - 8x^2 + 17x - 10$.

- **Degree 4 with zeros -3 and $2 - 5i$, with -3 a zero of multiplicity 2**

Because -3 is a zero of multiplicity 2, $(x+3)^2 = x^2 + 6x + 9$ is a factor. Because $2 - 5i$ is a zero, $2 + 5i$ is another zero. Another factor of the polynomial is

$$(x-(2-5i))(x-(2+5i)) = (x-2+5i)(x-2-5i)$$

$$= x^2 - 2x - 5ix - 2x + 4 + 10i + 5ix - 10i - 25i^2$$

$$= x^2 - 4x + 4 - 25(-1) = x^2 - 4x + 29$$

The polynomial has the form $a(x^2+6x+9)(x^2-4x+29)$, where a is any real number that makes all coefficients integers.

$$\begin{aligned}a(x^2+6x+9)(x^2-4x+29)\\ &= a(x^4-4x^3+29x^2+6x^3-24x^2+174x+9x^2-36x+261)\\ &= a(x^4+2x^3+14x^2+138x+261)\end{aligned}$$

Because the coefficients are already integers, we can let $a=1$. One polynomial that satisfies the given conditions is $x^4+2x^3+14x^2+138x+261$.

PRACTICE

Find a polynomial with integer coefficients having the given degree and zeros.

1. Degree 3 with zeros 0, -4, and 6
2. Degree 4 with zeros $-5i$ and $3i$
3. Degree 4 with zeros -1 and $6-7i$, where $x=-1$ has multiplicity 2

SOLUTIONS

1. One polynomial with integer coefficients, with degree 3 and zeros 0, -4, and 6 is

$$x(x+4)(x-6)=x(x^2-2x-24)=x^3-2x^2-24x$$

2. One polynomial with integer coefficieints, with degree 4 and zeros $-5i$ and $3i$ is

$$\begin{aligned}(x+5i)(x-5i)(x-3i)(x+3i) &= (x^2+25)(x^2+9)\\ &= x^4+34x^2+225\end{aligned}$$

3. One polynomial with integer coefficients, with degree 4 and zeros -1 and $6-7i$, where $x=-1$ has multiplicity 2 is

$$\begin{aligned}
&(x+1)^2(x-(6-7i))(x-(6+7i))\\
&\quad=(x+1)^2(x-6+7i)(x-6-7i)\\
&\quad=[(x+1)(x+1)][(x^2-6x-7ix-6x+36+42i\\
&\qquad+7ix-42i-49i^2)]\\
&\quad=(x^2+2x+1)(x^2-12x+85)\\
&\quad=x^4-12x^3+85x^2+2x^3-24x^2+170x+x^2-12x+85\\
&\quad=x^4-10x^3+62x^2+158x+85
\end{aligned}$$

In the previous problems, there were infinitely many answers because a could be any integer. In the following problems, there is exactly one polynomial that satisfies the given conditions. This means that a is likely to be a number other than 1.

EXAMPLE 10-33

Find a polynomial that satisfies the given conditions.

- **Degree 3 with zeros -1, -2, and 4, where the coefficient for x is -20**

$$\begin{aligned}
a(x+1)(x+2)(x-4)&=a(x+1)[(x+2)(x-4)]\\
&=a(x+1)(x^2-2x-8)\\
&=a(x^3-2x^2-8x+x^2-2x-8)\\
&=a(x^3-x^2-10x-8)\\
&=ax^3-ax^2-10ax-8a
\end{aligned}$$

Because we want the coefficient of x to be -20, we need $-10ax=-20x$, so $a=2$ (from $-10a=-20$). The polynomial that satisifies the conditions is $2x^3-2x^2-20x-16$.

- **Degree 3 with zeros $\frac{2}{3}$ and $-1-5i$, where the coefficient of x^2 is -4**

If $x = \frac{2}{3}$ is a zero, then $3x - 2$ is a factor.

$$x - \frac{2}{3} = 0$$

$$3\left(x - \frac{2}{3}\right) = 3(0)$$

$$3x - 2 = 0$$

The other factors are $x - (-1 - 5i) = x + 1 + 5i$ and $x - (-1 + 5i) = x + 1 - 5i$.

Factors involving complex zeros

$$\begin{aligned}(x+1+5i)(x+1-5i) &= x^2 + x - 5ix + x + 1 - 5i + 5ix + 5i - 25i^2 \\ &= x^2 + 2x + 26\end{aligned}$$

All factors

$$\begin{aligned}a(3x-2)(x^2+2x+26) &= a(3x^3 + 6x^2 + 78x - 2x^2 - 4x - 52) \\ &= a(3x^3 + 4x^2 + 74x - 52) \\ &= 3ax^3 + 4ax^2 + 74ax - 52a\end{aligned}$$

We want $4ax^2 = -4x^2$, so we need $a = -1$. The polynomial that satisifies the conditions is $-3x^3 - 4x^2 - 74x + 52$.

PRACTICE

Find the polynomial that satisifies the given conditions.

1. Degree 3 zeros 4 and ± 1, with leading coefficient 3
2. Degree 3 with zeros $-\frac{3}{5}$ and 1, where the multiplicity of 1 is 2, and the coefficient of x is 2
3. Degree 4 with zeros i and $4i$, with constant term 8

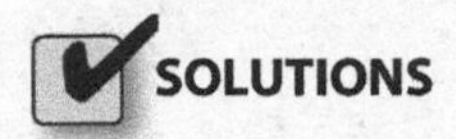

SOLUTIONS

1. **The factors are $x-4$, $x-1$, and $x+1$.**

$$\begin{aligned} a(x-4)(x-1)(x+1) &= a(x-4)[(x-1)(x+1)] \\ &= a(x-4)(x^2-1) = a[(x-4)(x^2-1)] \\ &= a(x^3-4x^2-x+4) \\ &= ax^3-4ax^2-ax+4a \end{aligned}$$

We want the leading coefficient to be 3, so $a=3$. The polynomial that satisifies the conditions is $3x^3-12x^2-3x+12$.

2. **Because $x=-\frac{3}{5}$ is a zero, $5x+3$ is a factor.**

$$x-\left(-\frac{3}{5}\right)=0$$

$$5\left(x+\frac{3}{5}\right)=5(0)$$

$$5x+3=0$$

The other factor is $(x-1)^2=(x-1)(x-1)=x^2-2x+1$.

$$\begin{aligned} a(5x+3)(x^2-2x+1) &= a(5x^3-10x^2+5x+3x^2-6x+3) \\ &= a(5x^3-7x^2-x+3) \\ &= 5ax^3-7ax^2-ax+3a \end{aligned}$$

We want $-ax=2x$, so $a=-2$. The polynomial that satisifies the conditions is $-10x^3+14x^2+2x-6$.

3. The factors are $x+i$, $x-i$, $x-4i$, and $x+4i$.

$$\begin{aligned}
a(x+i)(x-i)(x-4i)(x+4i) &= a[(x+i)(x-i)] \times [(x-4i)(x+4i)] \\
&= a(x^2+1)(x^2+16) \\
&= a(x^4+17x^2+16) \\
&= ax^4+17ax^2+16a
\end{aligned}$$

We want $16a=8$, so $a=\frac{1}{2}$. The polynomial that satisifies the conditions is $\frac{1}{2}x^4+\frac{17}{2}x^2+8$.

Summary

In this chapter, we learned how to

- *Use the leading term of a polynomial function to determine the function's end behavior.* The leading term of a polynomial function tells us whether both ends of its graph go up (the degree, n is even and the leading coefficient is positive), both ends go down (the degree is even and the leading coefficient is negative), the left end of the graph goes down and the right end go up (n is odd, the leading coefficient is positive), or the left end of the graph goes up and the right end goes down (n is odd, the leading coefficient is negative).
- *Use the factors of a polynomial to locate the x-intercepts (if any) for its graph.* For a real number a, the statements "$x=a$ is an x-intercept for the graph of $P(x)$" and "$x-a$ is a factor for $P(x)$" have the same meaning. That is, if we know the x-intercepts for a polynomial, then we know its factors, those involving real numbers, anyway.
- *Sketch the graph of a polynomial function.* We sketch the graph of a polynomial function by plotting the x-intercepts, a point between each pair of x-intercepts, and a point on each end to show the end behavior of the graph.
- *Perform polynomial division to find the zeros of a polynomial.* If we divide one polynomial by another (not the zero polynomial) and the remainder

is zero, then the quotient and divisor are factors of the polynomial. If we want to completely factor the polynomial, we then find the zeros of the quotient, by directly factoring it, using the quadratic formula, or carrying out the process again.

- *Perform synthetic division*. The many tedious steps of calculating the quotient and remainder for long polynomial division is reduced to only a few simple steps provided the denominator is in the form $x - a$.
- *Use the Rational Zero Theorem to locate an initial zero*. If a polynomial's coefficients are integers, then a rational number $\frac{p}{q}$ is a possible zero for the polynomial if the numerator divides the constant term and the denominator divides the leading coefficient. We use this theorem to locate a zero so that we can begin the factorization process. After listing all possible candidates for the rational zeros, we try them, one at at time, until we find a zero. Not all such polynomials have rational zeros, however.
- *Work with the Rule of Signs and the Upper and Lower Bounds Theorem*. These theorems help us to count and locate x-intercepts for the graph of a polynomial function. The Rule of Signs counts the possible number of positive zeros by counting the sign changes in the polynomial, $P(x)$, and the possible number of negative zeros by counting the number of sign changes in $P(-x)$. The Upper and Lower Bounds Theorem helps us to find an interval on the x-axis that contains every x-intercept.
- *Perform arithmetic with complex numbers*. A complex number is a number that can be written in the form $a + bi$, where a and b are real numbers and $i = \sqrt{-1}$. We add/subtract to complex numbers by combining like terms. We find the product of two complex numbers using the FOIL method and replacing i^2 with -1, if necessary. We can find the quotient of two complex numbers (usually written as a fraction) by multiplying the numerator and denominator by the denominator's conjugate and then simplifying. This works because the product of a complex number, $a + bi$, and its conjugate, $a - bi$, is the real number $a^2 + b^2$.
- *Find the complex zeros for a polynomial function*. A polynomial of degree $n \geq 1$ has n zeros according to the Fundamental Theorem of Algebra. Some of these zeros might be complex (i.e., complex but not real) and might be counted more than once. If a complex number is a zero for a polynomial, then its conjugate is also a zero. We can locate complex zeros using some of the techniques that we learned for locating the real zeros of a polynomial.

QUIZ

1. **The graph of a polynomial with leading term $-6x^4$**

   A. goes up on both ends.
   B. goes down on both ends.
   C. goes up on the left and down on the right.
   D. goes down on the left and up on the right.

2. **What are the zeros for the function $f(x) = x^3 - 7x^2 + 6x + 18$?**

   A. $3, -2 \pm \sqrt{10}$ B. $-3, -2 \pm \sqrt{10}$ C. $3, 2 \pm \sqrt{10}$ D. $-3, 2 \pm \sqrt{10}$

3. **The graph in Fig. 10-21 is the graph of which function?**

   A. $P(x) = \frac{1}{2}(x+1)^2(x-3)^2$
   B. $P(x) = -\frac{1}{2}(x+1)^2(x-3)^2$
   C. $P(x) = \frac{1}{2}(x+1)^2(x-3)$
   D. $P(x) = -\frac{1}{2}(x+1)^2(x-3)$

4. **For the polynomial function $f(x) = 4x^3 - 6x^2 + 9x - 5$,**

   A. the degree is 4, and the constant term is 3.
   B. the degree is 3, and the constant term is –5.
   C. the degree is 3, and the constant term is 4.
   D. the degree is 4, and the constant term is 5.

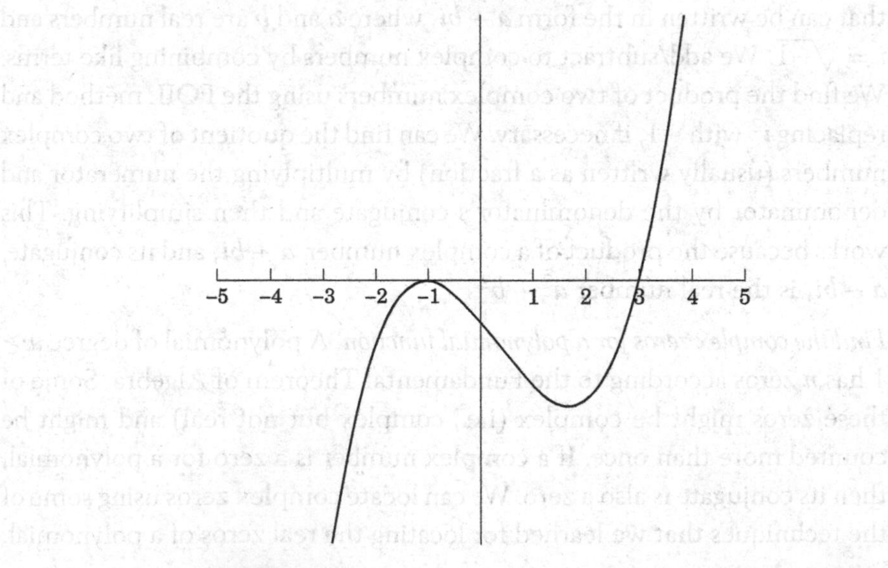

FIGURE 10-21

5. For the polynomial function $f(x) = (x + 6)(x + 2)(x - 4)$, which set of x-coordinates would be best to plot to sketch the graph of the function?

A. −7, −4, 1, 5 B. −4, 1, 2, 5 C. −7, −4, 5 D. −7, −1, 0, 5

6. Find the remainder for $\frac{x^3 - 6x^2 - 4x + 5}{x^2 + 2}$.

A. $-6x + 17$ B. $-3x - 34$ C. $6x - 17$ D. $3x + 18$

7. Find the quotient for $(x^3 - 6x^2 - 4x + 5) \div (x + 3)$.

A. $x^2 - 3x - 13$ B. $x^2 - 9x + 21$ C. $x^2 + 3x - 10$ D. $x^2 - 9x + 23$

8. Find the quotient and remainder for $(4x^3 + 6x^2 - x + 3) \div (2x + 1)$.

A. Quotient, $2x^2 + 4x + \frac{3}{2}$; remainder, $\frac{3}{2}$
B. Quotient, $2x^2 + 2x - \frac{3}{2}$; remainder, $\frac{9}{2}$
C. Quotient, $2x^2 + 4x + \frac{3}{2}$; remainder, $\frac{9}{2}$
D. Quotient, $2x^2 + 2x - \frac{3}{2}$; remainder, $\frac{3}{2}$

9. What are the solutions to the equation $x^3 - 2x^2 + 1 = 0$?

A. 1, $\frac{1 \pm \sqrt{5}}{2}$ B. 1, $\frac{1 \pm \sqrt{3}i}{2}$ C. 1, $\frac{-1 \pm \sqrt{5}}{2}$ D. 1, $\frac{-1 \pm \sqrt{3}i}{2}$

10. What are the zeros for the polynomial $x^3 - 10x^2 + 34x - 40$?

A. $-4, -3 \pm i$ B. $4, -3 \pm i$ C. $-4, 3 \pm i$ D. $4, 3 \pm i$

11. According to the Rational Zero Theorem, which of the following is NOT a possible zero for the function $f(x) = 6x^3 - 8x^2 + 5x + 35$?

A. $\frac{1}{6}$ B. $\frac{5}{3}$ C. $\frac{7}{2}$ D. −6

12. What are the zeros for the function $f(x) = x^4 - 4x^3 + 4x^2 - 36x - 45$?

A. $2 \pm i, \pm 3i$ B. $2 \pm i, \pm 3$ C. $-5, 1 \pm 3i$ D. $5, -1, \pm 3i$

13. Which of the following functions has $3 + 5i$ as a zero?

A. $P(x) = x^3 - 8x^2 - 4x + 32$ B. $P(x) = x^3 - 8x^2 + 46x - 68$
C. $P(x) = x^3 - 8x^2 - 22x - 32$ D. $P(x) = x^3 - 8x^2 - 22x - 68$

14. Find the quotient $\frac{3-7i}{2+4i}$.

A. $\frac{17}{10} - \frac{1}{10}i$ B. $\frac{13}{10} - \frac{3}{10}i$ C. $-\frac{11}{10} - \frac{13}{10}i$ D. $\frac{17}{10} + \frac{13}{10}i$

15. Use the Upper and Lower Bounds Theorem to determine if $a = -3$ is a lower bound and/or $b = 3$ is an upper bound for the real zeros for $f(x) = 2x^4 + 7x^3 - 4x^2 + x - 8$.

A. $a = -3$ is not a lower bound, but $b = 3$ is an upper bound.

B. $a = -3$ is a lower bound, but $b = 3$ is not an upper bound.

C. $a = -3$ is a lower bound, and $b = 3$ is an upper bound.

D. $a = -3$ is not a lower bound, and $b = 3$ is not an upper bound.

chapter 11

Systems of Equations and Inequalities

A system of equations is a collection of two or more equations whose graphs might or might not intersect (share a common point or points). If the graphs do intersect, then we say that the solution to the system is the point or points where the graphs intersect. In this chapter, we learn two strategies for solving systems of linear equations. Later, we will apply these strategies to systems containing nonlinear equations.

After working with systems of equations and their applications, we turn our attention to solving inequalities and systems of inequalities. Solving inequalities involves plotting graphs of equations and then shading one side or the other of the graph to represent the solution to the inequality.

CHAPTER OBJECTIVES

In this chapter, you will

- Solve a system of linear equations with substitution
- Solve a system of linear equations with elimination by addition

- Solve applied problems represented by a system of linear equations
- Solve systems containing nonlinear equations
- Solve inequalities and systems of inequalities

Systems of Linear Equations

We can use one of several methods for solving a system of linear equations. One strategy is to examine their graphs and see where, if anywhere, the graphs intersect. For example, the solution to the system $\begin{cases} x+y=4 \\ 3x-y=0 \end{cases}$ is $x=1, y=3$ because the graphs intersect at the point (1, 3). See Fig. 11-1.

We say that (1, 3) *satisfies* the system because if we let $x=1$ and $y=3$ in each equation, both are true.

$1+3=4$ This is a true statement.

$3(1)-3=0$ This is a true statement.

Even with a graphing calculator, though, we might only find approximate solutions. Other strategies involve *matrices*. Here, we concentrate on two

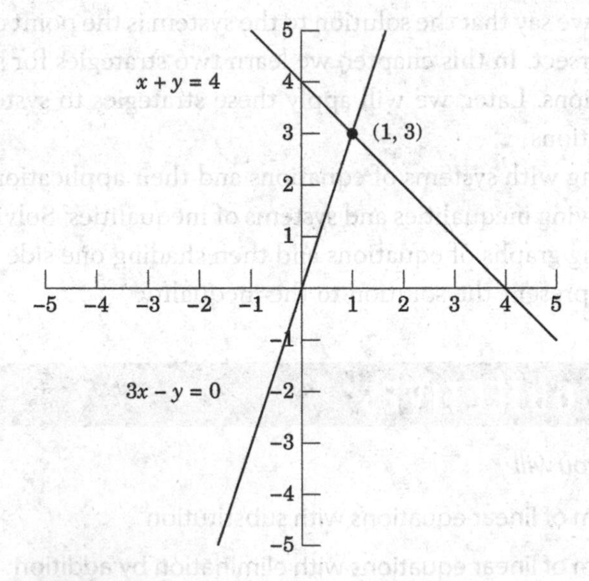

FIGURE 11-1

methods: one of them is called *substitution* and the other is called *elimination by addition*. Both methods work with many kinds of systems of equations, but we begin with systems of linear equations.

The substitution method works by solving for one variable in one equation and making a substitution in the other equation. Technically, it does not matter which variable we use or which equation we begin with, but some choices are easier than others.

EXAMPLE 11-1

Solve the system of equations. Write the solution in the form of a point, (x, y).

- $$\begin{cases} x + y = 5 \\ -2x + y = -1 \end{cases}$$

We have four places to start.

1. **Solve for x in the first equation: $x = 5 - y$.**
2. **Solve for y in the first equation: $y = 5 - x$.**
3. **Solve for x in the second equation: $x = \frac{1}{2} + \frac{1}{2}y$.**
4. **Solve for y in the second equation: $y = 2x - 1$.**

The third option looks like it would be the most trouble to use, so we use one of the others. We use the first option. Because $x = 5 - y$ came from the *first* equation, we substitute $5 - y$ for x in the *second* equation. The equation $-2x + y = -1$ becomes $-2(5 - y) + y = -1$. Now we can solve the equation $-2(5 - y) + y = -1$ because it has only one variable.

$$-2(5 - y) + y = -1$$

$$-10 + 2y + y = -1$$

$$3y = 9$$

$$y = 3$$

Now that we know $y = 3$, we could use any of the equations above to find x. We know that $x = 5 - y$, so we use this: $x = 5 - 3 = 2$.

The solution is $x = 2$ and $y = 3$ or the point $(2, 3)$. It is a good idea to check the solution.

$$x + y = 5: \quad 2 + 3 = 5 \qquad \text{This is true.}$$

$$-2x + y = -1: \quad -2(2) + 3 = -1 \qquad \text{This is true.}$$

- $\begin{cases} 4x - y = 12 & \text{A} \\ 3x + y = 2 & \text{B} \end{cases}$

We solve for y in equation B: $y = 2 - 3x$. Next we substitute $2 - 3x$ for y in equation A and solve for x.

$$4x - y = 12$$

$$4x - (2 - 3x) = 12$$

$$4x - 2 + 3x = 12$$

$$7x = 14$$

$$x = 2$$

Now that we know $x = 2$, we substitute $x = 2$ in one of the above equations. Here, we use $y = 2 - 3x$: $y = 2 - 3(2) = -4$. The solution is $x = 2$, $y = -4$, or $(2, -4)$. The graphs in Fig. 11-2 verify that the solution $(2, -4)$ is on both lines.

- $\begin{cases} y = 4x + 1 & \text{A} \\ y = 3x + 2 & \text{B} \end{cases}$

Both equations are already solved for y, so all we need to do is to set them equal to each other.

$$4x + 1 = 3x + 2$$

$$x = 1$$

We can use either equation A or equation B to find y when $x = 1$. We use A: $y = 4x + 1 = 4(1) + 1 = 5$. The solution is $x = 1$ and $y = 5$, or $(1, 5)$. We can see from the graphs in Fig. 11-3 that $(1, 5)$ is the solution to the system.

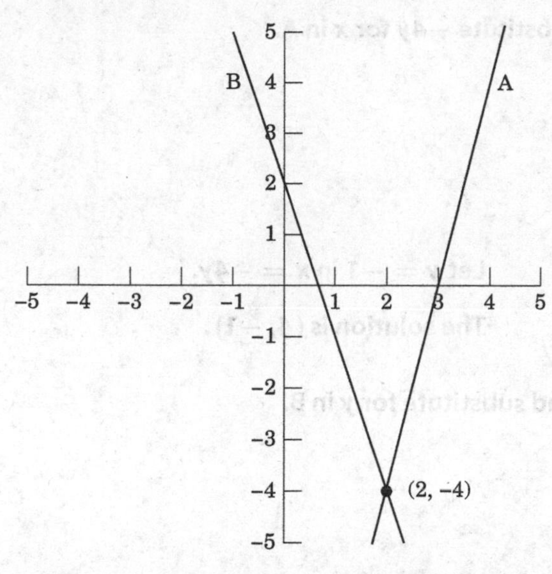

FIGURE 11-2

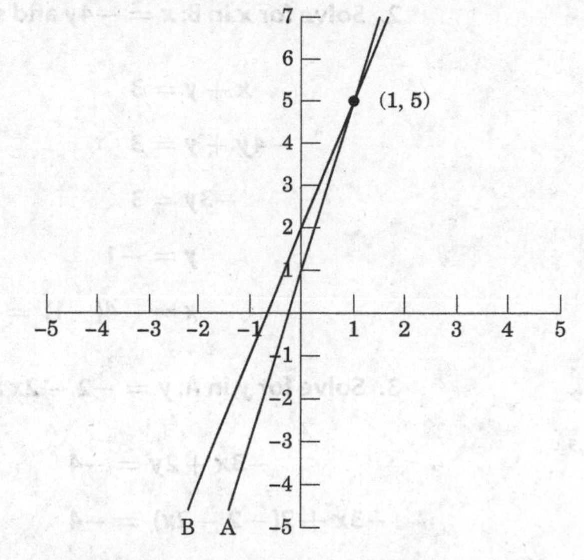

FIGURE 11-3

PRACTICE

Solve the system of equations. Write the solution in the form of a point, (x, y).

1. $\begin{cases} 2x + 3y = 1 & \text{A} \\ x - 2y = -3 & \text{B} \end{cases}$

2. $\begin{cases} x + y = 3 & \text{A} \\ x + 4y = 0 & \text{B} \end{cases}$

3. $\begin{cases} 2x + y = -2 & \text{A} \\ -3x + 2y = -4 & \text{B} \end{cases}$

4. $\begin{cases} y = x + 1 & \text{A} \\ -3x + 2y = -2 & \text{B} \end{cases}$

SOLUTIONS

1. Solve for x in B: $x = -3 + 2y$ and substitute this for x in A.

$$2x + 3y = 1$$

$$2(-3 + 2y) + 3y = 1$$

$$-6 + 4y + 3y = 1$$

$$7y = 7$$

$$y = 1 \qquad \text{Let } y = 1 \text{ in } x = -3 + 2y.$$

$$x = -3 + 2(1) = -1 \qquad \text{The solution is } (-1, 1).$$

2. Solve for x in B: $x = -4y$ and substitute $-4y$ for x in A.

$$x + y = 3$$
$$-4y + y = 3$$
$$-3y = 3$$
$$y = -1 \qquad \text{Let } y = -1 \text{ in } x = -4y.$$
$$x = -4(-1) = 4 \qquad \text{The solution is } (4, -1).$$

3. Solve for y in A: $y = -2 - 2x$ and substitute for y in B.

$$-3x + 2y = -4$$
$$-3x + 2(-2 - 2x) = -4$$
$$-3x - 4 - 4x = -4$$
$$-7x = 0$$
$$x = 0 \qquad \text{Let } x = 0 \text{ in } y = -2 - 2x.$$
$$y = -2 - 2(0) = -2 \qquad \text{The solution is } (0, -2).$$

4. Equation A is already solved for y. We substitute $x + 1$ for y in B.

$$-3x + 2y = -2$$
$$-3x + 2(x + 1) = -2$$
$$-3x + 2x + 2 = -2$$
$$-x = -4$$
$$x = 4 \qquad \text{Let } x = 4 \text{ in A.}$$
$$y = x + 1 = 4 + 1 = 5 \qquad \text{The solution is } (4, 5).$$

Elimination by Addition

Solving a system of equations by substitution can be messy when none of the coefficients is 1. Fortunately, we can use a more convenient strategy. We can always *add* the two equations to eliminate one of the variables. Sometimes, though, we might need to multiply one or both equations by a number to make this method work.

EXAMPLE 11-2

Solve the system of equations. Write the solution in the form of a point, (x, y).

- $\begin{cases} 2x - 3y = 16 & \text{A} \\ 5x + 3y = -2 & \text{B} \end{cases}$

We add the equations by adding like terms. Because we would be adding $-3y$ to $3y$, the y-terms cancel, leaving one equation with a single variable.

$$\begin{array}{r} 2x - 3y = 16 \\ \underline{5x + 3y = -2} \\ 7x + 0y = 14 \\ x = 2 \end{array}$$

We can let $x = 2$ in either A or B to find y. We let $x = 2$ in A.

$$2x - 3y = 16$$

$$2(2) - 3y = 16$$

$$-3y = 12$$

$$y = -4 \qquad \text{The solution is } (2, -4).$$

PRACTICE

Solve the system of equations. Write the solution in the form of a point, (x, y).

1. $\begin{cases} -2x + 7y = 19 & \text{A} \\ 2x - 4y = -10 & \text{B} \end{cases}$

2. $\begin{cases} 15x - y = 9 & \text{A} \\ 2x + y = 8 & \text{B} \end{cases}$

3. $\begin{cases} -5x + 4y = -3 & \text{A} \\ -3x - 4y = 11 & \text{B} \end{cases}$

SOLUTIONS

1.

$$-2x + 7y = 19 \quad A$$
$$2x - 4y = -10 \quad +B$$
$$3y = 9$$
$$y = 3$$
$$-2x + 7(3) = 19 \qquad \text{Let } y = 3 \text{ in A.}$$
$$x = 1 \qquad \text{The solution is } (1, 3).$$

2.

$$15x - y = 9 \quad A$$
$$2x + y = 8 \quad +B$$
$$17x = 17$$
$$x = 1$$
$$15(1) - y = 9 \qquad \text{Let } x = 1 \text{ in A.}$$
$$y = 6 \qquad \text{The solution is } (1, 6).$$

3.

$$-5x + 4y = -3 \quad A$$
$$-3x - 4y = 11 \quad +B$$
$$-8x = 8$$
$$x = -1$$
$$-5(-1) + 4y = -3 \qquad \text{Let } x = -1 \text{ in A.}$$
$$y = -2 \qquad \text{The solution is } (-1, -2).$$

Sometimes we must multiply one or both equations by some number or numbers so that one of the variables cancels.

EXAMPLE 11-3

- $\begin{cases} 3x + 6y = -12 & \text{A} \\ 2x + 6y = -14 & \text{B} \end{cases}$

Because the coefficients on *y* are the same, we only need to make one of them negative. Multiply either A or B by −1 and then add.

$$\begin{array}{rl} -3x - 6y = 12 & -\text{A} \\ 2x + 6y = -14 & +\text{B} \\ \hline -x = -2 & \\ x = 2 & \end{array}$$

$3(2) + 6y = -12$ Let $x = 2$ in A.

$y = -3$ The solution is $(2, -3)$.

- $\begin{cases} 2x + 7y = 1 & \text{A} \\ 4x - 2y = 18 & \text{B} \end{cases}$

We can use one of several options. We could multiply A by −2 so that we could add −4*x* (in −2A) to 4*x* in B. We could multiply A by 2 and multiply B by 7 so that we could add 14*y* (in 2A) to −14*y* (in 7B). We could also divide B by −2 so that we could add 2*x* (in A) to −2*x* (in $-\frac{1}{2}$B). Here, we add −2A + B.

$$\begin{array}{rl} -4x - 14y = -2 & -2\text{A} \\ 4x - 2y = 18 & +\text{B} \\ \hline -16y = 16 & \\ y = -1 & \end{array}$$

$2x + 7(-1) = 1$ Let $y = -1$ in A.

$x = 4$ The solution is $(4, -1)$.

PRACTICE

Solve the system of equations. Write the solution in the form of a point, (x, y).

1. $\begin{cases} -3x + 2y = 12 & \text{A} \\ 4x + 2y = -2 & \text{B} \end{cases}$

2. $\begin{cases} 6x - 5y = 1 & \text{A} \\ 3x - 2y = 1 & \text{B} \end{cases}$

3. $\begin{cases} 15x + 4y = -1 & \text{A} \\ 5x + 2y = -3 & \text{B} \end{cases}$

SOLUTIONS

1. We add $-\text{A} + \text{B}$.

$$\begin{array}{rl} 3x - 2y = -12 & -\text{A} \\ \underline{4x + 2y = \quad -2} & +\text{B} \\ 7x = -14 & \\ x = -2 & \end{array}$$

$$-3(-2) + 2y = 12 \qquad \text{Let } x = -2 \text{ in A.}$$

$$y = 3 \qquad \text{The solution is } (-2, 3).$$

2. We compute $\text{A} - 2\text{B}$.

$$\begin{array}{rl} 6x - 5y = 1 & \text{A} \\ \underline{-6x + 4y = -2} & -2\text{B} \\ -y = -1 & \\ y = 1 & \end{array}$$

$$6x - 5(1) = 1 \qquad \text{Let } y = 1 \text{ in A.}$$

$$x = 1 \qquad \text{The solution is } (1, 1).$$

3. We compute A −2B.

$$\begin{aligned} 15x + 4y &= -1 \quad \text{A} \\ -10x - 4y &= 6 \quad -2\text{B} \\ \hline 5x &= 5 \\ x &= 1 \\ 15(1) + 4y &= -1 \quad \text{Let } x = 1 \text{ in A.} \\ y &= -4 \quad \text{The solution is } (1, -4). \end{aligned}$$

In order to eliminate a variable in the next set of problems, we must change both equations in the system.

EXAMPLE 11-4

- $\begin{cases} 8x - 5y = -2 & \text{A} \\ 3x + 2y = 7 & \text{B} \end{cases}$

We have many options. Some are 3A −8B, −3A + 8B, and 2A + 5B. We compute 2A + 5B.

$$\begin{aligned} 16x - 10y &= -4 \quad 2\text{A} \\ 15x + 10y &= 35 \quad +5\text{B} \\ \hline 31x &= 31 \\ x &= 1 \\ 8(1) - 5y &= -2 \quad \text{Let } x = 1 \text{ in A.} \\ y &= 2 \quad \text{The solution is } (1, 2). \end{aligned}$$

- $\begin{cases} \frac{2}{3}x - \frac{1}{4}y = \frac{25}{72} & \text{A} \\ \frac{1}{2}x + \frac{2}{5}y = -\frac{1}{30} & \text{B} \end{cases}$

We begin by eliminating the fractions. The LCD for A is 72, and the LCD for B is 30.

$$48x - 18y = 25 \quad 72\text{A}$$

$$15x + 12y = -1 \quad 30\text{B}$$

Next, we multiply the first equation by 2 and the second by 3 so that we can eliminate *y*.

$$\begin{aligned} 96x - 36y &= 50 \\ \underline{45x + 36y} &\underline{= -3} \\ 141x &= 47 \\ x &= \frac{47}{141} = \frac{1}{3} \end{aligned}$$

$$96\left(\frac{1}{3}\right) - 36y = 50$$

$$y = -\frac{1}{2} \qquad \text{The solution is } \left(\frac{1}{3}, -\frac{1}{2}\right).$$

Two lines in the plane intersect at one point, are parallel, or are really the same line. Until now, our lines have intersected at one point. When solving a system of two linear equations whose graphs are parallel or are the same line, both variables cancel and we are left with a true statement such as "3 = 3" or a false statement such as "5 = 1." We have a true statement when the two lines are the same and a false statement when they are parallel.

EXAMPLE 11-5

- $\begin{cases} 2x - 3y = 6 & \text{A} \\ -4x + 6y = 8 & \text{B} \end{cases}$

We eliminate x by adding 2A + B.

$$\begin{array}{rl} 4x - 6y = 12 & \text{2A} \\ \underline{-4x + 6y = \ \ 8} & +\text{B} \\ 0 = 20 & \end{array}$$

This is a false statement, so the lines are parallel. They are sketched in Fig. 11-4

- $\begin{cases} y = \frac{2}{3}x - 1 \\ 2x - 3y = 3 \end{cases}$

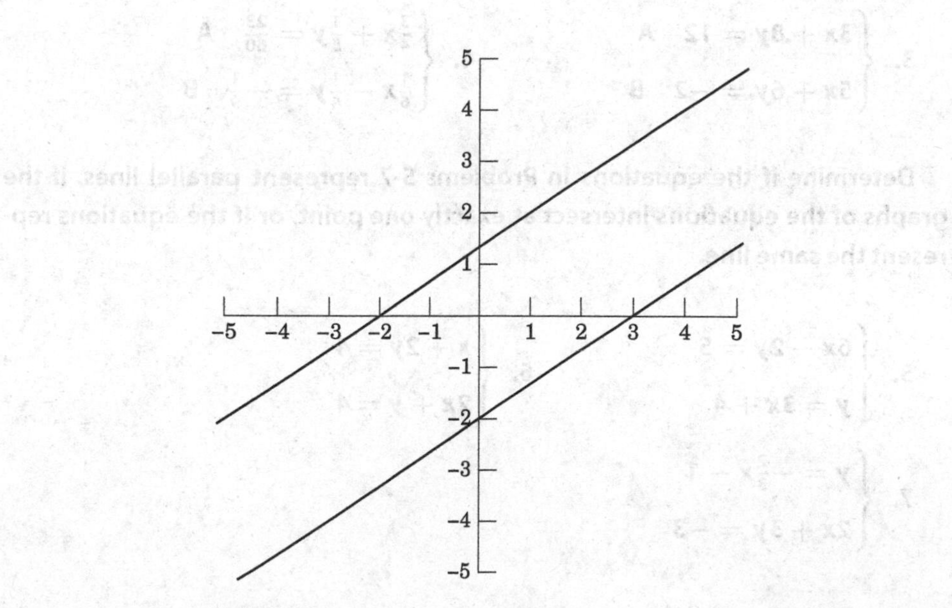

FIGURE 11-4

We use substitution.

$$2x - 3\left(\frac{2}{3}x - 1\right) = 3$$

$$2x - 2x + 3 = 3$$

$$0 = 0$$

Because $0 = 0$ is a true statement, these lines are the same.

PRACTICE

Solve the system of equations in Problems 1-4. Write the solutions in the form of a point, (x, y).

1. $\begin{cases} 5x - 9y = -26 & \text{A} \\ 3x + 2y = 14 & \text{B} \end{cases}$

2. $\begin{cases} 7x + 2y = 1 & \text{A} \\ 2x + 3y = -7 & \text{B} \end{cases}$

3. $\begin{cases} 3x + 8y = 12 & \text{A} \\ 5x + 6y = -2 & \text{B} \end{cases}$

4. $\begin{cases} \frac{3}{4}x + \frac{1}{5}y = \frac{23}{60} & \text{A} \\ \frac{1}{6}x - \frac{1}{4}y = -\frac{1}{9} & \text{B} \end{cases}$

Determine if the equations in Problems 5-7 represent parallel lines, if the graphs of the equations intersect at exactly one point, or if the equations represent the same line.

5. $\begin{cases} 6x - 2y = 5 \\ y = 3x + 4 \end{cases}$

6. $\begin{cases} x + 2y = 4 \\ 2x + y = 4 \end{cases}$

7. $\begin{cases} y = -\frac{2}{3}x - 1 \\ 2x + 3y = -3 \end{cases}$

SOLUTIONS

1. $15x - 27y = -78$ 3A

$-15x - 10y = -70$ $-$ 5B

$-37y = -148$

$y = 4$

$5x - 9(4) = -26$ Let $y = 4$ in A.

$x = 2$ The solution is $(2, 4)$.

2. $21x + 6y = 3$ 3A

$-4x - 6y = 14$ $-$ 2B

$17x = 17$

$x = 1$

$7(1) + 2y = 1$ Let $x = 1$ in A.

$y = -3$ The solution is $(1, -3)$.

3. $9x + 24y = 36$ 3A

$-20x - 24y = 8$ $-$ 4B

$-11x = 44$

$x = -4$

$3(-4) + 8y = 12$ Let $x = -4$ in A.

$y = 3$ The solution is $(-4, 3)$.

4. First, we clear the fractions.

$$45x + 12y = 23 \quad 60A$$

$$6x - 9y = -4 \quad 36B$$

We add 3 times the first to 4 times the second.

$$\begin{aligned} 135x + 36y &= 69 \\ \underline{24x - 36y} &\underline{= -16} \\ 159x &= 53 \\ x &= \frac{53}{159} = \frac{1}{3} \\ 45\left(\frac{1}{3}\right) + 12y &= 23 \\ y &= \frac{2}{3} \end{aligned}$$

The solution is $\left(\frac{1}{3}, \frac{2}{3}\right)$.

5. The lines in the system are parallel.
6. The lines in the system intersect at exactly one point.
7. The lines in the system are the same line.

Applications for Systems of Equations

Systems of two linear equations can be used to solve many kinds of word problems. In these problems, two facts will be given about two variables. Each pair of facts represents a linear equation, and the two linear equations form a system.

EXAMPLE 11-6

- A movie theater charges $4 for each children's ticket and $6.50 for each adult's ticket. One night 200 tickets were sold, amounting to $1100 in ticket sales. How many of each type of ticket was sold?

Let x represent the number of children's tickets sold and y the number of adult tickets sold. One equation comes from the fact that a total of 200 tickets were sold, giving us the equation $x + y = 200$. The other equation comes from the fact that the ticket revenue was $1100. The ticket revenue from children's tickets is $4x$, and the ticket revenue from adult tickets is $6.50y$. Their sum is 1100 giving us $4x + 6.50y = 1100$.

$$\begin{cases} 4x + 6.50y = 1100 & \text{A} \\ x + y = 200 & \text{B} \end{cases}$$

We could use either substitution or addition to solve this system. Substitution is a little faster. We solve for x in B.

$$x = 200 - y$$

$$4(200 - y) + 6.50y = 1100 \quad \text{Substitute } 200 - y \text{ for } x \text{ into A.}$$

$$800 - 4y + 6.50y = 1100$$

$$y = 120$$

$$x = 200 - y = 200 - 120 = 80$$

Eighty children's tickets were sold, and 120 adult tickets were sold.

- A farmer had a soil test performed. He was told that his field needed 1080 pounds of Mineral A and 920 pounds of Mineral B. Two mixtures of fertilizers provide these minerals. Each bag of Brand I provides 25 pounds of Mineral A and 15 pounds of Mineral B. Brand II provides 20 pounds of Mineral A and 20 pounds of Mineral B. How many bags of each brand should he buy?

Let x represent the number of bags of Brand I and y represent the number of bags of Brand II. Then the number of pounds of Mineral A he gets from Brand I is $25x$, and the number of pounds of Mineral B is $15x$. The number of pounds of Mineral A he gets from Brand II is $20y$, and the number of pounds of Mineral B is $20y$. He needs 1080 pounds of Mineral A: $25x$ pounds come from Brand I and $20y$ comes from Brand II. This gives us the equation $25x + 20y = 1080$. He needs 920 pounds of Mineral B, $15x$ comes from Brand I and $20y$ comes from Brand II. This gives us the equation $15x + 20y = 920$. We now solve the system $\begin{cases} 25x + 20y = 1080 & \text{A} \\ 15x + 20y = 920 & \text{B} \end{cases}$.

We compute A − B.

$$\begin{array}{rl} 25x + 20y = 1080 & \text{A} \\ \underline{-15x - 20y = -920} & -\text{B} \\ 10x = 160 & \\ x = 16 & \\ 25(16) + 20y = 1080 & \\ y = 34 & \end{array}$$

The farmer needs 16 bags of Brand I and 34 bags of Brand II.

- A furniture manufacturer has some discontinued fabric and trim in stock. The manufacturer can use them on sofas and chairs. In stock are 160 yards of fabric and 110 yards of trim. Each sofa takes 6 yards of fabric and 4.5 yards of trim. Each chair takes 4 yards of fabric and 2 yards of trim. How many sofas and chairs should the manufacturer produce in order to use all the fabric and trim?

Let x represent the number of sofas to be produced and y the number of chairs. The manufacturer needs to use 160 yards of fabric, $6x$ will be used on sofas and $4y$ yards on chairs. This gives us the equation $6x + 4y = 160$. There are 110 yards of trim, $4.5x$ yards will be used on the sofas and $2y$ on the chairs. This gives us the equation $4.5x + 2y = 110$.

$$\begin{cases} 6x + 4y = 160 & \text{F} \\ 4.5x + 2y = 110 & \text{T} \end{cases}$$

We compute F − 2T.

$$\begin{array}{rl} 6x + 4y = 160 & \text{F} \\ \underline{-9x - 4y = -220} & -\text{2T} \\ -3x = -60 & \\ x = 20 & \\ 6(20) + 4y = 160 & \\ y = 10 & \end{array}$$

The manufacturer should produce 20 sofas and 10 chairs.

PRACTICE

1. **A grocery store sells two different brands of milk. The price for the name brand is \$3.50 per gallon, and the price for the store's brand is \$2.25 per gallon. On one Saturday, 4500 gallons of milk were sold for sales of \$12,875. How many of each brand was sold?**
2. **A cable company offers two services—basic cable and premium cable. It charges \$25 per month for basic service and \$80 per month for premium service. Last month, it had 94,000 subscribers and had \$4,220,000 in billing. How many subscribers used the premium service?**
3. **Isaac wants to add 39 pounds of Nutrient A and 16 pounds of Nutrient B to a garden. Each bag of Brand X provides 3 pounds of Nutrient A and 2 pounds of Nutrient B. Each bag of Brand Y provides 4 pounds of Nutrient A and 1 pound of Nutrient B. How many bags of each brand should he buy?**
4. **A clothing manufacturer has 70 yards of a certain fabric and 156 buttons in stock. The manufacturer can make jackets and slacks that use this fabric and button. Each jacket requires $1\frac{1}{3}$ yards of fabric and 4 buttons. Each pair of slacks required $1\frac{3}{4}$ yards of fabric and 3 buttons. How many jackets and pairs of slacks should the manufacturer produce to use all the available fabric and buttons?**

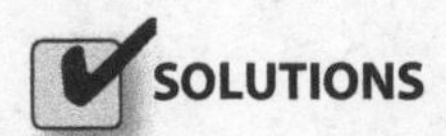

SOLUTIONS

1. Let x represent the number of gallons of the name brand sold and y the number of gallons of the store brand sold. The total number of gallons sold is 4500, giving us $x + y = 4500$. Revenue from the name brand is $3.50x$ and is $2.25y$ for the store brand. Total revenue is $12,875, giving us the equation $3.50x + 2.25y = 12{,}875$.

$$\begin{cases} x + y = 4{,}500 \\ 3.50x + 2.25y = 12{,}875 \end{cases}$$

We use substitution.

$$x = 4500 - y$$

$$3.50(4500 - y) + 2.25y = 12{,}875$$

$$y = 2300$$

$$x = 4500 - y = 4500 - 2300 = 2200$$

The store sold 2200 gallons of the name brand and 2300 gallons of the store brand.

2. Let x represent the number of basic service subscribers and y the number of premium service subscribers. The total number of subscribers is 94,000, so $x + y = 94{,}000$. Revenue from basic services is $25x$ and $80y$ from premium services. Billing was $4,220,000, giving us the equation $25x + 80y = 4{,}220{,}000$.

$$\begin{cases} x + y = 94{,}000 \\ 25x + 80y = 4{,}220{,}000 \end{cases}$$

We use substitution.

$$x = 94{,}000 - y$$

$$25(94{,}000 - y) + 80y = 4{,}220{,}000$$

$$y = 34{,}000$$

There are 34,000 premium service subscribers.

3. Let x represent the number of bags of Brand X and y the number of bags of Brand Y. Isaac will get $3x$ pounds of Nutrient A from x bags of Brand X and $4y$ pounds from y bags of Brand Y, so we need $3x + 4y = 39$. He gets $2x$ pounds of Nutrient B from x bags of Brand X and $1y$ pounds of Nutrient B from y bags of Brand Y, so we need $2x + y = 16$. We use substitution.

$$y = 16 - 2x$$

$$3x + 4(16 - 2x) = 39$$

$$x = 5$$

$$y = 16 - 2x = 16 - 2(5) = 6$$

Isaac should buy 5 bags of Brand X and 6 bags of Brand Y.

4. Let x represent the number of jackets to be produced and y the number of pairs of slacks. To use 70 yards of fabric, we need $1\frac{1}{3}x + 1\frac{3}{4}y = 70$. To use 156 buttons, we need $4x + 3y = 156$.

$$\frac{4}{3}x + \frac{7}{4}y = 70 \quad \text{F}$$

$$4x + 3y = 156 \quad \text{B}$$

$$16x + 21y = 840 \quad 12\text{F}$$

$$\underline{-16x - 12y = -624} \quad -4\text{B}$$

$$9y = 216$$

$$y = 24$$

$$4x + 3(24) = 156$$

$$x = 21$$

The manufacturer should produce 21 jackets and 24 pairs of slacks.

Systems Containing Nonlinear Equations

If a system of equations contains a nonlinear equation (such as a polynomial function), then the system might have more than one solution (or no solution at all). The strategies we used for linear systems work (sometimes) on nonlinear systems.

 **EXAMPLE 11-7**

- $\begin{cases} y = x^2 - 2x - 3 & \text{A} \\ 3x - y = 7 & \text{B} \end{cases}$

We cannot use elimination by addition to eliminate x^2 because B has no x^2 term to cancel x^2 in A. Solving for x in B and substituting it for x in A would work to eliminate x. We could use either substitution or elimination by addition to eliminate y. We use addition to eliminate y.

$$y = x^2 - 2x - 3 \quad \text{A}$$

$$\underline{3x - y = 7} \quad +\text{B}$$

$$3x = x^2 - 2x + 4$$

$$0 = x^2 - 5x + 4$$

$$0 = (x - 1)(x - 4)$$

The solutions occur when $x = 1$ or $x = 4$. We must find two y-values. We let $x = 1$ and $x = 4$ in A.

$$y = 1^2 - 2(1) - 3 = -4 \qquad (1, -4) \text{ is one solution.}$$

$$y = 4^2 - 2(4) - 3 = 5 \qquad (4, 5) \text{ is the other solution.}$$

We can see from the graphs in Fig. 11-5 that these solutions are correct.

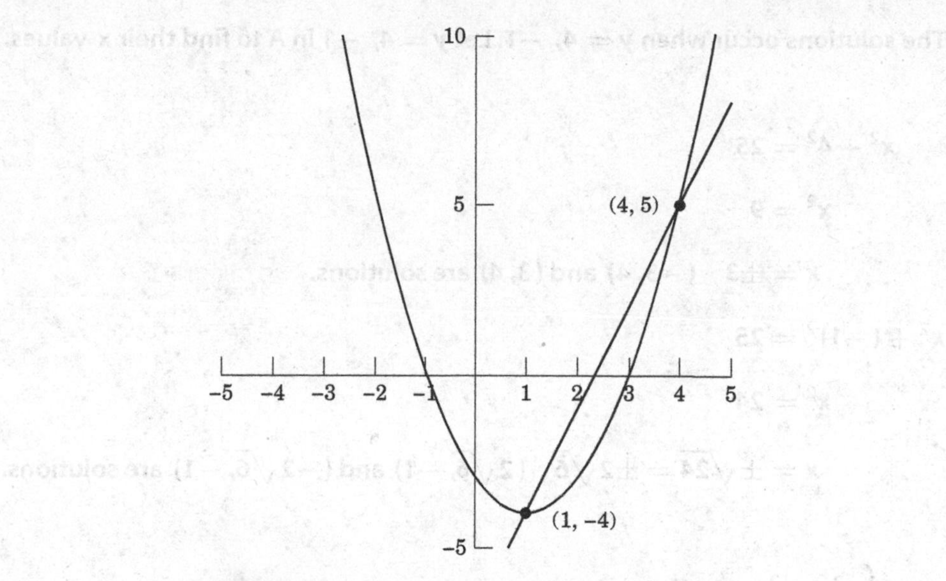

FIGURE 11-5

- $$\begin{cases} x^2 + y^2 = 25 & \text{A} \\ y = -\frac{1}{3}x^2 + 7 & \text{B} \end{cases}$$

We could solve for x^2 in A and substitute this in B. We cannot add the equations to eliminate y or y^2 because A does not have a y term to cancel y in B and B does not have a y^2 term to cancel y^2 in A. We move $-\frac{1}{3}x^2$ to the left side of equation B and multiply each side of B by -3. We can then add this to equation A to eliminate x^2.

$$\frac{1}{3}x^2 + y = 7 \quad \text{B}$$

$$x^2 + y^2 = 25 \quad +\text{A}$$

$$-x^2 - 3y = -21 \quad -3\text{B}$$

$$y^2 - 3y = 4$$

$$y^2 - 3y - 4 = 0$$

$$(y - 4)(y + 1) = 0$$

The solutions occur when $y = 4, -1$. Let $y = 4, -1$ in A to find their x-values.

$$x^2 + 4^2 = 25$$

$$x^2 = 9$$

$$x = \pm 3 \quad (-3, 4) \text{ and } (3, 4) \text{ are solutions.}$$

$$x^2 + (-1)^2 = 25$$

$$x^2 = 24$$

$$x = \pm\sqrt{24} = \pm 2\sqrt{6} \quad (2\sqrt{6}, -1) \text{ and } (-2\sqrt{6}, -1) \text{ are solutions.}$$

- $$\begin{cases} x^2 + y^2 = 4 & \text{A} \\ y = \frac{2}{x} & \text{B} \end{cases}$$

Elimination by addition will not help us with this system, but substitution will. We substitute $y = \frac{2}{x}$ for y in A. Doing so will give us an expression that factors the same way a quadratic expression can be factored.

$$x^2 + \left(\frac{2}{x}\right)^2 = 4$$

$$x^2 + \frac{4}{x^2} = 4$$

$$x^2\left(x^2 + \frac{4}{x^2}\right) = x^2(4) \quad \text{Clear the fraction.}$$

$$x^4 + 4 = 4x^2$$

$$x^4 - 4x^2 + 4 = 0 \quad \text{This is similar to a quadratic equation.}$$

$$(x^2 - 2)(x^2 - 2) = 0 \quad \text{Factor.}$$

$$x^2 = 2$$

$$x = \pm\sqrt{2}$$

We let $x = \sqrt{2}$ and $x = -\sqrt{2}$ in $y = \frac{2}{x}$ to find y.

$$y = \frac{2}{\sqrt{2}} = \frac{2\sqrt{2}}{\sqrt{2}\sqrt{2}} = \frac{2\sqrt{2}}{2} = \sqrt{2} \quad (\sqrt{2}, \sqrt{2}) \text{ is a solution.}$$

$$y = \frac{2}{-\sqrt{2}} = \frac{2\sqrt{2}}{-\sqrt{2}\sqrt{2}} = \frac{2\sqrt{2}}{-2} = -\sqrt{2} \quad (-\sqrt{2}, -\sqrt{2}) \text{ is a solution.}$$

PRACTICE

Solve the system of equations. Write the solutions in the form of a point, (x, y).

1. $\begin{cases} y = x^2 - 4 & \text{A} \\ x + y = 8 & \text{B} \end{cases}$

2. $\begin{cases} x^2 + y^2 + 6x - 2y = -5 & \text{A} \\ y = -2x - 5 & \text{B} \end{cases}$

3. $\begin{cases} x^2 - y^2 = 16 & \text{A} \\ x^2 + y^2 = 16 & \text{B} \end{cases}$

4. $\begin{cases} 4x^2 + y^2 = 5 & \text{A} \\ y = \frac{1}{x} & \text{B} \end{cases}$

SOLUTIONS

1.

$$\begin{aligned} y &= x^2 - 4 \quad \text{A} \\ -x - y &= -8 \quad -\text{B} \\ \hline -x &= x^2 - 12 \\ 0 &= x^2 + x - 12 = (x + 4)(x - 3) \end{aligned}$$

We have solutions for $x = -4$ and $x = 3$. We substitute these values in A.

$$y = (-4)^2 - 4 = 12; \; (-4, 12) \text{ is a solution.}$$

$$y = 3^2 - 4 = 5; \; (3, 5) \text{ is a solution.}$$

2. Substitute $-2x - 5$ for y in A.

$$x^2 + (-2x-5)^2 + 6x - 2(-2x-5) = -5$$
$$x^2 + 4x^2 + 20x + 25 + 6x + 4x + 10 = -5$$
$$5x^2 + 30x + 40 = 0 \quad \text{Divide by 5.}$$
$$x^2 + 6x + 8 = 0$$
$$(x+4)(x+2) = 0$$

The system has solutions for $x = -4$ and $x = -2$. We substitute these values in B instead of A because there is less computation to do in B.

$$y = -2(-4) - 5 = 3; \ (-4, 3) \text{ is a solution.}$$
$$y = -2(-2) - 5 = -1; \ (-2, -1) \text{ is a solution.}$$

3.

$$x^2 - y^2 = 16 \quad \text{A}$$
$$\underline{x^2 + y^2 = 16} \quad +\text{B}$$
$$2x^2 = 32$$
$$x^2 = 16$$
$$x = \pm 4$$

Let $x = 4$ and $x = -4$ in A.

$$(-4)^2 - y^2 = 16 \qquad 4^2 - y^2 = 16$$
$$16 - y^2 = 16 \qquad 16 - y^2 = 16$$
$$y^2 = 0 \qquad y^2 = 0$$
$$y = 0 \qquad y = 0$$

The solutions are $(-4, 0)$ and $(4, 0)$.

4. Substitute $\frac{1}{x}$ for y in A.

$$4x^2 + \left(\frac{1}{x}\right)^2 = 5$$

$$x^2\left(4x^2 + \frac{1}{x^2}\right) = x^2(5)$$

$$4x^4 + 1 = 5x^2$$

$$4x^4 - 5x^2 + 1 = 0$$

$$(4x^2 - 1)(x^2 - 1) = 0$$

$$(2x - 1)(2x + 1)(x - 1)(x + 1) = 0$$

The solutions are $x = \pm\frac{1}{2}$ (from $2x - 1 = 0$ and $2x + 1 = 0$) and $x = \pm 1$. We substitute these values in B to find y.

$$y = \frac{1}{\frac{1}{2}} = 2;\ \left(\frac{1}{2}, 2\right) \text{ is a solution.}$$

$$y = \frac{1}{-\frac{1}{2}} = -2;\ \left(-\frac{1}{2}, -2\right) \text{ is a solution.}$$

$$y = \frac{1}{1} = 1;\ (1, 1) \text{ is a solution.}$$

$$y = \frac{1}{-1} = -1;\ (-1, -1) \text{ is a solution.}$$

Inequalities and Systems of Inequalities

We represent the solution to a polynomial inequality (the only kind in this book) by shading the region above or below the curve. *Every* point in the shaded region is a solution to the inequality.

We begin with linear inequalities. We first sketch the graph of the equation using a solid graph for "≤" and "≥" inequalities and a dashed graph for "<" and ">" inequalities. We can decide which side of the graph to shade by choosing *any* point not on the graph itself. If the coordinates of the point make the inequality true, then we shade the side that contains that point. If the coordinates of the point make the inequality false, we shade the side that does *not* contain the point.

EXAMPLE 11-8

- $2x + 3y \leq 6$

We begin with the line $2x + 3y = 6$, using a solid line because the inequality is "$\leq$." See Fig. 11-6.

We always use the origin, $(0, 0)$, in our inequalities unless the graph goes through the origin. Does $(0, 0)$ make $2x + 3y \leq 6$ true? As $2(0) + 3(0) \leq 6$ is a true statement, we shade the side that has the origin. The solution to the inequality is the shaded region in Fig. 11-7.

- $x - 2y > 4$

We sketch the line $x - 2y = 4$ using a dashed line because the inequality is "$>$." See Fig. 11-8.

Now we need to decide which side of the line to shade. When we substitute the coordinates of $(0, 0)$ in $x - 2y > 4$, we have the false statement

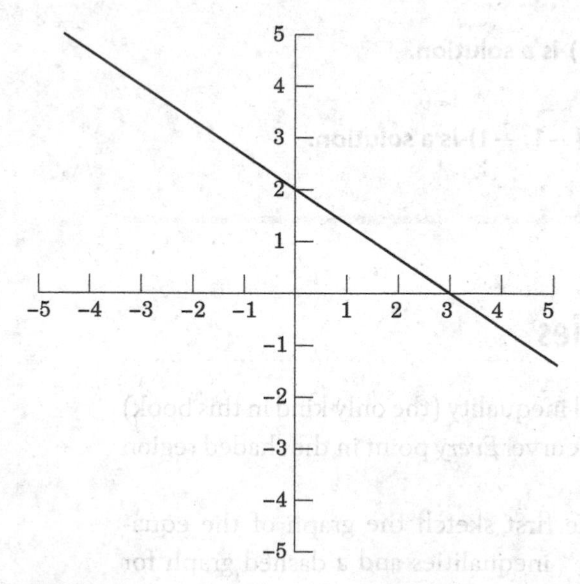

FIGURE 11-6

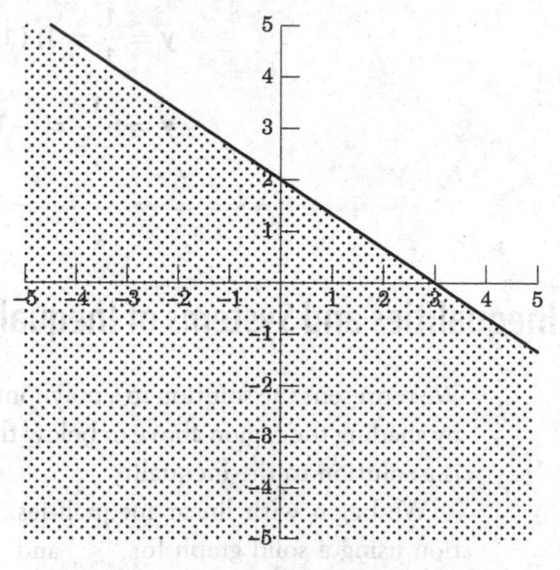

FIGURE 11-7

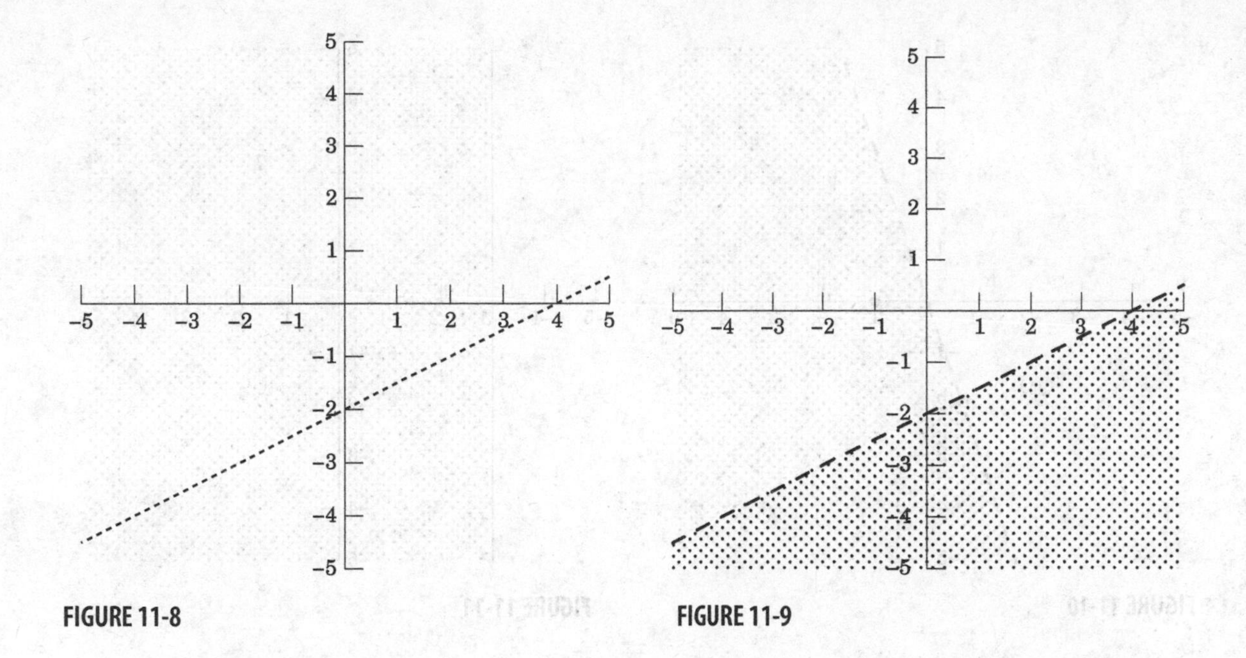

FIGURE 11-8

FIGURE 11-9

$0 - 2(0) > 4$. We need to shade the side of the line that does *not* have the origin. The solution to the inequality is the region shaded in Fig. 11-9.

- $y < 3x$

We use a dashed line to sketch the line $y = 3x$. Because the line goes through $(0, 0)$, we cannot use it to determine which side of the line to shade. This is because any point on the line makes the equation true. We want to know where the inequality is true. The point $(1, 0)$ is not on the line, so we can use it. Because the coordinates of this point make the inequality true, $0 < 3(1)$, we shade the side of the line that contains this point, which is the right side. The shaded region in Fig. 11-10 is the solution to the inequality.

- $x \geq -3$

The line $x = -3$ is a vertical line through $x = -3$. Because we want $x \geq -3$, we shade to the right of the line. See Fig. 11-11.

- $y < 2$

The line $y = 2$ is a horizontal line at $y = 2$. Because we want $y < 2$, we shade below the line. See Fig. 11-12.

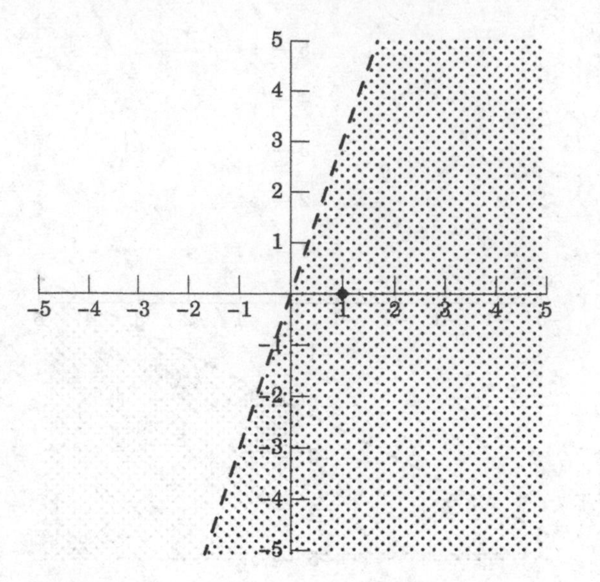

FIGURE 11-10

FIGURE 11-11

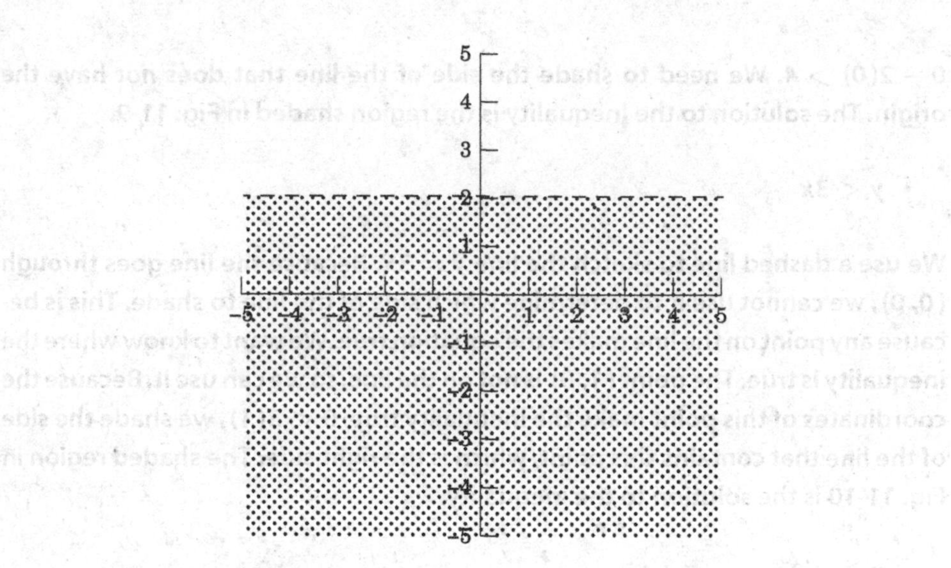

FIGURE 11-12

PRACTICE

Graph the solution to the inequality.

1. $x + y \geq 2$
2. $2x - 4y < 4$
3. $y \leq \frac{2}{3}x - 1$
4. $x > 1$
5. $y \leq -1$

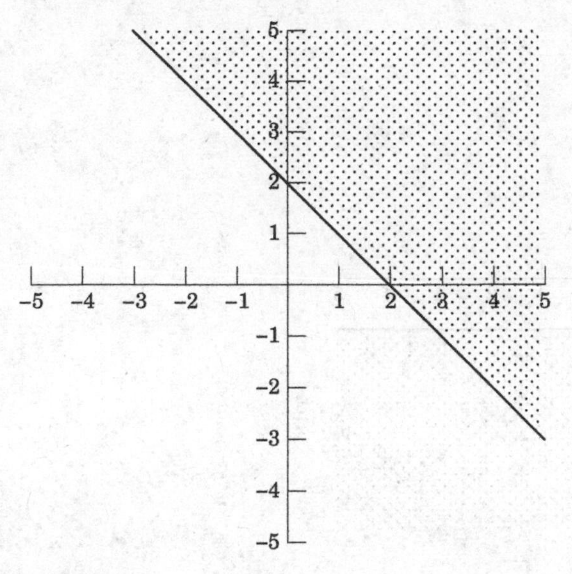

FIGURE 11-13

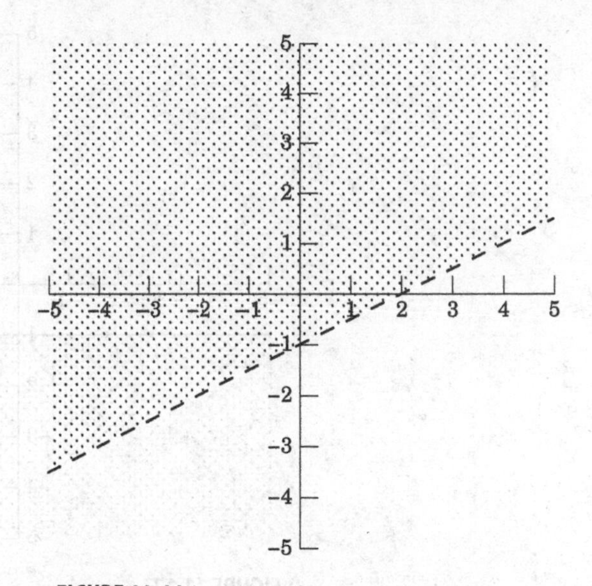

FIGURE 11-14

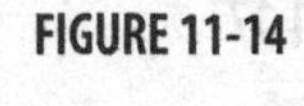

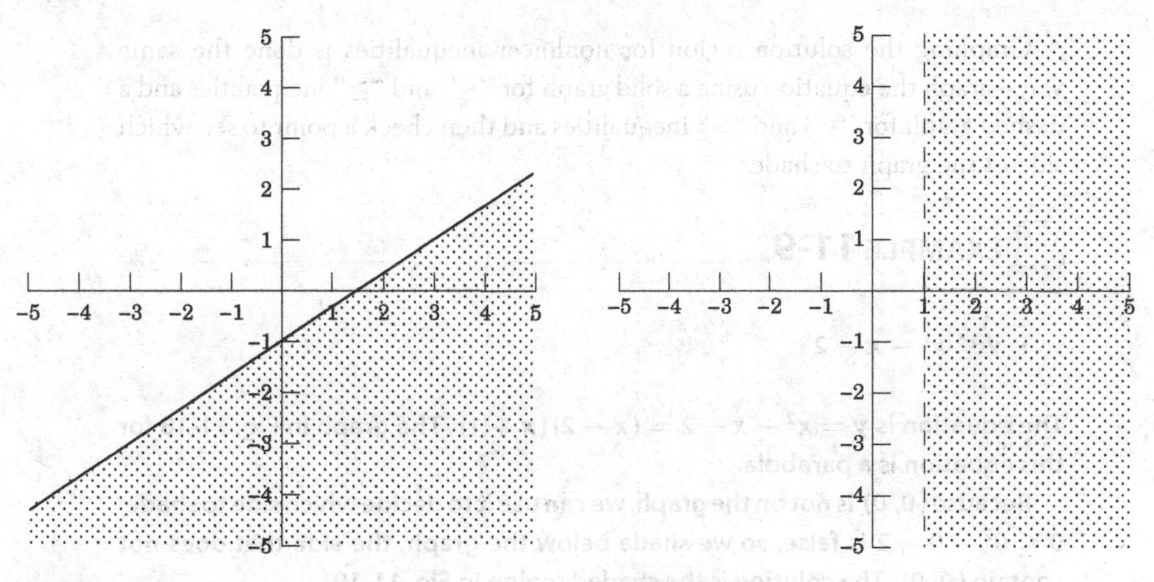

FIGURE 11-15

FIGURE 11-16

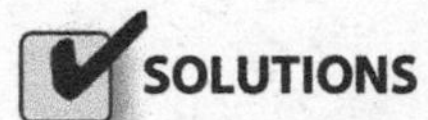

SOLUTIONS

1. See Fig. 11-13.
2. See Fig. 11-14.
3. See Fig. 11-15.
4. See Fig. 11-16.
5. See Fig. 11-17.

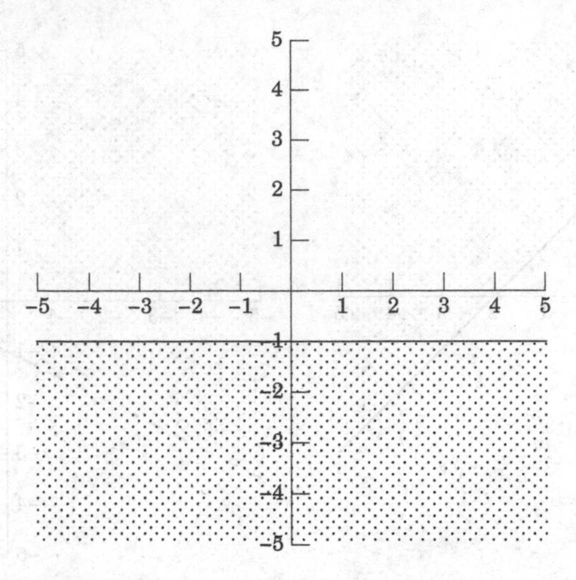

FIGURE 11-17

Graphing the solution region for nonlinear inequalities is done the same way—graph the equation using a solid graph for "≤" and "≥" inequalities and a dashed graph for "<" and ">" inequalities and then check a point to see which side of the graph to shade.

EXAMPLE 11-9

- $y \leq x^2 - x - 2$

The equation is $y = x^2 - x - 2 = (x - 2)(x + 1)$. The graph in Fig. 11-18 for this equation is a parabola.

Because $(0, 0)$ is not on the graph, we can use it to decide which side to shade; $0 \leq 0^2 - 0 - 2$ is false, so we shade below the graph, the side that does not contain $(0, 0)$. The solution is the shaded region in Fig. 11-19.

- $y > (x + 2)(x - 2)(x - 4)$

When we check $(0, 0)$ in the inequality, we get the false statement $0 > (0 + 2)(0 - 2)(x - 4)$. We shade above the graph, the region that does not contain $(0, 0)$. See Fig. 11-20.

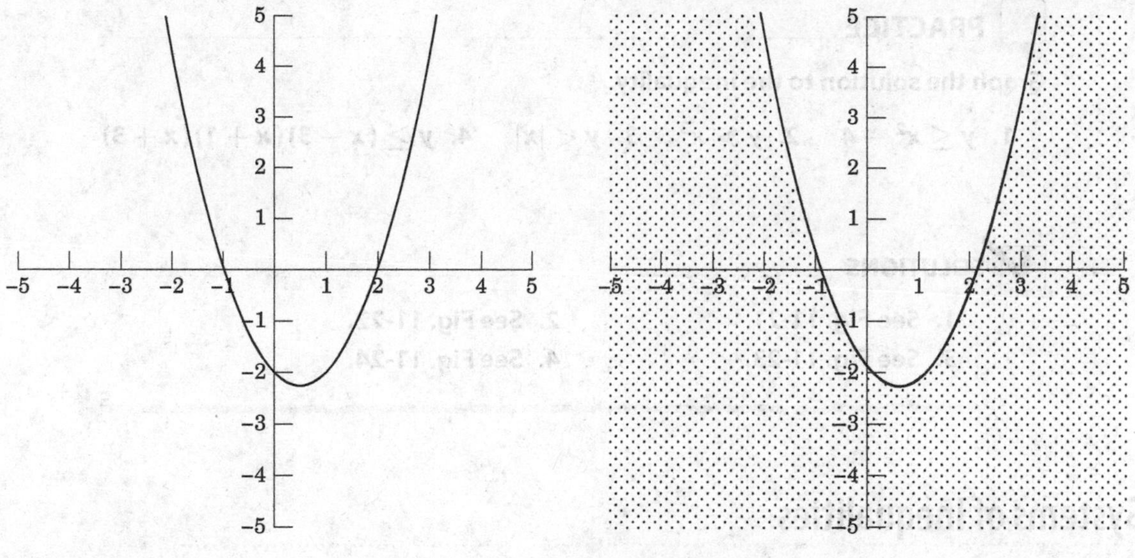

FIGURE 11-18

FIGURE 11-19

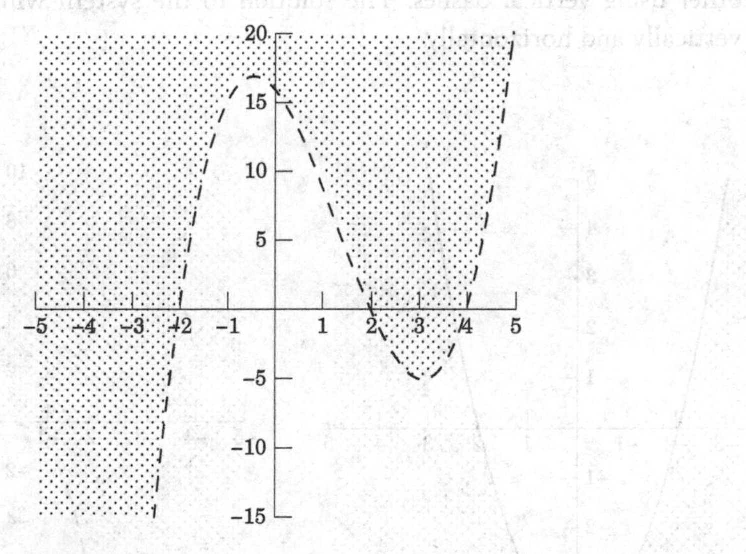

FIGURE 11-20

PRACTICE

Graph the solution to the inequality.

1. $y \leq x^2 - 4$ 2. $y > x^3$ 3. $y < |x|$ 4. $y \geq (x-3)(x+1)(x+3)$

SOLUTIONS

1. See Fig. 11-21.
2. See Fig. 11-22.
3. See Fig. 11-23.
4. See Fig. 11-24.

Systems of Inequalities

The solution (if there is one) to a system of two or more inequalities is the region where the shaded regions overlap. (In algebra, we say, the *intersection* of the solutions.) Most of the systems in the following problems contain two inequalities. We will shade the solution to one inequality using horizontal dashes and the other using vertical dashes. The solution to the system will be shaded both vertically and horizontally.

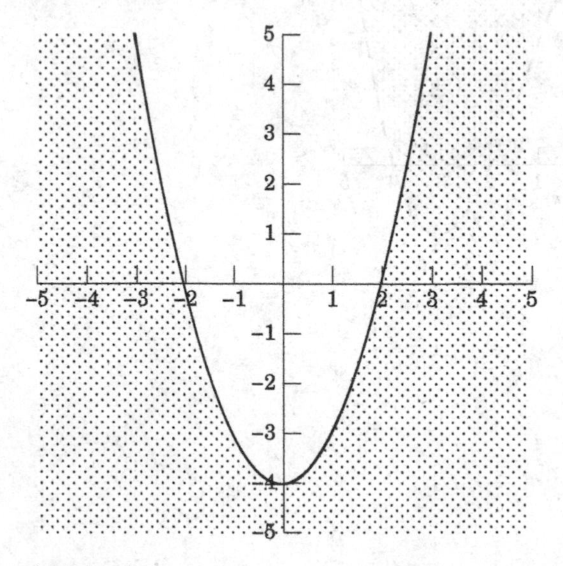

FIGURE 11-21

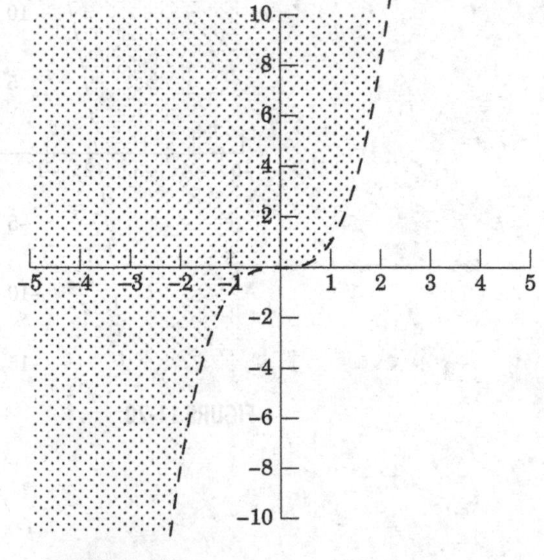

FIGURE 11-22

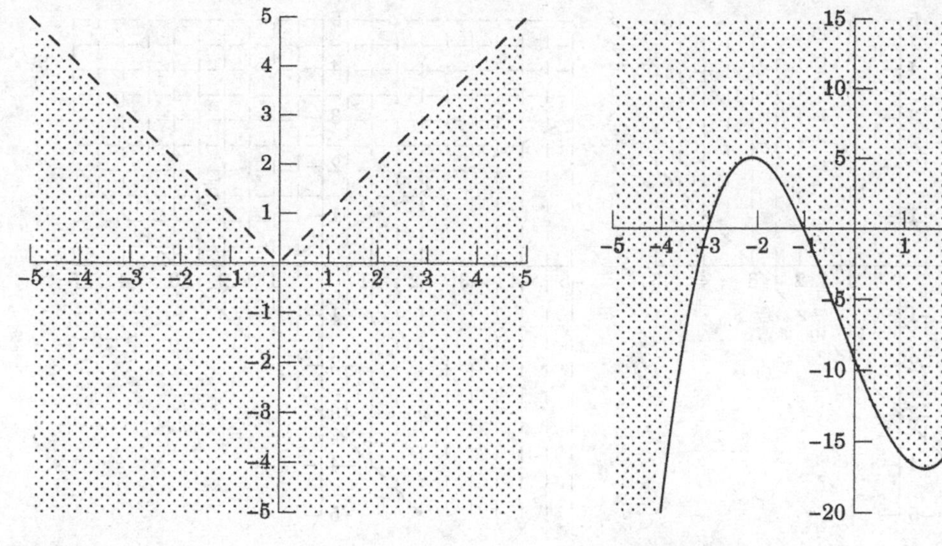

FIGURE 11-23

FIGURE 11-24

EXAMPLE 11-10

Graph the solution to the system of inequalities.

- $\begin{cases} x - y < 3 \\ x + 2y > 1 \end{cases}$

The solution to each individual inequality is shaded in Figs. 11-25 and 11-26.

The region that is shaded both vertically and horizontally is shown in Fig. 11-26. The solution to the system of inequalities is the shaded region in Fig. 11-27.

- $\begin{cases} y \leq 4 - x^2 \\ x - 7y \leq 4 \end{cases}$

The solution to the first inequality is the region shaded vertically in Fig. 11-28, and the solution to the second inequality is the region shaded horizontally.

The region that lies in both solutions is above the line and inside the parabola. See Fig. 11-29.

Because a solid graph indicates that points on the graph are also solutions, to be absolutely accurate, the correct solution uses dashed graphs for the part of the graphs that is not on the border of the shaded region. See Fig. 11-30.

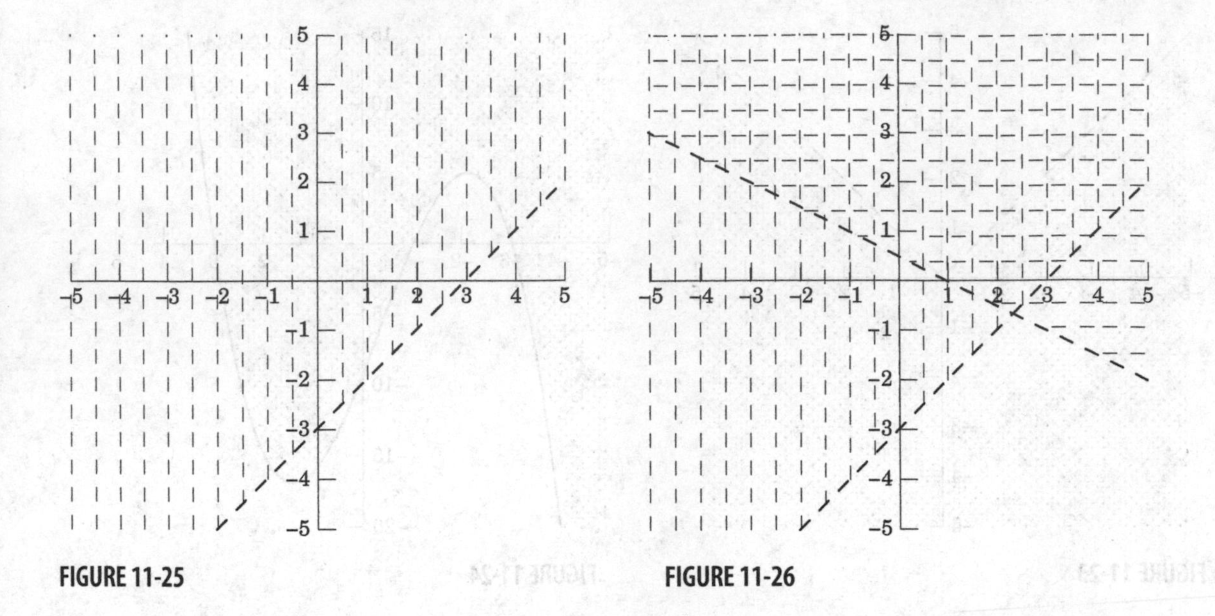

FIGURE 11-25

FIGURE 11-26

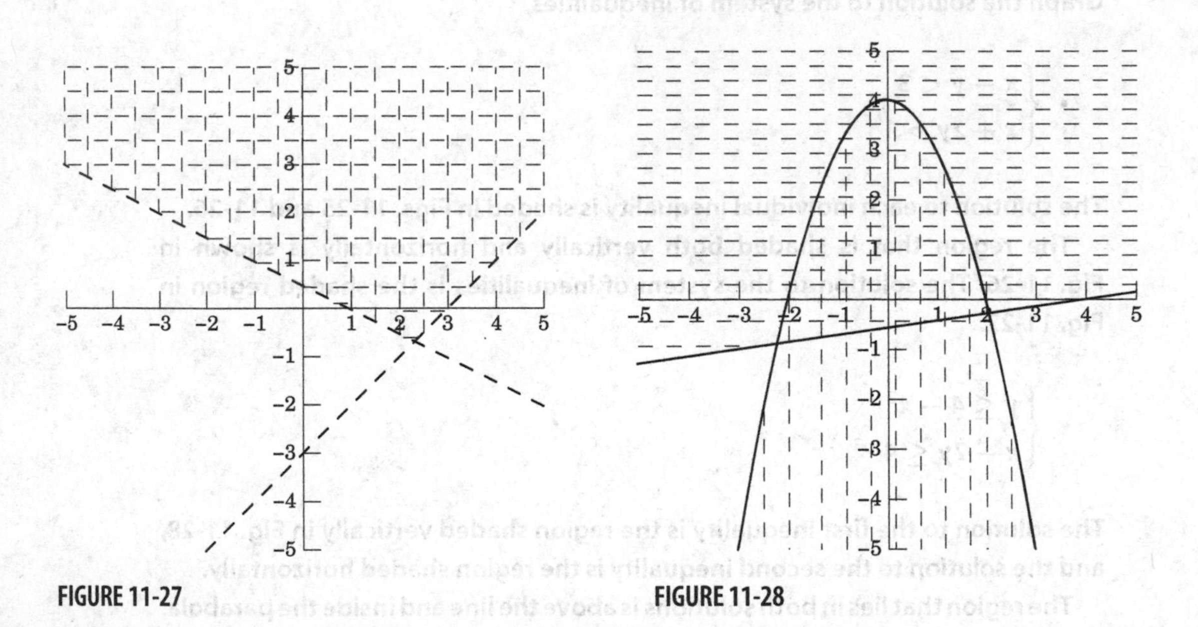

FIGURE 11-27

FIGURE 11-28

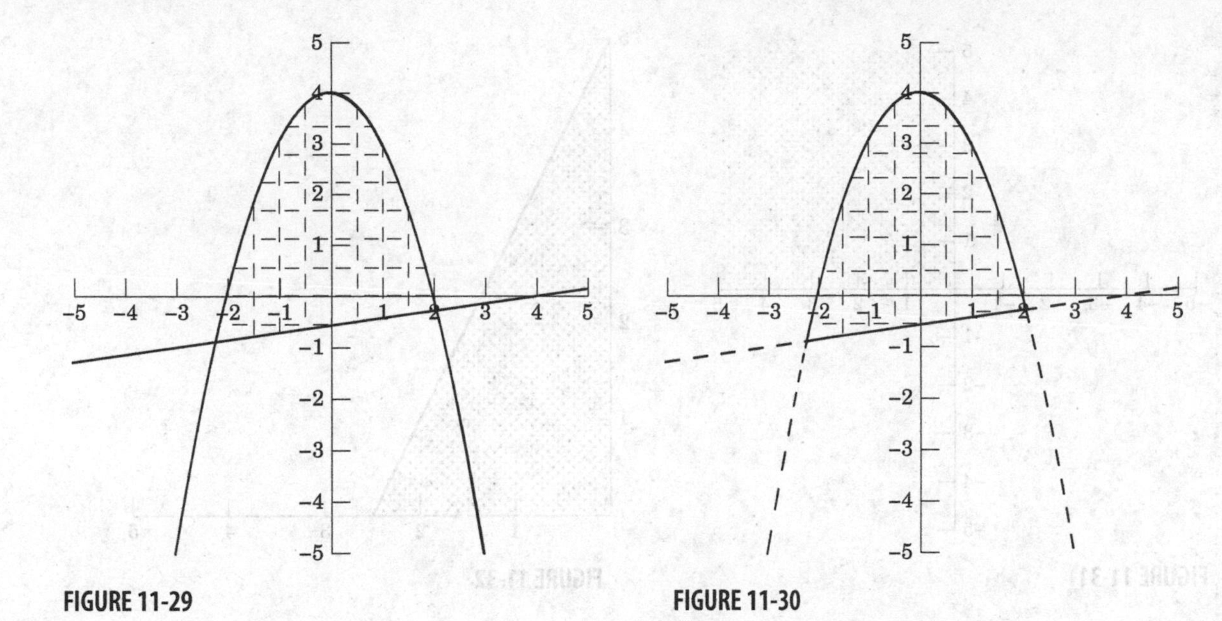

FIGURE 11-29

FIGURE 11-30

We will not quibble with this technicality here.

- $\begin{cases} 2x + y \leq 5 \\ x \geq 0 \\ y \geq 0 \end{cases}$

The inequalities $x \geq 0$ and $y \geq 0$ imply that we only need the top right corner of the graph, as in Fig. 11-31. These inequalities are common in word problems.

The solution to the system is the region in the top right corner of the graph below the line $2x + y = 5$. See Fig. 11-32.

PRACTICE

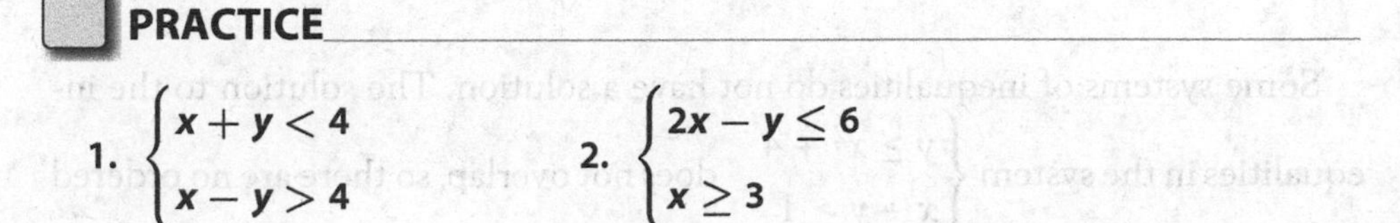

1. $\begin{cases} x + y < 4 \\ x - y > 4 \end{cases}$

2. $\begin{cases} 2x - y \leq 6 \\ x \geq 3 \end{cases}$

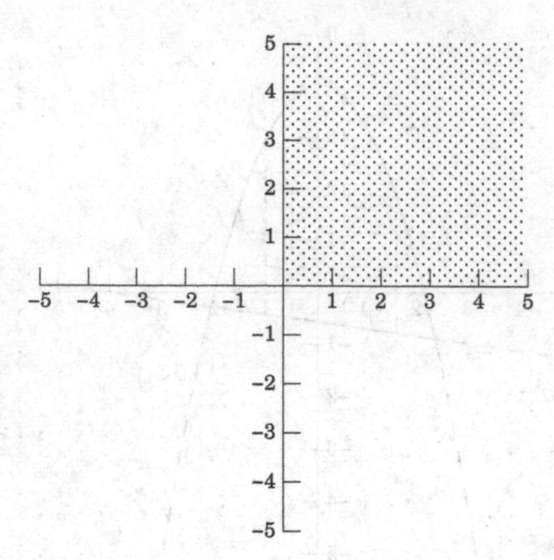

FIGURE 11-31

FIGURE 11-32

3. $\begin{cases} y > x^2 + 2x - 3 \\ x + y < 5 \end{cases}$

4. $\begin{cases} y \le 9 - x^2 \\ y \ge x^2 + 4x - 5 \end{cases}$

5. $\begin{cases} 2x + 3y \ge 6 \\ x \ge 0 \\ y \ge 0 \end{cases}$

SOLUTIONS

1. See Fig. 11-33. **2. See Fig. 11-34.** **3. See Fig. 11-35.**
4. See Fig. 11-36. **5. See Fig. 11-37.**

Some systems of inequalities do not have a solution. The solution to the inequalities in the system $\begin{cases} y \ge x^2 + 4 \\ x - y \ge 1 \end{cases}$ does not overlap, so there are no ordered pairs (points) that make both inequalities true. See Fig. 11-38.

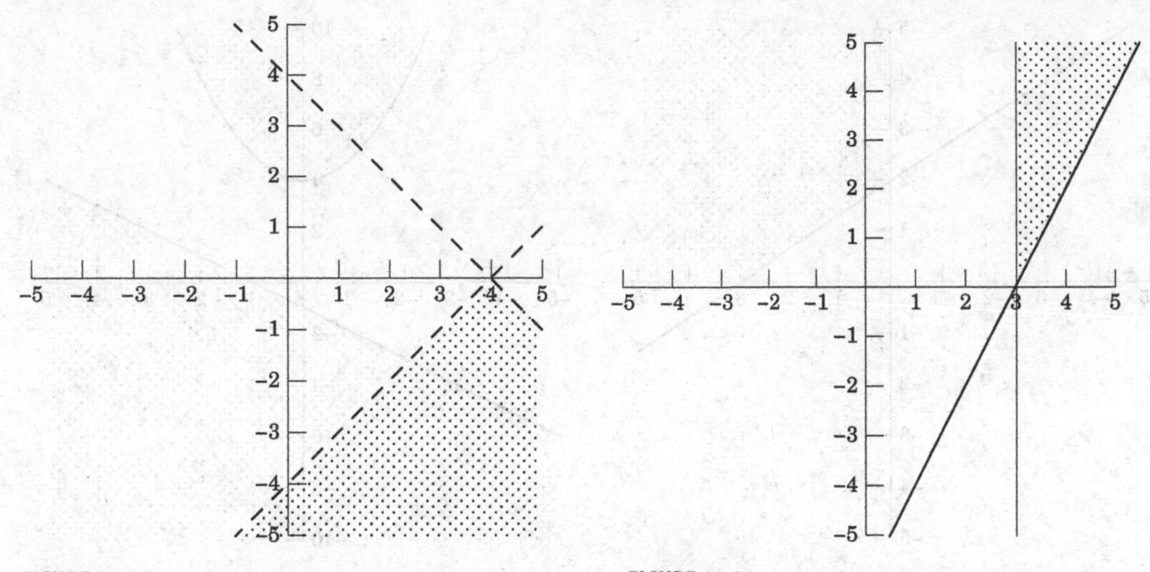

FIGURE 11-33

FIGURE 11-34

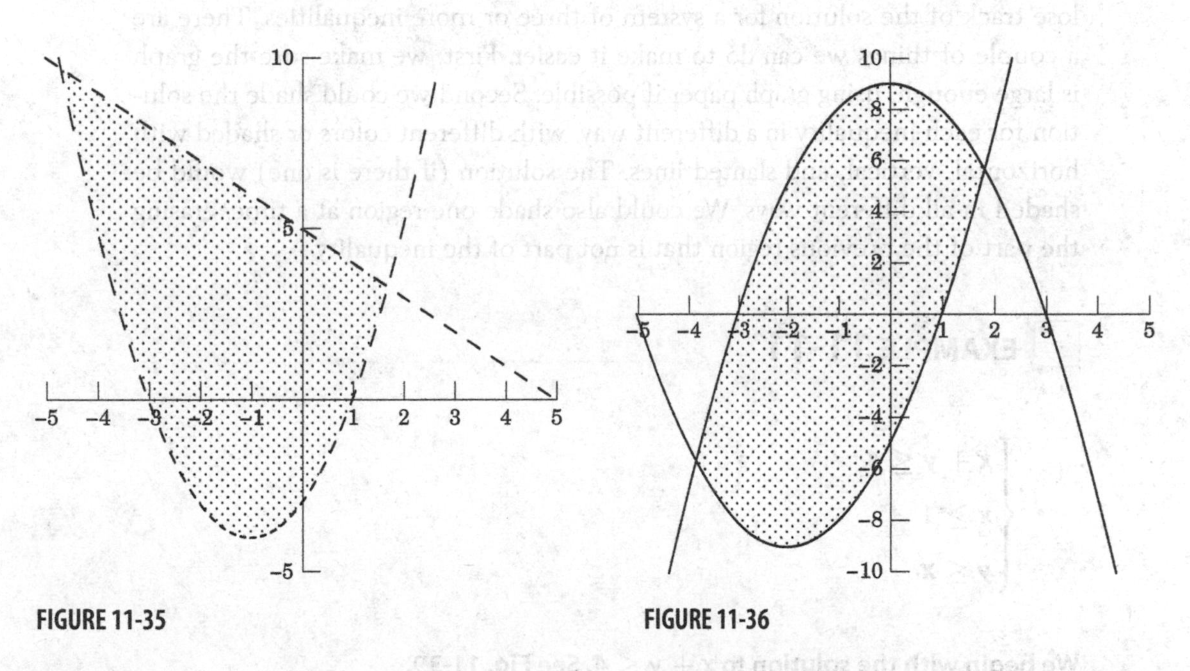

FIGURE 11-35

FIGURE 11-36

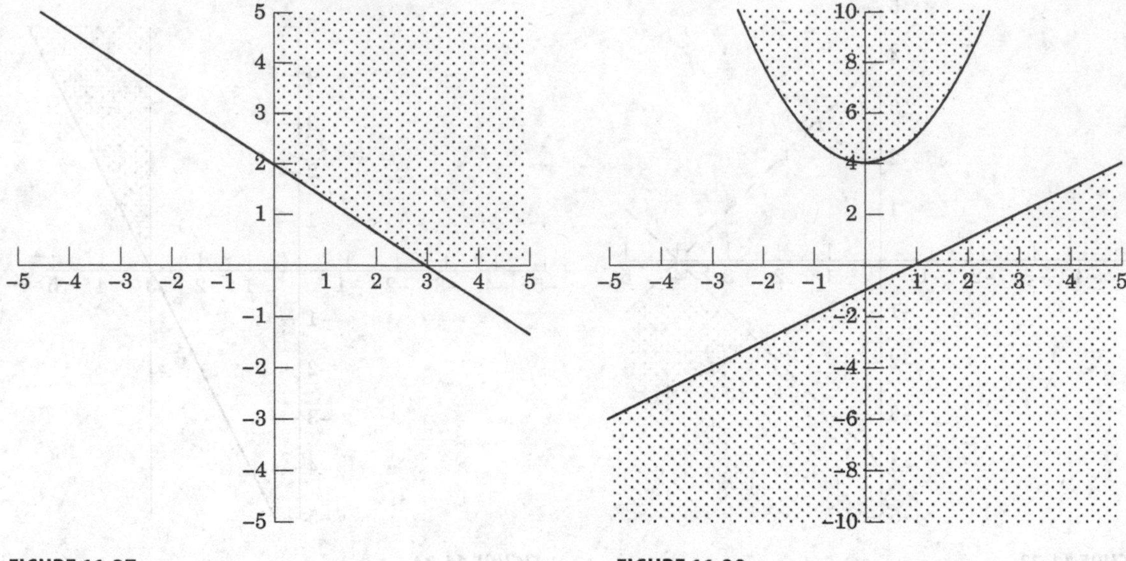

FIGURE 11-37

FIGURE 11-38

If a system of inequalities contains more than two inequalities, we can easily lose track of the solution for a system of three or more inequalities. There are a couple of things we can do to make it easier. First, we make sure the graph is large enough, using graph paper if possible. Second,we could shade the solution for each inequality in a different way, with different colors or shaded with horizontal, vertical, and slanted lines. The solution (if there is one) would be shaded in all different ways. We could also shade one region at a time, erasing the part of the previous region that is not part of the inequality.

EXAMPLE 11-11

- $$\begin{cases} x + y \leq 4 \\ x \geq 1 \\ y \leq x \end{cases}$$

We begin with the solution to $x + y \leq 4$. See Fig. 11-39.

The region for $x \geq 1$ is to the right of the line $x = 1$, so we erase the region to the *left* of $x = 1$. See Fig. 11-40.

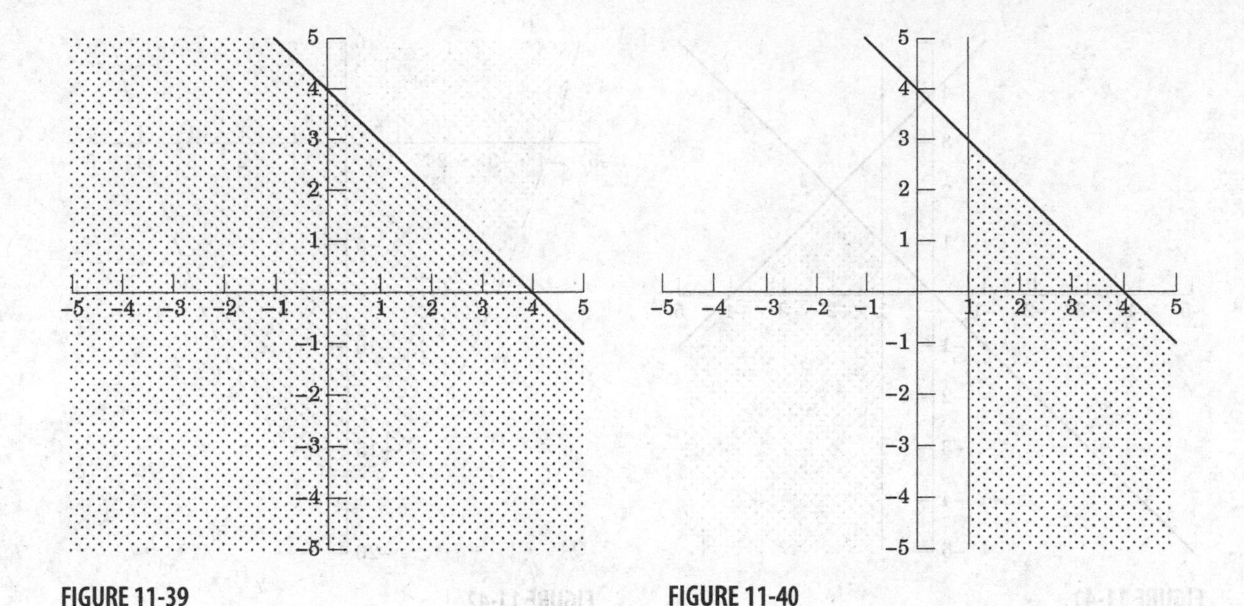

FIGURE 11-39

FIGURE 11-40

The solution to $y \leq x$ is the region below the line $y = x$, so we erase the shading *above* the line $y = x$.

The shaded region in Fig. 11-41 is the solution for the system.

- $$\begin{cases} y > x^2 - 16 \\ x < 2 \\ y < -5 \\ -x + y < -8 \end{cases}$$

We begin with the solution to $y > x^2 - 16$ in Fig. 11-42.

The solution to $x < 2$ is the region to the left of the line $x = 2$, so we erase the shading to the right of $x = 2$. See Fig. 11-43.

The solution to $y < -5$ is the region below the line $y = -5$. We erase the shading above the line $y = -5$. See Fig. 11-44.

The solution to $-x + y < -8$ is the region below the line $-x + y = -8$, so we erase the shading above the line. The solution to the system is in Fig. 11-45.

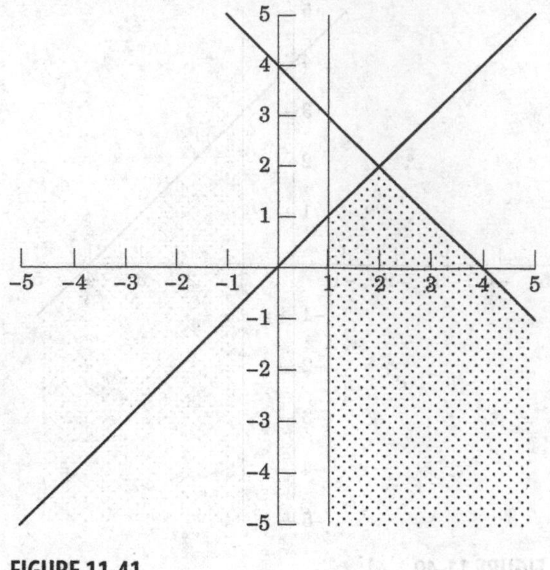

FIGURE 11-41

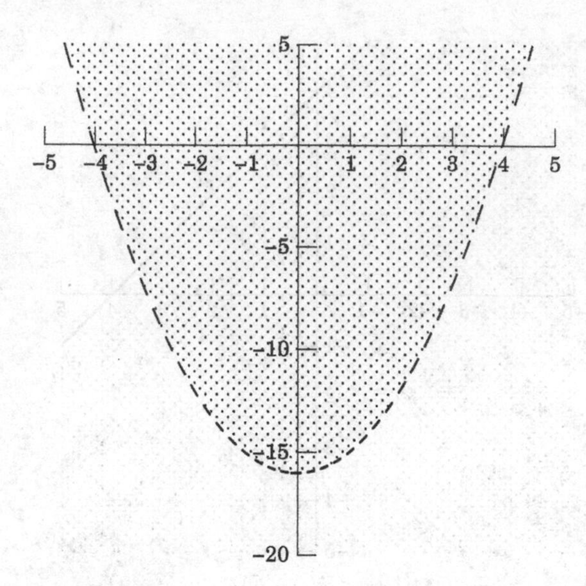

FIGURE 11-42

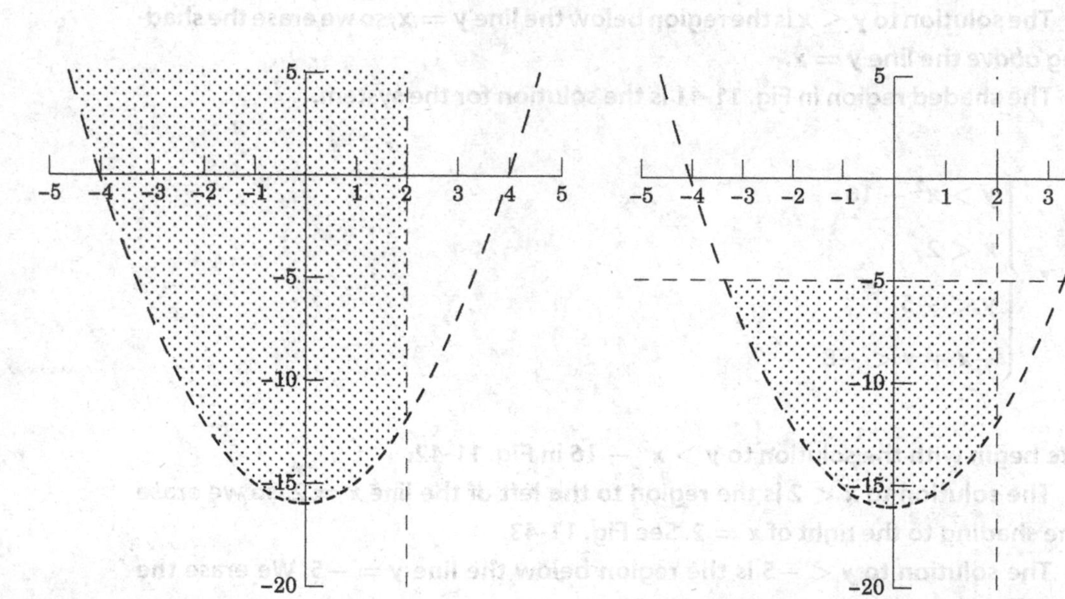

FIGURE 11-43

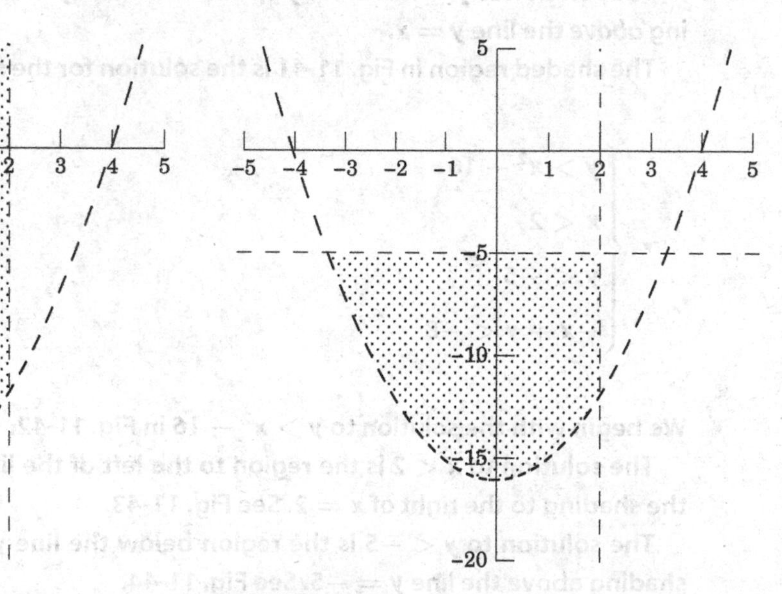

FIGURE 11-44

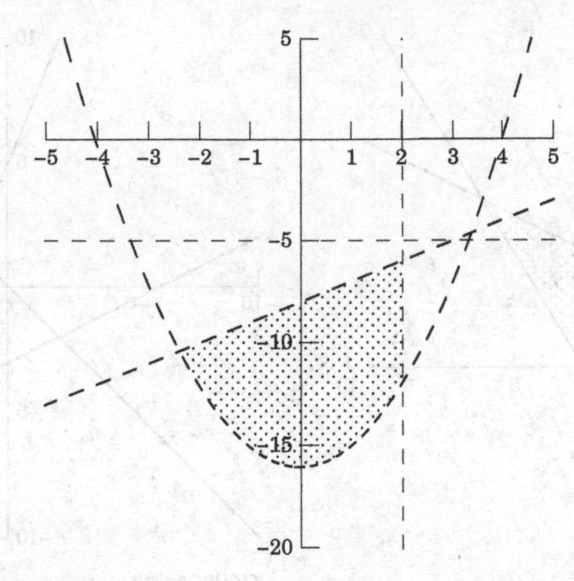

FIGURE 11-45

PRACTICE

1. $\begin{cases} 2x + y \geq 1 \\ -x + 2y \leq 4 \\ 5x - 3y \leq 15 \end{cases}$

2. $\begin{cases} x + 2y \geq -6 \\ y \leq x \\ 5x + 2y \geq 10 \end{cases}$

3. $\begin{cases} y \leq 9 - x^2 \\ x \leq 2 \\ y \leq -x \end{cases}$

4. $\begin{cases} 2x + y < 8 \\ y < x \\ x \geq 0 \\ y \geq 0 \end{cases}$

SOLUTIONS

1. See Fig. 11-46.
2. See Fig. 11-47.
3. See Fig. 11-48.
4. See Fig. 11-49.

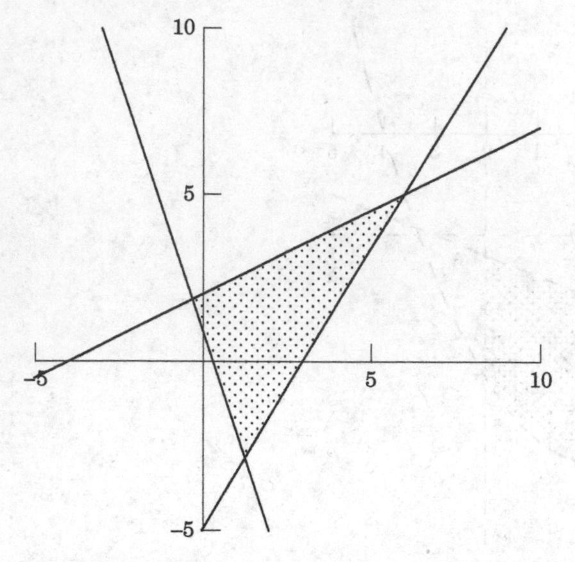

FIGURE 11-46

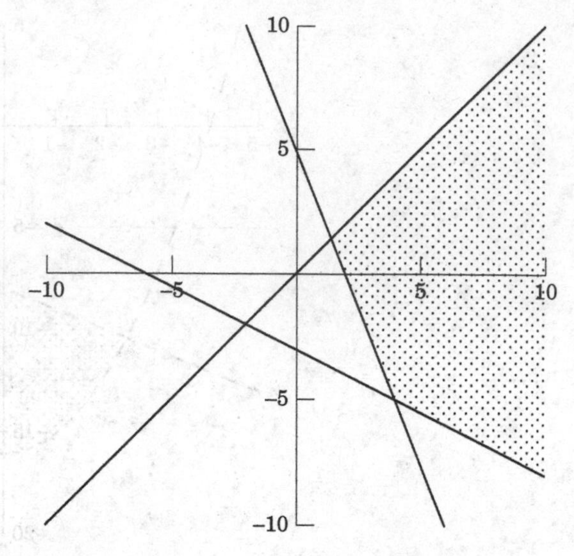

FIGURE 11-47

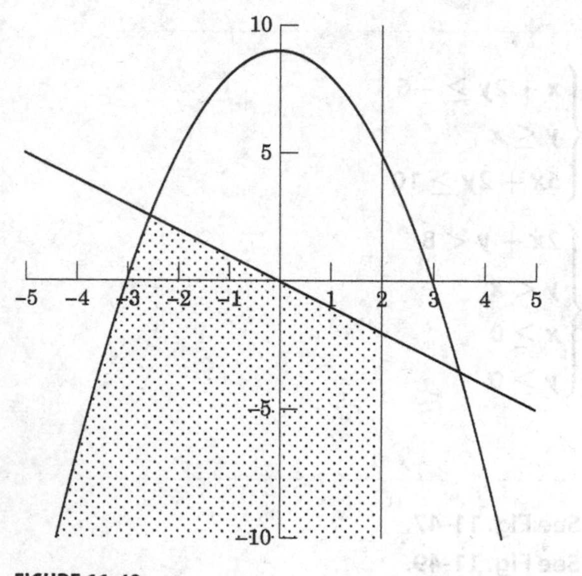

FIGURE 11-48

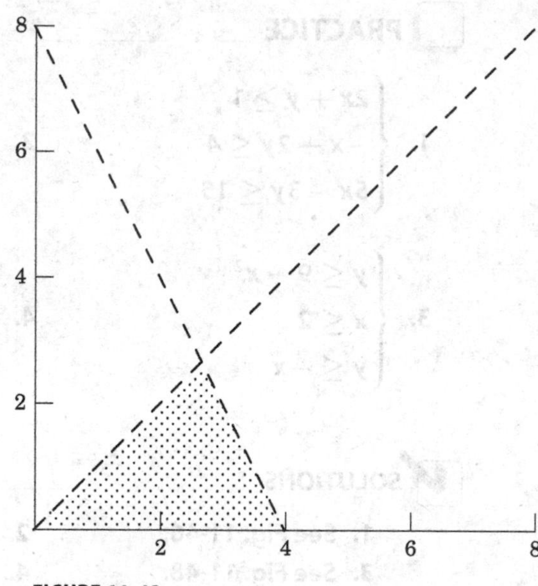

FIGURE 11-49

Summary

In this chapter, we learned how to

- *Solve a system of linear equations by substitution*. We solve for one variable in one equation and make a substitution in the other equation. This leaves us with a linear equation containing a single variable. We use ordinary algebra to solve this equation. Once we know the value of one variable, we use either equation in the system to find the other value. What we have found are the coordinates of the point where the graphs intersect.
- *Solve a system of linear equations with elimination by addition*. We can add two equations together (by combining like terms). If the coefficients for one of the variables are opposites, such as $4x$ in one equation and $-4x$ in the other, then when we add the equations, we are left with one equation containing a single variable. Again, we use ordinary algebra to solve this equation and then use one of the equations in the system to find the value of the other variable.
- *Determine if two equations in the system represent the same line or are parallel*. If a system of linear equations contains two parallel lines, then the system has no solution. If we try to solve such a system, we end up with a false statement such as "$0 = 5$." If the lines in the system are really the same line, then solving the system gives us a true statement such as "$12 = 12$."
- *Solve applied problems represented by a system of linear equations*. If an applied problem involves two quantities for which two sets of facts are true (such as the total number of movie tickets sold and the total revenue from the ticket sales), then we can represent the problem as a system of linear equations.
- *Solve a system of nonlinear equations*. We can solve many systems containing nonlinear equations using the same strategies. However, sometimes a strategy does not work. For instance, if one equation in the system has an x^2 term but the other does not, then the elimination by addition method might not work. The basic idea remains the same though: eliminate one of the variables, leaving us with one equation containing a single variable and using its solution(s) to find the value(s) of the other variable.
- *Solve an inequality*. We begin by sketching the graph of an equation, using a dashed line/curve if the inequality is "$<$" or "$>$" and a solid line/curve if the inequality is "$\leq$" or "$\geq$." We then choose any point that is not on

the graph and test it in the original inequality. If the coordinates of this point make the inequality true, we shade the side of the graph containing the point. Otherwise, we shade the other side.

- *Solve a system of inequalities*. The solution to a system of inequalities is the region that overlaps all of the solutions to the individual inequalities in the system. For example, if the system contains two inequalities and we shade one solution horizontally and the other vertically, then the solution to the system is shaded both horizontally and vertically.

QUIZ

In Problems 1-6, find $x + 2y$ for the system. That is, if $(-2, 5)$ is a solution, then the correct answer choice would be $-2 + 2(5) = 8$.

1. $\begin{cases} 2x - y = 5 \\ x + 4y = 7 \end{cases}$

   A. 4   B. 5   C. 6   D. 7

2. $\begin{cases} 3x + 5y = 4 \\ -2x + 4y = 12 \end{cases}$

   A. 2   B. 3   C. 4   D. 5

3. $\begin{cases} y = 3x + 10 \\ y = -x - 6 \end{cases}$

   A. −4   B. −6   C. −8   D. −10

4. $\begin{cases} -\frac{2}{3}x + \frac{1}{4}y = 1 \\ x - \frac{1}{3}y = 2 \end{cases}$

   A. 168   B. 198   C. 140   D. 145

5. $\begin{cases} 2x + y = 10 \\ y = x^2 - 2x - 15 \end{cases}$

   A. 10, −20   B. 10, 30   C. −5, 20   D. 5, 35

6. (Real solutions only) $\begin{cases} y = \frac{3}{x} \\ x^2 - y^2 = 8 \end{cases}$

   A. −2, 2   B. −1, 3   C. −5, 5   D. 0, 1

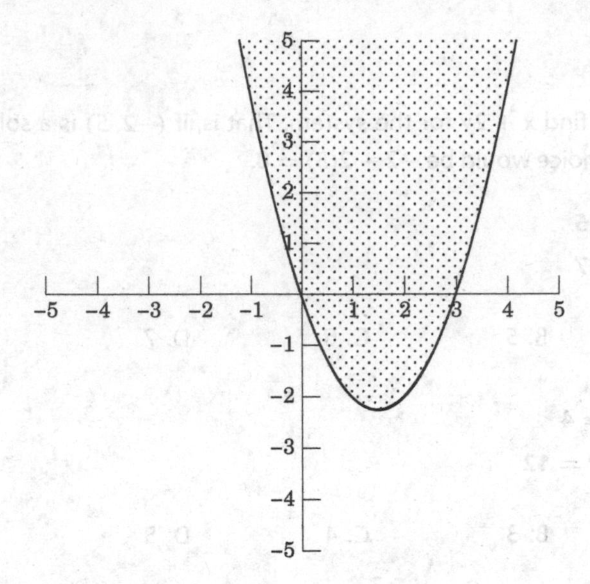

FIGURE 11-50

7. Determine whether the system has no solution or if the equations in the system represent the same line. $\begin{cases} -\frac{1}{2}x + y = 1 \\ x - 2y = -2 \end{cases}$

A. The system has no solution.

B. The equations in the system represent the same line.

8. The solution to which inequality is shaded in Fig. 11-50?

A. $y \geq x^2 + 3x$ B. $y \leq x^2 + 3x$

C. $y \leq x^2 - 3x$ D. $y \geq x^2 - 3x$

For Problems 9–12, match one of the graphs in Figs. 11-51–11-54 to its system of inequalities.

9. $\begin{cases} 2x - 3y > 6 \\ x + y < -1 \end{cases}$

A. Fig. 11-51 B. Fig. 11-52

C. Fig. 11-53 D. Fig. 11-54

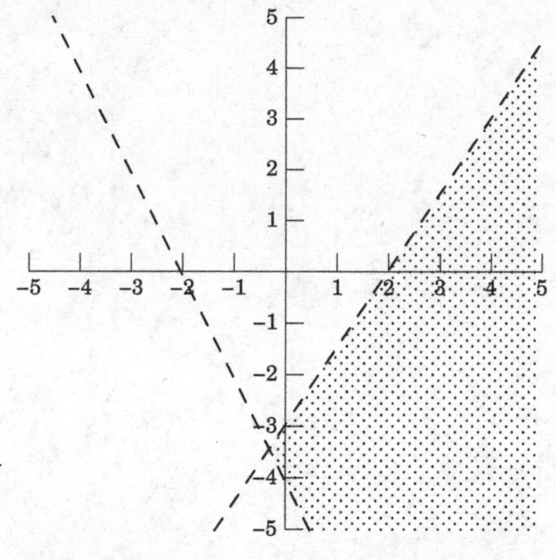

FIGURE 11-51

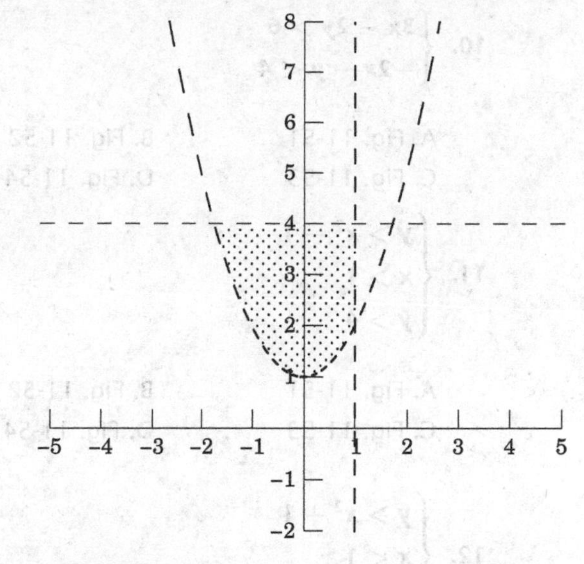

FIGURE 11-52

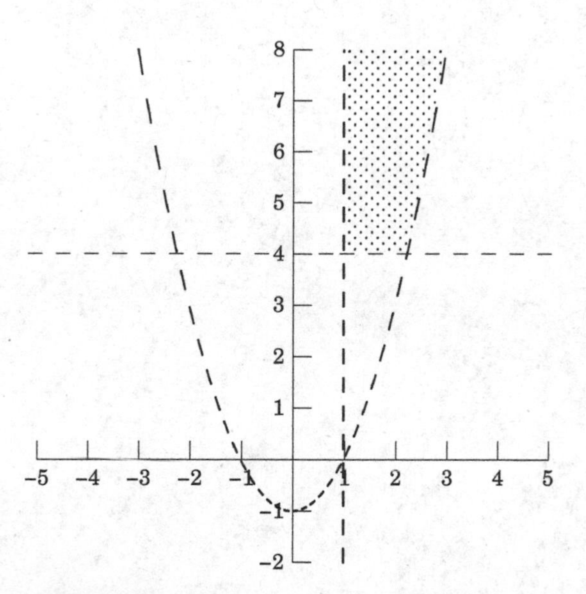

FIGURE 11-53

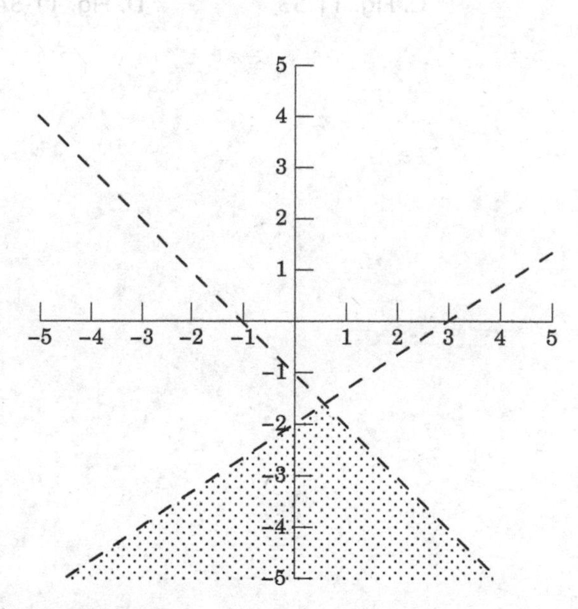

FIGURE 11-54

10. $\begin{cases} 3x - 2y > 6 \\ -2x - y < 4 \end{cases}$

A. Fig. 11-51 B. Fig. 11-52
C. Fig. 11-53 D. Fig. 11-54

11. $\begin{cases} y > x^2 - 1 \\ x > 1 \\ y > 4 \end{cases}$

A. Fig. 11-51 B. Fig. 11-52
C. Fig. 11-53 D. Fig. 11-54

12. $\begin{cases} y > x^2 + 1 \\ x < 1 \\ y < 4 \end{cases}$

A. Fig. 11-51 B. Fig. 11-52
C. Fig. 11-53 D. Fig. 11-54

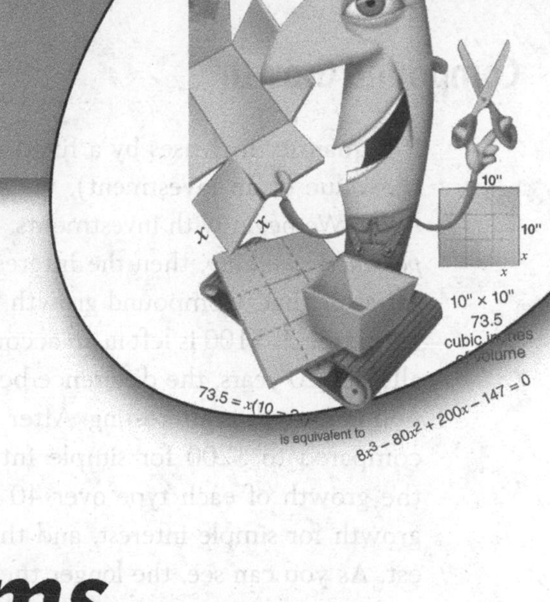

chapter 12

Exponents and Logarithms

Exponents and logarithms are used in many different fields: physics, chemistry, mathematics, and finance, to name a few. In addition to working with some of these applications, we will learn about properties of logarithms and how to solve equations involving exponents or logarithms and we will work with exponent and logarithm functions.

CHAPTER OBJECTIVES

In this chapter, you will

- Calculate the value of an investment earning compound interest
- Use the number "*e*" to represent quantities changing exponentially
- Rewrite exponent equations in logarithm form and vice versa
- Work with properties of logarithms
- Solve equations involving exponents or logarithms
- Sketch the graph of exponent functions and logarithm functions
- Find the domain of a logarithm function

Compound Growth

If a quantity increases by a fixed percent over time (such as a population or the value of an investment), we say that the quantity is increasing *exponentially*. We begin with investments. If an investment grows at 5% per year, *compounded annually*, then the interest earned in the first year itself earns interest after that. Compound growth is not dramatic over the short run but it is over time. If \$100 is left in an account earning 5% interest, compounded annually, for 20 years, the difference between the compound growth and noncompound growth is interesting. After 20 years, the compound amount is \$265.33 compared to \$200 for simple interest (noncompound growth). A graph of the growth of each type over 40 years is given in Fig. 12-1. The line is the growth for simple interest, and the curve is the growth for compound interest. As you can see, the longer the period, the greater the difference between an investment earning compound interest and an investment earning simple interest.

We begin with investments that earn interest that is compounded annually. We use the formula $A = P(1+r)^t$ to calculate the value of an investement earning compound interest. In this formula, A represents the *compound amount* (i.e., the value of the investment), P, the amount invested, r, the annual interest rate, and t, the number of years.

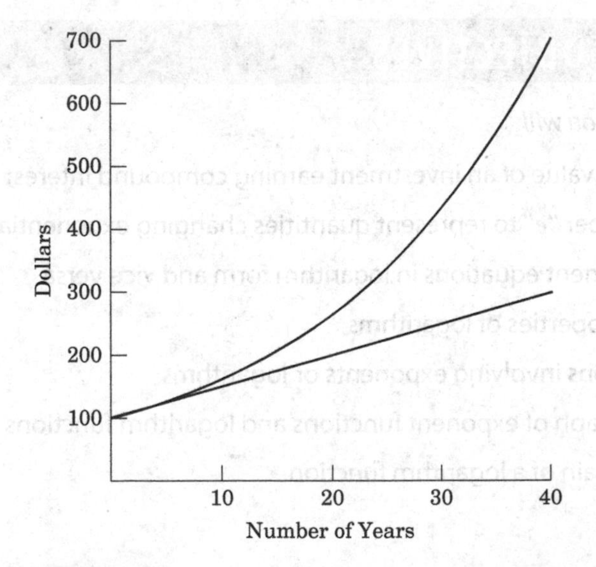

FIGURE 12-1

EXAMPLE 12-1

Find the compound amount.

- **$5000, after 3 years, earning 6% interest, compounded annually**

For the formula $A = P(1 + r)^t$, we have $P = 5000$, $r = 0.06$, and $t = 3$. We want to know A, the compound amount.

$$A = 5000(1 + 0.06)^3 = 5000(1.06)^3 = 5000(1.191016) \approx 5955.08$$

The compound amount is $5955.08.

- **$10,000 after 8 years, $7\frac{1}{4}$ % interest, compounded annually**

$$A = 10{,}000(1 + 0.0725)^8 = 10{,}000(1.0725)^8 \approx 10{,}000(1.7505656)$$
$$\approx 17{,}505.66$$

The compound amount is $17,505.66

Many investments pay more often than once a year, some paying interest daily. Instead of using the annual interest rate, we use the interest rate per period, and instead of using the number of years, we use the number of periods. If there are n compounding periods per year, then the interest rate per period is $\frac{r}{n}$ and the total number of periods is nt. The compound amount formula becomes

$$A = P\left(1 + \frac{r}{n}\right)^{nt}$$

EXAMPLE 12-2

Find the compound amount.

- **$5000, after 3 years, earning 6% annual interest**

  (a) Compounded semiannually

  (b) Compounded monthly

(a) **Interest that is compounded semiannually means that it is compounded twice each year, so $n = 2$.**

$$A = 5000\left(1+\frac{0.06}{2}\right)^{2(3)} = 5000(1.03)^6 \approx 5000(1.194052) \approx 5970.26$$

The compound amount is $5970.26.

(b) **Interest that is compounded semiannually means that it is compounded 12 times each year, so $n = 12$.**

$$A = 5000\left(1+\frac{0.06}{12}\right)^{12(3)} = 5000(1.005)^{36} \approx 5000(1.19668) \approx 5983.40$$

The compound amount is $5983.40.

- **$10,000, after 8 years, earning $7\frac{1}{4}$% annual interest, compounded weekly**

Interest that is paid weekly is paid 52 times each year, so $n = 52$.

$$A = 10{,}000\left(1+\frac{0.0725}{52}\right)^{52(8)} \approx 10{,}000(1.001394231)^{416}$$

$$\approx 10{,}000(1.785317) \approx 17{,}853.17$$

The compound amount is $17,853.17.

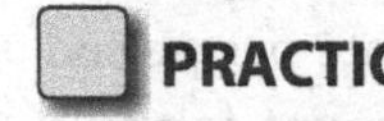

PRACTICE

Find the compound amount.

1. **$800, after 10 years, $6\frac{1}{2}$% interest, compounded annually**
2. **$1200, after 6 years, $9\frac{1}{2}$% interest, compounded annually**
3. **Christin, a 20-year-old college student, opens a retirement account with $2000. If her account pays $8\frac{1}{4}$% interest, compounded annually, how much will be in the account when she reaches age 65?**
4. **$800, after 10 years, earning $6\frac{1}{4}$% annual interest**
   (a) **Compounded quarterly**
   (b) **Compounded weekly**
5. **$9000, after 5 years, earning $6\frac{3}{4}$% annual interest, compounded daily (assume 365 days per year).**

SOLUTIONS

1. $A = 800(1 + 0.065)^{10} = 800(1.065)^{10} \approx 800(1.877137)$
≈ 1501.71
The compound amount is $1501.71.
2. $A = 1200(1 + 0.095)^{6} = 1200(1.095)^{6} \approx 1200(1.72379)$
≈ 2068.55
The compound amount is $2068.55.
3. $A = 2000(1 + 0.0825)^{45} = 2000(1.0825)^{45} \approx 2000(35.420585)$
$\approx 70{,}841.17$
Christin's account will be worth $70,841.17 when she is 65 years old.
4. (a) $n = 4$:

$$A = 800\left(1 + \frac{0.0625}{4}\right)^{4(10)} = 800(1.015625)^{40} \approx 800(1.85924)$$
$$\approx 1487.39$$

The compound amount is $1487.39.

(b) $n = 52$:

$$A = 800\left(1 + \frac{0.0625}{52}\right)^{52(10)} = 800(1.00120192)^{520}$$
$$\approx 800(1.86754) \approx 1494.04$$

The compound amount is $1494.04.

5. $n = 365$:

$$A = 9000\left(1 + \frac{0.0675}{365}\right)^{365(5)} \approx 9000(1.000184932)^{1825}$$
$$\approx 9000(1.4013959) \approx 12{,}612.56$$

The compound amount is $12,612.56.

The Number *e*

From Problem 4 in the previous set of Practice problems, you might have noticed that the same investment, for the same interest rate over the same time, earned *more* interest when it was compounded weekly instead of quarterly. This illustrates the fact that the more often the interest is compounded per year, the more interest is earned. An investment of \$1000 earning 8% annual interest, compounded annually, is worth \$1080 after 1 year. If interest is compounded quarterly, it is worth \$1082.43 after one year. And if interest is compounded daily, it is worth \$1083.28 after one year. What if interest is compounded each hour? Each second? It turns out that the most this investment could be worth (at 8% interest, at the end of one year) is \$1083.29, when interest is compounded at each and every instant of time. This is called *continuous* compounding. The formula for the compound amount for interest compounded continuously is $A = Pe^{rt}$, where A, P, r, and t are the same quantities as before. The letter e stands for a constant called Euler's number. It is approximately 2.718281828. You probably have an e or e^x key on your calculator. Although e is irrational, it can be approximated by rational numbers of the form

$$\left(1 + \frac{1}{m}\right)^m,$$

where m is a large integer. The larger m is, the better the approximation for e. If we make the substitution $m = \frac{n}{r}$ and use some algebra, we could see how $(1 + \frac{r}{n})^{nt}$ is very close to e^{rt}, for large values of n. If interest is compounded every minute, n would be about 525,600, a rather large number!

EXAMPLE 12-3

- **Find the compound amount of \$5000 after 8 years, earning 12% annual interest, compounded continuously.**

$$\mathbf{A = 5000e^{0.12(8)} = 5000e^{0.96} \approx 5000(2.611696) \approx 13{,}058.48}$$

The compound amount is \$13,058.48.

PRACTICE

Find the compound amount.

1. **\$800, after 10 years, earning $6\frac{1}{2}\%$ annual interest, compounded continuously**
2. **\$9000, after 5 years, earning $6\frac{3}{4}\%$ annual interest, compounded continuously**
3. **Christin, a 20-year-old college student, opens a retirement account with \$2000. If she earns $8\frac{1}{4}\%$ annual interest, compounded continuously, how much will the account be worth by the time she is 65? (Compare your answer to Problem 3 in the previous set of Practice problems when Christin's investment was only compounded annually.)**

SOLUTIONS

1. $A = 800e^{0.065(10)} = 800e^{0.65} \approx 800(1.915540829) \approx 1532.43$

   The compound amount is \$1532.43.

2. $A = 9000e^{0.0675(5)} = 9000e^{0.3375} \approx 9000(1.401439608)$

   $\approx 12{,}612.96$

   The compound amount is \$12,612.96.

3. $A = 2000e^{0.0825(45)} = 2000e^{3.7125} \approx 2000(40.95606882)$

   $\approx 81{,}912.14$

   Christin's account will be worth \$81,912.14 by the time she is 65.

Increasing Population

The compound growth formula for continuous compounded interest is used for other growth and decay problems. The general exponential growth model is $n(t) = n_0e^{rt}$, where $n(t)$ replaces A and n_0 replaces P. Their meanings are the same—$n(t)$ is still the compound growth, and n_0 is still the beginning amount. The variable t represents time in this formula; although, time is not always measured in years. The growth rate and t must have the same unit of measure. If the growth rate is in days, then t must be in days. If the growth rate is in hours, then t must be in hours, and so on. If the "population" is getting smaller, then the formula is $n(t) = n_0 e^{-rt}$. We now use these formulas when the rate of change of a quantity is given as a percent.

EXAMPLE 12-4

- **The population of a city is estimated to be growing at the rate of 10% per year. In 2010, its population was 160,000. Estimate its population in the year 2015.**

The year 2010 corresponds to $t = 0$, so the year 2015 corresponds to $t = 5$; n_0, the population in year $t = 0$, is 160,000. The population is growing at the rate of 10% per year, so $r = 0.10$. The formula $n(t) = n_0 e^{rt}$ becomes $n(t) = 160{,}000e^{0.10t}$. We want to find $n(t)$ for $t = 5$.

$$n(5) = 160{,}000e^{0.10(5)} \approx 263{,}795$$

The city's population is expected to be 264,000 in the year 2015 (estimates and projections are normally rounded off).

- **In an experiment, a culture of bacteria grew at the rate of 35% per hour. If 1000 bacteria were present at 10:00, how many were present at 10:45?**

$n_0 = 1000$, $r = 0.35$, t is the number of hours after 10:00.
The growth model becomes $n(t) = 1000e^{0.35t}$. We want to find $n(t)$ for 45 minutes or $t = 0.75$ hours.

$$n(0.75) = 1000e^{0.35(0.75)} = 1000e^{0.2625} \approx 1300$$

At 10:45, there were approximately 1300 bacteria present in the culture.

PRACTICE

1. **The population of a city in the year 2012 is 2,000,000 and is expected to grow 1.5% per year. Estimate the city's population for the year 2022.**
2. **A school is built for a capacity of 1500 students. The student population is projected to grow 6% per year. If 1000 students attend when it opens, will the school be at capacity in 7 years?**

3. **A construction company estimates that a piece of equipment is worth \$150,000 when new. If it loses value continuously at the annual rate of 10%, what will its value be in 10 years?**
4. **Under certain conditions, bacteria in a culture grow at the rate of about 200% per hour. If 8000 bacteria are present in a dish, how many are in the dish after 30 minutes?**

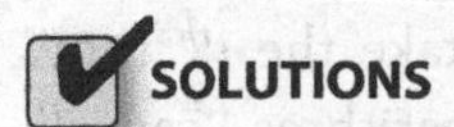

SOLUTIONS

1. $n_0 = 2{,}000{,}000, r = 0.015$
   The growth formula is $n(t) = 2{,}000{,}000e^{0.015t}$. We want to find $n(t)$ when $t = 10$.

   $$n(10) = 2{,}000{,}000e^{0.015(10)} \approx 2{,}323{,}668$$

   The population in the year 2022 is expected to be about 2.3 million.
2. $n_0 = 1000, r = 0.06$
   The growth formula is $n(t) = 1000e^{0.06t}$. We want to find $n(t)$ when $t = 7$.

   $$n(7) = 1000e^{0.06(7)} \approx 1522$$

   Yes, the school is projected to be at capacity in 7 years.
3. $n_0 = 150{,}000, r = 0.10$
   We use the decay formula because value is being lost. The formula is $n(t) = 150{,}000e^{-0.10t}$. We want to find $n(t)$ when $t = 10$.

   $$n(10) = 150{,}000e^{-0.10(10)} \approx 55{,}181.92$$

   The equipment will be worth about \$55,000 after 10 years.
4. $n_0 = 8000, r = 2$
   The growth formula is $n(t) = 8000e^{2t}$. We want to find $n(t)$ when $t = 0.5$.

   $$n(0.5) = 8000e^{2(0.5)} \approx 21{,}746$$

   About 21,700 bacteria are present after 30 minutes.

Logarithms

A common question for investors is, "How long will it take for my investment to double?" If $1000 is invested so that it earns 8% interest, compounded annually, how long will it take to grow to $2000? To answer the question using the compound growth formula, we need to solve for t in the equation $2000 = 1000(1.08)^t$. When we divide both sides of the equation by 1000 we have $2 = (1.08)^t$. Now what? It does not make sense to "take the tth root" of both sides of the equation. We must use logarithms. Logarithms "cancel" exponentiation in the same way subtraction "cancels" addition and division "cancels" multiplication. Logarithms (or *logs*) are very useful in solving many science and business problems.

The logarithm equation $\log_a x = y$ is another way of writing the exponent equation $a^y = x$. Verbally, we say, "log base a of x is (or equals) y." For "$\log_a x$," we say, "(the) log base a of x." We begin by rewriting exponent equations in logarithm form and logarithm equations in exponent form.

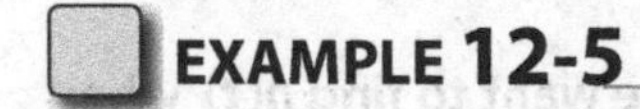

Rewrite the logarithm equation as an exponent equation.

- $\log_3 9 = 2$

The base of the logarithm is the base of the exponent, so 3 is raised to a power. The number that is equal to the log is the power, so the power on 3 is 2.

$$\log_3 9 = 2 \text{ rewritten as an exponent is } 3^2 = 9$$

- $\log_2 \frac{1}{8} = -3$

The base is 2 and the power is −3,

$$2^{-3} = \frac{1}{8}$$

- $\log_9 3 = \frac{1}{2}$

The base is 9, and the power is $\frac{1}{2}$.

$$9^{\frac{1}{2}} = 3$$

We also must work in the other direction, rewriting exponent equations as logarithm equations. Remember, the exponent equals the logarithm.

EXAMPLE 12-6

- **The equation $4^3 = 64$ written as a logarithm equation is $\log_4 64 = 3$.**
- **$5^2 = 25$ written in logarithm form is $\log_5 25 = 2$.**
- **$4^0 = 1$ written as a logarithm equation is $\log_4 1 = 0$.**

PRACTICE

Rewrite the logarithm equation in Problems 1-8 in exponent form.

1. $\log_4 16 = 2$
2. $\log_3 81 = 4$
3. $\log_{100} 10 = \frac{1}{2}$
4. $\log_a 4 = 3$
5. $\log_e 2 = 0.6931$
6. $\log_{(x+1)} 9 = 2$
7. $\log_7 \frac{1}{49} = -2$
8. $\log_8 4 = \frac{2}{3}$

Rewrite the exponent equation in Problems 9-13 in logarithm form.

9. $7^{-1} = \frac{1}{7}$
10. $125^{1/3} = 5$
11. $10^{-4} = 0.0001$
12. $e^{1/2} = 1.6487$
13. $8^x = 5$

SOLUTIONS

1. $4^2 = 16$
2. $3^4 = 81$
3. $100^{\frac{1}{2}} = 10$
4. $a^3 = 4$
5. $e^{0.6931} = 2$
6. $(x+1)^2 = 9$
7. $7^{-2} = \frac{1}{49}$
8. $8^{2/3} = 4$
9. $\log_7 \frac{1}{7} = -1$
10. $\log_{125} 5 = \frac{1}{3}$
11. $\log_{10} 0.0001 = -4$
12. $\log_e 1.6487 = \frac{1}{2}$
13. $\log_8 5 = x$

Properties of Logarithms

The first two logarithm properties we learn are the cancelation properties. They come directly from rewriting one form of an equation into the other form.

$$\log_a a^x = x \text{ and } a^{\log_a x} = x$$

When the bases of the exponent and logarithm are the same, they cancel. Let us see why these properties are true. What would the expression "$\log_a a^x$" be? We rewrite the equation "$\log_a a^x =?$" in exponent form: "$a^? = a^x$." Now we see that "?" is x. This is why "$\log_a a^x = x$." What would the expression "$a^{\log_a x}$" be? Rewriting "$a^{\log_a x} =?$" in logarithm form gives us "$\log_a ? = \log_a x$." From this equation, we see that "?" is x, so $a^{\log_a x} = x$.

EXAMPLE 12-7

- $5^{\log_5 2}$

The bases of the logarithm and exponent are both 5, so $5^{\log_5 2}$ simplifies to 2.

- $10^{\log_{10} 8} = 8$
- $4^{\log_4 x} = x$
- $e^{\log_e 6} = 6$
- $29^{\log_{29} 1} = 1$
- $\log_m m^r = r$
- $\log_7 7^{ab} = ab$
- $\log_6 6^5 = 5$
- $\log_{16} 16^{-4} = -4$
- $\log_{10} 10^x = x$

PRACTICE

Use logarithm properties to simplify the expression.

1. $9^{\log_9 3}$
2. $10^{\log_{10} 14}$
3. $5^{\log_5 x}$
4. $\log_{15} 15^2$
5. $\log_{10} 10^{-8}$
6. $\log_e e^x$

SOLUTIONS

1. $9^{\log_9 3} = 3$
2. $10^{\log_{10} 14} = 14$
3. $5^{\log_5 x} = x$
4. $\log_{15} 15^2 = 2$
5. $\log_{10} 10^{-8} = -8$
6. $\log_e e^x = x$

Sometimes we must use exponent properties before using the property $\log_a a^x = x$ to simplify the expression.

$$\sqrt[n]{a^m} = a^{\frac{m}{n}} \quad \text{and} \quad \frac{1}{a^m} = a^{-m}$$

For example, $16^{1/2} = \sqrt{16}$ and $\frac{1}{5} = 5^{-1}$. Our goal in the next set of problems is to write $\log_a m$ so that m is a power of a. This allows us to use a cancelation property.

EXAMPLE 12-8

- $\log_9 3 = \log_9 \sqrt{9} = \log_9 9^{1/2} = \frac{1}{2}$
- $\log_7 \frac{1}{49} = \log_7 \frac{1}{7^2} = \log_7 7^{-2} = -2$
- $\log_{10} \sqrt[4]{10} = \log_{10} 10^{1/4} = \frac{1}{4}$
- $\log_{10} \sqrt[5]{100} = \log_{10} \sqrt[5]{10^2} = \log_{10} 10^{2/5} = \frac{2}{5}$

Two types of logarithms occur frequently enough to have their own notation. They are $\log_e$ and $\log_{10}$. The notation for $\log_e$ is "ln," pronounced "ell-in," and is called the *natural logarithm*. The notation for $\log_{10}$ is "log" (no base is written) and is called the *common logarithm*. The cancelation properties for these special logarithms are

$$\ln e^x = x, \quad e^{\ln x} = x \quad \text{and} \quad \log 10^x = x, \quad 10^{\log x} = x$$

EXAMPLE 12-9

- $\ln e^{15} = 15$
- $10^{\log 5} = 5$
- $e^{\ln 14} = 14$
- $\log 10^{1/2} = \frac{1}{2}$
- $\ln e^{-4} = -4$
- $\log 10^{-4} = -4$

PRACTICE

Use the cancelation properties to evaluate the logarithm.

1. $\log_7 \sqrt{7}$
2. $\log_5 \frac{1}{5}$
3. $\log_3 \frac{1}{\sqrt{3}}$
4. $\log_4 \frac{1}{16}$
5. $\log_{25} \frac{1}{5}$
6. $\log_8 \frac{1}{2}$
7. $\log_{10} \sqrt{1000}$
8. $\ln e^5$
9. $\log 10^{\sqrt{x}}$
10. $10^{\log 9}$
11. $e^{\ln 6}$
12. $\log 10^{3x-1}$
13. $\ln e^{x+1}$

SOLUTIONS

1. $\log_7 \sqrt{7} = \log_7 7^{1/2} = \frac{1}{2}$
2. $\log_5 \frac{1}{5} = \log_5 5^{-1} = -1$
3. $\log_3 \frac{1}{\sqrt{3}} = \log_3 \frac{1}{3^{1/2}} = \log_3 3^{-1/2} = -\frac{1}{2}$
4. $\log_4 \frac{1}{16} = \log_4 \frac{1}{4^2} = \log_4 4^{-2} = -2$
5. $\log_{25} \frac{1}{5} = \log_{25} \frac{1}{\sqrt{25}} = \log_{25} \frac{1}{25^{1/2}} = \log_{25} 25^{-1/2} = -\frac{1}{2}$
6. $2 = \sqrt[3]{8}$

$$\log_8 \frac{1}{2} = \log_8 \frac{1}{\sqrt[3]{8}} = \log_8 \frac{1}{8^{1/3}} = \log_8 8^{-1/3} = -\frac{1}{3}$$

7. $1000 = 10^3$

$$\log_{10} \sqrt{1000} = \log_{10} \sqrt{10^3} = \log_{10} 10^{3/2} = \frac{3}{2}$$

8. $\ln e^5 = 5$
9. $\log 10^{\sqrt{x}} = \sqrt{x}$
10. $10^{\log 9} = 9$
11. $e^{\ln 6} = 6$
12. $\log 10^{3x-1} = 3x - 1$
13. $\ln e^{x+1} = x + 1$

Three More Important Logarithm Properties

The following three logarithm properties come directly from the exponent properties $a^m \cdot a^n = a^{m+n}$, $\frac{a^m}{a^n} = a^{m-n}$, and $a^{mn} = (a^m)^n$.

1. $\log_b mn = \log_b m + \log_b n$ We call this "the first log property."
2. $\log_b \frac{m}{n} = \log_b m - \log_b n$ We call this "the second log property."
3. $\log_b m^t = t \log_b m$ We call this "the third log property."

Let us see why the first log property works. Let $x = \log_b m$ and $y = \log_b n$. Rewriting these equations as exponent equations gives us $b^x = m$ and $b^y = n$. Multiplying m and n, we have $mn = b^x \cdot b^y = b^{x+y}$. Rewriting the equation $mn = b^{x+y}$ as a logarithm equation, we have $\log_b mn = x + y$. Because $x = \log_b m$ and $y = \log_b n$, $\log_b mn = x + y$ becomes $\log_b mn = \log_b m + \log_b n$.

EXAMPLE 12-10

Use the first log property to rewrite the logarithms.

- $\log_4 7x = \log_4 7 + \log_4 x$
- $\ln 15t = \ln 15 + \ln t$
- $\log_6 19t^2 = \log_6 19 + \log_6 t^2$
- $\log 100y^4 = \log 10^2 + \log y^4$
- $\log_9 3 + \log_9 27 = \log_9 3(27) = \log_9 81 = 2$
- $\ln x + \ln \sqrt{y} = \ln x\sqrt{y}$

EXAMPLE 12-11

Use the second log property to rewrite the logarithms.

- $\log \left(\frac{x}{4}\right) = \log x - \log 4$
- $\ln \left(\frac{5}{x}\right) = \ln 5 - \ln x$
- $\log_{15} 3 - \log_{15} 2 = \log_{15} \left(\frac{3}{2}\right)$
- $\ln 16 - \ln t = \ln \frac{16}{t}$
- $\log_4 \left(\frac{4}{3}\right) = \log_4 4 - \log_4 3 = 1 - \log 3$
- $\log_6 \frac{x^2}{y} = \log_6 x^2 - \log_6 y$

Still Struggling

Be carefull when applying these rules: $\log_a(x+y)$ and $\log_a(x-y)$ cannot be simplified in general.

The exponent property $\sqrt[n]{a^m}$ allows us to apply the third logarithm property to roots as well as to powers. The third logarithm property is especially useful in science and business applications. We will use the third logarithm property later when solving equations involving exponents.

EXAMPLE 12-12

Use the third log property to rewrite the logarithms.

- $\log_4 3^x = x \log_4 3$
- $\log x^2 = 2 \log x$
- $\frac{1}{3} \ln t = \ln t^{1/3}$
- $-3 \log 8 = \log 8^{-3}$
- $\log_6 \sqrt{2x} = \log_6 (2x)^{1/2} = \frac{1}{2} \log_6 2x$
- $\ln \sqrt[4]{t^3} = \ln t^{3/4} = \frac{3}{4} \ln t$

PRACTICE

Use logarithm properties to rewrite the logarithm.

1. $\ln 59t$
2. $\log 0.10y$
3. $\log_{30} 148x$
4. $\log_6 3 + \log_6 12$
5. $\log_5 9 + \log_5 10$
6. $\log_3 5 + \log_3 20$
7. $\log_4 \frac{10}{9x}$
8. $\log_2 \frac{7}{9}$
9. $\ln \frac{t}{4}$
10. $\log \frac{20}{x}$
11. $\log_7 2 - \log_7 4$
12. $\log_8 x - \log_8 3$
13. $\ln 5^x$
14. $\log_{12} \sqrt{3}$
15. $\log \sqrt{16x}$
16. $\log_5 6^{-t}$
17. $2 \log_8 3$
18. $(x+6) \log_4 3$
19. $\log_{16} 10^{2x}$
20. $-2 \log_4 5$

SOLUTIONS

1. $\ln 59t = \ln 59 + \ln t$
2. $\log 0.10y = \log 0.10 + \log y$
3. $\log_{30} 148x = \log_{30} 148 + \log_{30} x$
4. $\log_6 3 + \log_6 12 = \log_6(3 \cdot 12) = \log_6 36 = \log_6 6^2 = 2$
5. $\log_5 9 + \log_5 10 = \log_5(9 \cdot 10) = \log_5 90$
6. $\log_3 5 + \log_3 20 = \log_3(5 \cdot 20) = \log_3 100$
7. $\log_4 \dfrac{10}{9x} = \log_4 10 - \log_4 9x$
8. $\log_2 \dfrac{7}{9} = \log_2 7 - \log_2 9$
9. $\ln \dfrac{t}{4} = \ln t - \ln 4$
10. $\log \dfrac{20}{x} = \log 20 - \log x$
11. $\log_7 2 - \log_7 4 = \log_7 \dfrac{2}{4} = \log_7 \dfrac{1}{2}$
12. $\log_8 x - \log_8 3 = \log_8 \dfrac{x}{3}$
13. $\ln 5^x = x \ln 5$
14. $\log_{12} \sqrt{3} = \log_{12} 3^{1/2} = \dfrac{1}{2} \log_{12} 3$
15. $\log \sqrt{16x} = \log(16x)^{1/2} = \dfrac{1}{2} \log 16x$
16. $\log_5 6^{-t} = -t \log_5 6$
17. $2 \log_8 3 = \log_8 3^2 = \log_8 9$
18. $(x+6) \log_4 3 = \log_4 3^{x+6}$
19. $\log_{16} 10^{2x} = 2x \log_{16} 10$
20. $-2 \log_4 5 = \log_4 5^{-2} = \log_4 \dfrac{1}{5^2} = \log_4 \dfrac{1}{25}$

Using Multiple Logarithm Properties

Sometimes we must use several logarithm properties to rewrite more complicated logarithms. The hardest part of this is to use the properties in the correct order. For example, which property should be used first on $\log \frac{x}{y^3}$? Do we first use the third property or the second property? We use the second property first. For the expression $\log(\frac{x}{y})^3$, we would use the third property first.

Going in the other direction, we use all three properties in the expression $\log_2 9 - \log_2 x + 3 \log_2 y$. We use the second property to combine the first two terms.

$$\log_2 9 - \log_2 x + 3 \log_2 y = \log_2 \frac{9}{x} + 3 \log_2 y$$

We cannot use the first property on $\log_2 \frac{9}{x} + 3\log_2 y$ until we have used the third property to move 3.

Recall: $a\left(\frac{b}{c}\right) = \frac{ab}{c}$

$$\log_2 \frac{9}{x} + 3\log_2 y = \log_2 \frac{9}{x} + \log_2 y^3 = \log_2 y^3 \frac{9}{x} = \log_2 \frac{9y^3}{x}$$

EXAMPLE 12-13

Rewrite as a single logarithm.

- $\log_2 3x - 4\log_2 y$

We use the third property to move 4 and then we can use the second property.

$$\log_2 3x - 4\log_2 y = \log_2 3x - \log_2 y^4 = \log_2 \frac{3x}{y^4}$$

- $3\log 4x + 2\log 3 - 2\log y$

$$\begin{aligned} 3\log 4x + 2\log 3 - 2\log y &= \log(4x)^3 + \log 3^2 - \log y^2 && \text{Third property} \\ &= \log 4^3x^3 \cdot 3^2 - \log y^2 && \text{First property} \\ &= \log 576x^3 - \log y^2 \\ &= \log \frac{576x^3}{y^2} && \text{Second property} \end{aligned}$$

- $t\ln 4 + \ln 5$

$$t\ln 4 + \ln 5 = \ln 4^t + \ln 5 = \ln(5 \cdot 4^t) \quad \text{(not } \ln 20^t\text{)}$$

Expand each logarithm.

- $\ln \frac{3\sqrt{x}}{y^2}$

$$\ln \frac{3\sqrt{x}}{y^2} \overset{\text{Second property}}{=} \ln 3(x^{1/2}) - \ln y^2 \overset{\text{First property}}{=} \ln 3 + \ln x^{1/2} - \ln y^2 \overset{\text{Third property}}{=} \ln 3 + \frac{1}{2}\ln x - 2\ln y$$

- $\log_7 \frac{4}{10xy^2}$

$$\log_7 \frac{4}{10xy^2} = \log_7 4 - \log_7 10xy^2 = \log_7 4 - (\log_7 10 + \log_7 x + \log_7 y^2)$$

$$= \log_7 4 - (\log_7 10 + \log_7 x + 2\log_7 y)$$

$$\text{or } \log_7 4 - \log_7 10 - \log_7 x - 2\log_7 y$$

PRACTICE

For Problems 1-5, rewrite each as a single logarithm.

1. $2\log x + 3\log y$
2. $\log_6 2x - 2\log_6 3$
3. $3\ln t - \ln 4 + 2\ln 5$
4. $t\ln 6 + 2\ln 5$
5. $\frac{1}{2}\log x - 2\log 2y + 3\log z$

For Problems 6-10, expand each logarithm.

6. $\log \frac{4x}{y}$
7. $\ln \frac{6}{\sqrt{y}}$
8. $\log_4 \frac{10x}{\sqrt[3]{z}}$
9. $\ln \frac{\sqrt{4x}}{5y^2}$
10. $\log \sqrt{\frac{2y^3}{x}}$

SOLUTIONS

1. $2\log x + 3\log y = \log x^2 + \log y^3 = \log x^2y^3$
2. $\log_6 2x - 2\log_6 3 = \log_6 2x - \log_6 3^2 = \log_6 2x - \log_6 9 = \log_6 \frac{2x}{9}$
3. $3\ln t - \ln 4 + 2\ln 5 = \ln t^3 - \ln 4 + \ln 5^2 = \ln \frac{t^3}{4} + \ln 25$

   $$= \ln 25\frac{t^3}{4} = \ln \frac{25t^3}{4}$$
4. $t\ln 6 + 2\ln 5 = \ln 6^t + \ln 5^2 = \ln[25(6^t)]$

5. $$\frac{1}{2}\log x - 2\log 2y + 3\log z = \log x^{1/2} - \log(2y)^2 + \log z^3$$
$$= \log x^{1/2} - \log 2^2y^2 + \log z^3$$
$$= \log x^{1/2} - \log 4y^2 + \log z^3$$
$$= \log \frac{x^{1/2}}{4y^2} + \log z^3 = \log z^3\frac{x^{1/2}}{4y^2}$$
$$= \log \frac{z^3x^{1/2}}{4y^2} \text{ or } \log \frac{z^3\sqrt{x}}{4y^2}$$

6. $$\log \frac{4x}{y} = \log 4x - \log y = \log 4 + \log x - \log y$$

7. $$\ln \frac{6}{\sqrt{y}} = \ln 6 - \ln \sqrt{y} = \ln 6 - \ln y^{1/2} = \ln 6 - \frac{1}{2}\ln 6$$

8. $$\log_4 \frac{10x}{\sqrt[3]{z}} = \log_4 10x - \log_4 \sqrt[3]{z} = \log_4 10x - \log_4 z^{1/3}$$
$$= \log_4 10 + \log_4 x - \frac{1}{3}\log_4 z$$

9. $$\ln \frac{\sqrt{4x}}{5y^2} = \ln\sqrt{4x} - \ln 5y^2 = \ln(4x)^{1/2} - \ln 5y^2$$
$$= \frac{1}{2}\ln 4x - (\ln 5 + \ln y^2)$$
$$= \frac{1}{2}(\ln 4 + \ln x) - (\ln 5 + 2\ln y)$$
$$\text{or } \frac{1}{2}\ln 4 + \frac{1}{2}\ln x - \ln 5 - 2\ln y$$

10. $$\log\sqrt{\frac{2y^3}{x}} = \log\left(\frac{2y^3}{x}\right)^{1/2} = \frac{1}{2}\log\frac{2y^3}{x} = \frac{1}{2}(\log 2y^3 - \log x)$$
$$= \frac{1}{2}(\log 2 + \log y^3 - \log x)$$
$$= \frac{1}{2}(\log 2 + 3\log y - \log x)$$
$$\text{or } \frac{1}{2}\log 2 + \frac{3}{2}\log y - \frac{1}{2}\log x$$

Equations Involving Exponents and Logarithms

We now solve equations involving exponents and logarithms. If we have an equation involving an exponent, we use a logarithm to eliminate the exponent. Our goal is to rewrite the exponent equation in logarithm form, use algebra to solve for x, and then use a calculator (if necessary). We begin with equations involving base 10 and base e. Later, we will solve equations involving other bases.

EXAMPLE 12-14

Solve for x. Give solutions accurate to four decimal places.

- $e^{2x} = 3$

$e^{2x} = 3$ **Rewrite in logarithm form.**

$2x = \ln 3$ **Divide each side by 2.**

$x = \frac{\ln 3}{2} \approx \frac{1.0986}{2} \approx 0.5493$ **Use a calculator.**

- $10^{x+1} = 9$

$x + 1 = \log 9$ **Rewrite in logarithm form.**

$x = -1 + \log 9 \approx -1 + 0.9542 \approx -0.0458$

- $2500 = 1000e^{x-4}$

$2500 = 1000e^{x-4}$ **Divide both sides by 1000 before rewriting the equation.**

$e^{x-4} = 2.5$ **Reverse sides.**

$x - 4 = \ln 2.5$ **Rewrite in logarithm form.**

$x = 4 + \ln 2.5 \approx 4 + 0.9163 \approx 4.9163$

PRACTICE

Solve for x.

1. $10^{3x} = 7$
2. $e^{2x+5} = 15$
3. $5000 = 2500e^{4x}$
4. $32 = 8 \cdot 10^{6x-4}$
5. $200 = 400e^{-0.06x}$

SOLUTIONS

1.
$$10^{3x} = 7$$
$$3x = \log 7$$
$$x = \frac{\log 7}{3} \approx \frac{0.8451}{3} \approx 0.2817$$

2.
$$e^{2x+5} = 15$$
$$2x + 5 = \ln 15$$
$$2x = -5 + \ln 15$$
$$x = \frac{-5 + \ln 15}{2} \approx \frac{-5 + 2.7081}{2} \approx -1.1460$$

3.
$$5000 = 2500e^{4x}$$
$$\frac{5000}{2500} = e^{4x}$$
$$4x = \ln\left(\frac{5000}{2500}\right)$$
$$4x = \ln 2 \qquad \left(\frac{5000}{2500} = 2\right)$$
$$x = \frac{\ln 2}{4} \approx \frac{0.6931}{4} \approx 0.1733$$

4.
$$32 = 8 \cdot 10^{6x-4}$$ **Divide both sides by 8.**
$$4 = 10^{6x-4}$$
$$6x - 4 = \log 4$$
$$6x = 4 + \log 4$$
$$x = \frac{4 + \log 4}{6} \approx \frac{4 + 0.6021}{6} \approx 0.767$$

5.

$$200 = 400e^{-0.06x}$$

$$\frac{1}{2} = e^{-0.06x}$$

$$-0.06x = \ln\left(\frac{1}{2}\right)$$

$$x = \frac{\ln\left(\frac{1}{2}\right)}{-0.06} \approx \frac{-0.69315}{-0.06} \approx 11.5525$$

Most exponent equations involve a base other than e or 10, so we must adjust our strategy for solving them. After we have isolated the exponent term (this is usually already done), we take either the common logarithm or the natural logarithm of each side. We then use the third logarithm property to "bring down" the exponent. At this point, we use ordinary algebra to solve for x and a calculator to approximate the solution.

EXAMPLE 12-15

Solve for *x*.

- $3^{4x} = 16$

$$3^{4x} = 16$$

$$\ln 3^{4x} = \ln 16 \qquad \text{Take the natural logarithm of each side.}$$

$$(4x)\ln 3 = \ln 16 \qquad \text{Use the third logarithm property.}$$

$$4x = \frac{\ln 16}{\ln 3} \qquad \text{Divide each side by } \ln 3.$$

$$x = \frac{1}{4}\left(\frac{\ln 16}{\ln 3}\right) \approx 0.6309$$

- $7^{2x-3} = 50$

$$7^{2x-3} = 50$$

$$\ln 7^{2x-3} = \ln 50$$ Take the natural logarithm of each side.

$$(2x - 3)\ln 7 = \ln 50$$ Use the third logarithm property.

$$2x - 3 = \frac{\ln 50}{\ln 7}$$ Divide each side by ln 7.

$$2x = 3 + \frac{\ln 50}{\ln 7}$$ Add 3 to each side.

$$x = \frac{1}{2}\left(3 + \frac{\ln 50}{\ln 7}\right) \approx 2.5052$$ Divide each side by 2.

Still Struggling

If the exponent has more than one term, be sure to put parentheses around the entire power when using the third logarithm property.

PRACTICE

Solve for x.

1. $2^{3x+1} = 60$
2. $5^{1-x} = 21$
3. $6^{2x+1} = 32$

SOLUTIONS

1.

$$2^{3x+1} = 60$$

$$\ln 2^{3x+1} = \ln 60$$

$$(3x + 1)\ln 2 = \ln 60$$

$$3x + 1 = \frac{\ln 60}{\ln 2}$$

$$3x = -1 + \frac{\ln 60}{\ln 2}$$

$$x = \frac{1}{3}\left(-1 + \frac{\ln 60}{\ln 2}\right) \approx 1.6356$$

2.

$$5^{1-x} = 21$$

$$\ln 5^{1-x} = \ln 21$$

$$(1-x)\ln 5 = \ln 21$$

$$1 - x = \frac{\ln 21}{\ln 5}$$

$$-x = -1 + \frac{\ln 21}{\ln 5}$$

$$x = 1 - \frac{\ln 21}{\ln 5} \approx -0.8917$$

3.

$$6^{2x+1} = 32$$

$$\ln 6^{2x+1} = \ln 32$$

$$(2x+1)\ln 6 = \ln 32$$

$$2x + 1 = \frac{\ln 32}{\ln 6}$$

$$2x = -1 + \frac{\ln 32}{\ln 6}$$

$$x = \frac{1}{2}\left(-1 + \frac{\ln 32}{\ln 6}\right) \approx 0.4671$$

Now that we can solve equations involving exponents, we can answer the question at the beginning of this section: How long will it take for \$1000 to grow to \$2000 if it earns 8% annual interest, compounded annually? In the formula $A = P(1+r)^t$, we know $A = 2000$, $P = 1000$, and $r = 0.08$ but we do not know t. Here is where we use logarithms: to solve the exponent equation $2000 = 1000(1.08)^t$ for t.

$2000 = 1000(1.08)^t$	
$2 = 1.08^t$	Divide each side by 1000.
$\ln 2 = \ln 1.08^t$	Take the natural logarithm of each side.
$\ln 2 = t \ln 1.08$	Use the third logarithm property.
$\frac{\ln 2}{\ln 1.08} = t$	Divide each side by ln 1.08.
$9 \approx t$	In approximately 9 years, the value of the investment will double.

Equations involving logarithms come in several forms, and sometimes more than one strategy works to solve them. Here, we solve equations of the form "log = number" and "log = log." For an equation of the form "log = number," we rewrite the equation as an exponent equation, which eliminates the logarithm.

EXAMPLE 12-16

Solve for *x*.

- $\log_3(x + 1) = 4$

Rewrite the equation as an exponent equation.

$$3^4 = x + 1$$

$$81 = x + 1$$

$$80 = x$$

- $\log_2(3x - 4) = 5$

$$2^5 = 3x - 4$$

$$32 = 3x - 4$$

$$12 = x$$

PRACTICE

Solve for x.

1. $\log_7(2x+1)=2$
2. $\log_4(x+6)=2$
3. $\log 5x=1$
4. $\log_2(8x-1)=4$

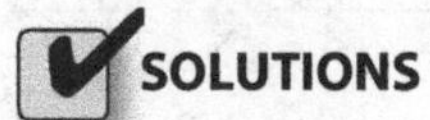

SOLUTIONS

1. $\log_7(2x+1)=2$ becomes $7^2=2x+1$, so $x=24$
2. $\log_4(x+6)=2$ becomes $4^2=x+6$, so $x=10$
3. $\log 5x=1$ becomes $10^1=5x$, so $x=2$
4. $\log_2(8x-1)=4$ becomes $2^4=8x-1$, so $x=\frac{17}{8}$

The logarithms "cancel" for equations in the form "log = log" as long as the bases are the same. For example, the solution to the equation $\log_8 x=\log_8 10$ is $x=10$. The cancelation law $a^{\log_a x}=x$ makes this work.

$\log_8 x=\log_8 10$	
$8^{\log_8 x}=8^{\log_8 10}$	Two equal numbers are equal as powers, too.
$x=10$	By the cancelation property

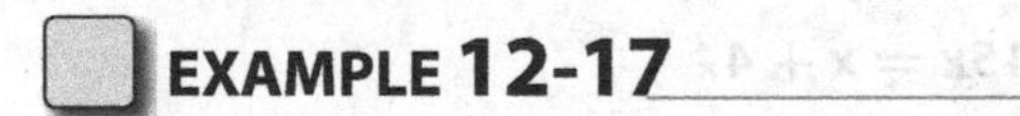

EXAMPLE 12-17

Solve for x.

- $\log_6(x+1)=\log_6 2x$

$\log_6(x+1)=\log_6 2x$	
$x+1=2x$	The logs cancel.
$1=x$	

- $\log 4 = \log(x - 1)$

$$\log 4 = \log(x-1)$$

$$4 = x - 1 \quad \text{The logs cancel.}$$

$$5 = x$$

PRACTICE

Solve for x.

1. $\log_3(4x-1) = \log_3 2$
2. $\log_2(3-x) = \log_2 17$
3. $\ln 15x = \ln(x+4)$
4. $\log \dfrac{x}{x-1} = \log \dfrac{1}{2}$

SOLUTIONS

1.
$$\log_3(4x-1) = \log_3 2$$
$$4x - 1 = 2$$
$$x = \frac{3}{4}$$

2.
$$\log_2(3-x) = \log_2 17$$
$$3 - x = 17$$
$$x = -14$$

3.
$$\ln 15x = \ln(x+4)$$
$$15x = x + 4$$
$$x = \frac{4}{14} = \frac{2}{7}$$

4.
$$\log \frac{x}{x-1} = \log \frac{1}{2}$$
$$\frac{x}{x-1} = \frac{1}{2} \quad \text{Cross-multiply}$$
$$2x = x - 1$$
$$x = -1$$

Exponent and Logarithm Functions

Exponent and logarithm functions are important for many subjects: mathematics, business, finance, chemistry, and physics, to name a few. In this section, we look at functions of the form $f(x) = a^x$ and $g(x) = \log_a x$. We will be concerned with sketching their graphs and identifying their domains.

A basic exponential function is of the form $f(x) = a^x$, where a is any positive number except 1. The graph of $f(x) = a^x$ comes in one of two shapes, depending on whether $0 < a < 1$ (a is positive but smaller than 1) or $a > 1$. Fig. 12-2 represents the first case, which is the graph of $f(x) = (\frac{1}{2})^x$, and Fig. 12-3 represents the second case, which is the graph of $f(x) = 2^x$.

We sketch the graph of $f(x) = a^x$ by plotting points for $x = -3$, $x = -2$, $x = -1$, $x = 0$, $x = 1$, $x = 2$, and $x = 3$. (If a is too large or too small, points for $x = -3$ and $x = 3$ might be too awkward to graph because their y-values are too large or too close to 0.) These points illustrate how fast the graph rises and how fast it approaches its horizontal asymptote.

Before we begin sketching graphs, let us review the following exponent laws.

$$a^{-n} = \frac{1}{a^n} \qquad \left(\frac{1}{a}\right)^{-n} = a^n$$

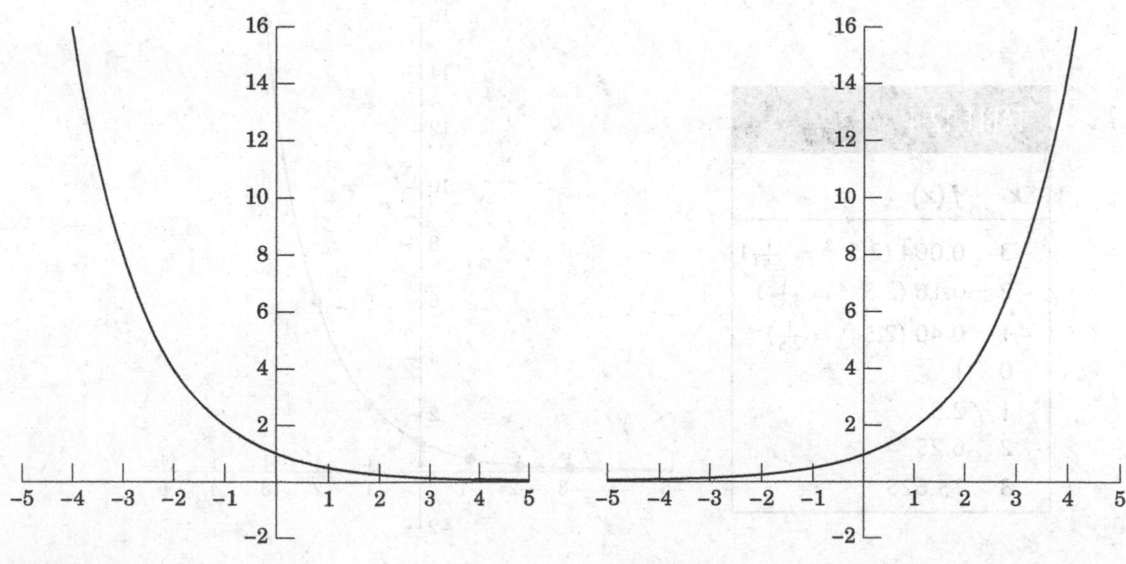

FIGURE 12-2

FIGURE 12-3

EXAMPLE 12-18

Sketch the graph of the exponential function.

- $f(x) = 2.5^x$

We begin with $x = -3, -2, -1, 0, 1, 2,$ and 3 in a table of values. See Table 12-1 and Fig 12-4.

- $g(x) = \left(\frac{1}{3}\right)^x$

See Table 12-2 and Fig 12-5.

PRACTICE

Sketch the graph of the exponential function.

1. $f(x) = \left(\frac{3}{2}\right)^x$
2. $g(x) = \left(\frac{2}{3}\right)^x$
3. $h(x) = e^x$ (Use the e^x key on a calculator.)

TABLE 12-1

x	$f(x)$
−3	0.064 ($2.5^{-3} = \frac{1}{2.5^3}$)
−2	0.16 ($2.5^{-2} = \frac{1}{2.5^2}$)
−1	0.40 ($2.5^{-1} = \frac{1}{2.5}$)
0	1
1	2.5
2	6.25
3	15.625

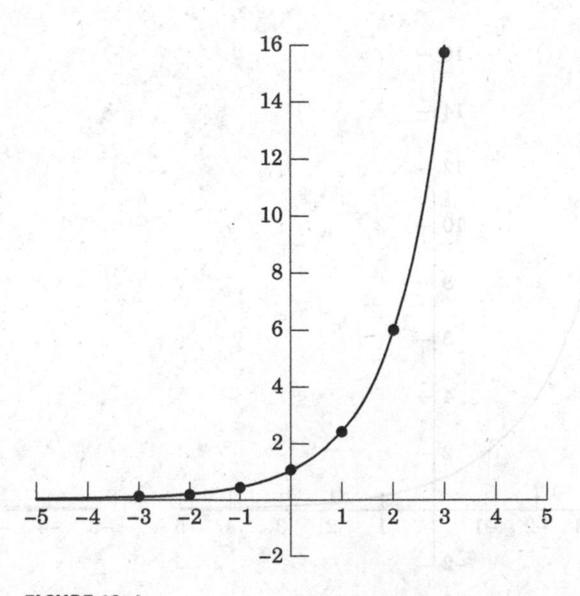

FIGURE 12-4

TABLE 12-2

x	$f(x)$
−3	27 $((\frac{1}{3})^{-3} = 3^3)$
−2	9 $((\frac{1}{3})^{-2} = 3^2)$
−1	3 $((\frac{1}{3})^{-1} = 3^1)$
0	1
1	0.33
2	0.11
3	0.037

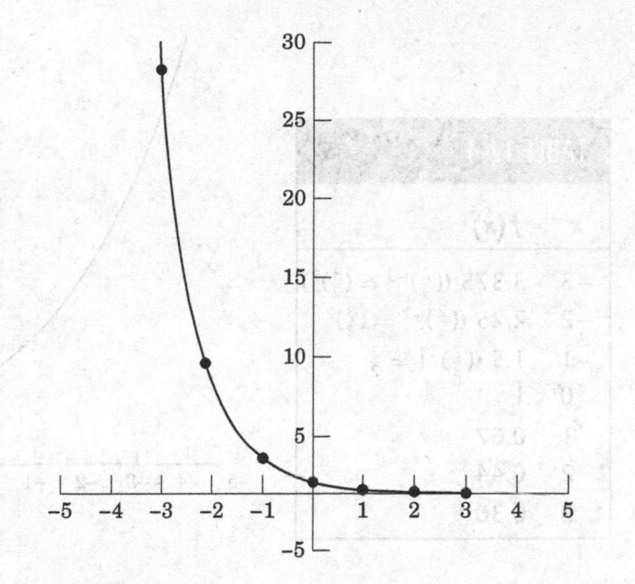

FIGURE 12-5

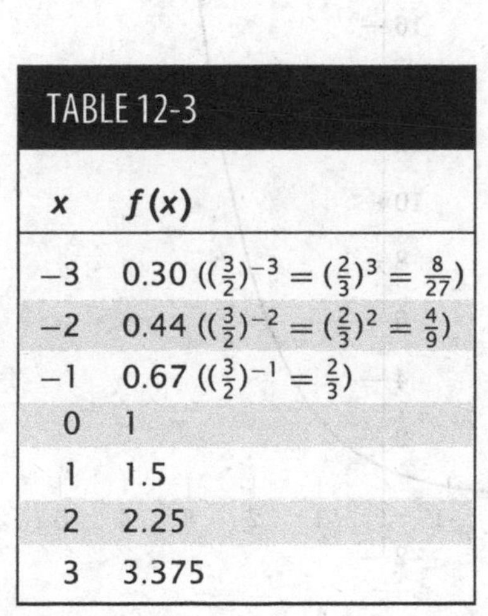

TABLE 12-3

x	$f(x)$
−3	0.30 $((\frac{3}{2})^{-3} = (\frac{2}{3})^3 = \frac{8}{27})$
−2	0.44 $((\frac{3}{2})^{-2} = (\frac{2}{3})^2 = \frac{4}{9})$
−1	0.67 $((\frac{3}{2})^{-1} = \frac{2}{3})$
0	1
1	1.5
2	2.25
3	3.375

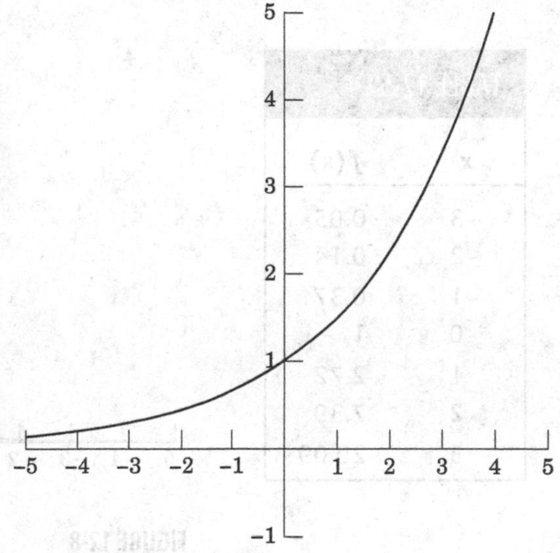

FIGURE 12-6

SOLUTIONS

1. See Fig. 12-6. 2. See Fig. 12-7. 3. See Fig. 12-8.

TABLE 12-4

x	$f(x)$
−3	3.375 $((\frac{2}{3})^{-3} = (\frac{3}{2})^3)$
−2	2.25 $((\frac{2}{3})^{-2} = (\frac{3}{2})^2$
−1	1.5 $((\frac{2}{3})^{-1} = \frac{3}{2}$
0	1
1	0.67
2	0.44
3	0.30

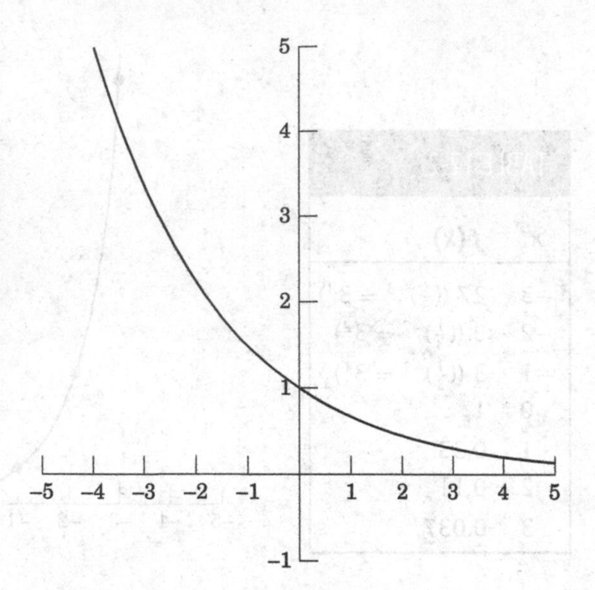

FIGURE 12-7

TABLE 12-5

x	$f(x)$
−3	0.05
−2	0.14
−1	0.37
0	1
1	2.72
2	7.39
3	20.09

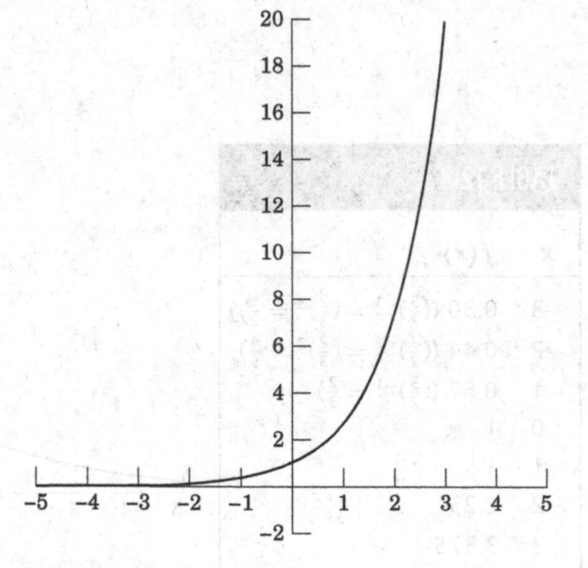

FIGURE 12-8

To sketch the graph of a logarithm function by hand, we rewrite the logarithm function as an exponent equation and graph the exponent equation. This strategy allows us to use what we know about graphs of exponent functions to sketch the graph of logarithm functions. After rewriting the logarithm equation in exponent form, we let the exponent (in the following problems, this is y) be the numbers −3, −2, −1, 0, 1, 2, and 3. These numbers show how fast the graph rises and how fast it approaches its vertical asymptote. Once we have computed the x-coordinates, we plot the points and draw a curve through them.

EXAMPLE 12-19

Sketch the graph of the logarithm function.

- $y = \log_2 x$

Rewrite the equation in exponential form, $x = 2^y$, and let the exponent, y, be the numbers −3, −2, −1, 0, 1, 2, and 3. See Table 12-6 and Fig 12-9.

- $y = \ln x$

Rewritten as an exponent equation, this is $x = e^y$. Let $y = -3, -2, -1, 0, 1, 2,$ and 3. See Table 12-7 and Fig 12-10.

PRACTICE

Sketch the graph of the logarithmic function.

1. $y = \log_{1.5} x$
2. $y = \log_3 x$

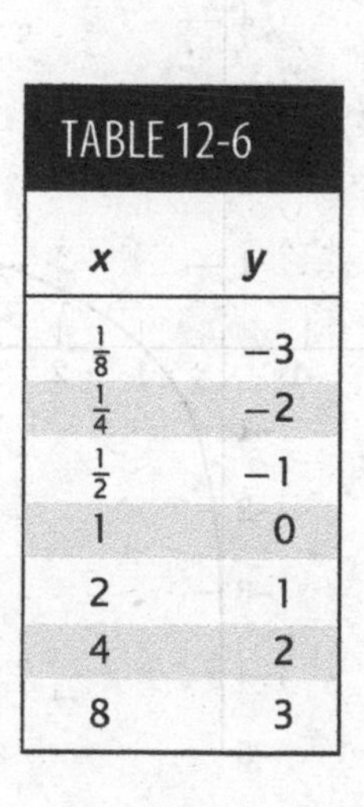

TABLE 12-6

x	y
$\frac{1}{8}$	−3
$\frac{1}{4}$	−2
$\frac{1}{2}$	−1
1	0
2	1
4	2
8	3

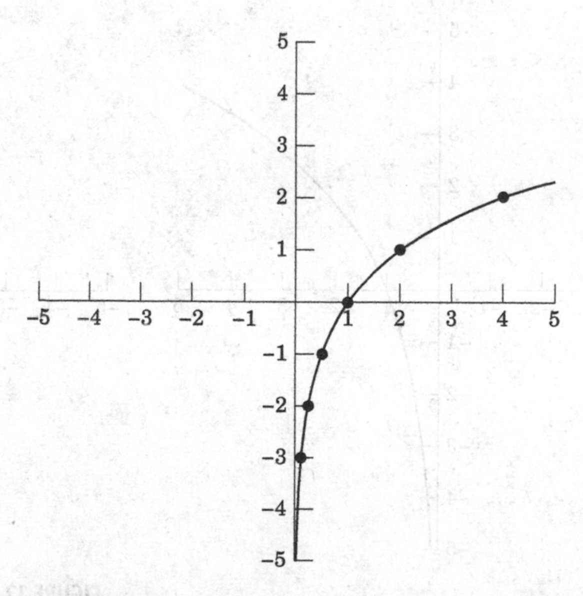

FIGURE 12-9

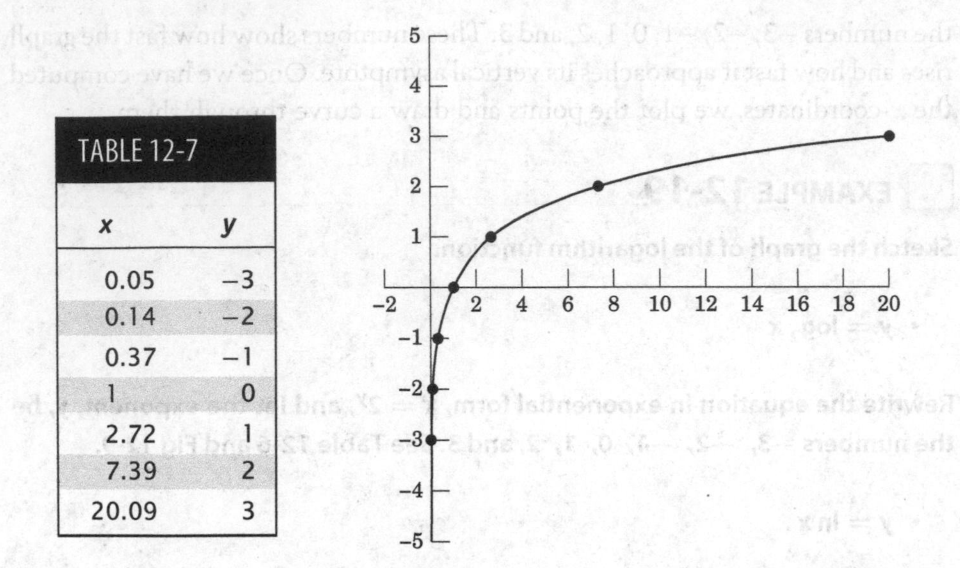

TABLE 12-7

x	y
0.05	−3
0.14	−2
0.37	−1
1	0
2.72	1
7.39	2
20.09	3

FIGURE 12-10

SOLUTIONS

1. **See Fig. 12-11.**
2. **See Fig. 12-12.**

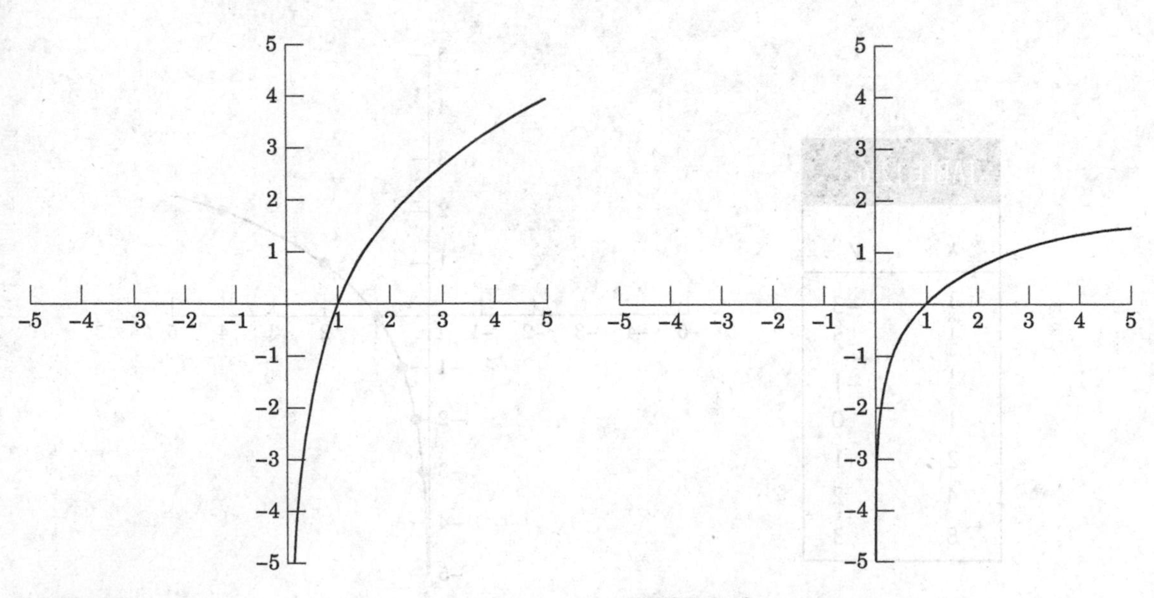

FIGURE 12-11

FIGURE 12-12

Transformations of the graphs of exponent and logarithm functions behave in the same way as transformations of other functions. That is, we can predict the change to a function's graph after making certain changes to its equation.

EXAMPLE 12-20

- **The graph of $f(x) = -2^x$ is the graph of $y = 2^x$ reflected about the x-axis (flipped upside down).**
- **The graph of $g(x) = 2^{-x}$ is the graph of $y = 2^x$ reflected about the y-axis (flipped sideways).**
- **The graph of $h(x) = 2^{x+1}$ is the graph of $y = 2^x$ shifted to the left 1 unit.**
- **The graph of $f(x) = -3 + 2^x$ is the graph of $y = 2^x$ shifted down 3 units.**
- **The graph of $f(x) = \log_2(x - 2)$ is the graph of $y = \log_2 x$ shifted to the right 2 units.**
- **The graph of $f(x) = -5 + \log_3 x$ is the graph of $y = \log_3 x$ shifted down 5 units.**
- **$f(x) = \frac{1}{3} \log x$ is the graph of $y = \log x$ flattened vertically by a factor of one-third.**

The Domain of a Logarithm Function

As you can see from their graphs, the domain of $f(x) = a^x$ is all real numbers, and the domain of $g(x) = \log_a x$ is all positive real numbers, $(0, \infty)$. This means that we cannot take a logarithm of 0 or the log of a negative number. The reason is that a is a positive number. Raising a positive number to *any* real number power always gives us another positive number. We find the domain by setting the quantity behind "log" (called the *argument*) greater than 0.

EXAMPLE 12-21

Find the domain. Give the answer in interval notation.

- **$f(x) = \log_5(2 - x)$**

Because we are taking the log of $2 - x$, we solve $2 - x > 0$.

$$2 - x > 0$$

$$-x > -2$$

$$x < 2 \qquad \text{The domain is } (-\infty, 2).$$

When solving equations involving logarithms, you should check to make sure the solution(s) lies in the domain of each logarithm expression. For example, the solution to $\log(2x) = \log(x - 5)$ does not exist because the solution to $2x = x - 5$ is not in the domain of either logarithm.

PRACTICE

Find the domain. Give the answer in interval notation.

1. $f(x) = \ln(10 - 2x)$ 2. $f(x) = \log(x^2 + 4)$ 3. $g(x) = \log_4(6x + 9)$

SOLUTIONS

1. **Solve $10 - 2x > 0$. The solution is $x < 5$, $(-\infty, 5)$.**
2. **Because $x^2 + 4 > 0$ is always positive, the domain is all real numbers, $(-\infty, \infty)$.**
3. **Solve $6x + 9 > 0$. The solution is $x > -\frac{3}{2}$. The domain is $(-\frac{3}{2}, \infty)$.**

Summary

In this chapter, we learned how to

- *Use the compound amount formula to calculate the value of an investment earning compound interest.* If interest is compounded annually, the formula is $A = P(1 + r)^t$, where A is the compound amount, P is the principle, r is the annual interest rate, and t is the number of years. If interest is compounded n times per year, then the compound amount formula becomes $A = P(1 + \frac{r}{n})^{nt}$, where A, P, r, and t are the same, and n is the number of times per year that interest is compounded. For example, if interest is compounded monthly, then $n = 12$.

- *Use an exponential growth formula to project the population for a population growing exponentially.* This formula involves the number e: $n(t) = n_0 e^{rt}$, where $n(t)$ is the population at time t (which might not be measured in years) and r is the growth rate. If interest is compounded continuously, we use a similar formula for the compound amount: $A = Pe^{rt}$. If a quantity is *decaying* instead of growing, the formula is $n(t) = n_0 e^{-rt}$.
- *Rewrite exponent equations in logarithm form and vice versa.* The equation $a^x = y$ written in logarithm form is $\log_a y = x$, provided a is a positive number not equal to 1. The number a is the base of the logarithm (and the exponent).
- *Work with natural and common logarithms.* The number e has its own notation as the base of a logarithm: $\ln x$ means $\log_e x$ and is called the natural logarithm. The common logarithm is a logarithm with base 10. That is, $\log_{10} x$ is normally written as $\log x$ (no base is written).
- *Work with cancelation properties of logarithms.* We first learned the cancelation properties $\log_a a^x = x$ and $a^{\log_a x} = x$. We used these properties (especially the first property) to simplify logarithms.
- *Use three other properties of logarithms to rewrite logarithms.* These properties are $\log_a mn = \log_a m + \log_a n$, $\log_a \frac{m}{n} = \log_a m - \log_a n$, and $\log_a m^t = t \log_a m$, where m and n are positive numbers.
- *Solve equations involving exponents or logarithms.* We solve an exponent equation by isolating the term with the exponent, taking the natural (or common) logarithm of each side and using the third logarithm property (above) to bring the exponent in front of the logarithm. From this point, we use ordinary algebra to solve for the variable. We had two strategies for solving equations containing one or more logarithms. If the equation can be written in the form "$\log_a$ Expression = number," then we eliminate the logarithm by rewriting the equation in exponent form and then using algebra to solve for the variable. If the equation can be written in the form "$\log_a$ Expression #1 = $\log_a$ Expression #2," we solve "Expression #1 = Expression #2," that is, the logarithms cancel.
- *Sketch the graph of an exponent function.* For the graph of $y = a^x$, we plot points for x to be -3, -2, -1, 0, 1, 2, and 3 (if a is rather large or rather close to 0, then the points for $x = -3$ and $x = 3$ might be too hard to plot).

- *Sketch the graph of a logarithm function.* We sketch the graph of the function $y = \log_a x$ by rewriting the equation in exponent form and then plotting points for the exponent (in this case, y) for $-3, -2, -1, 0, 1, 2$, and 3.
- *Determine the domain of a logarithm function.* The domain of a logarithm function is all positive numbers. To find the domain of the function $y = \log_a[f(x)]$, we solve the inequality $f(x) > 0$.

QUIZ

1. If $1500 earns 6% annual interest, compounded monthly, what is the investment worth after 8 years?

A. $2672 B. $2391 C. $2053 D. $2421

2. What would an investment of $12,000, earning 4% annual interest, compounded continuously, be worth after 9 years?

A. $16,817 B. $17,200 C. $18,401 D. $17,992

3. The graph in Fig. 12-13 is the graph of which function?

A. $y = (\frac{3}{4})^x$ B. $y = (\frac{4}{3})^x$ C. $y = (\frac{16}{9})^x$ D. $y = (\frac{9}{16})^x$

4. The graph in Fig. 12-14 is the graph of which function?

A. $y = \log_{3/4} x$ B. $y = \log_{4/3} x$ C. $y = \log_{9/16} x$ D. $y = \log_{16/9} x$

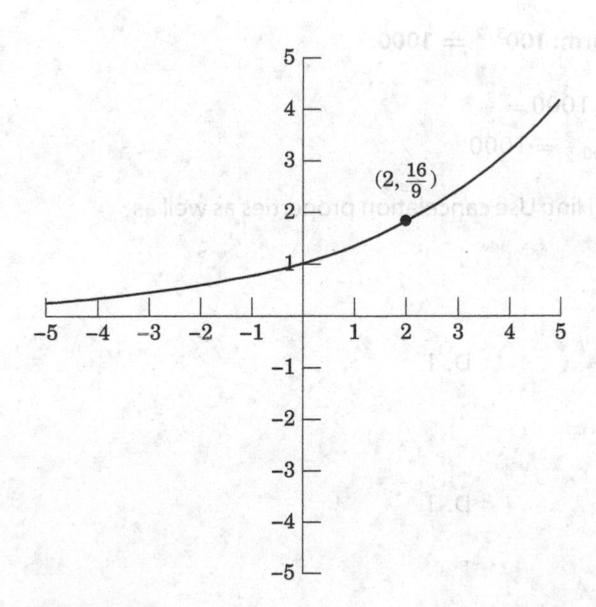

FIGURE 12-13

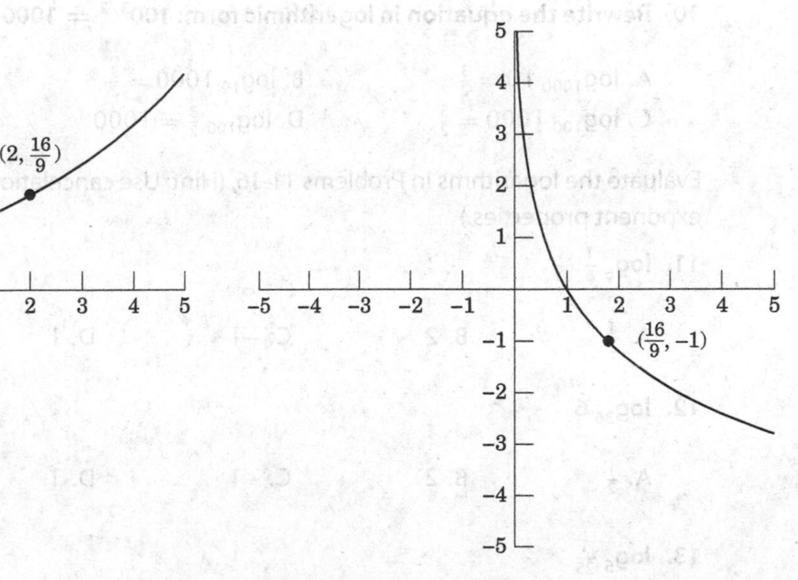

FIGURE 12-14

5. A wildlife organization tracks the population of feral hogs in a Texas county. The organization estimates that the population of feral hogs can be predicted by the function $n(t) = 260e^{0.04t}$, where t is the number of years after 2014. Use the model to predict the feral hog population for the year 2024.

A. 388 B. 390 C. 392 D. 394

6. The population of a city is growing at an annual rate of 1.25%. If the population in 2012 was 180,000, estimate the population for the year 2020.

A. 191,130 B. 194,990 C. 198,140 D. 198,930

7. Rewrite the equation in logarithmic form: $5^x = 18$.

A. $\ln 5x = 18$ B. $\log_{18} x = 5$ C. $\log_5 x = 18$ D. $\log_5 18 = x$

8. Rewrite the equation in exponential form: $\ln 7 = m$.

A. $e^m = 7$ B. $e^2 = m$ C. $7^e = m$ D. $7^m = 4$

9. Rewrite the equation in exponential form: $\log_{49} 7 = \frac{1}{2}$.

A. $7^{1/2} = 49$ B. $49^{1/2} = 7$ C. $49^{1/7} = \frac{1}{2}$ D. $\left(\frac{1}{2}\right)^7 = 49$

10. Rewrite the equation in logarithmic form: $100^{3/2} = 1000$.

A. $\log_{1000} 10 = \frac{3}{2}$ B. $\log_{10} 1000 = \frac{3}{2}$
C. $\log_{100} 1000 = \frac{3}{2}$ D. $\log_{100} \frac{3}{2} = 1000$

Evaluate the logarithms in Problems 11-16. (Hint: Use cancelation properties as well as exponent properties.)

11. $\log_6 \frac{1}{6}$

A. $\frac{1}{2}$ B. 2 C. -1 D. 1

12. $\log_{36} 6$

A. $\frac{1}{2}$ B. 2 C. -1 D. 1

13. $\log_5 \frac{1}{25}$

A. -2 B. $\frac{1}{2}$ C. $-\frac{1}{2}$ D. 0.2

14. $\ln e^4$

A. $\ln 4$ B. 4 C. e^4 D. $\frac{1}{4}$

15. $\log_4 \frac{1}{2}$

A. $-\frac{1}{2}$ B. -1 C. -2 D. $\frac{1}{2}$

16. $e^{\ln x}$

A. e^x B. $\ln x$ C. x D. $\frac{1}{2}$

Use properties of logarithms to rewrite the expression in Problems 17-20.

17. $\log_8 x - \log_8 y$

A. $\log_8(x - y)$ B. $\log_8 xy$ C. $\frac{\log_8 x}{\log_8 y}$ D. $\log_8 \frac{x}{y}$

18. $2 \log_5 x$

A. $\log_2 x^5$ B. $\log_5 x^2$ C. $\log_5 \sqrt{x}$ D. $\log_5 2x$

19. $3 \log_4 x - \log_4 y$

A. $(\frac{\log_4 x}{\log_4 y})^3$ B. $\frac{\log_4 x^3}{\log_4 y}$ C. $\log_4(\frac{x}{y})^3$ D. $\log_4 \frac{x^3}{y}$

20. $\ln x\sqrt{y}$

A. $\ln x + \frac{1}{2} \ln y$ B. $\frac{1}{2} \ln x + \ln y$ C. $\frac{1}{2} \ln x + \frac{1}{2} \ln y$ D. $\frac{\ln x}{\sqrt{\ln y}}$

21. Solve the equation: $6^{2x-4} = 10$.

A. 2.64 B. 2.26 C. 2.30 D. The equation has no solution.

22. Solve the equation: $\log_8 3x = 1$.

A. $x = -\frac{1}{2}$ B. $x = \frac{3}{8}$ C. $x = \frac{8}{3}$ D. The equation has no solution.

23. Solve the equation: $\log_5(1 - 2x) = \log_5 x$.

A. $x = \frac{1}{3}$ B. $x = -\frac{1}{3}$ C. $x = -3$ D. The equation has no solution.

Final Exam

1. **The graph of $y = 2f(x - 1)$ is the graph of $y = f(x)$**

   A. stretched vertically and shifted to the left 1 unit.

   B. stretched vertically and shifted to the right 1 unit.

   C. shifted to the right 1 unit and up 2 units.

   D. shifted to the left 1 unit and up 2 units.

2. **Solve: $\frac{4}{3}x + 2 \leq 6$**

   A. $(-\infty, \frac{16}{3}]$ B. $(-\infty, 3]$ C. $(-\infty, 6]$ D. $[6, \infty)$

3. **Rewrite as an exponential equation: $\log_a m = n$.**

   A. $m^a = n$ B. $a^n = m$ C. $m^n = a$ D. $a^m = n$

Problems 4-6 refer to Fig. F.E.-1.

4. **Evaluate $f(2)$.**

   A. -2 B. -1 C. 2 D. 1

5. **What is the range of the function $y = f(x)$?**

   A. $[1, \infty)$ B. $(-\infty, -2]$ C. $(-\infty, \infty)$ D. $[-2, \infty)$

6. **Where is the function increasing?**

   A. $(-\infty, \infty)$ B. $(-\infty, -2)$ C. $(1, \infty)$

   D. The function is never increasing.

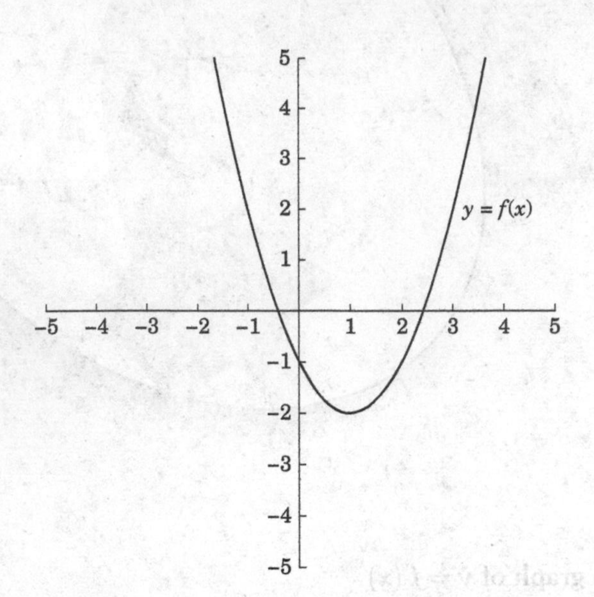

FIGURE F.E.-1

FIGURE F.E.-2

7. Find the quotient $\frac{6-i}{3+2i}$.

A. $\frac{20}{13}+\frac{9}{13}i$ B. $\frac{16}{13}-\frac{15}{13}i$ C. $\frac{20}{13}-\frac{9}{13}i$ D. $20-9i$

8. Factor: $36x^2-25=$

A. $(12x-5)(3x+5)$ B. $-(6x+5)^2$

C. $(6x-5)(6x+5)$ D. $(6x-5)^2$

9. Find $x+2y$ for the system: $\begin{cases} x-y=3 \\ x+y=5 \end{cases}$.

A. 5 B. 6 C. 7 D. 8

10. The solution to which inequality is shaded in Fig. F.E.-2?

A. $x-y\le 1$ B. $x+y\le 1$ C. $x-y\ge 1$ D. $x+y\ge 1$

11. Find the midpoint between the points $(\frac{3}{2},-1)$ and $(\frac{5}{2},2)$.

A. $(-2,\frac{3}{2})$ B. $(\frac{1}{2},-\frac{1}{2})$ C. $(2,-\frac{3}{2})$ D. $(2,\frac{1}{2})$

12. Rewrite as a logarithm equation: $e^{2x} = 15$.

A. $\ln 15 = 2x$ B. $\ln 2x = 15$ C. $\log_{15} e = 2x$ D. $\log_{15} 2x = e$

13. Solve the inequaltity: $x^2 + 49 > 0$.

A. $(-7, 7)$ B. $(-\infty, -7) \cup (7, \infty)$

C. $(-\infty, \infty)$ D. There is no solution.

14. What is the range of the function $f(x) = 0.025x^2 + 2x - 1$?

A. $[-40, \infty)$ B. $(-\infty, -40]$ C. $[-41, \infty)$ D. $(-\infty, -41]$

15. Find an equation of the line that goes through the points $(4, -6)$ and $(1, 3)$.

A. $3x - y = 30$ B. $3x - y = 6$

C. $3x + y = 6$ D. $x - 3y = 22$

16. Is the function $f(x) = x^4 - 3x^2 + 1$ even, odd, or neither?

A. Even B. Odd C. Neither

17. Solve: $x^2 - 8x - 20 = 0$

A. $x = -2,\ 10$ B. $x = 2,\ -10$

C. $x = 4 \pm 2\sqrt{5}$ D. $x = -4 \pm 2\sqrt{5}$

18. Add: $\frac{4}{x^2+7x+6} + \frac{x-1}{x^2+2x+1} =$

A. $\frac{x^2+13x-2}{(x+1)^2(x+6)}$ B. $\frac{x^2+9x-5}{(x+1)^2(x+6)}$ C. $\frac{x-3}{(x+1)^2(x+6)}$ D. $\frac{x^2+9x-2}{(x+1)^2(x+6)}$

19. Evaluate: $\log_{16} 4$.

A. $-\frac{1}{2}$ B. -1 C. $\frac{1}{4}$ D. $\frac{1}{2}$

20. Find $x + 2y$ for the system: $\begin{cases} 8x + y = 5 \\ 3x - 2y = 9 \end{cases}$.

A. -2 B. -3 C. -4 D. -5

21. The solution to which system of inequalites is shaded in Fig. F.E.-3?

A. $\begin{cases} y < \frac{1}{2}x + 1 \\ x > 1 \\ y > -1 \end{cases}$ B. $\begin{cases} y < \frac{1}{2}x + 1 \\ x < 1 \\ y > -1 \end{cases}$

C. $\begin{cases} y > \frac{1}{2}x + 1 \\ x > 1 \\ y < -1 \end{cases}$ D. $\begin{cases} y > \frac{1}{2}x + 1 \\ x < 1 \\ y > -1 \end{cases}$

22. Evaluate $\frac{f(a+h)-f(a)}{h}$ for $f(x) = 2x^2 + 3$.

A. $4a + 2h$ B. $4a + 2h^2$ C. $4ah + 2$ D. $4a + 2h^2 + 6$

23. Solve: $10x + 3 = 7$

A. $x = 1$ B. $x = \frac{5}{2}$ C. $x = -1$ D. $x = \frac{2}{5}$

24. Find $x + 2y$ for the system: $\begin{cases} \frac{2}{5}x - y = 3 \\ x + 3y = 13 \end{cases}$.

A. 8 B. 10 C. 12 D. 14

25. Evaluate: $\log_{12} \frac{1}{144}$.

A. $-\frac{1}{2}$ B. -2 C. $\frac{1}{4}$ D. -1

26. Find an equation for the line whose graph is shown in Fig. F.E.-4.

A. $4x + 5y = 10$ B. $5x + 4y = 8$

C. $5x - 4y = -8$ D. $4x - 5y = -10$

27. Find the distance between the points $(\frac{3}{2}, -1)$ and $(-\frac{5}{2}, 2)$.

A. $\frac{\sqrt{5}}{2}$ B. $\sqrt{17}$ C. 5 D. $\sqrt{2}$

28. For the function $f(x) = -20x^2 + 100x + 25$,

A. the maximum value is 2.5. B. the minimum value is 2.5.

C. the maximum value is 150. D. the minimum value is 150.

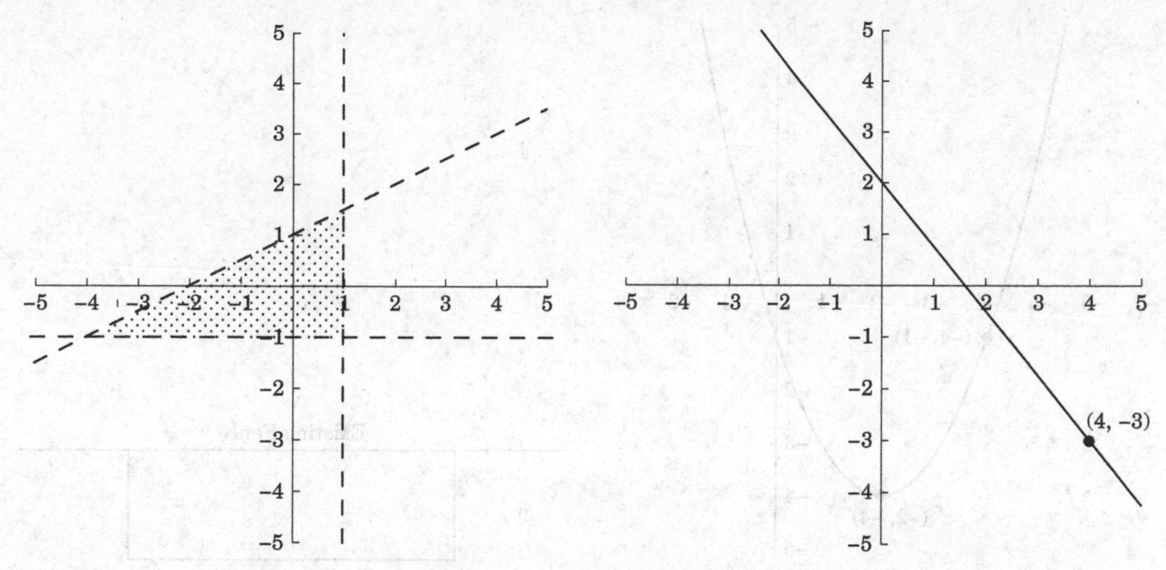

FIGURE F.E.-3

FIGURE F.E.-4

29. **Find $x + 2y$ for the system:** $\begin{cases} y = 10x - 9 \\ y = x + 18 \end{cases}$.

A. 35 B. 40 C. 45 D. 50

30. **Evaluate: $\log_{15} 15^t$.**

A. t B. 15^t C. 1^t D. $t \ln 15$

31. **Solve the inequality: $3x^2 + 2x \leq 8$.**

A. $[-\frac{4}{3}, 2]$ B. $(-\infty, -\frac{4}{3}] \cup [2, \infty)$

C. $[-2, \frac{4}{3}]$ D. $(-\infty, -2] \cup [\frac{4}{3}, \infty)$

32. **Let $f(x) = 3x - 4$ and $g(x) = x^2 + 1$. Evaluate $(f \circ g)(-2)$.**

A. 11 B. 101 C. −70 D. −13

33. **Solve: $x^2 + x - 10 = 0$**

A. $x = \frac{1 \pm \sqrt{41}}{2}$ B. $x = 1 \pm \frac{\sqrt{41}}{2}$

C. $x = \frac{-1 \pm \sqrt{41}}{2}$ D. $x = -1 \pm \frac{\sqrt{41}}{2}$

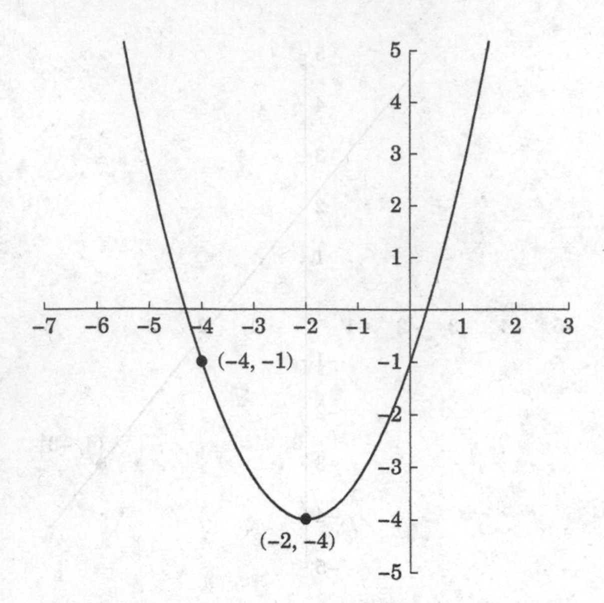

FIGURE F.E.-5

Existing Fence

FIGURE F.E.-6

34. Find an equation for the parabola whose graph is in Fig. F.E.-5.

A. $y = \frac{4}{3}x^2 + 2x - 1$ B. $y = \frac{4}{3}x^2 - 2x - 1$

C. $y = \frac{3}{4}x^2 - 3x - 1$ D. $y = \frac{3}{4}x^2 + 3x - 1$

35. Solve the equation: $\log_2(x + 4) = -1$.

A. $-\frac{7}{2}$ B. 2 C. −6 D. The equation has no solution.

36. Find the domain of $(f \circ g)(x)$ if $f(x) = \frac{1}{x-1}$ and $g(x) = 2x - 3$.

A. $x \neq 1$ B. $x \neq 2$ C. $x \neq 1,\ x \neq 2$ D. $x \neq 1,\ x \neq 0$

37. Evaluate: $e^{\ln 10}$.

A. ln 10 B. e^{10} C. 10 D. $e \ln 10$

38. Jonathon, a rancher, wishes to enclose a rectangular pen along the edge of his property, which is already fenced. If Jonathon has 800 feet of fencing available for the other three sides, what is the maximum area he can enclose? (See Fig. F.E.-6.)

A. 80,000 ft^2 B. 90,000 ft^2

C. 100,000 ft^2 D. 110,000 ft^2

39. Which of the following is true for the system? $\begin{cases} 3x - 4y = 1 \\ -6x + 8y = -2 \end{cases}$?

A. The system has no solution.

B. The solution has a unique solution.

C. The graph of each equation in the system is the same line.

40. Simplify: $(5xy^{-2})^2(2x)^3 =$

A. $\frac{10x^3}{y^4}$ B. $\frac{40x^4}{y^4}$ C. $\frac{40x^5}{y^4}$ D. $\frac{200x^5}{y^4}$

41. What is the domain for the function $f(x) = \frac{\sqrt[3]{x-1}}{x-5}$?

A. $(-\infty, 1) \cup (1, 5) \cup (5, \infty)$ B. $[1, 5) \cup (5, \infty)$

C. $(-\infty, 5) \cup (5, \infty)$ D. $(1, 5) \cup (5, \infty)$

42. What would an investment of $25,000 earning 5.25% annual interest, compounded monthly, be worth after 6 years?

A. $34,233 B. $25,663 C. $33,984 D. $35,463

43. Solve the inequality: $\frac{x^2-x-6}{x^2-25} \geq 0$.

A. $(-\infty, -5] \cup [-2, 3] \cup [5, \infty)$ B. $[-5, -2] \cup [3, 5]$

C. $(-5, -2] \cup [3, 5)$ D. $(-\infty, -5) \cup [-2, 3] \cup (5, \infty)$

44. Find all solutions, real or complex: $x^4 - 7x^3 + 19x^2 - 3x - 30 = 0$.

A. $1, \ -2, 3 \pm \sqrt{6}$ B. $-1, 2, 3 \pm \sqrt{6}$

C. $1, \ -2, 3 \pm \sqrt{6}i$ D. $-1, \ 2, 3 \pm \sqrt{6}i$

45. The graph of which function is sketched in Fig. F.E.-7?

A. $y = -(\frac{2}{3})^x$ B. $y = (\frac{2}{3})^x$ C. $y = (\frac{3}{2})^x$ D. $y = -(\frac{3}{2})^x$

46. The points $(5, -1)$, $(10, -8)$, and $(3, -3)$ are the vertices of which kind of triangle?

A. Equilateral (all three sides equal)

B. Isosceles (exactly two sides equal)

C. Right triangle (not isosceles) D. None of the above

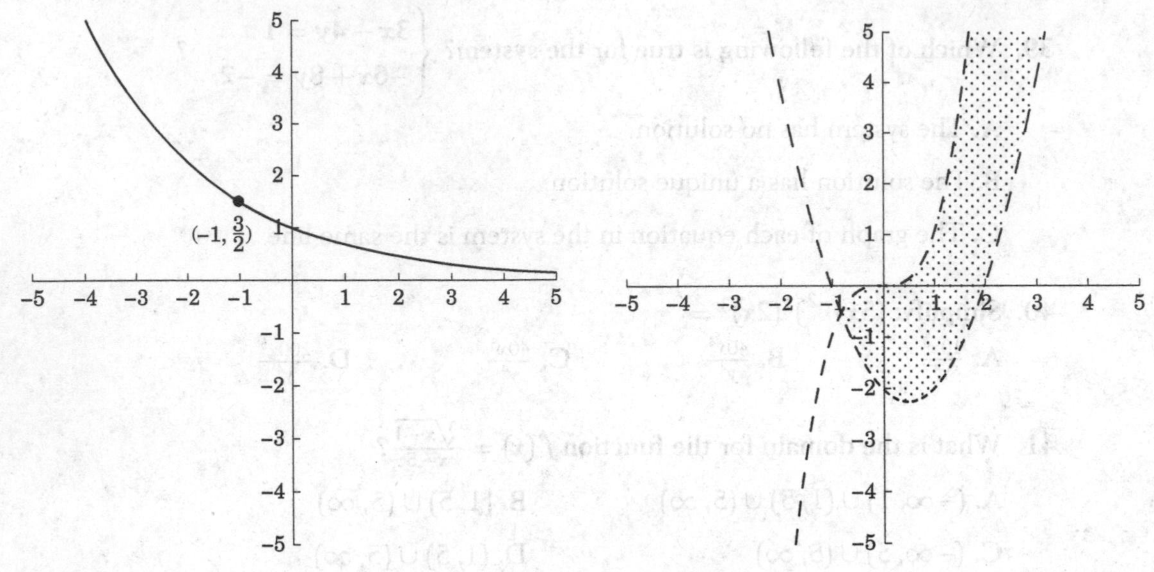

FIGURE F.E.-7

FIGURE F.E.-8

47. The solution to which system is shaded in Fig. F.E.-8?

A. $\begin{cases} y > x^3 \\ y < x^2 - x - 2 \end{cases}$ B. $\begin{cases} y > x^3 \\ y > x^2 - x - 2 \end{cases}$

C. $\begin{cases} y < x^3 \\ y < x^2 - x - 2 \end{cases}$ D. $\begin{cases} y < x^3 \\ y > x^2 - x - 2 \end{cases}$

48. Solve the equation: $4^{-2x+5} = 3$.

A. −2.644 B. −0.029 C. 0.058 D. 2.104

49. Solve the equation: $|4x - 9| = 7$.

A. $x = \frac{1}{2}, 4$ B. $x = -\frac{1}{2}, 4$ C. $x = -\frac{1}{2}, -4$ D. $x = \frac{1}{2}, -4$

50. The future student population of a large state university is estimated by the function $p(t) = 30e^{0.02t}$, where $p(t)$ is in thousands and t is the number of years after 2010. Project the university's student population for the year 2018.

A. 34,800 B. 32,500 C. 33,600 D. 35,200

51. **Find the intercepts for the graph of $y = \frac{1}{x-2}$.**

A. There is no x-intercept, and the y-intercept is $-\frac{1}{2}$.

B. The x-intercept is 2, and there is no y-intercept.

C. The x-intercept is 1, and the y-intercept is $-\frac{1}{2}$.

D. The graph has no intercepts.

52. **Simplify: $\sqrt{\frac{16x^4}{27y^3}} =$**

A. $\frac{4x^2}{3y\sqrt{3y}}$ B. $\frac{2x^2}{3y\sqrt{3y}}$ C. $\frac{2x\sqrt{2x}}{3y\sqrt{3y}}$ D. $\frac{2x}{3y\sqrt{3y}}$

53. **Evaluate $f(6)$ for $f(x) = 8$.**

A. 8 B. 48 C. 6 D. $f(6)$ does not exist.

54. **The graph of which function is given in Fig. F.E.-9?**

A. $f(x) = -\frac{1}{5}(x+2)^2(x-3)$ B. $f(x) = -\frac{1}{5}(x+2)(x-3)^2$

C. $f(x) = \frac{1}{5}(x+2)^2(x-3)^2$ D. $f(x) = \frac{1}{5}(x+2)^2(x+3)^2$

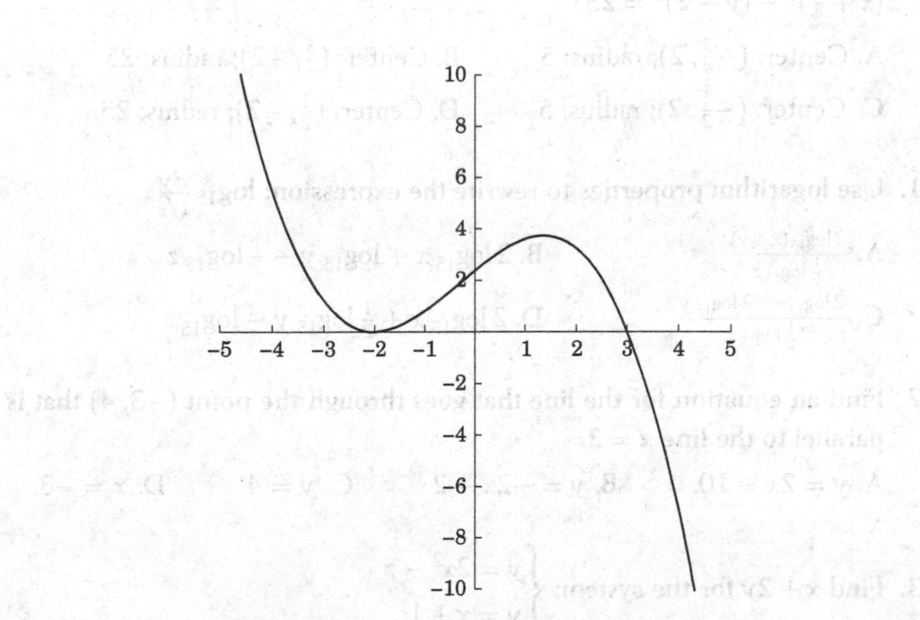

FIGURE F.E.-9

55. **Use logarithm properties to rewrite the expression: $2\ln x - 3\ln y$.**

A. $\ln(x^2 - y^3)$ B. $\ln(\frac{x^2}{y^3})$ C. $\frac{\ln x^2}{\ln y^3}$ D. $\frac{2\ln x}{3\ln y}$

56. **Find $x + 2y$ for the system:** $\begin{cases} \frac{1}{9}x - \frac{1}{2}y = 1 \\ \frac{1}{3}x + \frac{1}{4}y = -\frac{1}{2} \end{cases}$.

A. -4 B. $-\frac{3}{2}$ C. 2 D. $\frac{5}{2}$

57. **Find a polynomial that has zeros $-3, -2, 2$, and 5 and whose coefficient for x^2 is 19.**

A. $x^4 - 3x^3 + 19x^2 - 10x + 60$ B. $x^4 - 2x^3 + 19x^2 - 7x + 60$

C. $-x^4 + 2x^3 + 19x^2 - 8x - 60$ D. $-x^4 + 3x^2 + 19x^2 - 6x - 60$

58. **Is the function $f(x) = 5x^3 - 1$ even, odd, or neither?**

A. Even B. Odd C. Neither

59. **Solve the equation: $\log_{10}(x + 4) = \log_{10}(3x - 8)$.**

A. $\frac{8}{3}$, 4 B. $\frac{32}{7}$ C. 6 D. 4

60. **What is the center and radius of the circle whose equation is $(x + \frac{3}{4})^2 + (y - 2)^2 = 25$?**

A. Center: $(-\frac{3}{4}, 2)$; radius: 5 B. Center: $(\frac{3}{4}, -2)$; radius: 25

C. Center: $(-\frac{1}{2}, 2)$; radius: 5 D. Center: $(\frac{1}{2}, -2)$; radius: 25

61. **Use logarithm properties to rewrite the expression: $\log_{15}\frac{x^2y}{\sqrt{z}}$.**

A. $\frac{2\log_{15}(x+y)}{\frac{1}{2}\log_{15} z}$ B. $2\log_{15} x + \log_{15} y - \frac{1}{2}\log_{15} z$

C. $\frac{2\log_{15} x + 2\log_{15} y}{\frac{1}{2}\log_{15} z}$ D. $2\log_{15} x + \frac{1}{2}\log_{15} y - \log_{15} z$

62. **Find an equation for the line that goes through the point $(-3, 4)$ that is parallel to the line $x = 2$.**

A. $y = 2x + 10$ B. $y = -2x + 2$ C. $y = 4$ D. $x = -3$

63. **Find $x + 2y$ for the system:** $\begin{cases} y = 2x^2 - 5 \\ y = x + 1 \end{cases}$.

A. 8 and $-\frac{7}{2}$ B. -3 and $\frac{5}{2}$ C. 4 and $\frac{3}{2}$ D. 8 and $-\frac{5}{2}$

64. **Solve for x: $|8x - 5| = 7$.**

A. $x = \frac{1}{4}, \frac{3}{2}$ B. $x = -\frac{3}{2}, \frac{3}{2}$

C. $x = -\frac{1}{4}, \frac{3}{2}$ D. There is no solution.

65. **What is the domain for the function $f(x) = \sqrt{x^2 - 16}$?**

A. $[-4, 4]$ B. $(-4, 4)$

C. $(-\infty, -4] \cup [4, \infty)$ D. $(-\infty, -4) \cup (4, \infty)$

66. **Find the slope and the y-intercept for the graph of the line $6x - 4y = 12$.**

A. The slope is $\frac{3}{2}$, and the y-intercept is 3.

B. The slope is $-\frac{3}{2}$, and the y-intercept is 3.

C. The slope is $\frac{3}{2}$, and the y-intercept is -3.

D. The slope is $-\frac{3}{2}$, and the y-intercept is -3.

67. **Solve for x: $x^2 - 81 = 0$.**

A. $x = 9$ B. $x = -9, 9$

C. $x = \pm 3\sqrt{3}$ D. There is no solution.

68. **The solid graph in Fig. F.E.-10 is the graph of $y = f(x)$, and the dashed graph is the graph of which function?**

A. $y = -2f(x)$ B. $y = -f(-x)$

C. $y = -f(x)$ D. $y = f(-x)$

69. **Find the vertex for the graph of $y = 10x^2 - 20x + 4$.**

A. $(2, 4)$ B. $(1, -6)$ C. $(-2, 84)$ D. $(-1, 34)$

70. **Solve the inequality: $\frac{1}{x-5} < 1$.**

A. $(5, 6)$ B. $(-\infty, 5) \cup (6, \infty)$ C. $(6, \infty)$ D. $(-\infty, 6)$

71. **Find the zeros and multiplicity for the function $f(x) = x(x+4)^2(x-6)^3$.**

A. 4, multiplicity 2; 6, multiplicity 5

B. -4, multiplicity 2; -6, multiplicity 3

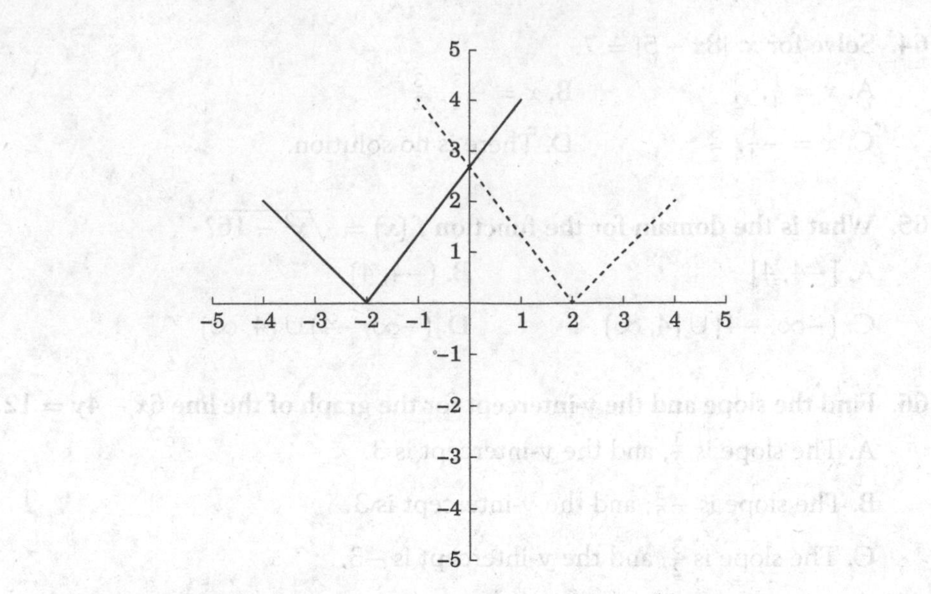

FIGURE F.E.-10

C. 0, multiplicity 1; −4, multiplicity 2; 6, multiplicity 3

D. 0, multiplicity 1; 4, multiplicity 2; −6, multiplicity 3

72. Find an equation for the circle that has a diameter between the points (12, 16) and (−2, −32).

A. $(x-5)^2+(y+8)^2=625$ B. $(x+5)^2+(y-2)^2=625$

C. $(x-5)^2+(y-8)^2=25$ D. $(x+5)^2+(y+8)^2=25$

73. Simplify: $\sqrt{12x^3}=$

A. $12x\sqrt{x}$ B. $2x\sqrt{3x}$ C. $6x$ D. $6x\sqrt{x}$

74. Solve: $\frac{4}{3x-2}=\frac{3}{2x+6}$

A. $\frac{5}{3}$ B. $x=6$ C. $x=\frac{5}{6}$ D. $x=30$

75. Instructions on a package of oatmeal call for $\frac{1}{2}$ cup of water for $\frac{2}{3}$ cup of oatmeal. Find an equation that gives the amount of water in terms of the amount of oatmeal.

A. $y=\frac{3}{2}x$ B. $y=\frac{4}{3}x$ C. $y=\frac{3}{4}x$

D. There is not enough information to answer the question.

FIGURE F.E.-11

76. Which of the following is NOT a candidate for the rational zeros for the function $f(x) = 4x^3 - x^2 + 5x + 15$.

A. $\frac{4}{3}$ B. $-\frac{5}{4}$ C. $\frac{3}{2}$ D. -1

77. The graph of which function is sketched in Fig. F.E.-11?

A. $y = \log_{4/5} x$ B. $y = -\log_{5/4} x$

C. $y = \log_{5/4}(-x)$ D. $y = \log_{5/4} x$

78. To raise money for a local animal shelter, a school group runs an annual car wash. From past experience, they expect to wash 40 cars if they charge \$6. After conducting a survey, the group's faculty advisor concludes that for every \$0.50 decrease in the price, the group should have 10 more customers. Under these conditions, what is the most revenue the group should expect?

A. \$305 B. \$310 C. \$315 D. \$320

79. Find $x + 2y$ for the system (real solutions only): $\begin{cases} y = x^2 - 8 \\ y = \frac{9}{x^2} \end{cases}$.

A. -1 and 5 B. -2 and 6 C. 1 and 2 D. 2 and 3

80. Solve for x: $3x^2+8x+4=0$.

A. $x=\frac{3}{2},\ 2$ B. $x=-2,\ \frac{3}{2}$

C. $x=-2,\ -\frac{2}{3}$ D. $x=2,\ -\frac{2}{3}$

81. What is the center and radius of the circle $(x+4)^2+(y-6)^2=15$.

A. Center: $(-4, 6)$; radius: 15 B. Center: $(-4, 6)$; radius: $\sqrt{15}$

C. Center: $(4, -6)$; radius: 15 D. Center: $(4, -6)$; radius: $\sqrt{15}$

82. According to Descartes' Rule of Signs, how many real zeros are possible for the function $f(x)=4x^3-x^2+5x+15$?

A. 1 or 2 positive, 1 or 2 negative zeros

B. 1 positive, 1 negative zeros

C. 0 or 2 positive, 1 negative zeros

D. 1 positive, 0 or 2 negative zeros

83. Are the lines $8x-5y=20$ and $25x+40y=-6$ parallel, perpendicular, or neither?

A. Parallel B. Perpendicular C. Neither

84. Solve for x: $8-5x\le 10$.

A. $(-\infty, \frac{18}{5}]$ B. $[\frac{18}{5}, \infty)$ C. $(-\infty, -\frac{8}{5}]$ D. $[-\frac{2}{5}, \infty)$

85. Solve for x: $x^2+\frac{1}{3}x=3$.

A. $x=-\frac{1}{6}\pm\frac{\sqrt{37}}{6}$ B. $x=-\frac{1}{6}\pm\sqrt{3}$

C. $x=\frac{5}{6},\ \frac{11}{6}$ D. $x=-\frac{1}{6}\pm\frac{\sqrt{109}}{6}$

86. Solve the inequality: $|3x+4|>1$.

A. $(-\frac{5}{3}, \infty)\cup(-1, \infty)$ B. $(-\frac{5}{3}, -1)$

C. $(-1, -\frac{5}{3})$ D. $(-\infty, -\frac{5}{3})\cup(-1, \infty)$

87. Find the intercepts for the graph of $y=x^2-9$.

A. The x-intercepts are -3, 3, and the y-intercept is -9.

B. The x-intercept is -3, and the y-intercept is -9.

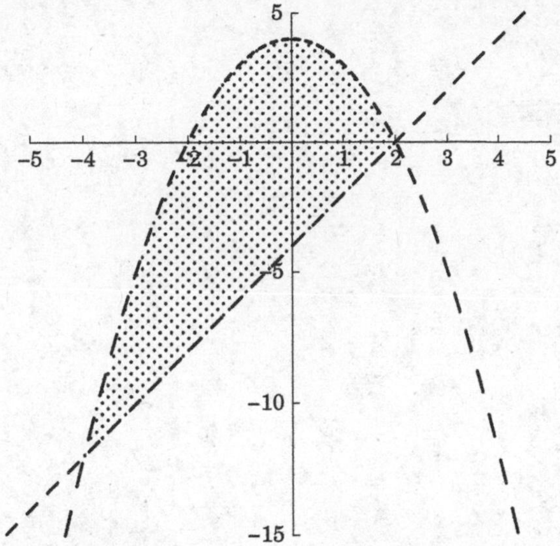

FIGURE F.E.-12

C. The graph has no x-intercepts, and the y-intercept is -9.

D. The x-intercept is 3, and the y-intercept is -9.

88. Find the center and radius of the circle whose equation is $x^2 - 12x + y^2 + 12y + 23 = 0$.

A. Center: $(6, -6)$; radius: 8

B. Center: $(6, -6)$; radius: 7

C. Center: $(6, -6)$; radius: $2\sqrt{2}$

D. Center: $(6, -6)$; radius: $\sqrt{7}$

89. The graph of which system of inequalities is shaded in Fig. F.E.-12?

A. $\begin{cases} 2x - y < 4 \\ y > 4 - x^2 \end{cases}$

B. $\begin{cases} 2x - y < 4 \\ y < 4 - x^2 \end{cases}$

C. $\begin{cases} 2x - y > 4 \\ y > 4 - x^2 \end{cases}$

D. $\begin{cases} 2x - y > 4 \\ y < 4 - x^2 \end{cases}$

90. Find the quotient for $(8x^3 - 6x^2 + 3x + 1) \div (2x - 5)$.

A. $4x^2 + 7x - 19$

B. $4x^2 - 13x + 31$

C. $4x^2 + 7x + 19$

D. $4x^2 + 13x - 34$

Answers to Quizzes and Final Exam

Chapter 1
1. C
2. B
3. A
4. B
5. D
6. B
7. B
8. C
9. D
10. D
11. A
12. A
13. B
14. C

Chapter 2
1. B
2. C
3. D
4. B
5. D
6. B
7. C
8. B
9. A
10. D
11. B
12. C
13. B
14. D

Chapter 3
1. D
2. C
3. D
4. A
5. D
6. B
7. A
8. C
9. A

Chapter 4
1. A
2. B
3. A
4. C
5. D
6. D
7. D
8. B
9. C
10. C

Chapter 5
1. C
2. A
3. B
4. D
5. C
6. A
7. D
8. B
9. D
10. A
11. B
12. C
13. A
14. C
15. D

Chapter 6
1. B
2. A
3. D
4. A
5. A
6. B
7. A
8. C

Chapter 7
1. B
2. C
3. D
4. A
5. D
6. B
7. D

8. A
9. D
10. D
11. A
12. B
13. C
14. B
15. D

Chapter 8

1. D
2. B
3. D
4. C
5. A
6. B
7. B
8. A

Chapter 9

1. C
2. A
3. A
4. C
5. B
6. B
7. B
8. D
9. D
10. A
11. B
12. D

Chapter 10

1. B
2. C
3. C
4. B
5. A
6. A
7. D
8. B
9. A
10. D
11. D
12. D
13. B
14. C
15. A

Chapter 11

1. B
2. A
3. C
4. B
5. D
6. C
7. B
8. D
9. D
10. A
11. C
12. B

Chapter 12

1. D
2. B
3. B
4. C
5. A
6. D
7. D
8. A
9. B
10. C
11. C
12. A
13. A
14. B
15. A
16. C
17. D
18. B
19. D
20. A
21. A
22. C
23. A

Final Exam

1. B
2. B
3. B
4. B
5. D
6. C
7. B
8. C
9. B
10. A
11. D
12. A
13. C
14. C
15. C
16. A
17. A
18. D
19. D
20. D
21. B
22. A
23. D
24. C
25. B
26. B
27. C
28. C
29. C
30. A
31. C
32. A
33. C
34. D
35. A
36. B
37. C
38. A
39. C
40. D
41. C
42. A
43. D
44. D
45. B
46. B
47. D
48. D
49. A
50. D
51. A
52. A
53. A
54. A
55. B
56. A
57. C
58. C
59. C
60. A
61. B
62. D
63. D
64. C
65. C
66. C

67. B
68. D
69. B
70. B
71. C
72. A
73. B
74. D
75. C
76. A
77. D
78. D
79. A
80. C
81. B
82. C
83. B
84. D
85. D
86. D
87. A
88. B
89. B
90. C

Index

A

absolute value inequalities, 39–46
absolute value
  definition of, 30
  equations, 30–36
  functions, 267, 270–275
  graphs of, 267
  interval notation and, 40–46
  in transformations, 270–275
addition
  of complex numbers, 347–348
  elimination by, 377, 396–398
  of rational expressions, 9, 11–12
  in systems of equations, 380–390
alike terms, defined, 2
applications
  linear. See linear applications
  of quadratic functions, 239–242
  for systems of equations, 390–395
arguments, defined, 459
averages
  midpoint formula for, 76–77

C

cancellation. See also elimination by addition
  FOIL method and, 26, 29
  multiplying rational expressions and, 9
cancellation (*Cont.*):
  properties, 436–438, 451–452
  simplifying rational expressions with, 8
  simplifying square roots and, 15
  in systems of equations, 381–382, 386
circles
  completing squares for, 85–87
  endpoints in diameter of, 90
  equations for, 247
  with given radius/center, 77–81
  sketching graphs of, generally, 77–78
  slope-intercept form of lines and, 112–113
  without known radius/center, 81–87
  xy-coordinate planes and, 77–85
closed dots, 202–210
coefficients
  definition of, 2, 306
  leading, 306–308, 313
  in synthetic division, 324–327
combinations of functions. See also transformations
  domains of functions in, 295–298
  function composition in, generally, 287–290
  introduction to, 247–248, 286–287
common denominators. See least common denominators (LCDs)
common logarithms, 437

completing squares
circles and, 85–87
locating vertices by, 141–147
quadratic equations with, 56–63
complex conjugates, 350–354
complex numbers
multiplying with FOIL method, 349–350
as zeros of a polynomial, 346–354
complex solutions, 354–356
complex zeros
Fundamental Theorem of Algebra and, 357–363, 369
introduction to, 343–346
quadratic equations and, 356
composition of functions. See also functions
domains in, 295–298
introduction to, 287–290
for single values, 290–295
compound amounts, 426–429, 431
compound inequalities, 35–39
compounding
annually, 426–429
continuous, 430–433
of growth, generally, 426–429
of population increase, 431–433
constant functions
graphs of, 267
intervals and, 199–200
constants, defined, 2
continuous compounding, 430–433
coordinates
definition of, 66
planes of. See *xy*-coordinate planes
x-. See *x*-coordinates
y-. See *y*-coordinates
cubic functions, 267

D

decreasing intervals, 197–200
denominators
conjugates of, 352–353
difference quotient and, 213
denominators (*Cont.*):
domains of functions and, 182–183, 187, 295–297
factoring, 1–7, 10–12
introduction to, 9–10
least common. See least common denominators (LCDs)
in rational inequalities, 165–172
Rational Zero Theorem and, 334–335
dependent variables, 175–176
Descartes' Rule of Signs, 340–346
diameter, 81–83, 90
difference
of coordinates, 67
of factors, 4
of two squares, 6
difference quotient, 212–217
discriminant, 356
distance formula, 67–76
Distributive Property
adding rational expressions with, 12
expanding expressions with, 2–4, 6–7
factoring expressions with, 4–7
FOIL method and, 2, 12
like terms and, 2
solving linear equations with, 22–25
division
Fundamental Theorem of Algebra and, 357–363
polynomial, 317–323, 357–363
synthetic, for factoring of polynomials, 331–340
synthetic, generally, 323–331
Division Algorithm
introduction to, 305
polynomial functions and, 317–324
Remainder Theorem and, 329
domains of functions. See also functions
in combinations of functions, 295–298
finding graphically, 193–197
interval notation and, 182–188, 194–197
logarithms and, 459–460
real numbers in, 182–188
sign graphs and, 185

dots
closed, 202–210
open, 193–194, 202–210

E

e (Euler's number), 430–431
elimination by addition
introduction to, 377
in systems of equations, generally, 380–390
in systems of nonlinear equations, 396–398
end behavior, 306–312
endpoints
definition of, 81
of diameters of circles, 81–83, 90
in graphing piecewise functions, 202
equations
absolute value, 30–36
exponents and, 445–452
linear. See linear equations
logarithms and, 445–452
quadratic. See quadratic equations
systems of. See systems of equations
Euler's number(*e*), 430–431
evaluation of functions, 178–182, 209–212
even multiplicity, 364
expanding expressions
with Distributive Property, 2–4, 6–7
with FOIL method, 3–4
exponent functions. See also exponents, 267
exponential growth, 426, 431–433
exponents
for compound growth, generally, 426–429
for continuous compounding, 430–431
base *e*, 430–431
equations with logarithms and, 445–452
functions of, 453–459
graphs of, 267
for increasing population, 431–433
exponents (*Cont.*):
logarithms and, generally, 434–435, 439–440
properties of, 13–16
radical properties of, 13–16
expressions
expanding, 2–4, 6–7
factoring, 2–7, 52–53
irreducible quadratics, 356
rational, 7–12, 14–15
extracting roots, 58–60
extraneous solutions, 25–27

F

factoring denominators, 1–7, 10–12
factoring expressions
with Distributive Property, 4–7
introduction to, 2–3
solving quadratic equations by, 52–53
factorization, 10–12, 52–53
FOIL method
adding rational expressions and, 10, 12
expanding expressions with, 3–4
factoring expressions with, 5–7
introduction to, 2
multiplying complex numbers with, 349–350, 352
multiplying rational expressions with, 9
solving linear equations with, 26
solving quadratic equations with, 56
fractions
addition of, 9, 11–12
arithmetic for, generally, 7
in equations for lines, 103, 107
in linear equations, 23, 25
multiplication of, 8–12
in polynomial division, 320
in rational expressions, generally, 7
in rational inequalities, 166–172
simplification of, 8, 10
subtraction of, 9, 11–12
in systems of equations, 386, 390
zero and, 100–102

function composition
domains in, 295–298
introduction to, 287–290
for single values, 290–295
functions
combining. See combinations of functions
composition of. See function composition
decreasing intervals of, 197–200
difference quotient and, 212–217
domains of. See domains of functions
evaluation of, 178–182, 209–212
of exponents, 453–459
graphs for domains/ranges of, 193–197
graphs of, generally, 188–193
increasing intervals of, 197–200
introduction to, 175–178
of logarithms, 453–459
piecewise-defined, generally, 180–182
piecewise-defined, graphs of, 200–209
quadratic. See quadratic functions
range of, generally, 182–188
ranges of, finding graphically, 193–197
sketching graphs of. See transformations
Fundamental Theorem of Algebra
finding zeros, real or complex, 357–363
given conditions in, 367–370
given zeros in, 365–370
introduction to, 305, 357
multiplicity of zeros in, 363–364
fundamentals
of Distributive Property, 2–6
of exponents, 13–16
rational expressions, 7–12

G

general exponential growth model, 431–433
general quadratic equations. See also quadratic equations, 141
graphs
above or below x-axis, 154–159
of absolute value, 267
of circles, 77–85
of constant functions, 267
for domains/ranges of functions, 193–197
of functions, generally, 188–193
of linear functions, 267
of lines with slope and y-intercept, 113–116
of logarithm functions, 267
for nonlinear inequalities, 154–159
of parabolas, 135–141
of piecewise-defined functions, 200–209
of polynomial functions, 314–317
of quadratic functions, 248–253, 267
of special functions, 264
of square root functions, 267
of transformations, 259–264

H

horizontal change, 102–104, 113–116
horizontal distance, 232–233
horizontal lines
definition of, 66
distance between points on, 67–76
finding equations for, 107–109, 111
in graphs, generally, 188–193
in inequalities, 403–404
parallel, 120–121
perpendicular, 120–121
plotting points on, 66–67
slope and, 102–104
slope-intercept form of, 112–113
values of, 105–107
horizontal reflection, 276
hypotenuse, 70–72

I

increasing intervals, 197–200
increasing population, 431–433

independent variables, 176
inequalities
  absolute value, 39–46
  compound, 35–39
  linear. See linear inequalities
  nonlinear. See nonlinear inequalities
  systems of. See systems of inequalities
intercepts. See also x-intercepts; y-intercepts
  graphing lines with slope and, 113–116
  graphing of polynomial functions and, 314–317
  in lines, generally, 95–102
  nonlinear inequalities and, 158–159
  in polynomial functions, generally, 309–312
  rational inequalities and, 165–166
  sign graphs and, 161
  in slope of lines, 112–113, 121–123
intersection
  in compound linear inequalities, 37
  definition of, 408
  in systems of inequalities, 408–412
interval notation
  in absolute value inequalities, 40–46
  in compound inequalities, 36–39
  domains in, 182–188, 194–197
  in functions, 197–200
  in linear inequalities, 35–36
  in logarithmic functions, 459–460
  nonlinear inequalities and, 160–163
  ranges in, 194–197
  rational inequalities and, 166–172
irreducible quadratic expressions, 356
isosceles triangles, 70–76

K

kilo-watt hours, 124, 131

L

LCDs (least common denominators). See least common denominators (LCDs)
leading coefficients, 306–308, 313
leading terms, 306–308
least common denominators (LCDs)
  introduction to, 9–10
  leading to linear equations, 25–26
  in point-slope formulas, 109, 111
  solving linear equations and, 23–25, 28–29
like terms, defined, 2
linear applications
  finding points on lines in, 124–127
  introduction to, 123
  points and rate of change in, 129–133
  points and slope of lines in, 127–129
linear equations
  absolute value equations and, 30–36
  basic, 22–25
  definition of, 92
  equations leading to, 25–30
  graphing equations on, 77
  introduction to, 21
  for lines, 92–95, 123–133
  systems of, 376–380, 396–401
linear factors, 357
linear functions, 267
linear inequalities
  absolute values in, 39–46
  compound, 35–39
  introduction to, 21
  solutions to, 38–39
linearly related variables, defined, 123
lines
  finding equations for, 107–111
  graphing with slope and y-intercept, 113–116
  horizontal. See horizontal lines
  intercepts in, 95–102
  introduction to, 91
  linear applications and, 123–133
  linear equations for, 92–95, 123–133
  parallel, 116–123
  perpendicular, 116–123
  slope-intercept form of, 112–113
  slope of, 102–104
  vertical. See vertical lines

logarithms
  domains of functions of, 459–460
  equations with exponents and, 445–452
  functions and, 453–459
  graphs of, 267
  introduction to, 425, 434–435
  multiple properties of, 441–444
  properties of, generally, 436–441
logs. See logarithms
lower bounds, 340–346
lowest terms, defined, 8

M

maximum number of zeros, 340–346
maximum values
  finding, 230–239
  in miscellaneous functions, 242–243
  in quadratic functions, generally, 228–230
midpoint formula, 76–77
minimum values
  finding, 230–239
  in miscellaneous functions, 242–243
  in quadratic functions, generally, 228–230
multiple logarithm properties, 441–444
multiplication
  of complex conjugates, 350–352
  of complex numbers, 348–349
  of fractions, 7, 10
  by least common denominators, 23–25, 28–29, 109–111
  of linear inequalities, 35–36, 45
  logarithm properties and, 439
  in polynomial division, 317–320
  of rational expressions, 8–9, 10–12
  in systems of equations, 381–383, 386
multiplicity of zeroes, 363–370

N

natural logarithms, 437, 447–448
negative reciprocals, 116–121
negative zeros, 343–346
nonlinear inequalities
  introduction to, 153
  rational inequalities and, 165–172
  sign graphs for, 159–165
  solving algebraically, 159–165
  solving graphically, 154–159
notation. See interval notation
numbers
  complex, 27, 346–354
  Euler's, 430–431
  real. See real numbers
  of zeros, 340–346

O

odd multiplicity, 364
open dots, 193–194, 202–210
optimization, 228
origin
  definition of, 66–67
origin symmetry, 276–281

P

parabolas
  introduction to, 91, 134–135
  locating vertex by completing squares, 141–147
  sketching graphs of, 135–141
  vertices of, 141–147, 224–226
parallel lines, 116–123
perpendicular lines, 116–123
piecewise-defined functions, 180–182, 200–209
plotting points, 66–67
point-slope formula, 107–109, 129–133
points
  distance between, 67–76
  end, 81–83, 90
  finding on lines, 124–127
  in linear applications, 124–133
  mid, 76–77
  plotting, 66–67

points (*Cont.*):
for slope of lines, 104–108, 121–123, 127–129
turning around, 198
polynomial division
Fundamental Theorem of Algebra and, 357–363
overview of, 317–324
Remainder Theorem and, 331–340
synthetic, for factoring of polynomials, 331–340
synthetic, generally, 324–331
polynomials
complex solutions to equations and, 354–356
division of. See polynomial division
finding zeros in, real or complex, 357–363
Fundamental Theorem of Algebra and, generally, 357
given conditions and, 367–370
given zeros and, 364–367
introduction to, 305–314
multiplicity of zeros in, 363–364
quadratic equations and, 354–356
Rational Zero Theorem and, 333–340
Rule of Signs and, 340–346
sketching graphs of, 314–317
synthetic division for factoring of, 331–340
synthetic division of, generally, 323–331
Upper and Lower Bounds Theorem and, 340–346
population growth, 431–433
positive zeros, 343–346
possible number of zeros, 340–346
properties
cancellation, 436–438, 451–452
Distributive. See Distributive Property
of exponents, 13–16
first logarithm property, 439–443
of logarithms, 436–444
multiple, 441–444
radical, 13–16
properties (*Cont.*):
second logarithm property, 439–443
third logarithm property, 440–443, 447–448
Pythagorean theorem, 69–70

Q

quadratic equations
completing squares for, 56–63
complex solutions for, 354–356
factoring for, 52–53
general vs. standard form, 141–142
introduction to, 51
quadratic formula for, 53–56
quadratic formula
for quadratic equations, 53–56
square roots and, 14
quadratic functions
graphs of, 248–253, 267
introduction to, 223–224
maximum/minimum of, finding, 230–239
maximum/minimum of, generally, 228–230
maximum/minimum of, in miscellaneous functions, 242–243
parabolas and, 224–226
range of, 226–227
transformation in, 248–253
vertices and, 224–226
quotients
in combining functions, 286
difference, 212–217
in division of complex numbers, 351–354
fractions as, 103
in polynomial functions, generally, 317–324
in slope of lines, 103–104
in synthetic division for factoring, 331–340
in synthetic division, generally, 324–331

R

radical properties, 13–16
radicals, defined. See also roots, 13
radius, defined, 78
range
  finding graphically, 193–197
  of functions, generally, 182–188
  in interval notation, 194–197
  of quadratic functions, 226–227
rational expressions
  addition of, 9, 11–12
  introduction to, 7
  multiplication of, 8–12
  simplification of, 8, 10
  subtraction of, 9, 11–12
rational inequalities, 165–172
Rational Zero Theorem, 333–340
real numbers
  complex numbers vs., 346–354
  Fundamental Theorem of Algebra and, 357
  interval notation for, 155
  in logarithmic functions, 459–460
  in nonlinear inequalities, 155
  in polynomial functions, 306, 309, 346–354
  in radical properties, 13
  square roots of, 27
real zeros
  complex numbers and, 346–356
  Fundamental Theorem of Algebra and, 357–363
  Upper and Lower Bounds Theorem and, 343–346
reciprocals, 116–121
reducing fractions, 8
Remainder Theorem, 329–333
remainders
  in polynomial functions, generally, 317–324
  in synthetic division for factoring, 331–340
  in synthetic division, generally, 324–331
roots. See also square roots, 13–16
Rule of Signs, 340–346

S

second logarithm property, 439–443
sign graphs, 159–165, 185
simplification
  in difference quotients, 212
  in equations for circles, 79
  in even/odd functions, 261–262, 263–264
  of fractions, 15–16, 25–29
  in linear equations, 22–25
  logarithm properties for, 436–437
  of rational expressions, 8, 10, 14–15
  of square roots, 14–15
sketching graphs. See also graphs
  of circles, 77–85
  of functions. See transformations
  of lines with slope and y-intercept, 113–116
  of parabolas, 135–141
  of polynomials, 314–317
  of quadratic functions, 248–253
  of transformations, 259–264
slope
  difference quotient and, 217
  graphing lines with y-intercepts and, 113–116
  of lines, generally, 102–104
  of parallel/perpendicular lines, 116
  rate of change as, 129–133
slope-intercept form of lines, 112–113, 121–123
square root functions, 267
square roots
  of constant terms, 58
  graphs of, 267
  of real numbers, 27
  of sides of equations, 58–60, 177, 355–356
  simplification of, 14–15

squares, completing
  circles and, 85–87
  locating vertices by, 141–147
  solving quadratic equations by, 56–63
squares, difference of, 6
standard quadratic equations, 141
substitution
  introduction to, 377–380
  in nonlinear equation systems, 396–401
  in systems of equations, 388, 391–395
subtraction
  of complex numbers, 347–348
  in compound inequalities, 37–38
  in linear equations, 23–27
  in nonlinear inequalities, 162
  of rational expressions, 9, 11–12
symmetry, 276–281
synthetic division
  for factoring of polynomials, 331–340
  of polynomials, generally, 323–331
  Remainder Theorem and, 331–340
systems of equations
  applications for, 390–395
  elimination by addition in, 380–390
  introduction to, 375–376
  linear equations in, 376–380, 396–401
systems of inequalities
  inequalities and, 401–408
  intersection in, 408–412
  introduction to, 375–376
  more than two inequalities in, 414–418
  without solution, 412–414

T

terms, defined, 2
theorems
  Fundamental. See Fundamental Theorem of Algebra
  Pythagorean, 69–70
  Rational Zero, 333–340
  Remainder, 329–333
  Upper and Lower Bounds, 340–346
third logarithm property, 440–443, 447–448
transformations. See also combinations of functions
  absolute value function in, 270–275
  compression in, 253–259
  even/odd functions in, 281–286
  of exponent functions, 459
  graphs of, 259–264
  introduction to, 247–253
  of logarithmic functions, 459
  origin symmetry in, 276–281
  reflections in, 253–259
  sketching graphs of, 259–264
  special functions in, generally, 264–265
  symmetry in, 276–281
  vertical stretching in, 253–259
  x/y-axis symmetry in, 276–281
triangles, 70–76
turning around points, 198

U

Upper and Lower Bounds Theorem, 340–346

V

variables
  constants *vs.*, 2
  in denominators, 165–166
  dependent, 175–176
  in elimination by addition, 381–386
  in equations leading to linear equations, 25
  in general exponential growth model, 431–433
  graphing on *xy*-planes and, 79
  independent, 176
  intercepts for graphs with, 98
  in linear applications, 123–133
  in linear equations, 23–25
  linearly related, 123
  in lines with given points, 108
  radical properties and, 14–15

variables (*Cont.*):
in rational expressions, 7
in substitution method, 377–380
Vertical Line Test, 189–190
vertical lines
definition of, 66
distance between points on, 67–76
finding equations for, 110–111
in graphing with slope and *y*-intercept, 112–113
in graphs, generally, 188–193
in inequalities, 403–404
parallel, 120–121
perpendicular, 120–121
plotting points on, 66–67
slope-intercept form *vs.*, 112
vertical reflection, 276
vertices
definition of, 134
locating by completing squares, 141–147
of parabolas, 141–147, 224–226
quadratic functions and, 224–226
of squares, 76
of triangles, 72–75

X

x-axis
definition of, 66
in exponents, 459
graphs above or below, 154–159
in graphs of transformations, 259–265
in logarithmic functions, 459
in polynomial functions, 309, 343
in reflections, 253–259
in transformations, 259–265, 276–281
Upper and Lower Bounds Theorem and, 343
x-axis symmetry, 276–281
x-coordinates. See also *xy*-coordinate planes
composition of functions and, 293–295
definition of, 66
distance between points and, 67–76
x-coordinates (*Cont.*):
in linear equations, 128
in logarithmic functions, 457
plotting points and, 66–67
in polynomial functions, 373
y-intercepts and, 95
x-intercepts
in graphs of polynomial functions, 314–317
in linear equations, 95–102
in nonlinear inequalities, 158–159
in polynomial functions, 309–312
in rational inequalities, 165–166
sign graphs and, 161
x-values
in constant functions, 178–179
in domains of functions, 193
function composition for, 290–295
in functions, generally, 176–178
in piecewise functions, 180–182
in vertical lines, 191
xy-coordinate planes
distance formula for, 67–76
graphing equations on, 77
introduction to, 65
midpoint formula for, 76–77
plotting points in, 66–67

Y

y-axis
definition of, 66
in domains/ranges of functions, 194
in exponents, 459
intercepts and, 95
in logarithmic functions, 459
in rational inequalities, 165
in special functions, 266–268
in transformations, 259
y-axis symmetry, 276–281
y-coordinates
in circles, 78
definition of, 66
in distance between points, 69
distance between points and, 67–76

y-coordinates (*Cont.*):
  in linear applications, 128
  in plotting points, 66–67
  plotting points and, 66–67
  in polynomial functions, 309
  in quadratic functions, 228
  as value of functions, 218, 293
  of vertices, 228
  x-intercepts and, 96, 309
y-intercepts
  in linear equations, 95–102
  in slope-intercept form of lines, 112–113
  in slope of lines, 113–116, 247
y-values
  in constant functions, 178–179
  in functions, generally, 176–178
  in graphs of polynomials, 315
  of horizontal lines, 105–106, 111
y-values (*Cont.*):
  in midpoint formulas, 76
  in nonlinear inequalities, 158–159
  in piecewise functions, 180–182
  in quadratic functions, 248–253
  in reflections, 253–259
  sign graphs and, 159–165
  in systems of nonlinear equations, 396
  in transformations, 259–265

Z

zeros (0s)
  complex, 343–346, 356–363, 369
  complex numbers and, 346–356
  finding, 357–363
  of functions, 309
  multiplicity of, 363–364
  real, 343–346, 357–363